Springer Handbook of Auditory Research

Series Editors: Richard R. Fay and Arthur N. Popper

Springer
New York
Berlin
Heidelberg
Barcelona
Budapest
Hong Kong
London
Milan
Paris
Santa Clara
Singapore
Tokyo

SPRINGER HANDBOOK OF AUDITORY RESEARCH

Volume 1: The Mammalian Auditory Pathway: Neuroanatomy
Edited by Douglas B. Webster, Arthur N. Popper, and Richard R. Fay

Volume 2: The Mammalian Auditory Pathway: Neurophysiology
Edited by Arthur N. Popper and Richard R. Fay

Volume 3: Human Psychophysics
Edited by William Yost, Arthur N. Popper, and Richard R. Fay

Volume 4: Comparative Hearing: Mammals
Edited by Richard R. Fay and Arthur N. Popper

Volume 5: Hearing by Bats
Edited by Arthur N. Popper and Richard R. Fay

Volume 6: Auditory Computation
Edited by Harold L. Hawkins, Teresa A. McMullen, Arthur N. Popper, and Richard R. Fay

Volume 7: Clinical Aspects of Hearing
Edited by Thomas R. Van de Water, Arthur N. Popper, and Richard R. Fay

Volume 8: The Cochlea
Edited by Peter Dallos, Arthur N. Popper, and Richard R. Fay

Forthcoming volumes (partial list)

Development of the Auditory System
Edited by Edwin W. Rubel, Arthur N. Popper, and Richard R. Fay

Plasticity in the Auditory System
Edited by Edwin W. Rubel, Arthur N. Popper, and Richard R. Fay

Comparative Hearing: Amphibians and Fish
Edited by Arthur N. Popper and Richard R. Fay

Comparative Hearing: Insects
Edited by Ronald R. Hoy, Arthur N. Popper, and Richard R. Fay

Comparative Hearing: Birds and Reptiles
Edited by Robert Dooling, Arthur N. Popper, and Richard R. Fay

Peter Dallos
Arthur N. Popper
Richard R. Fay
Editors

The Cochlea

With 134 Illustrations

 Springer

Peter Dallos
Auditory Physiology Laboratory and
Department of Neurobiology
Northwestern University
Evanston, IL 60208, USA

Arthur N. Popper
Department of Zoology
University of Maryland
College Park, MD 20742-9566, USA

Richard R. Fay
Department of Psychology and
Parmly Hearing Institute
Loyola University of Chicago
Chicago, IL 60626, USA

Series Editors: Richard R. Fay and Arthur N. Popper

Cover illustration: Schematic of the olivocochlear efferents to the right cochlea of a cat. This figure appears on p. 436 of the text.

Library of Congress Cataloging in Publication Data
The cochlea / Peter Dallos, Arthur N. Popper, Richard R. Fay, editors.
   p. cm. — (Springer handbook of auditory research : v. 8)
  Includes bibliographical references and index.
  ISBN 0-387-94449-4 (hardcover : alk. paper).
  1. Cochlea—Physiology. I. Dallos, Peter. II. Popper, Arthur N.
III. Fay, Richard R. IV. Series.
QP471.2.C63   1996
612.8′58—dc20   95-44634

Printed on acid-free paper.

© 1996 Springer-Verlag New York, Inc.
All rights reserved. This work may not be translated or copied in whole or in part without the written permission of the publisher (Springer-Verlag New York, Inc., 175 Fifth Avenue, New York, NY 10010, USA), except for brief excerpts in connection with reviews or scholarly analysis. Use in connection with any form of information storage and retrieval, electronic adaptation, computer software, or by similar or dissimilar methodology now known or hereafter developed is forbidden.
The use of general descriptive names, trade names, trademarks, etc., in this publication, even if the former are not especially identified, is not to be taken as a sign that such names, as understood by the Trade Marks and Merchandise Marks Act, may accordingly be used freely by anyone.
While the advice and information in this book are believed to be true and accurate at the date of going to press, neither the authors nor the editors nor the publisher can accept any legal responsibility for any errors or omissions that may be made. The publisher makes no warranty, express or implied, with respect to the material contained herein.

Production managed by Karina Gershkovich and Terry Kornak; manufacturing supervised by Jeffrey Taub.
Typeset by TechType, Inc., Ramsey, NJ.
Printed and bound by Maple-Vail, Binghamton, NY.
Printed in the United States of America.

9 8 7 6 5 4 3 2 1

ISBN 0-387-94449-4 Springer-Verlag New York Berlin Heidelberg SPIN 10491114

Series Preface

The *Springer Handbook of Auditory Research* presents a series of comprehensive and synthetic reviews of the fundamental topics in modern auditory research. The volumes are aimed at all individuals with interests in hearing research including advanced graduate students, postdoctoral researchers, and clinical investigators. The volumes are intended to introduce new investigators to important aspects of hearing science and to help established investigators to better understand the fundamental theories and data in fields of hearing that they may not normally follow closely.

Each volume is intended to present a particular topic comprehensively, and each chapter will serve as a synthetic overview and guide to the literature. As such, the chapters present neither exhaustive data reviews nor original research that has not yet appeared in peer-reviewed journals. The volumes focus on topics that have developed a solid data and conceptual foundation rather than on those for which a literature is only beginning to develop. New research areas will be covered on a timely basis in the series as they begin to mature.

Each volume in the series consists of five to ten substantial chapters on a particular topic. In some cases, the topics will be ones of traditional interest for which there is a substantial body of data and theory, such as auditory neuroanatomy (Vol. 1) and neurophysiology (Vol. 2). Other volumes in the series will deal with topics that have begun to mature more recently, such as development, plasticity, and computational models of neural processing. In many cases, the series editors will be joined by a co-editor having special expertise in the topic of the volume.

Richard R. Fay
Arthur N. Popper

Preface

Any consideration of vertebrate hearing must include a comprehensive understanding of the inner ear. In terms of both structure and function, the mammalian ear may be the most intriguing. It is certainly the most complex. The purpose of this volume is to present a comprehensive overview of the mammalian cochlea—from anatomy and physiology to biophysics and biochemistry. Together, the contributors to this volume provide a detailed and unified introduction to how sounds are processed in the cochlea and how the ensuing signals are prepared for the central nervous system.

In the first chapter, Dallos provides a general overview of cochlear structure and function that brings together the subsequent chapters. Slepecky (Chapter 2) gives the structural bases for cochlear function through a detailed description of its gross and ultrastructural features. Wangemann and Schacht (Chapter 3) describe the ways in which the inner ear maintains the proper environment for function. Patuzzi (Chapter 4) considers the very important issue of the mechanics and micromechanics of cochlear response to stimulation, and de Boer (Chapter 5) examines cochlear mechanics from the perspective of modeling theory. Chapter 6 by Kros is directed at questions of the physiology of hair cells; specific issues of hair cell motility are discussed by Holley in Chapter 7. The efferent system plays a significant role in cochlear physiology, and this is considered by Guinan in Chapter 8. Finally, Sewell, in Chapter 9, discusses the afferent and efferent transmitters involved in cochlear function.

This volume is closely tied to others in the *Springer Handbook of Auditory Research*. The anatomy and physiology of the auditory nerve are considered in detail by Ryugo in Vol. 1 of the series (*The Mammalian Auditory Pathway: Neuroanatomy*) and by Ruggero in Vol. 2 (*The Mammalian Auditory Pathway: Neurophysiology*). Volumes 1 and 2 also consider processing of auditory signals by the brain. Mammalian inner ear structures, as described by Slepecky in this volume, are dealt with, in a very comparative way, by Echteler, Fay, and Popper in Vol. 4 (*Comparative Hearing: Mammals*), and in a chapter on the cochlea of bats by Kössel and

Vater in Vol. 5 (*Hearing by Bats*). Cochlear mechanics (Patuzzi) and modeling (de Boer) in the present volume are considered from the standpoint of computational systems in chapters by Hubbard and Mountain in Vol. 6 (*Auditory Computation*). The physiology of the efferent system (Guinan) is paralleled by a detailed discussion of efferent system anatomy by Warr in Vol. 1.

Peter Dallos
Arthur N. Popper
Richard R. Fay

Contents

Series Preface		v
Preface		vii
Contributors		xi
Chapter 1	Overview: Cochlear Neurobiology PETER DALLOS	1
Chapter 2	Structure of the Mammalian Cochlea NORMA B. SLEPECKY	44
Chapter 3	Homeostatic Mechanisms in the Cochlea PHILINE WANGEMANN AND JOCHEN SCHACHT	130
Chapter 4	Cochlear Micromechanics and Macromechanics ROBERT PATUZZI	186
Chapter 5	Mechanics of the Cochlea: Modeling Efforts EGBERT DE BOER	258
Chapter 6	Physiology of Mammalian Cochlear Hair Cells CORNÉ J. KROS	318
Chapter 7	Outer Hair Cell Motility MATTHEW C. HOLLEY	386
Chapter 8	Physiology of Olivocochlear Efferents JOHN J. GUINAN JR.	435
Chapter 9	Neurotransmitters and Synaptic Transmission WILLIAM F. SEWELL	503
Index		535

Contributors

Peter Dallos
Auditory Physiology Laboratory, Northwestern University, Frances Searle Building, 2299 Sheridan Road, Evanston, IL 60208, USA

Egbert de Boer
Academic Medical Center, Meibergdreef 9 1105 AZ Amsterdam, The Netherlands

John J. Guinan Jr.
Eaton Peabody Laboratory, 243 Charles St., Boston, MA 02114, USA

Matthew C. Holley
Department of Physiology, Medical School, University Walk, Bristol BS8 1TD, U.K.

Corné J. Kros
School of Biological Sciences, University of Sussex, Falmer, Brighton, BN1 9QG, U.K.

Robert Patuzzi
Physiology Department, University of Western Australia, Nedlands 6009, Australia

Jochen Schacht
Kresge Hearing Research Institute, University of Michigan, Ann Arbor, MI 48109-0506, USA

William F. Sewell
Eaton-Peabody Laboratory, Massachusetts Eye and Ear Infirmary, 243 Charles St., Boston, MA 02114, USA

Norma B. Slepecky
Department of Bioengineering and Neuroscience, Institute for Sensory Research, Syracuse University, Syracuse, NY 13244-5290, USA

Philine Wangemann
Boys Town National Research Hospital, Omaha, NE 68131, USA

1
Overview: Cochlear Neurobiology

PETER DALLOS

1. Introduction

All environmental and biologically significant sounds change from frequency to frequency and from instant to instant. The hearing organ must be capable of analyzing time-varying frequency components and representing them in a spatiotemporal array of neural discharges in the fibers of the auditory nerve. When a signal is analyzed, its frequency spectrum and temporal characteristics are reciprocally related. The simplest signal in terms of its spectrum is an ongoing sine wave, corresponding to a pure tone in acoustics, and has only one spectral component, the frequency of the sound. This simple spectral line is achieved by having a signal that, at least in theory, lasts forever. In contrast, the simplest temporal signal, an impulse, or a click in acoustics, is infinitesimally short in duration but, in theory, contains all frequencies. Of course, natural sounds are in a continuum between these extremes; they have finite duration and finite spectral breadth. However, the reciprocal relation holds. The shorter the signal, the broader its frequency content and, conversely, the longer the signal, the narrower the band of frequencies that represents it. A device constructed to analyze signals is constrained in the same way. In order to ensure good frequency resolution of the signal, one must process in narrow-frequency bands and observe for long durations. In contrast, to resolve short temporal features in the signal, the analysis must be relatively broad band. These are conflicting requirements for which advanced ears must find a solution. Let us inquire about the severity of the requirements.

Under laboratory conditions, young adult humans can discriminate two sounds whose frequencies are only about 0.2% apart. To put this in an easily understood context, consider that two adjacent keys on the piano represent a tempered semitone and are approximately 6% apart in frequency, as the cartoon of Figure 1.1 depicts. The frequency resolution of the young ear then is 30 times greater than what is needed to discriminate a "musical unit." This indicates that the neural representations, prepared by the cochlea, of two sounds 0.2% apart are sufficiently different to enable

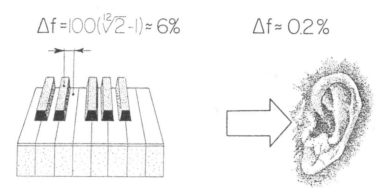

FIGURE 1.1 Diagram showing the difference in frequency between two adjacent keys on the piano and the ability of a young human to discriminate two pure tones. The former is about 6% whereas the latter is only 0.2%.

the cortex to distinguish between them. Then, engaging a relatively simple model of frequency discrimination (Zwicker 1970), one can estimate that filters used to perform such analysis would have to possess bandwidths of approximately 1.2% of the center frequency of the filter. Theory tells us that a filter of this sharpness, when excited by a brief transient, would persist in its ringing oscillatory response for a significant time period, corresponding to a time-constant equivalent to 26 cycles. Such long persistence would preclude, for all practical purposes, useful analysis of brief sounds. Yet, timing differences of 6–10 μs between signals presented to the two ears can be detected by human observers. Furthermore, at threshold, the hearing organ responds to sound with internal movements that are subatomic in dimensions. From recent measurements one can estimate these displacements to be 10^{-10}–10^{-11} m (Sellick, Patuzzi and Johnstone 1982). To gauge such motions, consider the comparison presented in Figure 1.2. The proximal input structure of the ear's sensory receptor cells is their stereocilia (Slepecky, Chapter 2). If one scales the dimensions of a single stereocilium to that of the world's tallest building, Chicago's Sears Tower, then the motion of the tip of the cilium at auditory threshold corresponds to an approximately 5-cm displacement of the top of the tower.

Broad-band root mean square (RMS) thermal noise motion of the sensory cell's input structure, its ciliary bundle, is estimated to be at least 10^{-9} m (Bialek 1987). After prefiltering by the cochlea's mechanical frequency analysis, the thermal noise reaching the cilia may be of the order of 10^{-10} m. We see that the cells detect signals that may be smaller than or of the same magnitude as the background noise.

The ear is also capable of processing sounds over a remarkably wide intensity range, encompassing at least a million-fold change in energy. To appreciate this range, in Figure 1.3 we represent a similar range of potential energies by contrasting the weight of a mouse with that of five elephants.

Another requirement is the ability to respond over a wide frequency

1. Overview 3

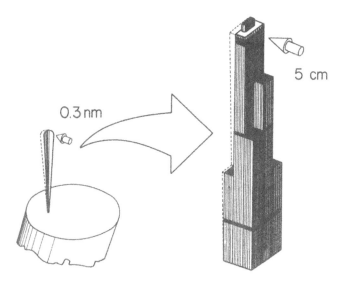

FIGURE 1.2 Illustrates the extraordinarily small motions of the sensory hairs (stereocilia) of cochlear hair cells at the threshold of hearing. If the volume of an individual cilium is scaled up to that of the Sears Tower, then the equivalent ciliary displacement is as if the tower moved 5 cm.

range. An ordinary mammal can process sounds that range over 10 octaves. Nonmammalian vertebrates generally respond to the much narrower bands of frequencies that encompass the bottom of the mammalian range. In fact, the extension of the frequency range may have been the primary pressure for developing mammalian specializations.

Among vertebrates the biological processes that subserve frequency analysis vary a great deal. In other words, there is no such thing as *the* cochlear mechanism. One well-understood and apparently widely utilized arrangement among lower vertebrates is spectral analysis by the sensory receptor cells themselves in a process of electrical resonance. An excellent review of this scheme is available (Fettiplace 1987). Electrical resonance relies on the interplay between inward (Ca^{2+}) and outward (K^+) membrane currents. It is a specialization that utilizes common cellular components and is ideally suited for regulation by changes in membrane potential. Setting of the cell's center frequency is determined by the activation rate of the K^+ current and by the number of K^+ channels. In the turtle cochlea electrical resonance frequencies are seen between 20 and 600 Hz, and the cells line up along a spatial axis of frequency in the auditory papilla (Crawford and Fettiplace 1980). Electrical resonance also determines frequency analysis in amphibian papilla (Pitchford and Ashmore 1978) and saccule (Lewis and Hudspeth 1983) and probably in some avian hair cells (Fuchs and Mann 1986). Mechanical resonance of the ciliary bundle itself may be the principal means of frequency analysis in some lizard cochleae (Weiss et al. 1978; Holton and Hudspeth 1983; Nielsen and Turner 1983). This mechanism

FIGURE 1.3 Represents the energy range encompassed between auditory threshold and the top of the physiological range of hearing (pain threshold). The range is similar to the ratios of weight of a mouse and five elephants.

affords analysis of higher frequencies than does electrical resonance. In the alligator lizard the best frequencies range between 900 an 4000 Hz (Weiss et al. 1978). The cochlea in all amniotes is remarkable for its spatial organization. Receptor cell and ciliary dimensions change along the length of the papilla (Lim 1980; also Slepecky, Chapter 2), and such gradients are thought to underlie physical alterations that control resonant frequency.

In mammals the hearing range is generally extended to higher frequencies and the process of spectral analysis is greatly elaborated. The means of reconciling the demands upon the mammalian cochlea appears to be a profoundly nonlinear local feedback process. The feedback is most commonly assumed to be manifested in a cycle-by-cycle boost in the vibratory amplitude within the cochlea and thereby the amplification of the signal (Gold 1948; Davis 1983; see Patuzzi, Chapter 4; de Boer, Chapter 5; Holley, Chapter 7). The amplification preferentially operates at low signal levels and, because of the saturating nonlinear nature of the process, becomes inconsequential at about 40 dB (100-fold increase in signal amplitude) above threshold (Zwicker 1979). The process confers high sensitivity and permits operation over a wide dynamic range. It is advantageous in effectively increasing signal-to-noise characteristics at the hair cell input (Bialek 1987) and is essential in producing the necessary degree and configuration of cochlear spectral filtering along with optimizing time-analysis capabilities. Finally, one consequence of the nature of the cochlear amplification process is the nonlinear processing of auditory signals over much of their intensity range.

This book considers cochlear function in mammalian ears only. It does not consider specializations such as those developed to subserve infra- or ultrasonic hearing. Readers interested in those topics may wish to consult Fay and Popper (1994) and Popper and Fay (1995). The emphasis of chapters dealing with mechanisms of sound analysis by the ear tends to be on spectral analysis, since this is a predominantly cochlear process. In contrast, peripheral representation of temporal features cannot, by itself, account for the time analysis capabilities of the mammalian auditory system. Temporal processing apparently relies on the central auditory nervous system's simultaneous analysis of signals presented in multiple narrow bands by the two cochleae.

Our understanding of cochlear frequency analysis progressed through three main epochs. The first was dominated by Helmholtz's suggestions that lightly damped, spatially ordered, resonant elements in the cochlea performed the spectral analysis (this early period was reviewed by Wever 1949; also see Patuzzi, Chapter 4). The second epoch, lasting from the late 1940s to the early 1970s, was dominated by von Békésy's description of the traveling wave (von Békésy 1960) and its incorporation into a theoretical schema of cochlear frequency analysis (Zwislocki 1953). The third period is now (an overview of its evolution is given in Dallos 1988; also see Patuzzi, Chapter 4; de Boer, Chapter 5) and it is an active period indeed during which a new "party line" has emerged on the basis of a wide array of experiments and theoretical considerations. According to this, von Békésy's traveling wave is boosted by a local electromechanical amplification process (Patuzzi, Chapter 4; de Boer, Chapter 5) in which outer hair cells function as both sensors and feedback elements (Kros, Chapter 6; Holley, Chapter 7). The local amplification process is under the control of the central nervous system (Guinan, Chapter 8), and information is conveyed to the central nervous system by the mechano-electro-chemical operation of inner hair cells and their afferent synapses (Kros, Chapter 6; Sewell, Chapter 9). The entire system is maintained by an elaborate homeostatic apparatus (Slepecky, Chapter 2; Wangemann and Schacht, Chapter 3). The resulting performance, noted above, is a remarkably achievement of this complex sensory system. This Overview visits the salient features of the cochlea's structure and operation. The purpose is to anticipate the details provided by the contributors of this volume, to offer a coherent overview, and to mention some topics that did not conveniently fit into other chapters.[1]

2. Anatomy and Homeostasis

The key elements for the understanding of the neurobiology of the cochlea are contained within the cochlear partition. This triangular-shaped duct (shown in Figs. 2.1 and 2.2, Slepecky, Chapter 2) is the cochlear portion of

---

[1]Portions of this material were included in the author's paper, The active cochlea (J Neurosci 12:4575–4595, 1992).

the membranous labyrinth that contains the sensory epithelia of the auditory and vestibular systems. The elaborate duct system roughly conforms to the shape of the bony labyrinth, a system of cavities in the temporal bone. The entire cavity is fluid-filled; between the bony and membranous labyrinths is a filtrate of cerebrospinal fluid or blood, the perilymph, while within the contiguous duct system the fluid is endolymph, about which more is said later (Slepecky, Chapter 2; Wangemann and Schacht, Chapter 3). The boundaries of the cochlear duct (or scala media) are the basilar membrane, which separates it from one of the perilymph-filled channels (the scala tympani); Reissner's membrane, which separates it from the other perilymphatic channel (the scala vestibuli); and the lateral wall of the cochlea. The auditory sensory epithelium, the spiral organ of Corti, is a cellular matrix situated on the scala media side of the basilar membrane. The endolymphatic space is separated from the surrounding structures by a lumenal layer of cells sealed by tight junctions (Jahnke 1975). This arrangement affords the biochemical isolation of the endolymphatic space from its surround. Lining the lateral wall is an important structure, the stria vascularis. The stria is a three-cell-layer epithelium incorporating a vascular bed. Both it and the organ of Corti may be thought of as typical ion-transporting epithelia (Wangemann and Schacht, Chapter 3).

Reissner's membrane is only two cell layers thick. Its only function is to separate endolymph from perilymph. In other words, it is "acoustically transparent" and does not influence the cochlea's mechanical functions. From the vantage point of cochlear fluid mechanics, the scala vestibuli and the scala media may be thought of as a single compartment. In contrast, the basilar membrane is one of the salient hydromechanical elements in the workings of the cochlea. Its mechanical properties control the passive, von Békésy-type traveling wave that it sustains upon sound stimulation (Patuzzi, Chapter 4; de Boer, Chapter 5). If the basilar membrane were flattened and straightened out, it would be wedge-shaped with its width gradually increasing from the cochlea's "input" end (its base) toward the far end (its apex). The change in width, generally severalfold, results in a highly significant reduction of the membrane's stiffness of more than 100-fold from base to apex.

The primary structure of the organ of Corti (Figs. 2.2 and 2.4, in Slepecky, Chapter 2) and its principal mechanical properties are conferred by the arrangement and structure of two rows of pillar cells and three rows of Deiters' cells. These cells have their nuclei located inferiorly, near the basilar membrane, and they send actin-packed phalangeal processes toward the roof of the organ. At the roof, the so-called reticular lamina, these phalangeal processes flatten into an intermeshed network that is closed to the endolymphatic space above by tight junctions between adjacent cells. The apices of sensory cells are incorporated into the reticular lamina, again sealed by tight junctions (Smith 1978). The cell bodies of the Deiters' cells also serve to anchor the bottoms of one group of sensory cells, the outer hair cells (OHCs). The other sensory cell group, the inner hair cells (IHCs), are completely surrounded by support cells. The physiological function of the several supporting cell types of the organ of Corti, such as Hensen's, Deiters', border, and inner phalangeal cells, is obscure. It is possible, however, that some of these are involved in K^+ and neurotransmitter buffering (Johnstone et al. 1989; Oesterle and Dallos 1989; 1990).

The remaining structure of great importance lying outside the organ of Corti, is the tectorial membrane. This structure is a collagenous acellular gel that floats above the reticular lamina. Its principal anchoring points are its intimate connection to the interdental cells of the spiral limbus and the firm attachment to the tips of the OHC cilia. A connection with the peripheral edge of the organ of Corti in the region of Hensen's cells via a thin marginal net is possible. The space between the tectorial membrane and the recticular lamina is open to endolymph. Consequently, the apical faces of the hair cells and the entire reticular lamina are bathed in endolymph. Relative motion between the tectorial membrane and the reticular lamina is the primary mechanical input to the hair cells of the organ of Corti (Patuzzi, Chapter 4; Kros, Chapter 6).

Afferent and efferent nerve fibers enter the cochlea from its center core, the modiolus, through a spiraling, hollow-cored, bony shelf, the osseous spiral lamina. This lamina is also the central anchor to the basilar membrane. The cell bodies of primary afferent neurons are collected at the junction of the osseous spiral lamina and the modiolus in Rosenthal's canal and form the spiral ganglion. The peripheral extensions of the bipolar spiral ganglion neurons enter the organ of Corti through small openings in the osseous spiral lamina, the habenulae perforata, and subsequently approach the IHCs and OHCs. Efferent neurons, after entering the modiolus, diverge from the afferent nerve trunk to travel with the vestibular portion of the 8th nerve to their cell bodies in the superior olivary complex (Slepecky, Chapter 2; Guinan, Chapter 8).

The cochlear endolymphatic space has two peculiarities. First, the medium filling it is similar to intracellular fluid in its major ion contents (reviewed by Bosher and Warren 1968; Anniko and Wróblewski 1986; Slepecky, Chapter 2; Wangemann and Schacht, Chapter 3). Second, within the cochlear duct, endolymph is polarized to a high positive potential, approximately $+80$ mV, with respect to indifferent tissue (von Békésy 1960). Formation of endolymph is the function of the cells of the stria vascularis in the cochlea and the dark cell epithelium around vestibular sensory structures. As the DC trace obtained by penetrating through the stria and dye-marked cell indicates (Fig. 1-4A,B), the resting polarization of marginal cells is about 10 mV more positive than that of endolymph and about 80–90 mV positive with respect to indifferent tissue (Offner, Dallos and Cheatham 1987).

An outstanding and detailed recent review of hair cell function is

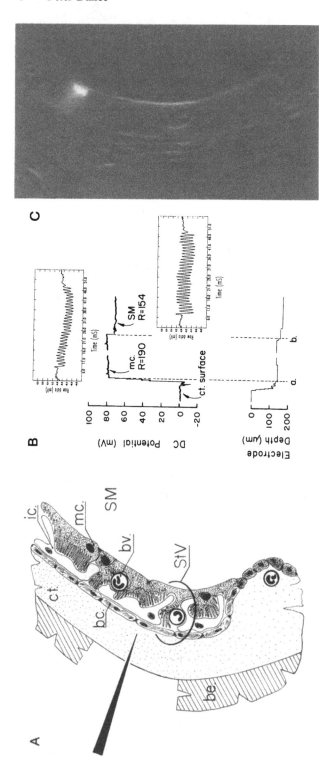

FIGURE 1.4 (A) Schematic drawing of the stria vascularis and an electrode penetrating through it. Abbreviations: ct, connecting tissue; bc, basal cell; ic, intermediate cell; mc, marginal cell; bv, blood vessel; StV, stria vascularis; SM, scala media; be, bone. (B) *Bottom trace*: Location of electrode tip if zero is the surface of the connective tissue underlying the bone. *Top trace*: DC potential recorded. Note that in the marginal cell (mc; electrode resistance 190 MΩ) the potential is about 10 mV more positive than in the endolymphatic (scala media) space (SM; electrode resistance 154 MΩ). *Insets* are recordings of electrical responses to tone burst stimuli in the marginal cell and in the scala media space. Response magnitude is somewhat larger in the scala media. (C) Image of fluorescently labeled marginal cell from which the recording is made. These data indicate that the intracellular resting potential of an identified marginal cell is about 10 mV more positive than the endocochlear potential. During penetration from the outside, there are no indications of positive resting potentials, except within the marginal cell. (Modified from Figs. 1 and 3 from Offner, Dallos and Cheatham 1987.)

available (Hudspeth 1989). Hair cells are the common receptors of the auditory and vestibular sensory organs. They are epithelial cells of somewhat variable morphology. IHCs are flask-shaped with a flat apical surface from which the stereociliary bundle, or sensory hairs, emerge. OHCs are cylindrical, with a ciliary bundle crowning their flattened apex. Ciliary bundles are strictly organized, both on a given cell and from cell to cell. Each bundle contains cilia in an array that has a distinct organization and an axis of symmetry. This axis is largely radial in the coiled cochlea. IHC cilia form a shallow "U" and OHC cilia present a "W" or "V." The foot of the configuration invariably points toward the periphery of the cochlear spiral. Within a bundle, the cilia form three parallel rows of nested "U"s or "W"s that are graded in height. The most peripheral row is invariably the tallest; the most central row is the shortest. There is a general gradation of ciliary height along the length of the cochlea, with the shortest dimensions found in the base and the longest in the apex. For example, the tallest row of cilia on OHCs changes from <1 μm in the base to ~ 6 μm in the apex, with a corresponding variation of ~ 2 to ~ 5 μm for IHCs (Lim 1980). Paralleling ciliary bundle change is a significant variation in OHC length, amounting to a fourfold increase toward the apex, reaching ~ 100 μm. OHC diameter (8–10 μm) and IHC soma shape are largely invariant. In addition to a longitudinal gradation of OHC height, OHC length also changes radially. The outermost row is the tallest and the innermost the shortest. This gradation is highly pronounced in the apex and gives a strong tilt to the reticular lamina with respect to the basilar membrane. In the base, the gradation is very modest and results in an almost parallel appearance of reticular lamina and basilar membrane. It is then clear that OHC height alone is an insufficient indicator of its longitudinal position. Cilia are exvaginations of the cell sheathed with the plasma membrane. When displaced at the tallest cilia, the entire bundle bends as a unit. Adjacent cilia in a bundle are connected to one another with filamentous material (Slepecky, Chapter 2). Whereas the apical aspect of the hair cell is dominated by the cilia, their basal, infranuclear end has the hallmarks of presynaptic regions (IHCs) and both pre- and postsynaptic regions (OHCs) (Slepecky, Chapter 2; Sewell, Chapter 9).

The IHCs are completely surrounded by supporting cells, allowing very narrow intercellular spaces. The OHCs are unique in that virtually their entire longitudinal extent is free of cellular neighbors and is bathed in perilymph that fills the spaces within the organ of Corti. The OHCs are only supported at their apices, where the cuticular plate is anchored to the reticular lamina by tight junctions, and at their bases, where the cell's bottom is cradled in the hollow of the Deiters' cells. The tips of the tallest row of cilia on OHCs are firmly embedded in the bottom layer of the tectorial membrane, whereas all their shorter cilia are free of attachment. In the adult mammal, it appears that IHC cilia are free-standing in the surrounding endolymph or at most are loosely coupled to the tectorial membrane (Lindemann et al. 1971; Lim 1980).

What may well be one of the most profound contemporary observations about the mammalian auditory system is the finding that the vast majority of afferent neurons innervate IHCs. Heinrich Spoendlin's 1969 discovery of the disparity in innervation patterns between IHCs and OHCs clearly paved the way to our present concept of cochlear function in which IHCs take the role of *the* sensory receptor in the hearing organ and in which the search is ongoing for assigning a primary function to OHCs (Dallos 1985a; Kim 1986). The peripheral processes of IHC afferents arise from approximately 30,000 (in the human) type I spiral ganglion cells that comprise 90%-95% of the afferent pool. Afferents destined for OHCs arise from type II pseudomonopolar ganglion cells. They are unmyelinated thin axons approximately 0.5 μm in diameter. As they enter the organ of Corti, they turn basally to project to a group of OHCs located approximately 0.6 mm away from the fiber's entrance. These afferents branch profusely and may innervate about 10 OHCs in the cochlear base and as many as 50 in the apex. In the adult ear, no fiber innervates both IHCs and OHCs. Central axons of both type I and II ganglion cells synapse in the ipsilateral cochlear nucleus.

It is not only the afferent innervation pattern that differs between the two receptor cell types; their efferent connections differ as well. The final descending path emerges in the superior olivary complex. Unmyelinated fibers originating in small cells around the lateral superior olivary (LSO) nucleus descend, mostly ipsilaterally, toward IHCs. They terminate on the afferent dendrites coming from IHCs and rarely on the cell bodies themselves. Myelinated fibers stemming from larger cell bodies around the medial superior olivary (MSO) nucleus travel mostly contralaterally toward OHCs. These fibers terminate in large granulated endings on OHC cell bodies and dominate their neural surround. An overview of efferent innervation is found in Warr and Guinan (1979), Warr (1992), Slepecky, Chapter 2, and Guinan, Chapter 8. The dominant neurotransmitter substance of efferents is acetylcholine; however, enkephalins, dynorphins and γ-aminobutyric acid (GABA) are also present (reviewed in Fex and Altschuler 1986; Klinke 1986; Eybalin 1993; Sewell, Chapter 9).

A simple summary of the neural connections between the organ of Corti and the central nervous system is that IHCs are almost exclusively innervated by afferents, whereas the dominant innervation of OHCs is efferent.

3. Mechanics and Micromechanics

3.1 Input to the Cochlea

All jawed vertebrates have evolved means to collect environmental sounds and funnel them to their cochleae. The external and middle ears jointly fulfill this role and their characteristics largely determine the frequency response properties (audiogram) of a given species (Dallos 1973a; Rosowski 1991, 1994). This implies that the cochlea performs as a detector of acoustic power at threshold and that the relationship between the sound power delivered to it by the middle ear and the sound power collected by the external ear is the principal determinant of the frequency profile of audibility (Khanna and Tonndorf 1969).

The resonance and diffractive effects of the head, earlobe and ear canal confer frequency-dependent pressure transformations between the external diffuse sound field and sound available at the eardrum (tympanic membrane). These linear sound transformations are dependent on the position of the head vis-à-vis the sound source (Shaw 1974) and generally produce a midfrequency boost. Sound-induced vibrations of the tympanic membrane are transmitted into the oval window of the fluid-filled cochlea by the middle ear ossicles, numbering from one to three, depending on species. The eardrum-to-oval window transformation accounts for a significant boost in sound pressure, especially at midfrequencies (between 1000 and 3000 Hz). The amplification derives from the lever action of the ossicular chain in some species, and in all species from the acoustic transformer action due to the different effective surface areas of the eardrum and the "footplate" of the innermost bone of the chain, as well as from possible force amplification by the eardrum itself. Not all power available at the eardrum enters the middle ear and not all power entering it is delivered to the cochlea. This is the consequence of shunting power away by parallel acoustic elements and losses occurring in the middle ear (Rosowski 1991). Since the middle ear is linear, sound transmission through it is quantified by a transfer function (Guinan and Peake 1967), which has been measured for several species.

The critical variable, average power into the cochlea (W_c), may be computed as $W_c = |U_s|^2 Re[Z_c]/2$, where U_s is the footplate volume velocity and Z_c is the input impedance of the cochlea. The pressure delivered to the cochlear fluid at the oval window (P_c) is obtained as $P_c = U_s Z_c$. We note that in the midfrequency range, where the input impedance of the cochlea is resistive (Zwislocki 1953), both W_c and P_c are directly determined by the volume velocity of the footplate of the innermost ossicle in mammals, the stapes. Inasmuch as the cochlea is driven by the pressure at the oval window, the consequence is that the temporal characteristics of this drive are controlled by the derivative of stapes motion.

The contemporary view of the operation of the mammalian hearing organ is that a hydromechanical event, von Békésy's traveling wave, inherent to the physical structure of the cochlea, provides the basis of frequency analysis. This rather crude analysis is augmented by a local cochlear amplification process that relies on OHCs as the feedback elements. The amplification operates effectively at low signal levels and is gradually disabled as sound input increases. When OHCs are prevented from fulfilling their feedback role, as in the case of loud sound input or injuries to the cochlea, the operation reverts to the analysis by the Békésy wave alone. Our task is to understand the physical genesis of this "passive" Békésy wave, to be followed by a discussion of the "active" feedback process that is thought to be mediated by OHCs (Patuzzi, Chapter 4; de Boer, Chapter 5; Kros, Chapter 6; Holley, Chapter 7).

3.2 Passive Mechanics

The following discussion is aided by the schematic diagram of Figure 1.5. Sound-induced, pistonlike motion of the stapes in the oval window produces pressure changes in the immediate vicinity of the footplate. The resulting acoustic events can be thought of as two waves (Peterson and Bogert 1950). The first is a pressure wave in both perilymphatic channels, traveling with the very high speed of sound waves in a liquid within rigid confines. This wave produces no forces upon the cochlear partition because it is the same in both scalae. The other wave develops as a pressure gradient across the partition. This gradient is made possible by the effective "grounding" of the scala tympani at the base by the flexible round window membrane that is placed at resting, atmospheric, pressure by its fronting on the air-filled middle ear space. The pressure difference exerts a force on the cochlear partition, setting it in motion. At low frequencies, the partition displacement is controlled by its stiffness, according to Hooke's law. Consequently, the motion is in phase with the pressure difference, which is in phase with fluid velocity, which, in turn, is determined by stapes velocity.

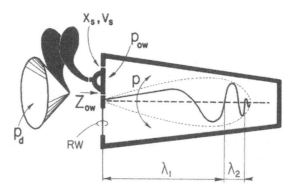

FIGURE 1.5 Schematic drawing of the eardrum, ossicular chain, and cochlea. Pressure at the drum (p_d) is transformed into stapes displacement (x_s) and stapes velocity (v_s) by the middle ear. Stapes motion creates a pressure in the vicinity of the footplate, p_{ow}, which is related to the cochlea's input impedance, Z_{ow}, and the stapes velocity. The pressure gradient across the cochlear partition is p and this pressure difference produces the traveling wave. Two full cycles of this wave are indicated by the wavelengths, λ_1 and λ_2. RW, round window.

One obvious result of this is that with a low-frequency tonal signal, the basilar membrane displacement leads in phase the stapes displacement, and thus the sound, by ~90°. At the far apical end of the cochlea, the endolymphatic space ends blindly and there is communication between the two perilymphatic scalae through an opening called the helicotrema. This opening provides an acoustic shunt across the cochlear partition for infrasonic frequencies. Therefore, extremely slow variations in sound pressure are entirely ineffective in displacing the cochlear partition.

To understand wave propagation in the cochlea, it is useful to consider a simple situation that is somewhat analogous and quite illuminating. When waves propagate in an open, fluid-filled channel, their properties are somewhat akin to those observed in the cochlea. The open water surface is the analog of the basilar membrane, whereas the rigid channel walls represent the bony boundaries of the cochlear scalae. If the wavelength is long in comparison to the depth of the channel, then the propagation velocity is independent of wavelength. Conversely, when the wavelength is short in relation to the channel depth, the propagation velocity depends directly on the wavelength. In the first, so-called long-wave case, the propagation is nondispersive. This means that in a complex multifrequency signal, all components travel together and a wave packet maintains its original shape. In the latter, short-wave case, propagation is dispersive. This means that different frequencies that start together, as in a complex acoustic signal, do not travel at the same speed. How does this relate to the cochlea?

As we saw, a pressure gradient across the basilar membrane–organ of Corti structure produces a force upon it, setting it in motion and originating a wave that propagates away from the window region. The wave's speed of travel and properties are determined by the physical characteristics of the flexible boundary between the two scalae, the basilar membrane–organ of Corti complex (Lighthill 1981). Since the stiffness of the partition decreases away from the window region, the speed of travel of this gradient wave also decreases while its amplitude increases. At a given frequency, f, as the pressure wave's velocity decreases, its wavelength (λ) becomes shorter, since $\lambda = v/f$. As a result of this shortening is that the relation between the depth of the scalae and the wavelength is altered, which produces important consequences. The depth of penetration of pressure into the canal away from the partition is roughly the reciprocal of the wave number, or $\lambda/2\pi$. Consequently, as λ decreases, the pressure across the scala becomes less and less uniform, with its increasing concentration near the partition. The less uniform the pressure, the more dispersive the wave. In other words, as a wave travels, it starts out as a nondispersive "long wave" in terms of the water-channel approximation and ends up as a dispersive "short wave." Moreover, the speed of propagation of acoustic energy (group velocity) decreases even more than that of the actual wavefronts (phase velocity). So, energy propagation progressively slows until the wave effectively halts, with the energy "piling up" at a particular, "characteristic," place (Lighthill 1991). As the energy is concentrated in a very narrow region, traveling at very low speed, any mechanical loss (damping) of the basilar membrane is sufficient to dissipate it, since this process of dissipation can take a long time. The location where this occurs depends on the frequency of the input.

Consider the local displacement of the basilar membrane at position x from the base to be $y(x)$. The local stiffness of the membrane is $k(x)$, while its local mass is $m_b(x)$. The fluid that moves with the membrane contributes an effective mass $m_f(x)$. Since the wave is concentrated more toward the basilar membrane as wavelength decreases, $m_f(x) \to 0$ as $\lambda \to 0$. Since at a given point the kinetic and potential energies associated with membrane motion are equal, we can write for the sinusoidal case (i.e., for $y = Ae^{j2\pi ft}$)

$$\frac{1}{2}k(x)A^2 = \frac{1}{2}[m_b(x) + m_f(x)]A^2(2\pi f)^2, \text{ or } (2\pi f)^2 = \frac{k(x)}{m_b(x) + m_f(x)}$$

We note that as $m_f(x) \to 0$, f approaches the membrane's own resonant frequency (f_r):

$$f_r = \frac{1}{2\pi}[\frac{k(x)}{m_b(x)}]^{1/2}$$

At this location, the membrane is purely dissipative and the wave is extinguished. Any location has a characteristic frequency, largely determined by the membrane's local stiffness. These frequencies are arranged in a spatial map so that they progressively decrease from cochlear base to apex, just as stiffness decreases along the length. One may envision this situation by noting that any location, x, has a resonant frequency that is the limiting frequency for waves propagating toward it. A wave whose frequency is the same as the resonance frequency will achieve zero wavelength at x; that is, it becomes extinguished. Waves with lower frequency propagate through this point.

The considerations above indicate that a particular basilar membrane motion pattern may be expected on simple physical grounds. With sinusoidal stapes movement, a sinusoidal pressure gradient is established across the basilar membrane. As a result, a wave motion is initiated at the basal end of the membrane. As the wave progresses away from its origin, it gradually increases in amplitude and slows down while its wavelength decreases. At a characteristic frequency-specific place, the wave energy is absorbed in the membrane's dissipation; consequently, the amplitude of vibrations rapidly diminishes. This "traveling wave" sustained on the basilar membrane was first measured by von Békésy (1960) in human cadaver ears. Some of his observed amplitude and phase patterns are reproduced in Figure 1.6. Note that as measured at the apical end of the cochlea for different low-frequency inputs, one sees at any frequency a gradual buildup of amplitude, a distinct maximum, and then a more rapid decline. Along with the amplitude changes, the phase accumulates. The lower the stimulus

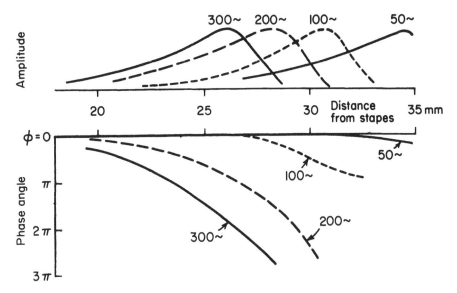

FIGURE 1.6 Von Békésy's classic data of "panoramic" amplitude and phase measurements of the traveling wave in the apical portion of the human cadaver cochlea for different stimulus frequencies, measured at different locations (abscissa) along the cochlear spiral. Normalized amplitude and phase data. The symbol ~ means Hz. (Fig. 11.58 from von Békésy 1960.)

frequency, the closer to the apex are these features apparent. This hydrodynamic behavior of the cochlea is the fundamental basis of its frequency-analyzing capability. Von Békésy measured the actual spatial pattern of vibrations by observing, at a given frequency, the amplitude and phase of basilar membrane motion at several locations ("panoramic view"; see Patuzzi, Chapter 4). An alternative method, one used exclusively in contemporary data gathering, is to observe the motion at a given point while varying frequency. This generates a so-called tuning curve or frequency response function. Some examples of this representation are given in Figure 4.11 (Patuzzi, Chapter 4).

3.3 Active Mechanics and Micromechanics

Von Békésy's traveling wave was linear; in others words, its shape was not dependent on stimulus level. It also showed shallow tuning. It is now known that von Békésy described the behavior of the "passive" or dead cochlea. In the living ear, the wave motion sustained by the basilar membrane is similar in its appearance to a traveling wave. It is, however, different both qualitatively and quantitatively from von Békésy's traveling wave, particularly in being exceedingly nonlinear and much more sharply tuned (Rhode 1971; Sellick, Patuzzi and Johnstone 1982). An example is shown in Figure 1.7 that

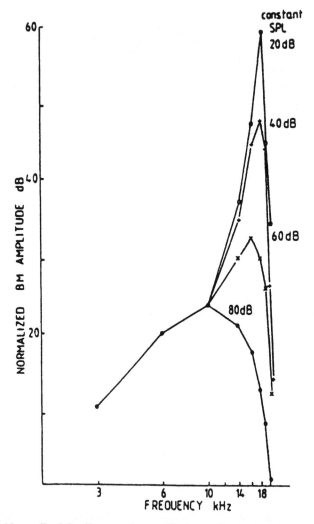

FIGURE 1.7 Normalized basilar membrane (BM) amplitude (gain) at a basal turn location in the guinea pig. Measurements are made at various constant sound pressure levels (SPL) as indicated. Note that the gain is independent of level below 10 kHz and is strongly dependent on level around the best frequency of ~18 kHz. Also note that the frequency of peak response shifts toward lower values as sound level increases. (Fig. 4B from Johnstone, Patuzzi and Yates 1986.)

highlights the increased sharpness of tuning. This plot is expressed as a gain function, that is, the ratio of basilar membrane and ossicular displacement. The gain is conspicuously dependent on stimulus level. At a given location along the basilar membrane, if measured at a high sound level, the displacement is similar to that found by von Békésy. As the sound level decreases, however, the gain functions become increasingly sharper. Note that the gain increases in the vicinity of the characteristic frequency (CF) only, and that for frequencies less than an octave below the CF the gain is independent of sound level. One may state that the response reflects a band-limited nonlinearity that is particularly pronounced at and around the CF. This phenomenon was first observed by Rhode (1971). This nonlinearity, and indeed the sharpness of tuning itself, depends on the physiological condition of the cochlea (Rhode 1973). This is convincingly demonstrated in Figure 4.14B (Patuzzi, Chapter 4). With increasing level and with cochlear deterioration, the best frequency shifts to lower values. This shift is just shy of an octave.

Some theoretical work intimates that the frequency response patterns are so sharp that they cannot be reproduced by a model in which all basilar membrane vibratory energy is derived from the original sound input (de Boer 1983; Neely and Kim 1983; Diependahl, deBoer and Viergever 1987; Geisler 1991; Zweig 1991; de Boer, Chapter 5). Passive linear models are well suited to characterize the dead cochlea. Passive nonlinear models, particularly those that allow for more than one degree of freedom in describing basilar membrane–related movements, are successful in representing many features of contemporary data but are generally incapable of quantitatively describing both amplitude and phase patterns (Viergever 1986). Active models incorporate a local supply of energy that may be utilized to selectively boost basilar membrane motion. Energy needs to be supplied to the traveling wave in a region that is basal to the best frequency. This is usually done with the formalism of providing "negative damping," meaning that there is some mechanism of counteracting the inherent damping (viscous friction) of the basilar membrane and cochlear fluids (Neely and Kim 1983). The basic idea is the following. If somehow a force is provided to the basilar membrane in opposite phase with the force produced by the passive damping, then the two forces cancel. Damping force is proportional to the velocity of motion. Augmentation requires a cycle-by-cycle force that lags displacement by 90°. We should note that expert opinion is not undivided as to the necessity of an active cycle-by-cycle feedback process. For example, Allen and Fahey (1992) maintain that the cochlear amplifier, as generally construed, is not necessary to explain the extant data.

Theoretical considerations, requiring amplification in the form of feedback, yield several predictions about possible vibratory patterns and also impose certain requirements on the system. Among the predictions is the possibility of instability, that is, oscillations due to excessive feedback. Interestingly, these may be observed in some situations. Acoustic energy has been shown to be produced in some ears where spontaneous oscillations, apparently of cochlear origin, are retransmitted by the middle ear and are detectable as sound in the ear canal (Wilson 1980). These so-called spontaneous otoacoustic emissions are the strongest evidence available that there is a possibility for the *production* of vibrations in the cochlea. It was demonstrated, by examining the amplitude spectrum of such emissions, that they are not merely noise filtered by a sharply tuned resonant element but the products of an active oscillatory source (Bialek and Wit 1984). One should note, however, that spontaneous otoacoustic emissions have been recorded from several nonmammalian vertebrates in which no evidence for cochlear amplification (active feedback process) exists (for review see Manley and Taschenberger 1993). Sound emissions from the ear that originate in the cochlea in response to acoustic inputs also support the validity of active processes providing amplification of the traveling wave (Kemp 1978). A requirement of any theoretical scheme of cochlear amplification is that energy is supplied locally and in a frequency-specific manner. This implies that the mechanism that provides the feedback needs to be tuned or at least possess some frequency-dependent characteristics. This requirement is considered after the identity and properties of the putative feedback element are discussed.

A great deal of evidence gathered in numerous laboratories during the 1970s (reviewed in Dallos 1988; Patuzzi, Chapter 4) indicated that OHCs were intimately associated with the sensitivity, frequency selectivity, and nonlinearity of the cochlea. Figure 1.8 summarizes the results of many investigators. It shows that damage to OHCs desensitizes, detunes, and linearizes the cochlear response. Fibers located near the lesion, while maintaining normal tuning curve shape, lose some two-tone suppression and have elevated $2f_1-f_2$ thresholds (middle panel). Fibers originating from the region of absent OHCs have abnormal tuning curves, no two-tone suppression, and do not show responses at $2f_1-f_2$. In many ways, without OHCs the cochlea behaves as in a dead animal. It is generally accepted today that OHCs function as the feedback elements in the cochlear amplification process. The key discovery was the demonstration that OHCs isolated from the cochlea are capable of shape changes at audio frequency rates upon electrical simulation (Brownell 1983; Brownell et al. 1985; Kachar et al. 1986; Zenner 1986; Ashmore 1987). These electromotile responses are primarily length changes of up to 3%-5% of the total length under maximal electrical stimulation. Depolarization of the cell results in contraction and hyperpolarization results in elongation (Ashmore 1987). Small diameter changes are also measurable. The controlling variable is transmembrane voltage change (Santos-Sacchi and Dilger 1988). The motile response is nonlinear and shows strong rectification in the contraction direction (Evans, Dallos and Hallworth 1989; Santos-Sacchi 1989). The sensitivity of the response is approximately 2-30 nm/mV at low frequencies. The correlated behavior under voltage clamp of motile response and nonlinear capacitive ("gating") currents suggests that motility involves the movement of membrane-bound charged molecules (Ashmore 1990; Santos-Sacchi 1991). Other work also intimates that OHC electromotility is produced by the concerted direct action of a large number of independent molecular motors that are closely associated with the cell's

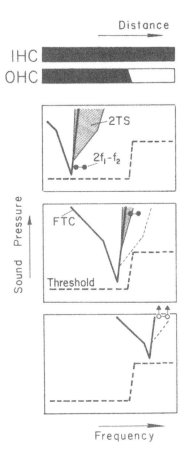

FIGURE 1.8 Schematic depicting the effects of OHC lesions on various characteristics of single auditory nerve fiber responses. *Bar graph* on the top depicts the presence (solid) or absence (open) of IHCs and OHCs along the length of the cochlea. Bottom three graphs are frequency response plots. *Dashed line* indicates neural threshold. At the boundary of the lesion the threshold becomes poorer by 40–50 dB (Ryan and Dallos 1975). For a fiber that originates far from the lesion (*top panel*), all responses are normal. Thus the tuning curve (FTC, *solid lines*) is sharply tuned with long tip segment. There is a prominent high-frequency two-tone suppression area (2TS) and the fiber responds to the $2f_1-f_2$ distortion component when the primaries f_1 and f_2 are presented at very low levels, even if the fiber does not respond to the primaries themselves. As the fiber's location of origin approaches the lesion, the FTC remains normal while the nonlinear responses, 2TS and $2f_1-f_2$, diminish. A fiber originating from the lesioned region (*bottom panel*) has a high threshold, a short or nonexistent tip, and does not produce nonlinear responses (Smoorenburg 1972; Dallos and Harris 1978; Harrison and Evans 1979; Dallos et al. 1980; Schmiedt, Zwislocki & Hamernik 1980).

basolateral membrane (Holley and Ashmore 1988; Dallos, Evans and Hallworth 1991; Holley, Chapter 7). This response is clearly a biophysical process and not dependent on ATP or Ca^{2+} (Kachar et al. 1986; Holley and Ashmore 1988). It is of such speed as to rule out all hitherto described motile mechanisms (Dallos and Evans 1995a).

In vivo, the electromotile response may be driven by the receptor potential produced in OHCs in response to acoustic stimulation (see below). Length changes resulting from alternating depolarization and hyperpolarization may feed back a cycle-by-cycle mechanical force upon the basilar membrane–tectorial membrane system. In addition, due to the rectifier properties of electromotility, a DC length change in the contraction direction would be produced as well. The AC force can act to counteract viscous damping, whereas the DC force may alter the basilar membrane-tectorial membrane relationship, thus setting the operating point of the micromechanical system (Dallos 1988). Aside from stimulus-evoked fast motile responses, a wide variety of agents are capable of stimulating slow contractions of OHCs (Brownell et al. 1985; Zenner, Zimmerman and Schmitt 1985; Flock, Flock and Ulfendahl 1986) or eliciting interactions with fast motility (Sziklai and Dallos 1993; Housley, Connor and Raybould 1995). Of particular interest is the modulatory effect elicited by focal application of acetylcholine, the transmitter substance for cochlear efferents (Sziklai and Dallos 1993). The possibility is thus open that the state of the organ of Corti may be mechanically modified by the central nervous system (see Guinan, Chapter 8) via alteration of motility gain mediated through efferent influence (Kim 1986; Zenner 1986) as well as stimulus-evoked DC contractions (Dallos 1988).

Energetic considerations intimate that the receptor potential is capable of supplying energy greatly in excess of that required to overcome the cell's own internal stiffness in producing extension–elongation cycles that are commensurate with basilar membrane movements in vivo. How this is affected by the mechanical load upon the cell in situ is unknown. The OHC motor itself, if driven by constant membrane voltage changes, is capable of AC displacements of undiminished amplitude up to at least 24 kHz (Dallos and Evans 1995a). However, an often-mentioned problem in associating OHC electromotile responses with a feedback role in vivo is the high-frequency attenuation of the receptor potential due to the cell's basolateral membrane capacitance. The high-frequency attenuation of the cell's receptor potential clearly reduces the cell's ability to produce sufficient motile feedback at the high end of the audio range. One can estimate that OHC electromotile displacement amplitude becomes less than basilar membrane motion at approximately 6 kHz. At higher frequencies, it is unlikely that electromotility, as commonly construed, would be capable of influencing cochlear mechanics. We have recently suggested as a partial remedy that the driving voltage for OHC motility at higher frequencies is an extracellular AC potential[2] gradient between scala media and intra-organ of Corti fluid spaces (Dallos and Evans 1995a,b). It was shown that the voltage gradient across the basolateral membranes of weakly excited cells basal to the CF location is greater (for the 6- to 20-kHz region) than that for strongly excited CF cells (Dallos 1996). This would allow basally located cell groups to provide electromotile feedback at frequencies where CF cells could no longer do so. Of course, it is also possible that in mammals some other hitherto unobserved aspects of motility, such as ciliary movement or change in ciliary stiffness (Kros, Chapter 7) are the critical feedback variables. Such phenomena have been detected in nonmammalian hair cells (Crawford and Fettiplace 1985; Howard and Hudspeth 1988; Assad, Hacohen and Corey 1989). Finally, the door still needs to be left open to the possibility that the feedback action in vivo is not produced by outright OHC motility, but that when these cells are constrained by the organ of Corti framework, the significant variable is a change in their axial stiffness and the consequent modulation of the total basilar membrane–organ of Corti stiffness (Kolston et al. 1989).

A locally active cochlea (see de Boer, Chapter 5) requires the presence of two frequency-analyzing systems. The second is obvious; this is the preanalysis by the traveling wave, which is then boosted by cochlear amplification. The first system is related to the selection of the cochlear region, for a given frequency, from which energy is provided to the traveling wave. Consider for argument's sake that the traveling wave peaks at 5 mm from the stapes for a pure tone f_o. Assume further that the active process operates about ⅓ octave basal from the traveling wave peak, or approximately at the 4.2 mm location in the guinea pig. In order for this location to be different from its neighbors and for it to "know" that it needs to provide feedback (OHC motility), it needs to be selective to f_o. This implies a second cochlear map that is systematically detuned from the "main" map by about ⅓–½ octave. It is generally assumed that this second filter can be found in a resonance between tectorial membrane and OHC cilia (Zwislocki and Kletsky 1979) or by resonance of the tectorial membrane itself (Allen 1980). There is some preliminary experimental evidence that indicates that the tectorial membrane may possess such tuned characteristics (ITER 1989; Allen and Fahey 1993; Gummer et al. 1993).

3.4 Transmission of Forces to Cilia

Forces exerted on the ciliary bundle may derive directly from the relative displacement between the reticular lamina and the tectorial membrane to which some of the cilia are attached. Alternatively, the flow of endolymph in the narrow gap between the two moving surfaces may be a consequence

---

[2]Extracellular potentials and their origins are discussed below.

of the aforementioned relative motion and deliver viscous forces to the cilia (Billone and Raynor 1973; Freeman and Weiss 1990). The latter forces are more efficient in stimulating cilia that are not attached to the tectorial membrane. In OHCs only the tallest row of cilia are connected to the tectorium, and it appears that IHC cilia are either entirely free-standing or only tenuously connected to the tectorial membrane (Lim 1980). There are important functional consequences of these anatomical differences.

As a result of the viscoelastic coupling of forces to IHC cilia, their displacement is likely to be proportional to basilar membrane velocity at low frequencies (Billone and Raynor 1973; Freeman and Weiss 1990). There is experimental evidence to support this claim, both indirectly through cochlear microphonic data (Dallos et al. 1972; Dallos 1973b) and directly from intracellular recordings from IHCs (Nuttall et al. 1981; Russell and Sellick 1981; Dallos and Santos-Sacchi 1983). At higher frequencies, above 300–500 Hz, ciliary displacement becomes proportional to basilar membrane displacement. The low-frequency velocity dependence, aside from imposing a dynamic characteristic upon the IHC response, has a more profound consequence. Because of the coupling properties of IHC cilia, very-low-frequency (DC) signals are probably ineffective in stimulating IHCs. In other words, the DC displacements of basilar and/or tectorial membranes, if any, that may arise from asymmetrical nonlinear processes are not likely to produce an IHC response (Dallos and Cheatham 1989).

4. Transduction and Receptor Potentials[3]

4.1 Transduction

The excitatory proximal stimulus to any hair cell is the bending of the ciliary tuft toward the kinocilium (Hudspeth and Corey 1977; Hudspeth and Jacobs 1979) or, in the case of mammalian hair cells, toward the tallest row of stereocilia (Russell and Richardson 1987). We have already considered how the forces that bend the bundle may arise. Extracellular current measurements indicate that the sink for transducer current is near the top of the ciliary bundle (Hudspeth 1982). An attractive and coherent hypothesis of hair cell transduction implicates mechanically gated channels located at or near the tips of stereocilia that are controlled by changing tension in attached elastic elements, termed gating springs (Hudspeth 1989). Slender filamentous tip links (Fig. 2.6D, Slepecky, Chapter 2) reach up from the tops of short cilia to the sides of adjacent tall cilia in a bundle, parallel to the mirror-symmetry axis (Pickles, Comis and Osborne 1984). The arrangement of tip links is consistent with the cell's directional sensitivity; stretching the links by displacing the bundle toward the tallest cilia is

---

[3]This section has been adapted from Dallos (1991).

excitatory, movement in the opposite direction is inhibitory, while orthogonal displacement is ineffective (Shotwell, Jacobs and Hudspeth 1981). Approximately 50% of the hair bundle's total stiffness (~ 300 μN/m) may be attributed to its basal pivot, while the rest resides in the gating springs. The latter stiffness depends on bundle position. It is minimal at displacements about the resting position where transduction gain is highest, suggesting an association between gating stiffness and open or closed states of transducer channels (Howard and Hudspeth 1988).

Figure 6.6 (Kros, Chapter 6) represents a version of the mechanical gating process of channels via the tip link (gating spring) connection. The number of channels per cilium is estimated to be few (Howard and Hudspeth 1988) and the "swing" of the gate about 4 nm. Because of its direct activation by the mechanical stimulus, the opening and closing times of the gate are very short, certainly less than 50 μs. The times are likely to be much shorter in animals in need of very-high-frequency hearing, such as chiropterans (Hudspeth 1989). There is strong evidence that during maintained bending of the cilia, the attachment point of the tip link is actively modified and travels up and down the cilium to produce a resetting of its elasticity and, thereby, adaptation of the receptor current (Eatock, Corey and Hudspeth 1987; Corey and Assad 1991).

The transducer channels themselves are nonselective aqueous pores that permit the passage of both monovalent and divalent cations (Corey and Hudspeth 1979) and probably possess single-channel conductance of ~ 50-100 pS (Ohmori 1984; Crawford, Evans and Fettiplace 1991). In all inner ear systems, the cilia, and hence the transducer channels, are in contact with endolymph, making K^+ the likely carrier of transducer current. The driving force for K^+, instead of an electrochemical potential, is the electrical gradient across the ciliary membrane produced by the sum of the cell's resting potential and the positive endocochlear potential. In the absence of stimulation, about 5%–15% of the transducer channels are open (Hudspeth and Corey 1977; Crawford and Fettiplace 1981; Russell, Cody and Richardson 1986). As a consequence, there is a standing current producing a "biased epithelium," in that the cell is normally somewhat depolarized to bring it closer to the range of activation of voltage-gated channels of the basolateral cell membrane. The resting potential of cochlear IHCs in vivo, measured with sharp microelectrodes, is between -40 and -50 mV (Russell and Sellick 1978; Dallos 1985b). A necessary caveat to these data is that the microelectrode almost certainly creates a significant leak conductance so that the cell appears more depolarized than it really is (see Kros, Chapter 6).

The transducer current in response to ciliary deflection has been measured for different hair cells. The relationship between current and ciliary rotation is sigmoidal and indicates great asymmetry for deflections in opposing directions. The gating process underlying transduction is probably describable as a three-state Boltzmann function (Corey and Hudspeth 1983;

Holton and Hudspeth 1986). Strong saturation is seen for bundle displacements in excess of 0.5 µm, which corresponds to a rotation of about 5°, and about 90% of the useful range is encompassed within ±1°. In the best hair cells, the transducer conductance is 2.5-5 nS and the sensitivity of the process is about 600 pA/µm (Crawford, Evans and Fettiplace 1989). This is similar to that found in frog saccular hair cells (Holton and Hudspeth 1986). Patch-clamp recordings reveal single-channel currents of ~9 pA and single-transducer-channel conductance of ~106 pS (Crawford, Evans and Fettiplace 1991). Assuming a hair cell input resistance in vivo of ~50 MΩ (Russell and Sellick 1978), the receptor potential produced may be around 0.1 mV at 1 nm ciliary displacement. This is of the same order as is seen experimentally around zero dB sound pressure level (Dallos 1985b).

It is noted that virtually all information about transducer channels and transducer gating comes from work on nonmammalian vertebrates. It is likely that the process of transduction is highly conserved for all hair cells and that the above information is applicable to the mammalian hearing organ and presumably to both IHCs and OHCs. In contrast, as noted below, the properties of the hair cell's basolateral membrane are likely to show considerable divergence that will not permit easy generalizations.

4.2 Basolateral Membrane and Receptor Potentials

Acoustic simulation and its mechanical consequences are AC signals. In other words, these are the sums of sinusoidal variations about a resting value. Nonlinearities at any stage of processing may distort the frequency composition of the signal and add DC components due to rectification and asymmetries. It is customary when analyzing receptor potentials to distinguish between their frequency-following, or AC, and their unidirectional, or DC, components.

Voltage drops produced by the receptor current may be shaped and altered by voltage-dependent conductances found in the cell's basolateral membrane. The measured receptor potential is the result of interaction among all active conductances. In some hair cells, turtle cochlear hair cells serving as a prime example, the basolateral conductances dominate the character of the cell's voltage response (Crawford and Fettiplace 1981). In contrast, the dominant character of the receptor potential of mammalian IHCs is likely determined by the transducer conductance itself (Kros and Crawford 1990). One form of ubiquitous influence by the basolateral membrane is due to the inevitable low-pass filtering, owing to the parallel resistance-capacitance (RC) of this membrane segment. The filter shunts those frequencies that are above its corner frequency $f_o = 1/2\pi RC$, and consequently limits the size of the AC receptor potential at high frequencies. It is widely assumed that limitations on neural phase-locking are largely a consequence of this presynaptic filtering action (Palmer and Russell 1986; Weiss and Rose 1988; Kidd and Weiss 1990). Inasmuch as R

(incorporating time- and voltage-dependent conductances) is dependent on the basolateral membrane voltage (Dallos and Cheatham 1990; Russell and Kössl 1991), the corner frequency changes with receptor potential (stimulus) level (Kros and Crawford 1990). The larger the input and hence the receptor potential, the higher the frequencies that may be transmitted by the hair cell. Even if these conductances are fully activated, it is unlikely that the corner frequency would much exceed 1 kHz, setting a maximum lower limit for the deterioration of phase-locked transmitter release.

Voltage-gated Ca^{2+} channels and Ca^{2+} gated K^+ channels have been identified in nonmammalian hair cells (Crawford and Fettiplace 1981; Lewis and Hudspeth 1983; Pitchford and Ashmore 1987; Fuchs and Evans 1988; Fuchs, Nagai and Evans 1988). In nonmammalian vertebrate hair cells, the rapidly activating Ca^{2+} inward current and delayed $K^+(Ca^{2+})$ outward current interact together and with the membrane capacitance to produce damped oscillations with high quality factor (Q). There is no direct evidence for resonant behavior in mammalian cochlear hair cells except for a very rapidly damped resonance in IHCs at driving current levels that are unlikely to be physiological (Kros and Crawford 1990; Kros, Chapter 6). In anuran hair cells, the Ca^{2+} channels and the $K^+(Ca^{2+})$ channels cluster together with preference to the synaptic region and probably correspond to presynaptic active zones (Roberts, Jacobs and Hudspeth 1990). In these same cells, the only conductances found in the cell's ciliated apex are those associated with transducer channels. The uneven division of ion channels between the apical and basolateral surfaces is a common characteristic of secretory and sensory epithelia. A preliminary report on OHCs indicates the possibility of the presence of conductances other than those of the transducer in the apical membrane (Gitter, Zenner and Frömter 1986). We now know that purinergic channels are present in the apex (Housley, Greenwood and Ashmore 1992).

Basolateral membrane conductances in IHCs (Kros and Crawford 1990) are dominated by two different voltage-gated K^+ channels, which are active in the -60 to -20 mV membrane potential range that encompasses all values seen in vivo. The two outward K^+ currents shape the receptor potential, in that a very rapid initial decline is a consequence of the activation of the tetraethylammonium (TEA)-sensitive conductance and a gradual, adaptation-like decline is produced by the activation of the 4AP-sensitive conductance. Neither K^+ current appears to be dependent on Ca^{2+} influx into the cell. Isolated IHCs have a zero-current membrane potential of approximately -67 mV and a conductance of about 2 nS. The conductance increases to 300–500 nS when fully activated; it decreases to as low as 0.5 nS when the cell is hyperpolarized. Figure 6.15 (Kros, Chapter 6) depicts the various currents and channels that dominate IHC electrical phenomena.

OHC basolateral membranes are endowed with two types of Ca^{2+}-activated K^+ channels (Ashmore and Meech 1986). It is likely that the basolateral membrane also possesses Ca^{2+} channels, inasmuch as blocking the K^+ current by internal Cs reveals a voltage-dependent inward current (Santos-Sacchi and Dilger 1988). The zero-current membrane potential of isolated OHCs ranges from -10 to -68 mV (Santos-Sacchi and Dilger 1988), "up to" -70 mV (Gitter, Zenner and Frömter 1986), and from -15 to -40 mV (Ashmore and Meech 1986). Apparently a rapid loss of internal K^+ and Na^+-loading occurs upon isolation and is responsible for the variable and low membrane potentials. If the cytoplasm is loaded with K^+, the resting potentials stabilize around -60 mV (Ashmore and Meech 1986). The mean conductance around this resting potential is 9.2 nS (Santos-Sacchi and Dilger 1988).

The first successful intracellular recordings from in vivo mammalian IHCs were obtained by Russell and Sellick (1977) and from OHCs by Dallos, Santos-Sacchi and Flock (1982). The properties of these receptor potentials have been extensively characterized (Russell and Sellick 1978, 1983; Russell 1983; Dallos 1985b, 1986; Cody and Russell 1987; Dallos and Cheatham 1989, 1991; Zwislocki 1989). Receptor potential measurements are also available from organotypic cultures of the mouse cochlea (Russell, Cody and Richardson 1986; Russell and Richardson 1987).

The schematic of Figure 1.9 depicts the two techniques employed for in vivo recordings from hair cells of the mammalian cochlea. In the original method of Russell and Sellick (1978), the organ of Corti is approached from the scala tympani. The method permits excellent control over the orientation and direction of electrodes, but thus far it has been applied to only that region of the cochlea having CFs between 16 and 20 kHz. With the technique developed by Dallos, Santos-Sacchi and Flock (1982), the electrode reaches the organ of Corti through an opening on the lateral bony wall of the cochlea. In theory, this approach is usable in all cochlear turns and has been applied to the fourth (CF $\approx$ 200 Hz), third (CF $\approx$ 1000 Hz), and second (CF $\approx$ 4000 Hz) turns. A drawback of the technique is the limited ability to visualize the target organ of Corti.

Both recording techniques reveal similar membrane potentials, approximately -40 mV for IHCs and -70 mV for OHCs. Similarly, both techniques yield IHC receptor potentials that imply invariance in the behavior of these cells along the length of the cochlea. In response to tones, all IHCs produce stereotyped receptor potentials. These contain a fundamental response, a DC component, and a harmonic series. At low sound levels, the fundamental response increases in proportion to the signal level, whereas the DC response reflects square-law behavior. Harmonics also rise faster than the fundamental. All responses saturate at moderately high sound levels. As the time patterns in Figure 1.10 show, the harmonic content is clearly manifested in the waveforms. The harmonic content plus the DC make the response waveforms exceedingly asymmetrical about the resting potential. The positive and negative peaks of the

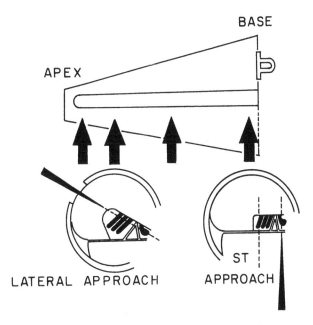

FIGURE 1.9 Schematic showing the two approaches to intracellular recording from cochlear hair cells in vivo. In the lateral approach (Dallos, Santos-Sacchi and Flock 1982), the electrode passes through a fenestra in the cochlear bone and approaches the organ of Corti through the scala media. The method has been used to collect data from the three low-frequency turns of the guinea pig cochlea (*three left arrows*). In the scala tympani (ST) approach (Russell and Sellick 1978), the electrode passes into the organ of Corti through the basilar membrane from the opened scala tympani. This approach is suitable to the high-frequency region of the cochlea (*right arrow*) (Fig. 2 from Dallos and Cheatham 1992.)

response plotted against sound pressure in the ±1 Pa range show the asymmetry of the receptor potential. Particularly severe saturation is always evident in the hyperpolarizing direction. The properties of the receptor potential versus sound pressure functions seem to be common among hair cell responses in various systems and may be quantified as rectangular hyperbolas or Boltzmann functions (Hudspeth and Corey 1977; Crawford and Fettiplace 1981; Russell and Sellick 1983; Dallos 1986).

The pattern is similar for a given IHC at all stimulus frequencies, but the quantitative properties of the response differ. In terms of sensitivity and many nonlinear response properties, the IHC reflects its mechanical input. Thus the tuning of its place along the cochlea is the prime determinant of the quantitative features of the IHC's response. As Figure 1.11 depicts for AC, and Figure 6.9 for DC, both fundamental and DC response components are tuned and the sharpness of tuning decreases with stimulus level. The plots of Figure 1.11 are gain plots, i.e., amplitude plots normalized for

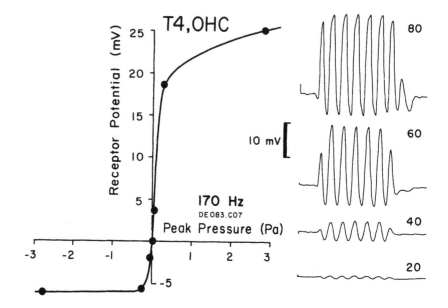

FIGURE 1.10 Intracellular recording from a fourth-turn OHC near the best frequency. *Left*: linear plot of peak receptor potential magnitude as a function of peak sound pressure at the eardrum. *Right*: waveforms of responses to sinusoidal tone burst stimuli at the indicated sound pressure levels. (Fig. 9 from Dallos 1986.)

sound level. Response saturation is most evident at the CF and more linear responses are found far from CF. All such features reflect the properties of basilar membrane vibrations (e.g., Ruggero and Rich 1991).

The low-level, low-frequency steepening of the AC plots reflects the velocity-dependence of the IHCs' low-frequency response. Concomitant with the slope change, there is an associated phase-lead that reaches 90° at the lowest frequencies (not shown). Aside from shape changes with level, there are also phase changes that depend on the relation between stimulus frequency and CF. This behavior also reflects the preceding mechanical nonlinearity. We note that the magnitude function of the DC component is sharper and reflects a square-law relation with the fundamental. It is also noted that no matter what the stimulus conditions, the DC response of IHCs is always depolarizing. When tuning curves (iso-response functions) are compared for AC and DC components, they have the same shape both at low frequencies (Dallos 1985b) and at high frequencies, after correction for the filtering of the AC (Russell and Sellick 1978). Inasmuch as high-frequency receptor potentials are shunted by the IHCs' basolateral RC filter (Russell and Sellick 1978), they cannot be effective in stimulating synaptic transmitter release. It follows that stimulus-related discharge-rate changes in primary auditory afferents and changes in their precursor transmitter exocytosis must be governed by DC receptor potentials pro-

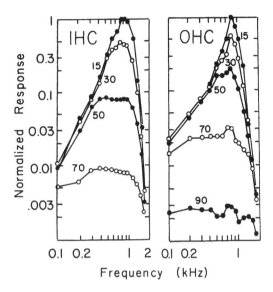

FIGURE 1.11 Normalized magnitude plots for AC receptor potentials recorded intracellularly in vivo from third-turn IHC and OHC from the same cochlea. Plots are obtained by shifting the response plots down to compensate for increases in sound pressure (shown next to the plots). (Modified from Figs. 10 and 11 from Dallos 1985b.)

duced in the nonlinear response of the IHC itself. This DC is likely to arise in the rectifier properties of the IHC transduction process (Russell and Sellick 1983; Dallos 1986; Dallos and Cheatham 1989), but may be enhanced by nonlinear properties of the cell's basolateral membrane (Dallos and Cheatham 1990). IHC receptor potentials reflect all well-studied nonlinear features of cochlear function, such as intermodulation distortion, harmonic production, and two-tone suppression (Sellick and Russell 1979; Cheatham and Dallos 1990).

Aside from two primary differences to be considered below, the previous description of IHC behavior could be applied to OHCs as well. In other words, OHC and IHC receptor potentials are more similar than different (Fig. 1.11). The first generally observed difference is seen in recordings of AC and DC responses from low-frequency IHCs and OHCs. There is a ~6 dB/octave low-frequency slope difference coupled with a ~90° phase difference between frequency responses of OHCs and IHCs. Both features reflect the basilar membrane displacement-dependence of OHCs and the velocity-dependence of IHCs at low frequencies. Second, whereas IHCs produce only depolarizing DC responses, OHCs generally show a frequency- and level-dependent transition between hyperpolarizing and depolarizing asymmetries. The frequency response plots of Figure 1.11 show AC responses, while those of Figure 6.9 (Kros, Chapter 6) depict DC responses at several levels, and the below-CF, low-level hyperpolarizing responses are amply evident. Such hyperpolarizing responses are more prominent in basal-turn OHCs below the CF (Cody and Russell 1987).

Certain features of OHC receptor potentials, however, are dependent on cochlear location, technique, or both. Recordings from OHCs in the high-frequency end of the cochlea via the scala tympani approach reveal a fundamental difference from that seen in lower-frequency regions. It is found that at the CF, and at low and moderate sound levels, the receptor potential versus sound pressure functions are symmetrical. In other words, little or no DC is produced. The comparison between the two types of recording is highlighted in Figure 1.12A. First, note that because of the basolateral membrane filters, the AC component is negligibly small in the basal turn recording for either IHCs or OHCs. The DC response does not appear in OHCs until the sound level reaches ~90 dB, whereas from an IHC located at the same place, the DC component is already very prominent at 10 dB sound pressure level. In contrast, at the 1-kHz location, both IHC and OHC DC responses are measurable at 30 dB. When measured in organotypic cultures of mouse cochleas (Russell, Cody, and Richardson 1986), both IHCs and OHCs exhibit the type of receptor potential versus input level functions as exemplified in Figure 1.12B. In other words, in the culture both cells show asymmetry.

All intracellular recordings in vivo suffer from a potential problem. We have already alluded to the artifact produced by imperfect sealing of the cell membrane around the recording microelectrode and the leakage conductance produced thereby. Aside from altering the cell's membrane potential, the leak conductance shunts out the effects of voltage-gated conductances of the cell's basolateral membrane (see Kros, Chapter 6). As a consequence, in vivo recordings of receptor potentials may not reflect the behavior of these potentials as they would appear in the absence of the recording electrode. What these recordings do reflect is the receptor current as it produces a voltage drop on the (largely ohmic and linear) basolateral leak resistance. The receptor current, in turn, renders ciliary deflection, as driven by macro- and micromechanical processes, through the nonlinearity of the transducer itself. At low signal levels, well below transducer saturation, the intracellular receptor potential is certainly a good representation of the preceding mechanical processes.

Some representations of the hair cell receptor currents can be measured extracellularly within various fluid compartments of the cochlea and even at more remote extracochlear locations. One can consider hair cells as sources of current that spreads through the complex electrical network surrounding them and that produces voltage drops, according to Ohm's law, on any electrical impedance across which the measurements are made. Aggregate receptor currents produced by groups of hair cells generate voltage drops on any external impedance in the current path. Consequently, these voltage drops (the extracellular receptor potentials) also reflect the transducer

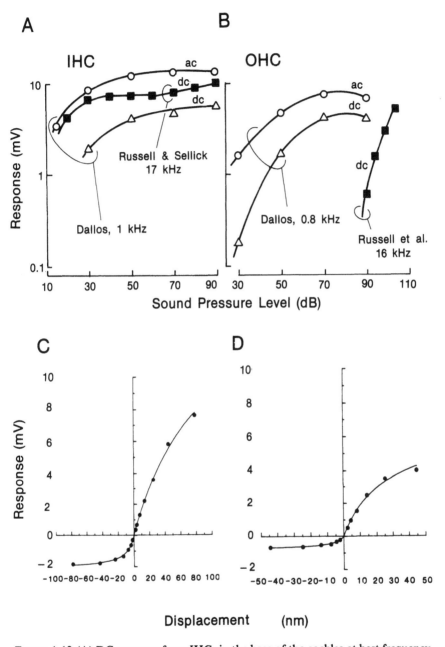

FIGURE 1.12 (A) DC response from IHCs in the base of the cochlea at best frequency (after Russell and Sellick 1978) and AC and DC responses from an IHC in the third turn at best frequency. (Dallos unpublished.) (B) DC response from an OHC in the base of the cochlea at best frequency (after Russell, Cody and Richardson 1986) and AC and DC responses from OHCs in the third turn (after Fig. 6 from Dallos 1985b). (C) Peak response magnitude versus ciliary bundle displacement recorded intracellularly from an IHC in a mouse organotypic cochlear culture. (D) As in C, but from an OHC. (From Fig. 5 of Russell, Cody and Richardson 1986.)

process. This reflection, however, is weighted and distorted by a spatial integration process over all active, current-producing hair cells (Whitfield and Ross 1965; Dallos 1973a). Connection between the intracellular and extracellular receptor potentials is made via a knowledge of cochlear "electroanatomy," the study of electrical impedance patterns in the cochlea (von Békésy 1960).

Indeed, the electrical activity of the cochlea was identified from such remote measurement of voltage drops, first of the AC component (cochlear microphonic, CM; Wever and Bray 1930), then of the DC component (summating potential, SP; Davis, Fernández and McAuliffe 1950). Aside from their limited clinical utility, contemporary interest in extracellular receptor potentials stems from some of their properties. The principal one is that the production of the extracellular AC potential is dominated by OHC receptor currents (Davis et al. 1958; Dallos and Cheatham 1976; Oesterle and Dallos 1989). The theoretical reason for this is examined below, but the utility of the relationship is noted here. In spite of its 14-year history (Dallos, Santos-Sacchi and Flock 1982), it is now apparent that intracellular recording from OHCs will never be routine. Consequently, there is a significant motivation to evaluate how well various extracellular responses can be used to assess OHC operation (Patuzzi, Yates and Johnstone 1989). Furthermore, we recently proposed that extracellular AC voltage gradients may have a direct function in stimulating OHC electromotility in vivo at high frequencies (Dallos and Evans 1995a,b). To evaluate this proposal, it is necessary to obtain detailed measurements of extracellular receptor potentials and electrical impedance patterns (electroanatomy) within the cochlea and, specifically, within the organ of Corti. Such information is as yet lacking.

Although there appears to be nearly universal agreement that OHCs dominate extracellular AC responses (Patuzzi, Yates and Johnstone 1989; Russell and Kössl 1991), the source of the extracellular DC is controversial. Recordings from high-frequency OHCs in vivo at their best frequency do not reveal a significant DC component except at very high sound levels. However, IHCs produce pronounced tonic response (Cody and Russell 1987; Kössl and Russell 1992). Consequently, it is quite commonly stated that the extracellular DC is of IHC origin (Geisler et al. 1990). In contrast, intracellular recordings from OHCs in the apical, low-frequency half of the cochlea show that they do produce large depolarizing DC responses at their best frequencies, and therefore do not differ substantially in their behavior from IHCs (Dallos and Cheatham 1990). Experiments with hair cell damage also point to OHCs as sources for extracellular DC responses, at least in the low-frequency region of the cochlea (Dallos and Cheatham 1976). Measured in organ cultures, both IHCs and OHCs generate DC receptor potentials (Fig. 1.12B). It is only in vivo that the DC component is lacking (Russell and Richardson 1987). The proposed explanation of the different OHC behavior in vitro and in vivo is the lack of a tectorial membrane in the culture. This would produce "open loop" operation for the OHC, inasmuch as the presumed motile OHC could not feed energy back in the absence of the tectorial membrane and in the presence of direct ciliary stimulation. The suggestion is that the fundamental transducer asymmetry of OHCs and IHCs is similar and is expressed in the culture. However, in vivo the OHCs, via their cilia and the tectorial membrane, feed energy back and alter the response so as to minimize the DC component (Russell, Cody and Richardson 1986). In this view, low-frequency OHCs either do not participate in the feedback process (the cochlear amplifier) or, less charitably, the lateral approach produces a mechanical bias that results in an artificial DC component. The converse, noncharitable view of the lack of a DC in high-frequency OHCs is that the scala tympani approach produces a mechanical bias that eliminates the DC response. However, there are no direct proofs extant that would show the reality of either of the above. Undoubtedly, an electrode dwelling in the organ of Corti will introduce some mechanical bias, no matter which approach is used. It remains to be determined which method produces less deleterious conditions. If a longitudinal dichotomy in the cochlea between high- and low-frequency OHC behavior does exist, then the low-frequency pattern extends at least to the 4-kHz location. We note in passing that OHCs from any cochlear location show electromotility, the presumed feedback link in cochlear amplification. It is mentioned that the distortion of the organ of Corti that may arise from either recording method has a relatively minimal effect on the AC response. These distortions probably affect the operating point of the OHC transducer and, thereby, the response asymmetry, without significantly altering the phasic response.

Why would OHCs produce more of the extracellular response? On the basis of relatively simple analysis of the organ of Corti circuit (Dallos 1983), it was shown that the ratio (ρ) of extracellular potential due to OHCs versus IHCs can be approximated as

$$\rho \approx k \frac{R_i^{IHC} \alpha (1 + \beta)^3}{R_i^{OHC} \beta (1 + \alpha)^3}$$

where $k = 3.8$ in the ratio of the numbers of OHCs and IHCs in a cochlear region, and the so-called shape factors are defined from the basolateral (R_b) and apical (R_a) membrane resistances of OHCs and IHCs as

$$\alpha = \frac{R_b^{OHC}}{R_a^{OHC}}; \beta = \frac{R_b^{IHC}}{R_a^{IHC}}$$

while the cells' input resistance, R_i, is obtained as

$$R_i^{HC} = \frac{R_a^{HC} R_b^{HC}}{R_a^{HC} + R_b^{HC}}$$

The input resistance of an average OHC is estimated as 40 MΩ (Housley and Ashmore 1992) and that of an IHC as 16 MΩ (Russell and Kössl 1991). Assuming a transducer conductance per cilium of 100 pS (Crawford, Evans and Fettiplace 1991), 100 cilia per OHC and 40 per IHC yield $\alpha = 0.66$ and $\beta = 0.07$. Combining all the numbers results in $\rho \approx 4$. Although the estimated numbers are quite uncertain (see Kros, Chapter 6), one may hazard the theoretical estimate of a 12-dB dominance of OHCs over IHCs in the production of extracellular potentials. This is in the direction of the experimental results (Dallos and Cheatham 1976).

5. Summary

Instead of providing a capsule summary of the above Summary, attention is called to a variety of unsolved problems, uncertain results, and opportunities for inquiry. The list is sketchy and is not in any particular order of significance. Most importantly, it is idiosyncratic, representing the questions that are of the greatest interest to this author.

The presence of spontaneous otoacoustic emissions obtained from different mammals has been construed as proof of an active cochlear process. One should note, however, that spontaneous otoacoustic emissions have been recorded from several nonmammalian vertebrates (Manley and Taschenberger 1993) in which no evidence for cochlear amplification (active feedback process) exists. How these internal oscillations arise and what general cochlear property they might signify is somewhat of a mystery.

No matter how much is written about the cochlear amplifier and no matter how clever its modeling descriptions are becoming, there are still some fundamental uncertainties about this process. The most obvious need is to resolve the controversy concerning whether the cochlear amplifier actually exists (Allen and Neely 1992; de Boer 1995). Although the evidence in favor appears overwhelming, the issue is not yet closed. Assuming that the cochlea is indeed locally active, additional questions relate to the exact location of this activity vis-à-vis the peak of the traveling wave and, most importantly, to the means whereby this location is established. In other words, is a tuned tectorial membrane (Zwislocki and Kletsy 1979; Allen 1980), a tuned OHC (Brundin, Flock and Canlon 1989), or some other distributed resonance that establishes the second cochlear map? Can the local activity be in the form of a reactance change (Kolston et al. 1989) or is negative damping imperative? A related question is whether it is OHC displacement or stiffness change that represents the feedback parameter of importance. Or, could it be that OHC somatic motility is an epiphenomenon, merely accompanying ciliary motility or stiffness change? Whatever it is, how does it work at high frequencies where the cell's self-generated receptor potential is drastically filtered? Could the extracellular potential field be of some help (Dallos and Evans 1995a,b)? What are the real properties of in vivo receptor potentials — without mechanical interference and electrical leak-conductance produced by the microelectrode? What are all the different classes of ion channels doing in OHCs if, seemingly, they are inactive around the cell's normal resting potential (Kros, Chapter 6)?

Whatever aspect of OHC motility-related phenomena will turn out to be relevant in cochlear amplification, the motor process proper is certainly novel and apparently unique in the animal kingdom. Identification of the putative motor protein is one of the most interesting current problems in molecular biology.

Probably the most productive research of the next decade will address the measurement of detailed micromechanical movements within the organ of Corti. It is obvious that micromechanics is one of the remaining frontiers of cochlear neurobiology. Thus far, due to the unavailability of suitable techniques, we have no substantive knowledge of how the many parts of the organ of Corti move during acoustic stimulation. We do have models, but they are pure fantasy: actual measurements are needed.

How do the medial efferents interact with the cochlear amplification process? Do they fine-tune the cochlea according to recent acoustic history or according to attentional and environmental needs? Do the type II afferents report on the mechanical state of OHCs and thereby participate in the feedback process?

Why are there so many well-established differences between the functioning of the apical and the basal cochlea? Why are there differences in efferent innervation pattern and neurotransmitters (Guinan, Chapter 8; Sewell, Chapter 9)? Does the cochlear amplifier even work in the apex? After all, tuning characteristics at low frequencies are not different between mammals and nonmammalian vertebrates. But then, why are apical OHCs motile? Are there real differences between DC receptor potentials in apical and basal OHCs?

The energy source of the organ of Corti, the stria vascularis, is the subject of a great deal of theoretical and experimental work (Wangemann and Schacht, Chapter 3). Yet it is still unclear how the three cell types in the stria interact, using what cell biological machinery, to pump K^+ into and Na^+ out of the scala media and produce the endocochlear potential. The explosion of cell and molecular biology has already had a salutary effect on cochlear research and promises to have an even greater effect in the future. What was the hunting ground for engineers and physicists is also rapidly becoming the legitimate province of biologists. The hallmark of contemporary neuroscience, the integration of disciplines and techniques brought to bear on a complex problem, is now exemplified in cochlear research. The chapters that follow clearly illustrate this.

Acknowledgments This work was supported by the National Institute of Deafness and Other Communication Disorders, National Institutes of Health.

References

Allen JB (1990) Cochlear micromechanics—a physical model of transduction. J Acoust Soc Am 68:1660-1679.

Allen JB, Fahey PF (1992) Using acoustic distortion products to measure the cochlear amplifier gain on the basilar membrane. J Acoust Soc Am 92:178-188.

Allen JB, Fahey PF (1993) Evidence for a second cochlear map. In: Duifhuis D, Horst JW, van Dijk P, van Netten SM (eds) Biophysics of Hair Cell Sensory Systems. Singapore: World Scientific, pp. 296-303.

Allen JB, Neely ST (1992) Micromechanical models of the cochlea. Physics Today 45:40-47.

Anniko M, Wróblewski R (1986) Ionic environment of cochlear hair cells. Hear Res 22:279-293.

Ashmore JF (1987) A fast motile event in outer hair cells isolated from the guinea pig cochlea. J Physiol 388:323-347.

Ashmore JF (1990) Forward and reverse transduction in the mammalian cochlea. Neurosci Res Suppl 12:S39-S50.

Ashmore JF, Meech RW (1986) Ionic basis of the resting potential in outer hair cells isolated from the guinea pig cochlea. Nature 322:368-371.

Assad JA, Hacohen N, Corey DP (1989) Voltage dependence of adaptation and active bundle movements in bullfrog saccular hair cells. Proc Natl Acad Sci USA 86:2918-2922.

Bialek W (1987) Physical limits to sensation and perception. Annu Rev Biophys Biophys Chem 16:455-478.

Bialek WS, Wit HP (1984) Quantum limits to oscillator stability: theory and experiments on acoustic emissions from the human ear. Phys Lett 104A:1973-1978.

Billone MC, Raynor S (1973) Transmission of radial shear forces to cochlear hair cells. J Acoust Soc Am 54:1143-1156.

Bosher SK, Warren RL (1968) Observations on the electrochemistry of the cochlear endolymph of the rat: a quantitative study of its electrical potential and ionic composition determined by flame spectrophotometry. Proc R Soc Lond B 171:227-247.

Brownell WE (1983) Observations on a motile response in isolated outer hair cells. In: Webster WR, Aitkin LM (eds) Mechanisms of Hearing. Clayton, Australia: Monash University Press, pp. 5-10.

Brownell WE, Bader CR, Bertrand D, de Ribaupierre Y (1985) Evoked mechanical responses of isolated cochlear outer hair cells. Science 227:194-196.

Brundin L, Flock Å, Canlon B (1989) Sound-induced motility of isolated cochlear outer hair cells is frequency-specific. Nature 342:814-816.

Cheatham MA, Dallos P (1990) Comparison of low- and high-side two-tone suppression in inner hair cell and organ of Corti responses. Hear Res 50:193-210.

Cody AR, Russell IJ (1987) The responses of hair cells in the basal turn of the guinea-pig cochlea to tones. J Physiol (Lond) 383:551-569.

Corey DP, Assad JA (1991) Transduction and adaptation in vertebrate hair cells: correlating structure with function. In: Corey DP, Roper SD (eds) Sensory Transduction. New York: Rockefeller University Press, pp. 325-342.

Corey DP, Hudspeth AJ (1979) Ionic basis of the receptor potential in a vertebrate hair cell. Nature 281:675-677.

Corey DP, Hudspeth AJ (1983) Kinetics of the receptor current in bullfrog saccular hair cells. J Neurosci 3:962-976.
Crawford AC, Fettiplace R (1980) The frequency selectivity of auditory nerve fibers and hair cells in the cochlea of the turtle. J Physiol (Lond) 306:79-125.
Crawford AC, Fettiplace R (1981) Non-linearities in the responses of turtle hair cells. J Physiol (Lond) 315:317-338.
Crawford AC, Fettiplace R (1985) The mechanical properties of ciliary bundles of turtle cochlear hair cells. J Physiol (Lond) 364:359-379.
Crawford AC, Evans MG, Fettiplace R (1989) Activation and adaptation of transducer currents in turtle hair cells. J Physiol (Lond) 419:405-434.
Crawford AC, Evans MG, Fettiplace R (1991) The actions of calcium on the mechano-electrical transducer current of turtle hair cells. J Physiol (Lond) 434:369-398.
Dallos P (1973a) The Auditory Periphery: Biophysics and Physiology. New York: Academic Press.
Dallos P (1973b) Cochlear potentials and cochlear mechanics. In: Møller AR (ed) Basic Mechanisms in Hearing. New York: Academic Press, pp. 335-372.
Dallos P (1983) Some electrical circuit properties of the organ of Corti. I. Analysis without reactive elements. Hear Res 12:89-119.
Dallos P (1985a) The role of outer hair cells in cochlear function. In: Correia MJ, Perachio AA (eds) Contemporary Sensory Neurobiology. New York: Alan R. Liss, pp. 207-230.
Dallos P (1985b) Response characteristics of mammalian cochlear hair cells. J Neurosci 5:1591-1608.
Dallos P (1986) Neurobiology of cochlear inner and outer hair cells: intracellular recordings. Hear Res 22:185-198.
Dallos P (1988) Cochlear neurobiology: some key experiments and concepts of the past two decades. In: Edelman GM, Gall EW, Cowan WM (eds) Functions of the Auditory System. New York: J. Wiley & Sons, pp. 153-188.
Dallos P (1991) Neurobiology of cochlear hair cells. In: Cazals Y, Demany L, Horner K (eds) Auditory Physiology and Perception. London: Pergamon Press, pp. 3-16.
Dallos P (1992) The active cochlea. J Neurosci 12:4575-4585.
Dallos P (1996) Driving voltage for cochlear outer hair cell motility in vivo and the cochlear amplifier. Audit Neurosci (to be published).
Dallos P, Cheatham MA (1976) Production of cochlear potentials by inner and outer hair cells. J Acoust Soc Am 60:510-512.
Dallos P, Cheatham MA (1989) Nonlinearities in cochlear receptor potentials and their origins. J Acoust Soc Am 86:1790-1796.
Dallos P, Cheatham MA (1990) Effects of electrical polarization on inner hair cell receptor potentials. J Acoust Soc Am 87:1636-1647.
Dallos P, Cheatham MA (1992) Cochlear hair cell function reflected in intracellular recordings in vivo. In: Corey DP, Roper SD (eds) Sensory Transduction. New York: Rockefeller University Press, pp. 372-393.
Dallos P, Evans BN (1995a) High-frequency motility of outer hair cells and the cochlear amplifier. Science 267:2006-2009.
Dallos P, Evans BN (1995b) High-frequency motility: corrections and addendum. Science 268:1420-1421.
Dallos P, Harris DM (1978) Properties of auditory nerve responses in the absence of

outer hair cells. J Neurophysiol 41:365–383.

Dallos P, Santos-Sacchi J (1983) AC receptor potentials from hair cells in the low-frequency region of the guinea pig cochlea. In: Webster WR, Aitkin LM (eds) Mechanisms of Hearing. Clayton, Australia: Monash University Press, pp. 11–16.

Dallos P, Billone MC, Durrant JD, Wang CY, Raynor S (1972) Cochlear inner and outer hair cells: functional differences. Science 177:356–358.

Dallos P, Harris DM, Relkin E, Cheatham MA (1980) Two-tone suppression and intermodulation distortion in the cochlea: effect of outer hair cell lesions. In: van den Brink G, Bilsen FA (eds) Delft University Press, Delft, The Netherlands, pp. 242–249.

Dallos P, Santos-Sacchi J, Flock Å (1982) Cochlear outer hair cells: intracellular recordings. Science 218:582–585.

Dallos P, Evans BN, Hallworth R (1991) Nature of the motor element in electrokinetic shape changes of cochlear outer hair cells. Nature 350:155–157.

Davis H (1983) An active process in cochlear mechanics. Hear Res 9:79–90.

Davis H, Fernández C, McAuliffe DR (1950) The excitatory process in the cochlea. Proc Natl Acad Sci USA 36:580–587.

Davis H, Deatherage BH, Rosenblut B, Fernández C, Kimura R, Smith CA (1958) Modification of cochlear potentials produced by streptomycin poisoning and extensive venous obstruction. Laryngoscope 68:596–627.

de Boer E (1983) No sharpening? A challenge to cochlear mechanics. J Acoust Soc Am 73:567–573.

de Boer E (1995) The "inverse problem" solved for a three-dimensional model of the cochlea. II. Application to experimental data sets. J Acoust Soc Am 98:904–910.

Diependahl RJ, de Boer E, Viergever MA (1987) Cochlear power flux as an indicator of mechanical activity. J Acoust Soc Am 82:917–926.

Eatock RA, Corey DP, Hudspeth AJ (1987) Adaptation of mechanoelectrical transduction in hair cells of the bullfrog sacculus. J Neurosci 7:2821–2836.

Evans BN, Dallos P, Hallworth R (1989) Asymmetries in motile response of outer hair cells in simulated in vivo conditions. In: Wilson JP, Kemp DT (eds) Cochlear Mechanisms. New York: Plenum Press, pp. 205–206.

Eybalin M (1993) Neurotransmitters and neuromodulators of the mammalian cochlea. Physiol Rev 73:309–373.

Fay RR, Popper AN (1994) Comparative Hearing: Mammals. New York: Springer-Verlag.

Fettiplace R (1987) Electrical tuning of hair cells in the inner ear. Trends Neurosci 10:421–425.

Fex J, Altschuler RA (1986) Neurotransmitter-related immunocytochemistry of the organ of Corti. Hear Res 22:249–263.

Flock Å, Flock B, Ulfendahl M (1986) Mechanisms of movement in outer hair cells and a possible structural basis. Arch Otorhinolaryngol 243:83–90.

Freeman, DM, Weiss TF (1990) Hydrodynamic forces on hair bundles at low frequencies. Hear Res 48:17–30.

Fuchs PA, Evans MG (1988) Voltage oscillations and ionic conductances in hair cells isolated from the alligator cochlea. J Comp Physiol A 164:151–163.

Fuchs PA, Mann AC (1986) Voltage oscillations and ionic currents in hair cells isolated from the apex of the chick's cochlea. J Physiol (Lond) 371:31P.

Fuchs PA, Nagai T, Evans MG (1988) Electrical tuning in hair cells isolated from

the chick cochlea. J Neurosci 8:2460-2467.
Geisler CD (1991) A cochlear model using feedback from motile outer hair cells. Hear Res 54:105-117.
Geisler CD, Yates GK, Patuzzi RB, Johnstone BM (1990) Saturation of outer hair cell receptor currents causes two-tone suppression. Hear Res 44:241-256.
Gitter AH, Zenner HP, Frömter E (1986) Membrane potential and ion channels in isolated outer hair cells of guinea pig cochlea. Otorhinolaryngol Relat Spec 48:68-75.
Gold T (1948) Hearing. The physical basis of the action of the cochlea. Proc R Soc Edinb (B) 135:492-498.
Guinan JJ, Jr, Peake WT (1967) Middle-ear characteristics in anesthetized cats. J Acoust Soc Am 41:1237-1261.
Gummer AW, Hemmert W, Morioka I, Reis P, Reuter G, Zenner HP (1993) Cellular motility in the guinea-pig cochlea. In: Duifhuis D, Horst JW, van Dijk P, van Netten SM (eds) Biophysics of Hair Cell Sensory Systems. Singapore: World Scientific, pp. 229-236.
Harrison RV, Evans EF (1979) Cochlear fibre responses in guinea pigs with well defined cochlear lesions. Acta Otolaryngol Suppl 9:83-92.
Holley MC, Ashmore JF (1988) On the mechanism of a high-frequency force generator in outer hair cells isolated from the guinea pig cochlea. Proc R Soc Lond B 232:413-429.
Holton T, Hudspeth AJ (1983) A micromechanical contribution to cochlear tuning and tonotopic organization. Science 222:508-510.
Holton T, Hudspeth AJ (1986) The transduction channel of hair cells from the bull-frog characterized by noise analysis. J Physiol (Lond) 375:195-227.
Housley GD, Ashmore JF (1992) Ionic currents of outer hair cells isolated from the guinea-pig cochlea. J Physiol (Lond) 448:73-98.
Housley GD, Connor BJ, Raybould NP (1995) Purinergic modulation of outer hair cell electromotility. In: Flock Å, Ottoson D, Ulfendahl M (eds) Active Hearing. Oxford: Elsevier, pp. 221-238.
Housley GD, Greenwood D, Ashmore JF (1992) Localisation of cholinergic and purinergic receptors on outer hair cells isolated from the guinea-pig cochlea. Proc R Soc Lond B 249:265-273.
Howard J, Hudspeth AJ (1988) Compliance of hair bundle associated with gating of the mechanoelectric transducer channels in the bullfrog's saccular hair cell. Neuron 1:189-199.
Hudspeth AJ (1982) Extracellular current flow and the site of transduction by vertebrate hair cells. J Neurosci 2:1-10.
Hudspeth AJ (1989) How the ear's works work. Nature 341:397-404.
Hudspeth AJ, Corey DP (1977) Sensitivity, polarity and conductance change in the response of vertebrate hair cells to controlled mechanical stimuli. Proc Natl Acad Sci USA 74:2407-2411.
Hudspeth AJ, Jacobs R (1979) Stereocilia mediate transduction in vertebrate hair cells. Proc Natl Acad Sci USA 76:1506-1509.
ITER (International Team for Ear Research) (1989) Cellular vibration and motility in the organ of Corti. Acta Otolaryngol Suppl 467:1-279.
Jahnke K (1975) The fine structure of freeze fractured intercellular junctions in the guinea pig inner ear. Acta Otolaryngol Suppl 336:1-40.
Johnstone BM, Patuzzi R, Yates GK (1986) Basilar membrane measurements and

the traveling wave. Hear Res 22:147-153.
Johnstone BM, Patuzzi R, Syka J, Syková E (1989) Stimulus-related potassium changes in the organ of Corti of guinea-pig. J Physiol (Lond) 408:77-92.
Kachar B, Brownell WE, Altschuler RA, Fex J (1986) Electrokinetic shape changes of cochlear outer hair cells. Nature 322:365-368.
Kemp DT (1978) Stimulated acoustic emissions from the human auditory system. J Acoust Soc Am 64:1386-1391.
Khanna SM, Tonndorf J (1969) Middle ear power transfer. Arch Klin Exp Ohr Nas Kehlk Heilk 193:78-88.
Kidd RC, Weiss TF (1990) Mechanisms that degrade timing information in the cochlea. Hear Res 49:181-207.
Kim DO (1986) Active and nonlinear cochlear biomechanics and the role of the outer hair cell sub-system in the mammalian auditory system. Hear Res 22:105-114.
Klinke R (1986) Neurotransmission in the inner ear. Hear Res 22:235-243.
Kolston PJ, Viergever MA, de Boer E, Diependaal RJ (1989) Realistic mechanical tuning in a micromechanical cochlear model. J Acoust Soc Am 86:133-140.
Kössl M, Russell IJ (1992) The phase and magnitude of hair cell receptor potentials and frequency tuning in the guinea pig cochlea. J Neurosci 12:1575-1586.
Kros CJ, Crawford AC (1990) Potassium currents in inner hair cells isolated from the guinea-pig cochlea. J Physiol (Lond) 421:263-291.
Lewis RS, Hudspeth AJ (1983) Voltage- and ion-dependent conductances in solitary vertebrate hair cells. Nature 304:538-541.
Lighthill J (1981) Energy flow in the cochlea. J Fluid Mech 106:149-213.
Lighthill J (1991) Biomechanics of hearing sensitivity. J Vibrat Acoust 113:1-13.
Lim DJ (1980) Cochlear anatomy related to cochlear micromechanics. A review. J Acoust Soc Am 67:1686-1695.
Lindemann HH, Ades HW, Bredberg G, Engström H (1971) The sensory hairs and the tectorial membrane in the development of the cat's organ of Corti: a scanning electron microscopic study. Acta Otolarngol 72:229-242.
Manley G, Taschenberger G (1993) Spontaneous otoacoustic emissions from a bird: a preliminary report. In: Duifhuis D, Horst JW, van Dijk P, van Netten SM (eds) Biophysics of Hair Cell Sensory Systems. Singapore: World Scientific, pp. 33-39.
Neely ST, Kim DO (1983) An active cochlear model showing sharp tuning and high sensitivity. Hear Res 9:123-130.
Nielsen DW, Turner RG (1983) Micromechanics of the reptilian ear. Audiology 22:530-544.
Nuttall AL, Brown MC, Masta RI, Lawrence M (1981) Inner hair cell responses to the velocity of basilar membrane motion in the guinea pig. Brain Res 211:171-174.
Oesterle EC, Dallos P (1989) Intracellular recordings from supporting cells in the guinea-pig cochlea: AC potentials. J Acoust Soc Am 86:1013-1032.
Oesterle EC, Dallos P (1990) Intracellular recordings from supporting cells in the guinea pig cochlea: DC potentials. J Neurophysiol 64:617-636.
Offner FF, Dallos P, Cheatham MA (1987) Positive endocochlear potential: mechanism of production by marginal cells of the stria vascularis. Hear Res 29:117-124.
Ohmori H (1984) Mechano-electrical transduction currents in isolated vestibular

hair cells of the chick. J Physiol (Lond) 359:189–217.

Palmer AR, Russell IJ (1986) Phase-locking in the cochlear nerve of the guinea pig and its relation to the receptor potentials of the inner hair cells. Hear Res 24:1–15.

Patuzzi RB, Yates GK, Johnstone BM (1989) Outer hair cell receptor current and its effect on cochlear mechanics. In: Wilson JP, Kemp DT (eds) Cochlear Mechanisms. New York: Plenum Press, pp. 169–176.

Peterson LC, Bogert BP (1950) A dynamical theory of the cochlea. J Acoust Soc Am 22:369–381.

Pickles JO, Comis SD, Osborne MP (1984) Cross-links between stereocilia in the guinea pig organ of Corti, and their possible relation to sensory transduction. Hear Res 15:103–112.

Pitchford S, Ashmore JF (1987) An electrical resonance in hair cells of the amphibian papilla of the frog *Rana temporaria*. Hear Res 27:75–83.

Popper AN, Fay RR (1995) Hearing by Bats. New York: Springer-Verlag.

Rhode WS (1971) Observation of the vibration of the basilar membrane in squirrel monkeys using the Mössbauer technique. J Acoust Soc Am 49:1218–1231.

Rhode WS (1973) An investigation of postmortem cochlear mechanics using the Mössbauer effect. In: Møller AR (ed) Basic Mechanisms of Hearing. New York: Academic Press, pp. 49–67.

Roberts WM, Jacobs RA, Hudspeth AJ (1990) Colocalization of ion channels involved in frequency selectivity and synaptic transmission at presynaptic active zones of hair cells. J Neurosci 10:3664–3684.

Rosowski JJ (1991) The effects of external- and middle-ear filtering on auditory threshold and noise induced hearing loss. J Acoust Soc Am 90:124–135.

Rosowski JJ (1994) Outer and middle ears. In: Fay RR, Popper AN (eds) Comparative Hearing: Mammals. New York: Springer-Verlag, pp. 172–247.

Ruggero MA, Rich N (1991) Application of a commercially-manufactured Doppler-shift laser velocimeter to the measurement of basilar-membrane motion. Hear Res 51:215–230.

Russell IJ (1983) Origin of the receptor potential in inner hair cells of the mammalian cochlea: evidence for Davis' theory. Nature 301:334–336.

Russell IJ, Kössl M (1991) The voltage response of hair cells in the basal turn of the guinea pig cochlea. J Physiol (Lond) 435:493–511.

Russell IJ, Richardson GP (1987) The morphology and physiology of hair cells in organotypic cultures of the mouse cochlea. Hear Res 31:9–24.

Russell IJ, Sellick PM (1977) Tuning properties of cochlear hair cells. Nature 267:858–860.

Russell IJ, Sellick PM (1978) Intracellular studies of hair cells in the mammalian cochlea. J Physiol (Lond) 284:261–290.

Russell IJ, Sellick PM (1981) The responses of hair cells to low frequency tones and their relationship to the extracellular receptor potentials and sound pressure level in the guinea pig cochlea. In: Syka J, Aitkin L (eds) Neuronal Mechanisms of Hearing. New York: Plenum Press, pp. 3–15.

Russell IJ, Sellick PM (1983) Low frequency characteristics of intracellularly recorded receptor potentials in mammalian hair cells. J Physiol (Lond) 338:179–206.

Russell IJ, Cody AR, Richardson GP (1986) The responses of inner and outer hair cells in the basal turn of the guinea-pig cochlea and in the mouse cochlea grown

in vitro. Hear Res 22:199–216.
Ryan A, Dallos P (1975) Absence of cochlear outer hair cells: effect on behavioural auditory threshold. Nature 253:44–46.
Santos-Sacchi J (1989) Asymmetry in voltage-dependent movements of isolated outer hair cells from the organ of Corti. J Neurosci 9:2954–2962.
Santos-Sacchi J (1991) Reversible inhibition of voltage-dependent outer hair cell motility and capacitance. J Neurosci 11:3096–3110.
Santos-Sacchi J, Dilger DP (1988) Whole cell currents and mechanical responses of isolated outer hair cells. Hear Res 35:143–150.
Schmiedt RA, Zwislocki JJ, Hamernik RP (1980) Effects of hair cell lesions on responses of cochlear nerve fibers. I. Lesions, tuning curves, two-tone inhibition, and responses to trapezoidal wave patterns. J Neurophysiol 43:1367–1389.
Sellick PM, Russell IJ (1979) Two-tone suppression in cochlear hair cells. Hear Res 1:227–236.
Sellick PM, Patuzzi RB, Johnstone BM (1982) Measurement of basilar membrane motion in the guinea-pig using the Mössbauer technique. J Acoust Soc Am 72:131–141.
Shaw EA (1974) Transformation of sound pressure level from the free-field to the eardrum in the horizontal plane. J Acoust Soc Am 56:1848–1861.
Shotwell SL, Jacobs R, Hudspeth AJ (1981) Directional sensitivity of individual vertebrate hair cells to controlled deflection of their hair bundles. Ann NY Acad Sci 374:1–10.
Smith CA (1978) Structure of the cochlear duct. In: Naunton RF, Fernández C (eds) Evoked Electrical Activity in the Auditory Nervous System. New York: Academic Press, pp. 3–19.
Smoorenburg GF (1972) Combination tones and their origin. J Acoust Soc Am 52:615–632.
Spoendlin H (1969) Innervation patterns in the organ of Corti in the cat. Acta Otolaryngol 67:239–254.
Sziklai I, Dallos P (1993) Acetylcholine controls the gain of the voltage-to-movement converter in isolated outer hair cells. Acta Otolaryngol (Stockholm) 113:326–329.
Viergever MA (1986) Cochlear macromechanics—a review. In: Allen JB, Hall JL, Hubbard A, Neely ST, Tubis A (eds) Peripheral Auditory Mechanisms. Berlin: Springer-Verlag, pp. 63–72.
von Békésy G (1960) Experiments in Hearing. New York: McGraw-Hill.
Warr WB (1992) Organization of olivocochlear efferent systems in mammals. In: Webster DB, Popper AN, Fay RR (eds) The Mammalian Auditory Pathway: Neuroanatomy. New York: Springer-Verlag, pp. 410–448.
Warr WB, Guinan JJ, Jr (1979) Efferent innervation of the organ of Corti: two separate systems. Brain Res 173:152–155.
Weiss TF, Rose C (1988) Stages of degradation of timing information in the cochlea: a comparison of hair cell and nerve fiber responses in the alligator lizard. Hear Res 33:167–174.
Weiss TF, Peake WT, Ling A, Holton T (1978) Which structures determine frequency selectivity and tonotopic organization of vertebrate cochlear nerve fibers? Evidence from the alligator lizard. In: Naunton RF, Fernández C (eds) Evoked Electrical Activity of the Auditory Nervous System. New York: Academic Press, pp. 91–112.

Wever EG (1949) Theory of Hearing. New York: John Wiley & Sons.

Wever EG, Bray C (1930) Action currents in the auditory nerve in response to acoustic stimulation. Proc Natl Acad Sci USA 16:344-350.

Whitfield IC, Ross HF (1965) Cochlear microphonics and summating potentials and the outputs of individual hair cell generators. J. Acoust Soc Am 38:126-131.

Wilson JP (1980) Evidence for cochlear origin for acoustic re-emissions, threshold fine-structure and tonal tinnitus. Hear Res 2:233-252.

Zenner HP (1986) Motile responses in outer hair cells. Hear Res 22:83-90.

Zenner HP, Zimmerman U, Schmitt U (1985) Reversible contraction of isolated mammalian cochlear hair cells. Hear Res 18:127-133.

Zweig G (1991) Finding the impedance of the organ of Corti. J Acoust Soc Am 89:1229-1254.

Zwicker E (1970) Masking and psychological excitation as consequences of the ear's frequency analysis. In: Plomp R, Smoorenburg G (eds) Frequency Analysis and Periodicity Detection in Hearing. Leiden: A.W. Sijthoff, pp. 376-394.

Zwicker E (1979) A model describing nonlinearities in hearing by active processes with saturation at 40 dB. Biol Cybernet 35:243-250.

Zwislocki JJ (1953) Review of recent mechanical theories of cochlear dynamics. J Acoust Soc Am 25:743-751.

Zwislocki JJ (1989) Phase reversal of OHC response at high sound intensities. In: Wilson JP, Kemp DT (eds) Cochlear Mechanics. New York: Plenum Press, pp. 163-168.

Zwislocki JJ and Kletsky EJ (1979) Tectorial membrane: a possible effect on frequency analysis in the cochlea. Science 204:639-641.

2
Structure of the Mammalian Cochlea

Norma B. Slepecky

1. Introduction

The cochlea within the inner ear contains the cells responsible for the perception of sound. Unfortunately for researchers, the structures of interest are housed in a rather inaccessible part of the skull, totally embedded in bone. In spite of this, the anatomy was well described in the mid-nineteenth century by Retzius, Huschke, Reissner, Kolliker, Deiters, Hensen, and Corti, names familiar even to present-day cochlear anatomists. From their studies, it was known that the cochlea is composed of a bony labyrinth, within which is found the cellular structures comprising the membranous labyrinth. These are easily seen in a section taken through the cochlea in a plane parallel to its long axis (Fig. 2.1).

The bony labyrinth includes the otic capsule, as the external boundary of the cochlea, and the modiolus, a bony tube that forms the central axis of the cochlea. Spiraling around the modiolus from base to apex, the interior of the bony labyrinth is partitioned into three tubes or spaces (scalae). The upper, more apical space (scala vestibuli) is separated from the middle (scala media) by Reissner's membrane. The scala media is separated from the lower space (scala tympani) by parts of the osseous spiral lamina and the basilar membrane. The scala tympani and scala vestibuli are actually continuous at the apical tip of the cochlea, connected by a narrow opening called the helicotrema. At the base of the cochlea, the scala tympani appears to end at the round window, a membrane-covered hole in the otic capsule. In actuality, a small, fluid-filled tube (the cochlear aqueduct) connects perilymph from the base of the cochlea with cerebrospinal fluid. The scala media and scala vestibuli both continue beyond the base of the cochlea into the vestibular portion of the inner ear. The scala media connects to the saccule via the ductus reuniens, and endolymph within it is thought to reach the endolymphatic sac through the endolymphatic duct. The stapes inserts into the oval window of the cochlea at a hole in the otic capsule over the scala vestibuli.

2. Cochlear Structure 45

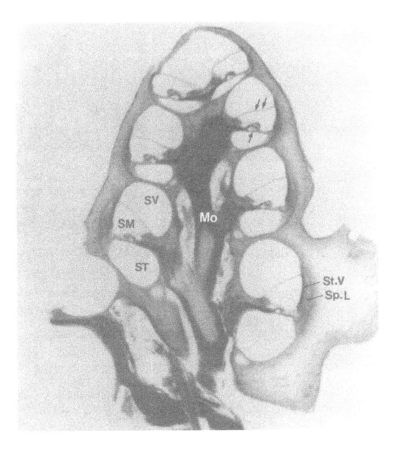

FIGURE 2.1 In a midmodiolar section through the guinea pig cochlea, the membranous labyrinth can be seen to be encased in bone. The three fluid-filled spaces of scala vestibuli (SV), scala tympani (ST), and scala media (SM) are separated from each other by the basilar membrane (*single arrow*) and Reissner's membrane (*double arrow*). The stria vascularis (StV) and the spiral ligament (SpL) lie close to the bone along the lateral wall of the cochlea. The organ of Corti can be seen resting on the basilar membrane. Auditory nerve fibers run in the modiolus (Mo).

The cochlea contains several fluid-filled compartments (Figs. 2.1, 2.2, 2.3). Two of these (or one if the scala tympani and scala vestibuli are considered to be continuous) are filled with a high-sodium solution called perilymph, and another is filled with a high-potassium solution called endolymph. Initially it was thought that the boundaries of these fluids correspond to the membrane-bound scalae. Thus, the term "cochlear duct" labels a compartment bounded by and including Reissner's membrane, the spiral ligament, the stria vascularis, the basilar membrane (BM), and the spiral limbus (Fig. 2.3A). It is now known that the anatomical boundaries

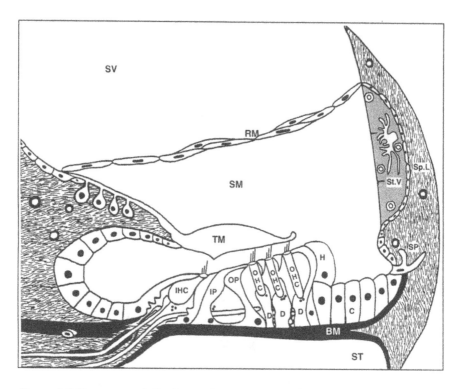

FIGURE 2.2 The organ of Corti contains sensory and supporting cells. The inner (IHC) and three rows of outer (OHC) sensory hair cells, along with the inner pillar (IP), outer pillar (OP), Deiters (D), Hensen (H), and Claudius (C) cells rest on the basilar membrane (BM). The tectorial membrane (TM) covers the apical surface of the sensory and supporting cells. The scala vestibuli (SV), scala media (SM), and scala tympani (ST) are three fluid-filled compartments. The scala media, containing endolymph, has as its boundaries the reticular lamina of the organ of Corti, Reissner's membrane (RM), and the lateral wall made up of the stria vascularis (StV), the spiral ligament (SpL), and the spiral prominence (SP).

of the scalae are not necessarily the boundaries of the fluids. For endolymph, the fluid boundaries consist of tight junctions between adjacent epithelial cells that inhibit free diffusion between them.

The boundaries of the endolymph-containing compartment are highlighted in Figure 2.3B. Tight junctions between the cells of Reissner's membrane separate the endolymph from the perilymph of the scala vestibuli. Tight junctions at the apical surfaces of the sensory and supporting cells, at the reticular lamina (Fig. 2.10), make up a second boundary; thus, perilymph in the scala tympani can diffuse beyond the BM, and perilymph-like fluid bathes the cell bodies of the sensory and supporting cells of the organ of Corti. In the stria vascularis along the lateral wall of the cochlea, tight junctions between adjacent marginal cells facing

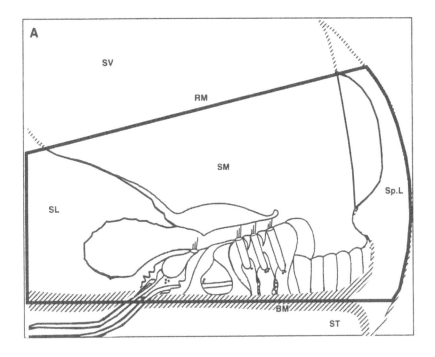

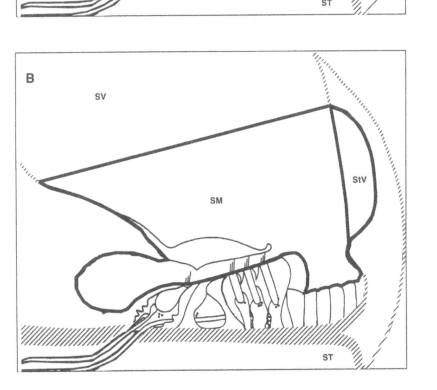

FIGURE 2.3 Boundaries of the cochlea. (A) Diagram showing the boundaries of the cochlear duct portion of the membranous labyrinth. Included in the cochlear duct are Reissner's membrane (RM), the spiral ligament (SpL), the basilar membrane (BM), and the spiral limbus (SL). (B) Diagram of the boundaries of the fluid-filled spaces in the cochlea. The areas with tight junctions separating the different fluid compartments are outlined showing the perilymphatic spaces (ST and SV), the endolymphatic space (SM), and the intrastrial space (StV).

the endolymph keep it from entering this vascularized epithelium. Within the lateral wall, adjacent cells coupled by tight junctions also form an intrastrial compartment. Thus, it appears that perilymph diffuses freely through the spiral ligament, but the stria vascularis is contained in a compartment that is formed and maintained by tight junctions among marginal cells on the endolymphatic side and among basal cells on the perilymphatic side (Fig. 2.3).

In certain regions within the membranous labyrinth, highly differentiated epithelial cells form the organ of Corti (Fig. 2.4), the receptor organ made up of specialized sensory and supporting cells that rest on the BM. Even early studies of the cochlea described the cells within the organ of Corti as anatomically polarized. With light microscopy, the hair cells can be characterized by stereocilia at one end (the apical surface) and nerve fibers clustered near the other end (the basolateral surface). These nerve fibers must pass through the bony holes, or habenulae perforatae, to traverse the distance between the sensory cells and the central nervous system.

Early studies also recognized that cochlear structures could be described in "spiral" and "radial" terms. The organ of Corti spirals along the basilar membrane, from the apex to the base of the cochlea. Along the spiral, the

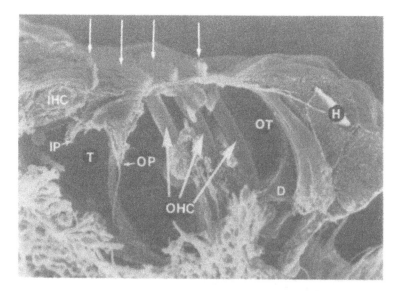

FIGURE 2.4 The three-dimensional arrangement of the organ of Corti is best seen in a scanning electron micrograph. The inner (IHC) and outer (OHC) hair cells each have stereocilia along their apical surface (arrows). The tunnel (T) is formed by the inner (IP) and outer (OP) pillar cells. The outer tunnel (OT) is the space between the third row of OHCs on one side and the third row Deiters (D) and Hensen (H) cells on the other. (Scanning electron micrograph of the guinea pig cochlea provided by Dr. Göran Bredberg, Sodersjukhuset, Stockholm, Sweden, from Bredberg et al. 1970.)

sensory epithelium contains (depending on the species studied) approximately 20,000 sensory hair cells, which interdigitate in a regular manner with various types of supporting cells. In radial terms, sensory cells are either "inner" hair cells (IHCs), which form one row along the length of the sensory epithelium closest to the modiolus, or "outer" hair cells (OHCs), which form three rows and are separated radially from the IHCs by the pillar supporting cells (Figs. 2.2, 2.4). In a cross section, the organ of Corti is described with the IHCs as medial and the OHCs as lateral. Spiral cochlear structures include spiral blood vessels (the spiral modiolar artery and vein shown in Fig. 2.19) and spiral nerves (efferent fibers that spiral through the ganglion cell bodies in Rosenthal's canal to the periphery of the cochlea, as shown in Figs. 2.13 and 2.14). Radial cochlear structures include radial blood vessels (radiating arterioles and venules shown in Fig. 2.19) and radial nerves (afferent fibers that run from the organ of Corti, through Rosenthal's canal, into the modiolus shown in Figs. 2.13 and 2.14). Unfortunately for cochlear anatomists, most structures have both a spiral and a radial component, often making the names of the structures ambiguous and always making the three-dimensional organization of the cochlea difficult to represent in diagrams on paper.

The goal of this chapter is to describe in detail what is known about the structures of the "typical" mammalian cochlea. This is a difficult task because of species differences. Although much of the early anatomy was conducted on human specimens, subsequent observations have been made on rodents (mouse, guinea pig, and gerbil). Studies of the normal cochlea were followed by experimental studies of the effects of noise trauma, ototoxic drugs, and aging. Recently, important work has been performed on cat and again on human inner ears. With comparative data across species (Echteler, Fay, and Popper 1994), it can be shown which similarities exist in structure and organization. In this chapter, these similarities have been stressed; yet note has been made of quantitative and qualitative differences among species.

A description of the entire cochlea traditionally requires division of the cochlea into either structurally related or functionally related regions, yet in such a complex structure the functionally related regions are often separated spatially. This review treats the organ of Corti by cell type: sensory and supporting cells, their accessory structures related to hair cell stimulation, and their nerve supply within the cochlea. Other aspects of the cochlea presented here are its vasculature and the cells of the spiral ligament and stria vascularis along the lateral wall. This approach provides structural descriptions organized by cell type, which implicates their functions.

Since it is not possible to include here all of the studies conducted over the years, the emphasis of this review is to focus on the original findings (so we do not forget our beginnings), recent reviews comparing specific areas in different species, and references for the most current data and interpretations of mammalian cochlear structure. Many structural similarities exist between the mammalian cochlea and other mechano-receptor organs, such as the mammalian vestibular system, the avian basilar papilla, the lizard basilar papilla, the frog saccule, and the fish otolithic organs and lateral line. Although these systems have been used to study structure–function relationships and the mechanisms involved with sensory transduction, this chapter is restricted to mammalian cochlear structure, and for the most part, nonmammalian inner ears will not be discussed.

2. Organ of Corti

2.1 Structure

The organ of Corti (Fig. 2.4), containing both sensory and supporting cells, spirals on the BM. It is loosely structured to permit movement of the sensory epithelium in response to mechanical stimuli, yet it is rigid enough to transmit vibrations from the BM to the stereocilia. Although the diameter of the bony labyrinth decreases from base to apex, most of the structures of the membranous labyrinth display a longitudinal gradient where there is an increase in size from base to apex (Smith 1968; Cabezudo 1978; Strelioff and Flock 1984; Wright 1984; Bohne and Carr 1985; Lim 1986). Thus, for the organ of Corti, the cells in the apex are larger than those in the base, the stereocilia on the hair cells are longer and less stiff, the BM is wider, and the tectorial membrane (TM) has a greater mass. The actual size of these structures varies from species to species (Echteler, Fay, and Popper 1994), and these morphological characteristics are probably related to the tuning capacities and sensitivities of the receptor organs, which differ in the absolute frequency range detected and the frequency range of best sensitivity.

In most species, the size of the organ of Corti changes continuously along the length of the sensory epithelium. In other species, exaggerated structural features appear suddenly in specific regions and correlate with functional importance. These can be seen in many bats, where one or more acoustic foveae are marked by specializations of the BM, TM, sensory and supporting cells, and nerve fiber density (Bruns 1980; Bruns and Schmieszek 1980; Henson and Henson 1991; Vater and Lenoir 1992; Vater, Lenoir, and Pujol 1992).

Fine structural analysis of the organ of Corti has allowed us to identify various cell types, as well as to speculate on and experimentally test their function. Apical specializations on sensory cells are responsible for detection of stimuli and sensory transduction. Adjacent to the sensory hair cells are cells that are specialized for mechanical and metabolic support. On the endolymphatic side of the spiral limbus, cells appear to have a secretory function and to be responsible for the formation of the TM. Some cells separate the compartments containing endolymph and perilymph, whereas others transport ions and may participate in the formation and maintenance of these two fluids. With characteristic synaptic specializations at their basal ends, hair cells contact both afferent and efferent nerve fibers to permit transfer of information to the central nervous system and feedback to the periphery.

2.2 Nature of Contact Between Cells Within the Organ of Corti

Since the cells in the organ of Corti separate endolymph from perilymph, and since they must withstand movement during sound stimulation, it is not surprising to find that an elaborate set of structures links them together. Their intercellular junctional specializations, characteristic of most epithelial cells, include tight junctions, adherens junctions, desmosomes, and gap junctions (Gulley and Reese 1976; Iurato et al. 1976a,b; Nadol et al. 1976; Nadol 1978). Tight junctions and adherens junctions are present at the apical surface of cells, usually where they face a lumen. Desmosomes and gap junctions are found along the basolateral surfaces when and if they are present.

2.2.1 Tight Junctions

A tight junction (zonula occludens) between cells serves at least two important functions. It creates and maintains a barrier against lateral diffusion of proteins within the plane of the lipid bilayer of the cell membrane. It also seals the apical intercellular space and prevents the diffusion of electrolytes and other molecules between cells. Tight junctions surround the entire apical circumference of epithelial cells, effectively sealing all cells within an epithelium together. The tight junctions do not withstand mechanical stress, and adhesive forces that maintain cell coupling under these conditions are provided by the adherens junctions.

The barrier separating endolymph from perilymph in the organ of Corti is formed by tight junctions between hair cells and supporting cells and among supporting cells. The tight junctions in the cochlea are extensive, as characterized by transmission electron microscopy and freeze-fracture studies. Freeze-fracture electron microscopy shows the functional fusion of the two cell membranes as a network of branching and anastomosing ridges on one fracture face, with complementary grooves on the opposite face. The number of junctional strands is positively correlated with the degree of transepithelial resistance, and tight junctions can range from very tight to leaky. The barrier between endolymph and perilymph is formed by very tight junctions. To date, two different tight junction–specific proteins have been characterized, ZO1 and cingulin. When the organ of Corti is immunostained with antibodies to cingulin, all the intercellular contacts at the reticular lamina are labeled (Raphael and Altschuler 1991).

All the other cells lining the scala media endolymphatic space are also joined to one another by tight junctions. This distribution is shown in

Figure 2.3B by the dark line along the reticular lamina, the spiral limbus, Reissner's membrane, and the marginal cells of the stria vascularis. Tight junctions are not found between the cells lining the scala vestibuli and scala tympani, the subepithelial cells of the spiral limbus and the spiral prominence, or the cells of the spiral ligament. This suggests that the intercellular spaces of these tissues are probably functionally continuous with the perilymph.

2.2.2 Adherens Junctions

Beneath the region of the tight junctions, cells are interconnected on their lateral surfaces by adherens junctions. These junctions physically hold the cells together and mediate intercellular communication related to cell–cell recognition. Adherens junctions may appear as belts (zonula adherens) encircling each cell, thus connecting the apical surface of one cell to the apical surfaces of all of its neighbors in an epithelial sheet. The junctions have three domains: cytoplasmic, intramembrane, and extracellular. At the cytoplasmic domain, the junction is linked to the intracellular actin cytoskeleton through vinculin and α-actinin. The actin filaments that associate with the adherens junction form a circumferential belt or ring within the cell. The intramembrane domains, present in the membrane of both cells on either side of the adherens junction, appear electron-dense when viewed with transmission electron microscopy. Bridging filaments run between the two cells in the extracellular area, but the cell-adhesion molecules in this region of adherens junctions have not yet been identified.

In the reticular lamina of the organ of Corti, adherens junctions are present between all cells in the organ of Corti, holding the cells together during the movement of the BM. The actin belt is especially prominent, although in heterotypic junctions (hair cell–supporting cell junctions), it is asymmetrical, with more actin on the supporting cell side. This can be seen in transmission electron microscopy (Engström 1958; Kimura 1975) and in surface preparations of the organ of Corti (Raphael and Altschuler 1991). The large number of actin filaments on the supporting cell side of the adherens junction indirectly suggests that these cells play a major role in maintaining the surface tension of the reticular lamina. They may also play an additional role during response to trauma, when two adjacent supporting cells make a scar to replace a damaged or lost hair cell (Raphael and Altschuler 1992).

In focal contact areas between some epithelial cells and the extracellular matrix (or between cells grown in culture and the surface of the culture dish), a small and focal junctional complex similar in structure and biochemical composition to the adherens junction is found. There are prominent actin filaments, anchored to the membrane by vinculin and α-actinin. This type of junction is thought to provide a mechanism for coupling the intracellular actin filament cytoskeleton to the extracellular matrix rather than to an adjacent cell. The adhesion molecule in these focal contacts is usually a protein of the integrin family. In the organ of Corti, these types of focal contacts have not yet been seen; however, this might be the type of connection expected between the stereocilia and the TM.

2.2.3 Desmosomes

A desmosome is an area where the lateral surfaces of adjacent cells appear to be connected by a small "spot" junction (a macula adherens or spot desmosome) that acts like a snap or small pieces of Velcro to hold cells together. Prominent intercellular bridging filaments meet to form an electron-dense plaque midway between the two cells. At their intracellular cytoplasmic surface, the filaments attach to bundles of cytokeratin-type intermediate filaments. Several proteins identified in desmosomal junctions include desmogleins, desmoplakins, and desmoglobins. A hemidesmosome is found at a point of contact between a cell and the extracellular matrix. In this region of contact, there is an accumulation of intermediate filaments within a cell.

Unlike the zones of tight junctions and adherens junctions, the distribution of desmosomes is not restricted to the apical region in inter-epithelial cell contacts. Rather, desmosomes can be found anywhere along the lateral membranes of cells. Some types of supporting cells have a dense line of desmosomes connecting supporting cell to supporting cell (homotypic junctions) along the entire length of their lateral surfaces. Desmosomal contacts between the head plates of inner and outer pillar cells are especially prominent. In the organ of Corti, desmosomes are not found in junctions between hair cells and supporting cells (heterotypic junctions). Thus, both the IHCs and OHCs of the mature organ of Corti have no desmosomal junctions (Gulley and Reese 1976) and no known intermediate filament cytoskeleton (Raphael et al. 1987).

2.2.4 Gap Junctions

Gap junctions are paired channels on adjacent cells, through which small molecules and electrical impulses can pass directly from one cell to another. Their component proteins are connexins, and they are arranged in six subunits per channel on each cell. The subunits on one cell line up with the subunits on an adjacent cell, and conformational changes, induced by electrical and biochemical events within the cell, cause the channels in both cells to switch between the open and closed positions.

Many of the supporting cells in the organ of Corti are connected to other supporting cells by gap junctions. In some nonmammalian species, dye-coupling experiments suggest that communicating junctions (gap junctions) are present between the sensory and supporting cells. Although gap junctions have not been found between sensory and supporting cells in the mammalian cochlea, gap junction–like structures have been noted at the lower boundaries of the tight junctions between sensory and supporting cells at the reticular lamina (Gulley and Reese 1976; Nadol 1978). Gap junctions occur between all cells in the stria vascularis and between the basal cells of the stria vascularis and fibrocytes in the spiral ligament. The role that gap junctions play in cochlear function is discussed further in this chapter in sections related to the cell types involved, and also in Wangemann and Schacht, Chapter 3.

2.3 Sensory Hair Cells

Two types of sensory cells, IHCs and OHCs, can be differentiated by their position within the organ of Corti, ultrastructure, arrangement of stereocilia, and innervation. Although the sensory cells differ both structurally and functionally, there are many similarities. Both types display stereocilia, stiff, fingerlike extensions that project from the apical surface of the sensory cells up into the endolymphatic space (Fig. 2.5). At the basal end of any hair cell, synaptic specializations are characteristic of interactions with nerve terminals (Fig. 2.15).

Hensen first described the stereocilia on some cells in the organ of Corti in the mid-nineteenth century, and suggested that the hair cells were sensory and the stereocilia were responsible for transduction. Direct confirmation of this hypothesis was not obtained until more than one hundred years later (Hudspeth and Corey 1977; Russell, Richardson, and Cody 1986). On the apical surface of each hair cell, the stereocilia are present in several rows arranged in a "U" or "W" shape (Fig. 2.5). Each row contains stereocilia that are similar in length, where the row of the shortest stereocilia faces the modiolus and the row of the tallest faces the lateral wall and is adjacent to the basal body of the cell (Flock et al. 1962). Deflection of the bundle of stereocilia in the direction of the tallest stereocilia is excitatory and deflection in the opposite direction is inhibitory.

The length of both the IHC and the OHC stereocilia increases along the length of the cochlear duct, from base to apex (Wright 1984; Bohne and Carr 1985; Lim 1986; Roth and Bruns 1992). Many of the properties of normal stereocilia in both mammalian and nonmammalian species have recently been reviewed (Nielsen and Slepecky 1986), and the most elegant studies on the internal arrangement of filaments, development of length differences, and arrangement of stereocilia within the bundle have been conducted on hair cells from the lizard and chicken basilar papilla (Tilney, DeRosier, and Mulroy 1980; Hirokawa and Tilney 1982; Tilney and DeRosier 1985; Tilney, Tilney, and DeRosier 1992).

Each stereocilium is covered by a plasma membrane with cell coat material that stains with cationic dyes (Spoendlin 1968; Flock, Flock, and Murray 1977; Slepecky and Chamberlain 1985a; Lim 1986; Prieto and Merchan 1987; Santi and Anderson 1987). This coat contributes a net negative charge to the external surface of each stereocilium and maintains

2. Cochlear Structure   55

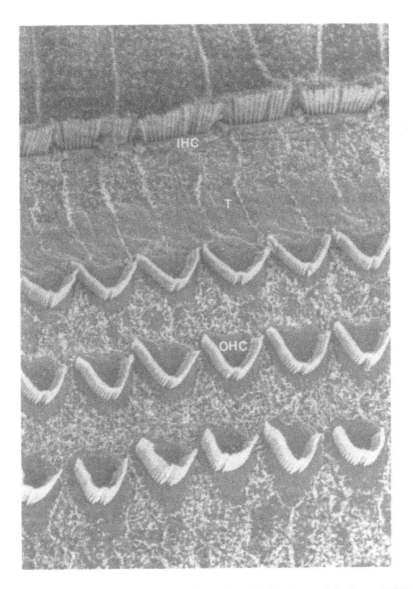

FIGURE 2.5 Comparison of the stereocilia on the apical surface of the inner (IHC) and outer (OHC) hair cells. The one row of IHCs and the three rows of OHCs are separated by the tunnel (T) composed of inner and outer pillar cells. (Scanning electron micrographs from guinea pig cochlea provided by Dr. G. Reiss, University of Hanover, Germany. From Reiss © 1992, reprinted by permission of John Wiley and Sons, Inc.)

a small distance between adjacent stereocilia in the bundle (Neugebauer and Thurm 1986). Fibrillar material extends out from different regions of each stereocilium and cross-links stereocilia within the hair bundle. "Within-row" links connect adjacent stereocilia of the same row and similar size along their lateral surfaces. "Across-row" links connect the lateral surfaces of stereocilia in one row with the taller stereocilia in the row behind. Both types of links appear to be robust and are thought to cause all stereocilia within a bundle to move as a group when only the tops of the longest stereocilia are deflected.

A single "tip link" (Fig. 2.6) runs upward from the tip of each shorter stereocilium to its taller neighbor (Pickles, Comis, and Osborne 1984). These tip links are extremely sensitive to histological processing and are difficult to preserve (Comis, Pickles, and Osborne 1985; Furness and Hackney 1985; Osborne and Comis 1990). A tip link has two components: a fine filamentous core surrounded by amorphous material, and an electron-dense submembrane specialization at the end of the link where it contacts the stereocilia (Furness and Hackney 1985; Osborne, Comis, and Pickles 1988). In transduction, it is thought that tip links transmit stimulus-induced movements to the transducer channels of the stereocilia. Stretch of the links could open ion channels at one or both points of attachment (Pickles, Comis, and Osborne 1984).

The tip link theory of transduction is supported by many observations on mammalian and nonmammalian mechanoreceptor systems: (1) Morphological polarization of hair cells, in which stereocilia, in rows of increasing length, are connected by tip links (Comis, Pickles and Osborne 1985), occurs in all species that have been examined (human, guinea pig, chinchilla, rat, bird, reptile, fish, and frog). (2) Transduction does not require the presence of a kinocilium, which is present in some sensory epithelia but is lacking in the organ of Corti of most adult mammals (Kimura 1966). (3) The relatively few ion channels involved in transduction (Russell 1983; Howard and Hudspeth 1988) appear to be located near the tips of the stereocilia (Hudspeth 1982; Jaramillo and Hudspeth 1990; Hackney, Furness, and Benos 1991). (4) Only small displacements of the tip of the bundle are required for near-maximum channel opening (Russell, Richardson, and Cody 1986). (5) The kinetics derived from electrophysiological experiments suggest that force applied to the transducer channel is transmitted via a link with elastic properties (Corey and Hudspeth 1983).

Lengthwise within each stereocilium, a bundle of parallel actin filaments are cross-linked to each other and to the plasma membrane (Flock, Flock, and Murray 1977; Tilney, DeRosier, and Mulroy 1980). Where the stereocilium inserts into the cuticular plate, the stereocilium tapers. Most of the actin filaments end there, and only a small number, coated with electron-dense material (Itoh 1982), continue as the rootlet into the meshwork of filaments of the cuticular plate (Fig. 2.6). Although the rootlet filaments are often not decorated with myosin S-1 (a specific probe for actin) because another material coats the filaments, the filaments have been identified as

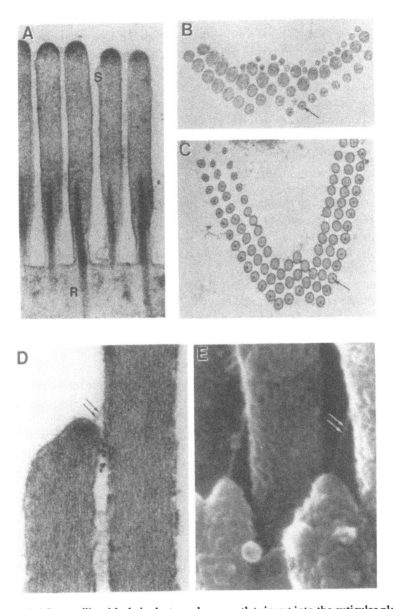

FIGURE 2.6 Stereocilia with their electron-dense rootlets insert into the cuticular plate of each IHC and OHC. The stereocilia are connected by side-to-side cross-links and tip links. (A) Transmission electron micrograph showing stereocilia (S) and rootlets (R). In cross section the different arrangements of the inner (B) and outer (C) hair cell stereocilia are seen, with the dark-staining rootlets (single arrow) in the center of each stereocilium. (D,E) Tip links (*double arrow*) are seen at higher magnification by both transmission (D) and scanning (E) electron microscopy. (D) Immunogold labeling with antibodies to an amiloride-sensitive epithelial sodium channel suggests that the stretch-sensitive ion channels responsible for transduction are located near the tip links. (Figs. 2.6A, B, and C from Slapecky NB et al. (1982); Figs. 2.6D and E provided by Drs. Carol Hackney and David Furness, University of Keele, United Kingdom.)

actin by other methods (Slepecky and Chamberlain 1985b). In mammals the rootlet extends well into the cuticular plate, unlike other animals, in which the filaments fan out as they enter (Tilney, DeRosier, and Mulroy 1980).

Given that actin filaments are oriented with the same polarity and that the two actin-cross-linking proteins present are fimbrin (in the stereocilium) and tropomyosin (in the rootlet), it was thought that the stereocilia themselves were not motile. However, the recent finding of calmodulin in stereocilia (Shepherd, Barres, and Corey 1989; Slepecky and Ulfendahl 1993) opens the possibility that the bridges between the core actin filaments and the plasma membrane may be composed, at least in part, of a protein complex of myosin I and calmodulin, similar to that seen in the microvilli of brush border epithelial cells. The possibility of a motor molecule along the length of the stereocilia has important implications for adaptation in hair cells. Adaptation to continued deflection of the stereocilia is both sensitive to calcium and blocked by calmodulin antagonists (Assad and Corey 1992).

The cuticular plate is a region made of filaments and proteins that may contribute to both support and motility functions. Randomly oriented actin filaments (Tilney, DeRosier, and Mulroy 1980; Flock et al. 1981; Slepecky and Chamberlain 1982) form a meshwork in the central region of the cuticular plate. The cuticular plate is discontinuous and there are gaps between the cuticular plate and the hair cell periphery through which vesicles appear to be transported through the cuticular plate to the stereocilia. Within the cuticular plate near the points of insertion of the stereocilia, there is also an area free of densely packed actin filaments that contains normal cytoplasmic components and a basal body composed of microtubules. The basal body is the origin of a kinocilium that is present during development but absent in the adult mammalian cochlea. Following loss of the kinocilium, the basal body remains in this region as the organizing center for cytoplasmic microtubules and indicates the functional polarity of the cell.

Actin filaments in the cuticular plate interact with actin filaments in the stereocilia through thin cross-links that extend radially from each stereocilium rootlet. Cuticular plate actin filaments also interact with actin filaments that run circumferentially at the periphery of the hair cell, parallel to the lateral cell membrane at the junctional complex (Hirokawa and Tilney 1982).

2.3.1 Structural Differences Between Inner and Outer Hair Cells

Initially, the two types of sensory cells in the organ of Corti were thought to differ in shape and number but to have similar functions. With ultrastructural studies on the sensory and supporting cells (Engström 1958; Kimura 1966; Angelborg and Engström 1973; Lim 1986) and on the innervation (Engström 1958; Smith and Sjostrand 1961a; Spoendlin 1973), it became clear that these two cell types differ dramatically. IHCs play a primarily sensory role in the cochlea, based on the fact that most of the afferent nerve fibers synapse with them. OHCs, although providing some direct sensory input to the central nervous system, more likely modify the mechanical properties of the organ of Corti and the BM. This chapter provides detailed descriptions of the structural and biochemical organization of these two cell types; information on their physiological responses and mechanical properties is presented in the chapters that follow.

2.3.2 Inner Hair Cells

IHCs are goblet-shaped with a centrally placed nucleus (Fig. 2.7) and form one continuous row along the length of the spiraling sensory epithelium. Each IHC has stereocilia arranged in a characteristic flattened "U" shape (Figs. 2.5 and 2.6) on its apical surface. It is thought that the tips of the longest row of stereocilia are only weakly attached to the bottom surface of the TM. The IHC stereocilia rootlets insert into the cuticular plate, where they traverse almost the entire depth (Takasaka et al. 1983), and have been observed to project into the apical cytoplasm, where they often end near subcuticular mitochondria.

The IHC cuticular plate is composed of randomly oriented actin filaments and obvious electron-dense bodies identified as the actin cross-linking protein α-actinin (Slepecky and Chamberlain 1985b). Circumferen-

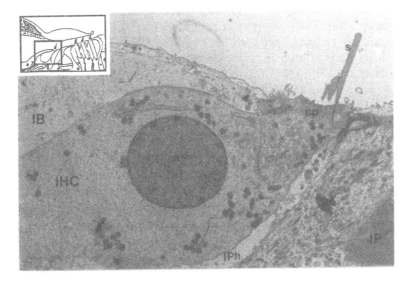

FIGURE 2.7 Each inner hair cell (IHC) is chalice-shaped with a large centrally placed nucleus (n). Along the apical surface, the stereocilia (S) insert into the cuticular plate (cp). Each IHC is surrounded on the modiolar side by an inner border (IB) cell and on the pillar side by an inner phalangeal cell (IPh) and an inner pillar cell (IP).

tial rings of actin filaments run parallel to the lateral cell membrane at the junctional complex. A striated region composed of alternating filaments and dense bodies is found lining the base of the cuticular plate. It was originally thought to be pathological in origin, but has now been observed in IHCs from normal specimens (Slepecky, Hamernik, and Henderson 1981). Although this structure has been described only in IHCs, it appears similar in composition to the cross-striated regions seen in the infracuticular network and/or the extracted cortical lattice observed in OHCs.

The IHC body is filled with round mitochondria, Golgi apparatus, multivesicular bodies, lysosomes, and a well-developed endoplasmic reticulum. Lining the plasma membrane, one single layer of discontinuous membrane cisterns extends to a region close to the nucleus. Cisterns are connected to the plasma membrane by cross-links or pillars (Forge 1987; Furness and Hackney 1990). Microtubules are present and more concentrated in the apex of the cell where they are nucleated by the pericentriolar material of the basal body (Furness, Hackney, and Steyger 1990). They rim the cuticular-free region and form a prominent meshwork underlying the cytoplasmic surface of the cuticular plate. In the cell body, microtubules line the single layer of membrane cisterns along the plasma membrane and in addition extend as a distinct longitudinal tubulovesicular bundle running between the apical and basal surfaces of the cell. The basal end of the cell synapses directly with afferent fibers projecting to the central nervous system.

Supporting cells and nerve fibers (Figs. 2.2, 2.7, 2.10) touch and surround IHCs on most of their basolateral surfaces so that IHCs do not contact the BM. The supporting cells span the distance between the BM and the reticular lamina. They have many microvilli on their apical surface, which suggests a role in ion exchange with endolymph or a site of attachment for the TM. On their basal surfaces, cytoplasmic processes envelop the nerve fibers below the IHCs and in the inner spiral bundle (Iurato 1967). Unlike the supporting cells that surround the OHCs, the cells surrounding the IHCs lack organized bundles of microtubules.

The inner border cells form a thin line on the modiolar side of the IHCs. Their basal portions are adjacent to nerve fibers entering and leaving the organ of Corti, and unlike the other supporting cells, the inner border cells have many mitochondria. They do not reach up to the reticular lamina. The inner phalangeal cells form a thin line between the IHCs and the inner pillar cells, and one process from each inner phalangeal cell projects between and separates the apical surfaces of two adjacent IHCs at the reticular lamina. The contact continues about halfway down the IHC, but in large areas from midcell down to the synaptic terminals, adjacent IHCs are in contact with each other. Thus, IHCs are like OHCs at the reticular lamina, where they form tight junctions with supporting cells (for IHCs, there are tight junctions between IHCs and both inner pillar cells and inner phalangeal cells) and their apical surfaces are separated by supporting cells. However, unlike OHCs, the IHCs are not separated from one another where the cell bodies become enlarged in the region of the nucleus. Although there is the possibility for direct communication between the IHCs, no junctional specializations between them have been found.

2.3.3 Outer Hair Cells

OHCs are long and cylindrical (Fig. 2.8) with a more basally placed nucleus. OHCs have a smaller diameter than IHCs, and there are almost four times as many OHCs as IHCs. At their apical surface, the OHCs contact supporting cells with tight junctions and link to them with adherens junctions. Their basal surface is cupped by adjacent supporting cells, but unlike IHCs, in most species they contact only fluid along their lateral surface.

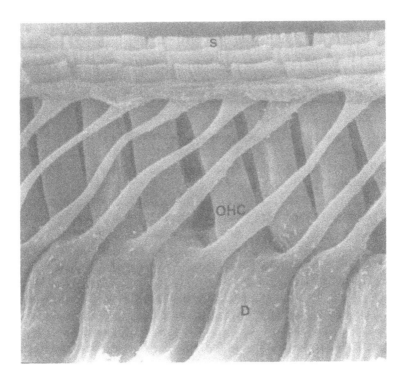

FIGURE 2.8 Each outer hair cell (OHC) is long and cylindrical, with stereocilia (S) projecting out from the cuticular plate at the apex. Each OHC is cupped at its base by a Deiters cell (D) (for example, the third-row OHC and Deiters cell seen here). Each Deiters cell has a stalk that extends up at an angle (toward the apex of the cochlea) and expands into a phalangeal process that inserts into the reticular lamina at a distance from the base of the Deiters cell. (Scanning electron micrograph of the mole rate cochlea provided by Dr. Y. Raphael, University of Michigan, Ann Arbor MI. From Raphael et al. © 1991, reprinted by permission of John Wiley and Sons, Inc.)

Since adjacent OHCs do not touch one another, there is no possibility for direct communication between them; however, they may be coupled to supporting cells through gap junctions. Morphological studies of the junctional complex near the reticular lamina have shown membrane specializations similar to, but not exactly like, gap junctions (Gulley and Reese 1976; Nadol 1978). Functional evidence is also equivocal, and although some say that cell coupling is unlikely (Oesterle and Dallas 1989) and that dye transfer has not occurred (Santos-Sacchi 1986), other studies have shown electrical connection and dye transfer (Zwislocki et al. 1992).

The OHC stereocilia are arranged in a "W" (Figs. 2.5, 2.6), and although graduated in length, they are longer and thinner than their IHC counterparts. In most species, they are arranged in three parallel rows; however, some species contain four or five rows (Kimura 1966; Raphael et al. 1991). At their apical tips, stereocilia from the longest rows are thought to be firmly attached to the TM. The stereocilia taper at their bases, and the rootlets project about halfway into the cuticular plate (Takasaka et al. 1983). The cuticular plate of the OHCs is composed of randomly and circumferentially oriented actin filaments; however, the body of the cuticular plate of the OHCs appears more homogeneous than that of the IHCs. The filamentous meshwork has fewer dense bodies (Slepecky and Chamberlain 1985b), and the large striated bodies seen running horizontally beneath the cuticular plate of IHCs are not present in OHCs.

In the apical turns of the guinea pig cochlea, cuticular plate-like material projects down from the cuticular plate toward the nucleus as an infracuticular network. This has dense bands running longitudinally (Kimura 1966), and the entire structure is associated with microtubules, membrane vesicles, and mitochondria. In spite of early theories that the presence, formation, or contraction of this network might be responsible for the slow shortening seen in OHCs, it cannot control this event, since it is not found in all OHCs that are known to be motile (Carlisle et al. 1988; Slepecky 1989b).

Above the nucleus, the OHC cytoplasm is homogeneous with ribosomes, mitochondria, and endoplasmic reticulum. Mitochondria are organized along the subsurface cisterns (SSCs) parallel to the lateral wall, unlike IHCs, in which mitochondria are scattered throughout the cytoplasm. Microtubules form a network in the cuticular-free region and beneath the cuticular plate (Furness, Hackney, and Steyger 1990) in a manner similar to, but less dense than, that seen in IHCs. By electron microscopy, few microtubules appear to be present in the OHC body, but some appear to terminate on the SSCs (Ekström von Lubitz 1981; Raphael and Wroblewski 1986). However, immunocytochemical analysis with antibodies to tubulin reveals an impressive network that runs the length of the cell (Slepecky, Ulfendahl, and Flock 1988; Steyger et al. 1989; Furness, Hackney, and Steyger 1990), and many are seen in the subnuclear region of the cell. Occasionally, a lamellar structure called a Hensen body is seen. It consists of concentric layers of membrane similar in structure to the SSCs and is sometimes seen in continuation with the SSCs, but its relationship with the SSCs is not known.

Below the nucleus microfilaments and microtubules are abundant, and at the base of the cell afferent and efferent synaptic specializations are present. It is thought that neural signals from the efferent nervous system activate a change in the OHCs that modifies the mechanical properties of the OHCs and hence the BM. This is supported by the presence of acetylcholine and γ-aminobutyric acid (GABA) receptors at the base of the OHCs (Canlon, Cartaud, and Changeux 1989; Zenner et al. 1989; Plinkert et al. 1990). However, the effect of efferents on OHCs is not known and can perhaps be elucidated by the study of species in which there are either no medial efferents innervating OHCs (Bruns and Schmieszek 1980; Raphael et al. 1991), or no base-to-apex gradient in the number or size of efferent terminals (Xie et al. 1993). No obvious relation appears between efferent innervation and frequency sensitivity since OHC efferents are absent in mammals with both high (horseshoe bat, *Rhinolophus rouxi*) and low (mole rat, *Spalax ehrenbergi*) audible frequency ranges.

The lateral walls of the OHC body below the cuticular plate (Fig. 2.9) are linked with parallel layers of membranous sacs called subsurface cisterns (SSCs) (Engström 1958; Smith and Sjostrand 1961a; Kimura 1975; Ekström von Lubitz 1981; Saito 1983), the inner layers of which are linked with elongated mitochondria and microtubules. The SSCs are single-layered in the human, rat, and cat, doubled-layered in the squirrel monkey, and multilayered in the guinea pig. Based on freeze-fracture analysis and sensitivity to different detergents, the SSC membrane appears to be chemically different from the plasma membrane (Forge 1991). When well preserved, the SSCs are composed of parallel, tightly packed lamellae in which the intracisternal lumen (25 nm) is wider than the intercisternal space (15–20 nm).

However, the morphological appearance of the SSCs is sensitive to the fixatives used for hair cell preservation and to trauma (Evans 1990; Dieler, Shehata-Dieler, and Brownell 1991; Slepecky and Ligotti 1992). Under these adverse conditions, the SSCs appear as a network of anastomosing and branching tubules composed of loosely stacked and swollen membranous vesicles. In spite of these problems, analysis of rapidly frozen preparations suggests that apical and basal hair cells may differ in SSC organization (Forge, Davies, and Zajic 1991), and further, that differences in SSC organization between cells in the same region of the cochlea may reflect membrane turnover (Forge et al. 1993).

The space between the SSCs and the plasma membrane (Fig. 2.9) is bridged by rows of pillarlike structures (Gulley and Reese 1977; Saito 1983; Flock, Flock, and Ulfendahl 1986; Raphael and Wroblewski 1986; Arnold and Anniko 1990). The pillars are connected to each other by a dense layer of material lining the SSCs, parallel to the cell membrane (Saito 1983; Bannister et al. 1988) and arranged in a lattice of circumferential filaments

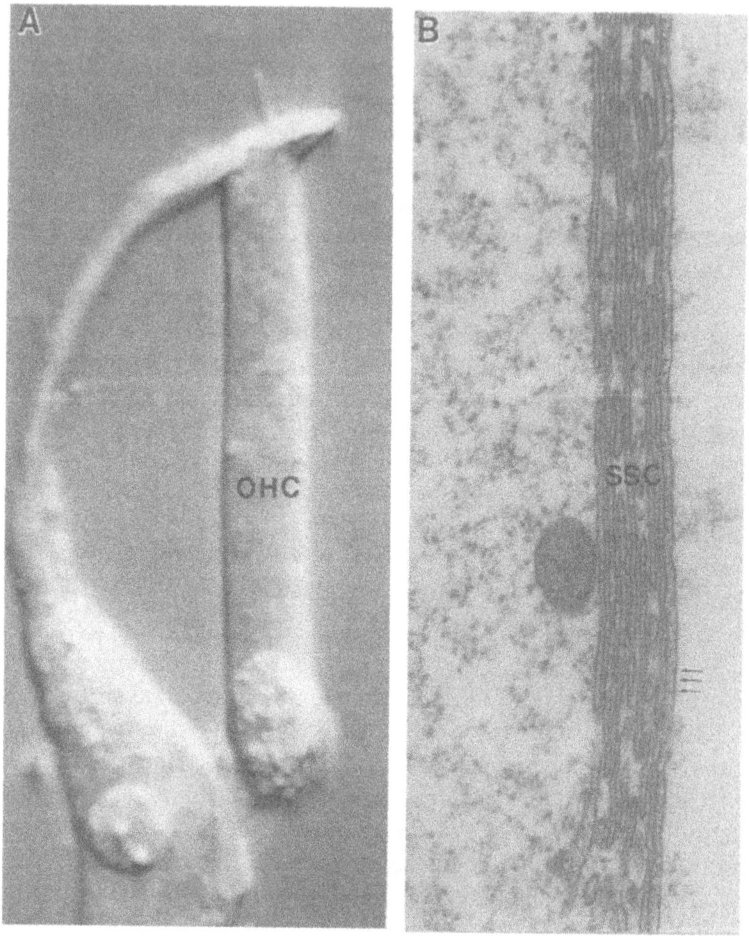

FIGURE 2.9 Outer hair cells have a unique system of subsurface cisterns (SSCs) along their lateral wall. (A) Hoffman modulation optics micrograph of an isolated outer hair cell (OHC) with attached Deiters and Hensen cells. (B) Transmission electron micrograph of a thin section through an OHC. The flattened membranous sacs of the SSCs are seen, as well as the periodic distribution of the pillars (*arrows*) connecting them to the cell plasma membrane.

and thin cross-linking filaments (Holley and Ashmore 1988, 1990; Arima et al. 1991; Forge 1991). Although originally modeled as a continuous helix (Holley and Ashmore 1988), it now appears that the cortical lattice is discontinuous with discrete domains that abut each other at various angles (Holley, Kalinec, and Kachar 1992). Immunofluorescent labeling suggests that they are composed of actin (Flock, Flock, and Ulfendahl 1986; Holley and Ashmore 1990; Holley, Kalinec, and Kachar 1992) and spectrin (Holley and Ashmore 1990; Nishida et al. 1993).

The interconnections between the SSCs, the pillars, and the plasma membrane in OHCs were initially thought to be similar to the "triad" structure in muscle, where junctional feet couple the transverse tubules of the cell membrane with the sarcoplasmic reticulum, and electrical information and ions flow between the two membranous layers. If present in OHCs, this structure could play an important role in motility.

The dependence of electromotility on an intact SSC system has been documented (Dieler, Shehata-Dieler, and Brownell 1991). However, morphological evidence argues against the analogy with muscle cells. First, the SSC-pillar-plasma membrane complex occurs, although less extensively studied, in IHCs (Forge 1987; Furness and Hackney 1990) where no motility has been demonstrated. Second, morphological differences exist between the SSC-pillar-plasma membrane complex in hair cells and the sarcoplasmic reticulum-pillar-T-tubule membrane complex forming the triad in muscle. These include: (1) The pillars in the two cell types differ in diameter (Arima et al. 1991). (2) The pillars in the two cell types differ in the way they are related to and interconnect the two membranous systems. In muscle cells, each pillar is made up of a protein composed of four subunits. On the muscle cell membrane end, the pillar subunit proteins are in direct contact with the transmembrane voltage-sensing protein at the T-tubule. On the sarcoplasmic reticulum membrane, the transmembrane domains of the four subunits constitute the calcium release channel. In hair cells, freeze-fracture analysis suggests an intimate relationship between the pillars and large particles in the plasma membrane (Arima et al. 1991), but there are no intramembranous particles to indicate that the pillars insert directly into either the SSC or the plasma membrane (Forge 1991).

Thus, the pillars only connect the plasma membrane (via short filaments) to the cortical lattice and do not form a transmembrane channel into the SSC. It has been suggested that the cortical lattice (filament-pillar complex) stabilizes the cell surface and contains elastic or motile components that could be involved in generating or transmitting acoustically driven changes in cell shape.

2.4 Supporting Cells with Filaments

The inner tunnel pillar cells, outer tunnel pillar cells, and Deiters cells are structurally similar in that they form a rigid scaffolding (Engström 1958; Iurato 1967; Angelborg and Engström 1972) adjacent to and surrounding the OHCs. Because of the position of the OHCs over the portion of the BM not supported by bone, this region of the organ of Corti undergoes significant movement during vibratory stimulation, and these supporting cells are specialized to withstand mechanical stress, to maintain the integrity of the reticular lamina, to transmit stimulus-induced motion of the BM to the reticular lamina, and to transmit stimulus-induced motion of the hair cells between the reticular lamina and the BM.

Structural rigidity is accomplished by microfilaments and microtubules within the cells. Flexibility is enhanced by the brace-and-strut-type arrangement of the different cell types. Inner tunnel pillar cells contact the BM over the osseous spiral lamina and reach in a relatively straight line from the BM to the reticular lamina. However, the outer pillar cells and Deiters cells slant in opposite directions, forming a crisscross pattern, like a folding babygate. With regard to one OHC, the outer pillars at its apical surface in the reticular lamina have their foot processes at a location distal to the OHC, two or three cells away in the apical direction of the cochlea. The Deiters cells that surround the apical surface of the same OHC have their cell bodies under other OHCs at a location two or three cells away, toward the base of the cochlea.

Anatomically, all three cell types have a broad base resting on the BM and an expanded apical process that makes a significant contribution to the reticular lamina (Fig. 2.10). The intracellular region between these two surfaces is spanned by rigid bundles of microtubules and actin microfilaments. Although the microfilaments appear to be of opposite polarity in these bundles (Slepecky and Chamberlain 1983; Arima, Uemura, and Yamamoto 1986), suggesting the possibility of motility, several lines of evidence argue against this: (1) The actin filaments do not interdigitate with each other, but rather are cross-linked to and interdigitate with microtubules, suggesting that there is no possibility for adjacent filaments to interact (Angelborg and Engström 1972). (2) The actin filaments in the bundles are surrounded by tropomyosin (Slepecky and Chamberlain 1987), as are filaments in muscle sarcomeres, an arrangement that is thought to maintain filament rigidity. (3) Myosin has not been found in the supporting cells of the mammalian inner ear. (4) The microtubules appear to be stable, based on the absence of newly synthesized forms of tubulin (Slepecky and Ulfendahl 1992) and the presence of more long-lived and stable modified forms of tubulin (Slepecky, Henderson, and Saha 1995). (5) Intermediate filaments, which are thought to play a role in maintenance of stable cell shape, are present (Raphael et al. 1987; Schulte and Adams 1989b; Bauwens et al. 1991; Kuijpers et al. 1991) in the supporting cells.

2.4.1 Inner and Outer Tunnel Pillar Cells

The inner and outer tunnel pillar cells separate the IHCs from the OHCs, provide a triangular basis of support (Figs. 2.10, 2.11), and form the fluid-filled tunnel of Corti. Each has a footplate that rests on the BM, filled with randomly oriented filaments consisting of several kinds of fibrous elements, into which numerous microtubules and actin filaments are inserted. A middle portion spans the distance between the BM and the reticular lamina and contains a large number of straight, interdigitating microfilaments and microtubules arranged in bundles. The microtubules and microfilaments insert into randomly oriented microfilaments in the head portions of the cells. At the reticular lamina, flattened processes, filled

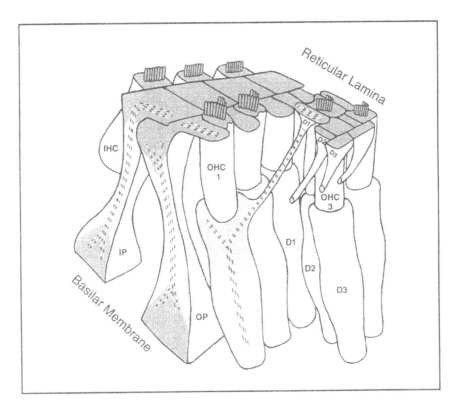

FIGURE 2.10 Diagram showing the complex relationship between the sensory hair cells and the supporting cells. The inner (IHC) and outer (OHC) hair cells are separated by the inner (IP) and outer (OP) pillar cells. The IHCs do not rest on the basilar membrane but are surrounded by inner border and inner phalangeal cells. The OHCs rest on the Deiters (D) cells. At the reticular lamina, the first-row OHCs are separated from each other laterally by the phalangeal processes of the OP cells. The second-row OHCs are separated from each other laterally by the phalangeal processes of the first row of Deiters cells (D1). The third-row OHCs are separated from each other laterally by the phalangeal processes of the second row of Deiters cells (D2). The phalangeal processes of the third row of Deiters cells (D3) separate the third-row OHCs from the Hensen cells. All of the supporting cells have bundles of microfilaments and microtubules that are thought to add to their ability to provide mechanical support.

with bundles of filaments and tubules, extend along the apical surface.

The inner pillar cells form a continuous wall of cells that separates the fluid in the tunnel from the IHCs. At the reticular lamina, they are coupled by tight junctions to IHCs, inner phalangeal cells, and outer pillar cells. The apical processes of the inner pillar cells are the rectangular surfaces that make up the tunnel in a surface preparation view, and there is approxi-

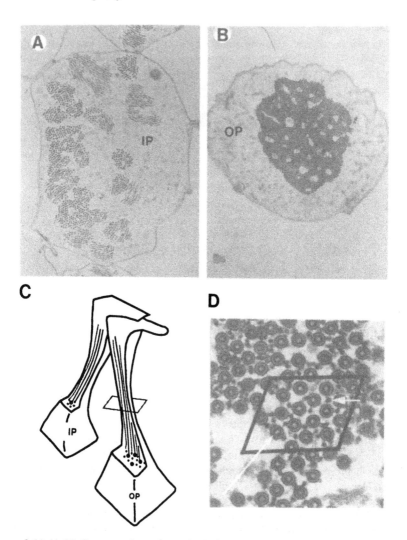

FIGURE 2.11 (A,B) Cross sections through the inner (IP) and outer (OP) pillar cells of the organ of Corti from a guinea pig cochlea. (C) Diagram showing the parallel array of microtubules and microfilaments in these cells. (D) High-power transmission electron micrograph showing the arrangement of the actin microfilaments (*short arrow*) and the microtubules (*long arrow*).

mately one inner pillar cell for each IHC. Below the reticular lamina, there are no tight junctions between these cells, so presumably the perilymph in the tunnel of Corti can diffuse past them, the adjacent inner phalangeal cells, and the IHCs. Nerve fibers travel between them on their way to the OHCs.

The outer pillar cells are more numerous than the inner pillar cells, with one outer pillar cell for each OHC in the first row. The apical processes of the outer pillar cells are tucked under the inner pillar cells in the tunnel regions, but reach up to the surface to interdigitate with, and separate, the lateral surfaces of the cuticular plates of the first row of OHCs. Thus, at the reticular lamina, the outer pillar cell is connected by tight junctions to the inner pillar cells, OHCs, and Deiters cells. The cylinder-shaped middle region of the outer pillar cell is entirely surrounded by fluid and divides the tunnel of Corti from the space of Nuel around the OHCs. The footplates of the outer pillars do not rest on the BM directly beneath their head plates, but at a place more apical in the cochlea.

These cochlear supporting cells are unique in that they are the only mammalian cells with microtubules composed of 15 protofilaments (Saito and Hama 1982). Consistent with this finding is the fact that the microtubules in these cells are not organized by centriole-containing centrosomes (as are 13 protofilament microtubules) but are organized by pericentriolar material, separated from the centriole and organized by densely staining material found at several subcellular locations just below and associated with the cell membrane (Tucker et al. 1992).

Evidence from histochemical and ultrastructural data suggests that in addition to structural support, the pillar cells are capable of transporting substances between fluid compartments by pinocytosis (Preston and Wright 1974). Microvilli line the surface of the pillar cells facing the tunnel fluid, as do holes that indicate pinocytotic activity. Electron micrographs show the presence of mitochondria in adequate numbers for oxidative metabolism in the cytoplasm of the pillar feet and clustered in the stalks and heads between the filaments and the cell membrane.

2.4.2 Deiters Cells

The Deiters cells span from the BM to the reticular lamina and are closely associated with the OHCs. These cells are also characterized by filament and microtubule bundles. The Deiters cells are compartmentalized even at the light microscopic level. The lower portion is adjacent to the BM, and the middle portion is at the base of the OHC. A thin stalk extends up to the reticular lamina and forms a phalangeal process that expands to fill the spaces between OHCs in the reticular lamina. The basal portion contains a basal cone of randomly oriented microfilaments close to the BM, a nucleus, mitochondria, and a bundle of filaments and microtubules extending from the BM to the portion of the cell at the base of one OHC. The basolateral membranes of adjacent Deiters cells envelope individual nerve fibers that contact OHCs as they run toward the apex and/or the base of the cochlea in the outer spiral bundles (Engström 1958).

In the region under the OHC, the cytoplasm of the Deiters cell becomes increasingly dense and organized, filled with filaments, organelles, and endoplasmic reticulum suggesting transport activity, and mitochondria with longitudinally oriented cristae (Vater, Lenoir, and Pujol 1992; Spicer and Schulte 1993). The randomly oriented filamentous material in this region is biochemically similar to the filamentous material at the footplates and headplates of the tunnel pillar cells (see Section 7.3) and probably plays a role in the structural support of the cell, since this portion of the Deiters cell forms a cup surrounding the hair cell and the nerve endings.

Each cell also has a stalk portion that runs obliquely from the base of one OHC to the reticular lamina, and a flattened phalangeal process adjacent to the cuticular plate of another (more apical) hair cell in the next row (Figs. 2.8, 2.9). The slender process is filled with a thin bundle of microtubules and filaments. At the reticular lamina it enlarges to form a flattened process, filled with all three cytoskeletal elements: microfilaments, intermediate filaments, and microtubules. In most species, processes from the third-row Deiters cells arch out (adjacent to the Hensen cells) rather than proceeding straight up to the reticular lamina, thus forming the fluid-filled outer tunnel (Figs. 2.4, 2.9). However it must be noted that in gerbils and bats, the third-row Deiters cell passes directly up from the base of the third OHC (Henson, Jenkins, and Henson 1983; Spicer and Schulte 1994b), and is free-standing in the fluid-filled space. In these species, a modified first row of Henson cells (the cover or tectal cell) forms the boundary of the outer tunnel. The marginal net of the TM extends processes that attach to the apical surface of the third-row Deiters cells.

Deiters cells increase in size from base to apex, and from the first-row Deiters cells to the more lateral third-row Deiters cells. They further show longitudinal gradients with regard to the number and location of mitochondria, the number and location of microtubules, and the presence or absence of the filamentous rosette under the base of the OHCs (Engström 1958; Iurato 1961; Kimura 1975; Harding, Baggot, and Bohne 1992; Spicer and Schulte 1993). These adaptations are probably related to their functions in different frequency-specific regions. The finding of intermediate filaments and tropomyosin in these cells suggests that they are not motile, but provide a scaffolding to support the organ of Corti. However, the presence of abundant creatine kinase (Spicer and Schulte 1992) and a tubulovesicular membrane system (Spicer and Schulte 1993) indicates a high level of energy consumption. Thus, Deiters cells may have roles in the cochlea in addition to simple mechanical support.

2.5 Surrounding Cells Without Filaments

Less is known about the supporting cells that lack organized filament bundles. These include the inner border and inner phalangeal cells surrounding the IHC discussed previously and the cells that span the distance from the OHC region to the lateral wall. The latter cells have recently been studied in detail in the gerbil (Spicer and Schulte 1994a). Morphologically they appear similar to each other (Iurato 1962; Kimura 1975) and are considered part of the supportive system of the organ of Corti, although their support may consist of transport of ions rather than mechanical support. They are characterized by relatively featureless and electron-lucent cytoplasm, scattered mitochondria, and prominent apical microvilli (Engström 1958), and display lateral intercellular spaces and junctions (tight junctions, adherens junctions, and desmosomes). On the basis of the cytological features of these cells, they are thought to participate in the formation of the fluid surrounding the hair cells. In culture, these cells appear to secrete extracellular fluid that fills the lateral intercellular spaces.

2.5.1 Hensen Cells

Hensen cells are high columnar cells that in most species are directly adjacent to the third row of Deiters cells (Figs. 2.2 and 2.4) and are still considered a part of the organ of Corti. They may reach from the reticular lamina to the BM, or they may rest on the Boettcher cells (at the base). The microvilli of their apical surface probably increase their surface area to facilitate ion exchange or attachment of the TM; they contain no filament or tubule bundles and have little developed endoplasmic reticulum. However, they may be distinguished even in living, unstained, preparations of the organ of Corti by their prominent lipid droplets, which glisten like pearls. In the gerbil and the bat, the first Hensen cell appears morphologically different from the other Hensen cells. On the basis of its structure and location, it has been called a cover cell or tectal cell of Corti's outer tunnel (Henson, Jenkins, and Henson 1983; Spicer and Schulte 1994b). These cells are characterized by fingerlike processes on their basolateral surface and are though to influence fluid and ion content in the outer tunnel.

2.5.2 Claudius Cells

Claudius cells are cuboidal cells that rest on the BM and, as a group, fill the distance between the Deiters cells and the lateral wall. They resemble Hensen cells in that their apical surfaces are in contact with endolymph and that they stain very lightly, with no morphological evidence for specialized function. They differ from Hensen cells in that they are shorter than Hensen cells, possess short apical microvilli, and have straight lateral cell borders (Duvall and Sutherland 1970; Kimura 1975). The Claudius cells are not part of the organ of Corti; rather, the Claudius cells extend from the organ of Corti laterally (Fig 2.2), becoming progressively shorter and flatter where they cover portions of the external sulcus cells (Fig. 2.21).

2.5.3 Boettcher Cells

In the basal turns of the cochlea, the Boettcher cells are found between the Claudius cells and the BM (Duvall and Sutherland 1970; Ishiyama, Cutt, and Keels 1970; Kimura 1975; Henson, Jenkins, and Henson 1982) so that their apical surfaces are never in contact with endolymph. Their cytoplasm is denser than that of the outer cells on the BM, and their basolateral surface membranes form elaborate interdigitations with each other. Thus, the surface area of a Boettcher cell is great relative to its volume, suggesting a transport function. At the very base of the cochlea, where the volume and number of the Boettcher cells are greatest, they fill the radial distance from the base of the third Deiters cell to the basilar crest of the spiral ligament. Toward the apex they decrease in number and eventually disappear.

2.5.4 External Sulcus Cells

The external sulcus cells, or root cells, are located at the junction of the BM with the cells of the lateral wall (Fig. 2.21). Some parts of their apical surface are in contact with endolymph, while other parts are covered by Claudius cells (Duvall and Sutherland 1970). Fingerlike projections or pegs from these cells penetrate the spiral ligament behind the spiral prominence, but are separated from the cells of the lateral wall by a basement membrane.

3. Tectorial Membrane

The TM is a gel-like structure composed of extracellular matrix material that overlies the reticular lamina of the organ of Corti. Because of its relationship to the stereocilia (Fig. 2.12), it is thought to play a crucial role in sensory transduction. In general, the relative mass of the TM increases along the length of the cochlea, from base to apex. The TM is anchored at the spiral limbus where the interdental cells secrete material that contributes to the TM during fetal and postnatal life (Iurato 1960; Kimura 1966; Lim 1972). The longest of the three rows of OHC stereocilia is thought to be attached to the TM along the entire length of the cochlea in all species, based on the consistent observation of stereocilia imprints on the TM and the presence of stereocilia attached to the underside of the TM when it has been detached during processing (Kimura 1966; Lim 1972).

The attachment of the IHC stereocilia to the TM is more controversial. In the horseshoe bat, clear imprints have been observed along the entire length of the lower surface of the TM (Vater and Lenoir 1992). In other species, such as the rat, imprints occur only in the basal turn (Lim 1986; Lenoir, Puel, and Pujol 1987), and in human and monkey, only fine grooves and fibrils appear on the undersurface in the basal turn (Hoshino 1977). In guinea pigs that have been treated with glycerol, there is some rearrangement of the fibers in the TM, and scanning electron microscopy shows attachment of the IHC stereocilia to Hensen's stripe (Meyer zum Gottesberge and Tsujikawa 1993). However, the imprints are quite durable and remain for months or years after hair cell loss, and actual attachment of the IHC stereocilia to the TM in a normal cochlea has not been observed. Thus, until convincing morphological evidence appears, there is no sure way to tell if there is weak attachment or no attachment between IHC stereocilia and the TM in adult animals.

The TM appears as a dense meshwork of fibers embedded in a microfi-

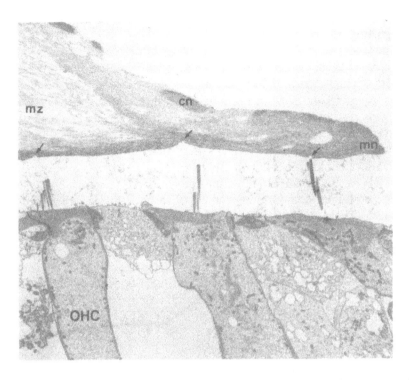

FIGURE 2.12 The tectorial membrane overlies the stereocilia, and the longest row of stereocilia in each bundle on the outer hair cells (OHC) insert into notches (arrows) on its lower surface. In mammals it is composed of collagen fibers and proteoglycans arranged in zones and layers. Shown are the middle zone (mz), cover net (cn), and marginal net (mn) of the tectorial membrane from a guinea pig cochlea.

brillar matrix (Kimura 1966; Lim 1972). These two fiber types have been characterized both ultrastructurally (Kronester-Frei 1978) and immunocytochemically. Type A fibers are straight, unbranched fibers, 10 nm in diameter, with obvious cross-striations. They run radially in parallel bundles within the body of the TM. Type B fibers, which lie among the type A fibers, are tightly coiled filaments that are wavy and highly branched. They are composed of fibrils 4 nm in diameter that are laterally interconnected to form striated sheets and that occur in both loosely packed and densely packed arrangements (Hasko and Richardson 1988). The distribution of the type A and type B fibers, and the different packing arrangement of the type B fibers (tightly packed and weakly hydrated vs loosely packed and highly hydrated), divide the TM into three zones and layers (Lim 1972).

The zones are marked in Figure 2.12 by lines drawn perpendicular to the reticular lamina. The limbal zone lies on top of and attaches to the spiral limbus. The middle zone is the thickest part of the TM and contains obvious thick fiber bundles that radiate outward from the inner limbal zone and slant toward the apex of the cochlea; this region covers the sensory cells.

The marginal zone is at the free end of the TM and connects the TM to the supporting cells below.

The layers run horizontal to the reticular lamina. The outer surface layer is nearest the endolymph and is made of densely packed type B fibers. It covers the limbal area and forms the cover net (over the body of the TM), the marginal band, and the net of the marginal zone. Beneath it, the homogeneous filament layer forms the body of the TM, and is composed of bundles of type A fibrils embedded in loosely packed type B fibrils. The layer closest to the reticular lamina is thin and amorphous and is composed of densely packed type B fibrils; this layer forms Hardesty's membrane over the OHC region and Hensen's stripe above the IHC region.

The molecular composition of the TM includes at least three types of collagen (Thalmann et a. 1985, 1987; Richardson et al. 1987) and various glycoproteins. Type II collagen is the major collagen in the cochlea and is localized only in the thick type A protofibrils (Slepecky et al. 1992). Type IX collagen co-localizes with type II collagen in the type A filaments (Slepecky, Cefaratti, and Yoo 1992). These heterotypic fibers are embedded in a reticular network of type B fibers containing type V collagen (Gil-Loyzaga, Gabrion, and Uziel 1985; Steel 1986; Khalkhali-Ellis, Hemming, and Steel 1987; Hasko and Richardson 1988; Prieto, Rubio, and Merchan 1990; Munyer and Schulte 1991; Sugiyama et al. 1991; Slepecky et al. 1992). Noncollagenous, glycosylated polypeptides and proteoglycans (Santi et al. 1990; Sugiyama et al. 1992; Suzuki et al. 1992; Munyer and Schulte 1994) can account for more than 50% of the total TM protein. The glycoconjugates are not homogeneous throughout the TM, and selective staining with lectins and colloidal iron have shown specific regional distribution as well as species differences.

The molecular components and their arrangement in the TM determine its unique biomechanical properties, and in light of the mechanical stress that the TM may undergo in the inner ear during fluid motion and movement of the stereocilia, it is not surprising that cartilage-type collagens and proteoglycans are present. Type II and IX collagen fibers in the TM are thought to contribute to its tensile strength and to resist stretch. The proteoglycans, because of their high negative charge density, create a highly hydrated, loose gel matrix that resists compressive forces and controls the radial expansion and interaction of the collagens.

4. Basilar Membrane

The BM, on which the organ of Corti rests, is composed mainly of extracellular matrix material, with fibers embedded in a homogeneous ground substance (Iurato 1962; Angelborg and Engström 1974; Cabezudo 1978). Although the BM was previously thought not to contain collagen, recently type II and IX collagen have been detected (Thalmann 1993).

Fibronectin is present in the amorphous ground substance (Santi et al. 1989; Woolf, Koern, and Ryan 1992; Keithley, Ryan, and Woolf 1993), where it may bind to and stabilize other connective tissue proteins.

Spindle-shaped tympanic border cells, with their axes oriented on the cochlear spiral, cover the side of the BM facing the scala tympani. Although the cells are connected by desmosomes and do not separate from each other during vibratory stimulation, the areas between the cells can be large and perilymph may be in direct contact with the BM. The lack of tight junctions in this region and the discontinuity of the basement membrane provide ultrastructural evidence that the BM is permeable to perilymph. A basement membrane separates the BM from the cells of the organ of Corti, the Claudius cells, and the Boettcher cells.

The structure of the BM contributes to the stiffness and mass of the cochlear partition. In normal adult animals, the stiffness decreases progressively toward the apex while the mass increases. In the base of the cochlea it is narrow and thick, while at the apex it becomes wide and thin. There are two zones of the BM. The pars tecta (arcuate zone) extends from the tympanic lip of the spiral limbus to the region under the outer pillar cells; its filaments are arranged transversely and not grouped into bundles. The pars pectinata (pectinate zone) extends from the region under the outer pillar cells to the basilar crest of the spiral ligament; its fibers are grouped into bundles and run radially, continuous with the fibers of the spiral limbus and the spiral ligament. Thus, the spiral limbus, the basilar membrane, and the spiral ligament form a single morphological and functional unit. The tension may be modulated through the action of type III fibrocytes in the spiral ligament that contain cytoskeletal proteins specialized for contraction. Such fibrocytes have been described in the mouse and two species of bat (Henson, Henson, and Jenkins 1984; Henson and Henson 1988).

The total thickness of the BM depends on the combined thickness of the homogeneous ground substance, the filaments, and the tympanic border cells. Although the thickness of the homogeneous ground substance and the filaments decreases from the base to the apex of the cochlea, the area of the mesothelial cells (tympanic border cells) increases. Thus, the total thickness of the BM shows a progressive increase from the base to the apex due to an increase in the population of mesothelial cells.

Although in most species (Echteler, Fay, and Popper 1994), the increase in thickness is gradual and the thickening of the arcuate and pectinate zones occurs only on the scala tympani side of the BM, this is not true for some species of bat (Bruns 1980). In the bat cochlea, where there is reason to have enhanced detection of a sound frequency used for echolocation, there can be one or more specializations of the BM. Often, thickening occurs on both the scala tympani and the scala media sides, close to the spiral ligament. The scala media portion contains a bundle of spirally oriented filaments, lateral to the base of the third-row Deiters cells, while the scala tympani area contains the radial and randomly oriented filaments. The secondary bony spiral lamina is thickened, and the spiral ligament, where the BM is anchored, is enlarged.

5. Reissner's Membrane

Reissner's membrane is composed of two cell layers, epithelial and mesothelial (Duvall and Rhodes 1967; Iurato 1967; Watanuki 1968; Hunter-Davis 1978), which separate the scala media from the scala vestibuli. Adjacent to the endolymph, the epithelial layer consists of flattened squamous cells linked at the endolympathic surface by tight junctions and adherens junctions; on their apical surface are microvilli. The cells do not contain many intracellular organelles other than a tubulocisternal endoplasmic reticulum (Qvortrup and Rostgaard 1990b). The epithelial cells are separated from the mesothelial cells adjacent to perilymph by a basement membrane. The presence of anionic sites, both on the endolympatic surface of Reissner's membrane (Arima, Uemura, and Yamamoto 1985; Torihara, Morimitsu, and Suganuma 1995) and on the basement membrane, is thought to provide a size- and charge-selective barrier between endolymph and perilymph.

The mesothelial cells are flat, elongated cells that are usually described as forming a discontinuous layer so that there are gaps or pores between them (Duvall and Rhodes 1967; Hunter-Duvar 1978). This seems consistent with the fact that the mesothelial cells are continuous with the cells lining the scala vestibuli. The presence of a continuous layer of mesothelial cells, coupled by traditional junctional complexes, has been reported in the guinea pig when special efforts have been made to improve fixation and minimize shrinkage of the tissue (Qvortrup and Rostgaard 1990a). In spite of its high metabolic rate, Reissner's membrane is avascular (Duvall and Rhodes 1967; Axelsson 1968). The role of Reissner's membrane in maintaining endolympathic and perilymphatic fluid composition is discussed in detail in Wangemann and Schacht, Chapter 3.

6. Innervation

Early studies on the cochlea differentiated several types of nerve fibers with processes running between the organ of Corti and the central nervous system. Anatomical studies were instrumental in identifying the fibers as afferent (toward the brain) or efferent (toward the periphery) and in determining the fine structure and the number of afferent and efferent nerve endings for each type of hair cell and each turn of the cochlea (Rasmussen 1940, 1953; Engström 1958; Smith and Sjostrand 1961a,b).

Since then, many reports on the innervation pattern and the fine structure of synapses at the hair cells have appeared in original articles for each species and in reviews (Spoendlin 1966; Iurato 1967; Angelborg and Engström 1973; Kimura 1975; Pujol and Lenoir 1986; Nadol 1988; Eybalin 1993).

Two common structural features of a hair cell synapse with an afferent fiber are a thickening of the pre- and postsynaptic membranes, and a presynaptic rod with vesicles in the hair cell. An efferent synapse on a hair cell is characterized by an accumulation of synaptic vesicles in the nerve terminal and a subsynaptic cistern postsynaptically in the hair cell. There are, however, some discrepancies depending on species; for example, synaptic rods differ in IHCs and OHCs and are not always present at the afferent synapse of the OHCs.

In regions of the cochlea away from the hair cells, the two populations of nerve fibers (afferent and efferent) are difficult to distinguish because the nerve fiber bundles contain both types. Early investigators studied degeneration following nerve section or hair cell loss. It is not possible to totally differentiate afferents from efferents by these methods. Sectioning of the efferent nerves to the cochlea, running with the vestibular nerve, causes the efferent nerve fibers and nerve endings to degenerate (Pujol and Carlier 1981). However, cutting only the crossed component (an operation that is easier to perform than cutting the entire efferent bundle innervating one cochlea) in different species will affect differing numbers of the medial and lateral subsystems (Warr 1992). Sectioning of the cochlear nerve in the internal acoustic meatus is followed by incomplete degeneration of the spiral ganglion cells (Spoendlin 1973). Thus, although it is possible to get a cochlea with only the afferent terminals present, it is not possible to get a cochlea with only the efferent fibers and terminals. Conclusions about the efferents have to be made by comparing normal cochleae with those in which the efferents have been cut. Thus, more accurate information on the innervation pattern of the organ of Corti can be obtained only with techniques that selectively identify the fiber types.

Afferent cells and fibers have been observed and traced using serial thin-section analysis (Kimura 1986), Golgi impregnation (Ginzberg and Morest 1983), intracellular labeling with horseradish peroxidase (Kiang et al. 1982; Berglund and Ryugo 1987; Brown 1987; Simmons and Liberman 1988), and immunocytochemistry for intermediate filaments (Berglund and Ryugo 1986; Romand, Hafidi, and Despres 1988). These results have confirmed that the afferent fibers are subdivided into at least two systems, with thick myelinated fibers originating from type I ganglion cells contacting IHCs, and thin unmyelinated fibers originating from type II ganglion cells contacting OHCs (Rosenbluth 1962; Thomasen 1966; Kellerhals, Engström, and Ades 1967). There is some evidence that these two groups can be further subdivided, and that up to five subgroups can be distinguished (Ross and Burkel 1973; Kimura, Bongiorno, and Iverson 1987; Nadol 1989; Romand et al. 1990).

Efferent cells and fibers have been selectively stained using acetylcholinesterase histochemistry, immunocytochemistry for neurotransmitters and their synthesizing enzymes, and synaptic vesicle proteins. As with the afferents, the efferents were initially divided into two systems: medial efferents, which synapse with the OHCs, and lateral efferents, which synapse on the IHC afferents (Warr and Guinan 1979). Further subdivisions have been suggested by immunocytochemical staining for neurotransmitter substances (Eybalin 1993).

On the basis of these studies, it is possible to describe the innervation of the cochlea in general terms (Figs. 2.13, 2.14), but it must be noted that there are species differences with regard to the absolute number of cells and fibers as well as their size and degree of myelination. Good reviews on the afferent (Schwartz 1986; Nadol 1988; Eybalin 1993) and efferent (Warr 1992; Eybalin 1993) systems have recently been published. Further discussion of hair cell synaptic ultrastructure and neurotransmitters is found in Sewell, Chapter 9.

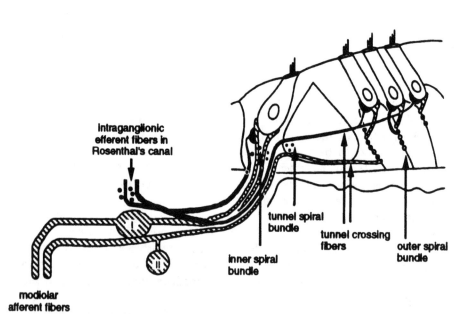

FIGURE 2.13 Comparison of the pathways taken by the afferent (*hatched*) and efferent (*solid black*) nerve fibers within the organ of Corti, the position of the afferent type I and type II ganglion cells and the spiral efferents in the osseous spiral lamina, and the afferent processes running toward the cochlear nucleus in the modiolus.

6.1 Inner Hair Cell Innervation

6.1.1 Afferent Fibers and Type I Spiral Ganglion Cells

Early studies in the cat (Spoendlin 1973) and guinea pig (Morrison, Schindler, and Wersäll 1975) showed that 90%-95% of the fibers in the auditory nerve contact only IHCs. In this region the nerves are unbranched (Spoendlin and Schrott 1988) in most species, so that each fiber terminates on one IHC and there is only one asymmetrical synaptic specialization per fiber (Fig. 2.15), and each IHC contacts approximately 20 different afferent nerves. The number of nerve fibers contacting each IHC is not constant and does not increase on a gradient, and the number may be enriched in frequency regions that are of functional importance. In the cat and human, the neural density is highest in the middle region of the cochlea, while in the bat, there may be multiple regions of increased density. Synapses between IHCs and afferent nerve terminals are characterized by a thin presynaptic membrane density and a thick postsynaptic membrane density, separated from each other by a synaptic cleft containing filamentous and granular material (Smith and Sjostrand 1961b; Dunn and Moret 1975; Saito 1980; Bodian 1983; Sobkowicz et al. 1986). In the active zone, the presynaptic cytoplasm contains a synaptic bar surrounded by membrane vesicles. In the inactive zone, there are presynaptic dense projections, coated vesicles, and membrane-bound tubular structures. The size and shape of the synaptic rods vary considerably among hair cells, and these variations in the appearance of the synaptic body are thought to be related to the functional state of the hair cell (Jorgensen and Flock 1976) and physiological characteristics of the postsynaptic nerve fiber (Liberman 1980b).

The afferent fibers of the 8th nerve transmit sensory information directly to the central nervous system. In the cat, postsynaptic afferent nerve terminals and fibers can be divided into three morphological groups according to the mitochondrial content, the site of the synapse on the IHC, the morphology of the synapse, and the number of contacts by efferent nerve terminals (Liberman and Oliver 1984; Liberman, Dodds, and Pierce 1990). The morphological characteristics correlate with fiber threshold sensitivity and spontaneous rate. Fibers with a large diameter, low threshold, and high spontaneous rate are located predominantly on the pillar side of the IHC. Thin fibers with a high threshold and low spontaneous rate are located only on the modiolar side of the IHC. It is not clear if this segregation is common to all species.

Each nerve runs as a thin, unbranched fiber directly out of the organ of Corti through a region with holes in the tympanic lip of the osseous spiral lamina called the habenulae perforatae. At this point, the radial afferent fiber becomes thicker and myelinated (depending on the species). As this dendrite of the IHC afferent approaches its cell soma, it becomes thinner

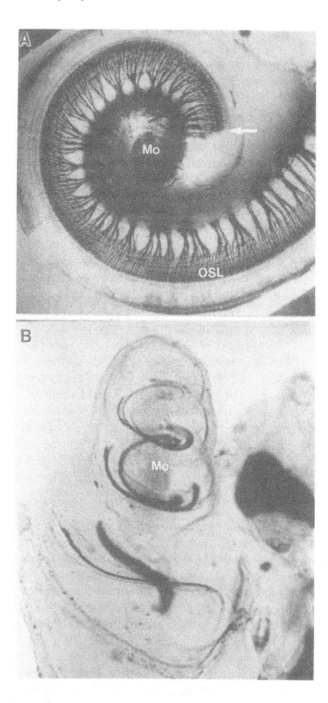

FIGURE 2.14 Comparison of the pathways taken by the afferent and efferent nerve fibers within the cochlea. (A) Myelinated afferent and efferent fibers stained black with osmium run radially from the organ of Corti (*arrow*) in the osseous spiral

(*continued*)

again. This decrease in diameter (and the subsequent unequal caliber of the axon and dendrite of the ganglion cell) is a characteristic that can be used to differentiate type I fibers and ganglion cells from type II fibers and ganglion cells when they are not myelinated.

Type I ganglion cells (Fig. 2.15) are the most numerous, comprising 95% of the cells found in the spiral ganglion. Depending on the species, the perikarya may be myelinated with 4–25 lamellae contributed by Schwann cells and surrounded by a basement membrane. The peripheral and central processes emerge from opposite ends of the cell, and the first node of Ranvier occurs on the proximal part of each neurite some distance away from the cell soma. The large, round nucleus contains a prominent nucleolus and is located toward one pole of the cell. The cell contains many large mitochondria, Golgi apparatus, ribosomes, SSCs, microtubules, some neurofilaments, dense bodies, and rough and smooth endoplasmic reticulum.

Myelinated fibers from these bipolar cells innervate the cochlear nucleus. In some species, the size of the spiral ganglion cells is correlated with the characteristic frequency (and thus the location of the neuron in the apex, or the base of the cochlea). In the cat (Romand, Romand, and Marty 1981) and guinea pig (Friede 1984), the apical fibers are thinner; in the mouse, the apical fibers are somewhat large (Anniko and Arnesen 1988); yet in the human, there appears to be no size difference. Although this may be of consequence physiologically, it is not related to the distance traversed by the neuron, since in the cat the total length from synapse to cochlear nucleus is similar for all nerve fibers (Arnesen and Osen 1978).

6.1.2 Efferent Innervation of Inner Hair Cells

Myelinated and unmyelinated efferent fibers coming from the central nervous system (a part of the lateral efferent system) enter the cochlea near the basal and middle turns of the cochlea (Rasmussen 1953) and spiral in both apical and basal directions as the intraganglionic spiral bundle through the spiral ganglion cells in Rosenthal's canal (Figs. 2.13 and 2.14). At a given point, depending on the frequency specificity of the individual cell bodies of origin, the efferent fibers turn out toward the organ of Corti in a

---

FIGURE 2.14 (*continued*) lamina (OSL). Afferent fibers continue past the spiral ganglion and enter the modiolus (Mo) where they travel as a group to the cochlear nucleus. (Micrograph of human cochlea provided by Dr. Göran Bredberg from Bredberg 1968.) (B) Efferent fibers stained darkly for acetylcholinesterase do not run in the modiolus (Mo), but rather spiral between the afferent ganglion cells in the intraganglionic spiral bundle. At the base of the cochlea they join the vestibular nerve and travel to the brain stem. (Micrograph of the guinea pig cochlea provided by Prof. Yasuya Nomura, Showa University, Tokyo, Japan, from Nomura 1970, reprinted by permission of Igaku Shoin Ltd.)

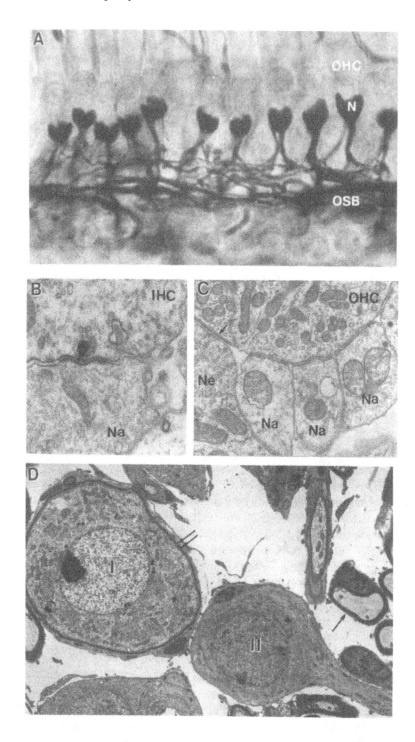

radial direction and run with the afferent fibers in the osseous spiral lamina. They pass through the habenulae perforatae as do the afferents, and like them are all unmyelinated within the organ of Corti.

The efferent fibers that remain in the IHC region are thin and travel in the inner spiral bundle or the tunnel spiral bundle, where they branch and form varicosities prior to making "en passant" synapses with the radial afferent dendrites beneath the IHCs. The fibers, comprising approximately 50%–100% of the total number of efferent fibers entering the cochlea, can originate from cell bodies associated with the lateral superior olivary complex, on either the ipsilateral or contralateral side of the brain stem. Projections from the ipsilateral side (75%–100% of the lateral efferent fibers) are distributed homogeneously along the length of the cochlea. The projections from lateral efferent cell bodies on the contralateral side are distributed preferentially to the apical part of the cochlea. In the cat, the lateral IHC efferents are most often found synapsing with afferent fibers that have low spontaneous rate and high threshold (Liberman 1980a; Liberman, Dodds, and Pierce 1990). The synaptic specializations are characterized by accumulations of small, clear and large, dense-cored synaptic vesicles and electron-dense conical spicules on the membrane. The pre- and postsynaptic elements are separated by an intercellular gap, and contact is made with an area of the afferent nerve fiber that does not appear specialized.

There are some sites of direct contact between efferent varicosities and IHCs where synaptic specializations are not found. In the few cases where synapses are present, they are characterized by small, bulbous varicosities containing synaptic vesicles and postsynaptic SSCs. In adult animals, such synaptic specializations are rare and depend on the species examined (Spoendlin 1966; Angelborg and Engström 1973; Kimura 1975; Saito 1980). Efferent contact with IHCs is common during development, so the

---

FIGURE 2.15 Comparison of inner and outer hair cell nerve terminals. (A) Maillet stained nerve fibers (N) synapse on outer hair cells (OHC) and run in the outer spiral bundle (OSB) prior to exiting the organ of Corti. (Micrograph provided by Dr. Göran Bredberg from Bredberg 1968.) (B) Inner hair cells (IHCs) synapse onto unbranched afferent nerve terminals (Na) with synaptic ribbons surrounded by vesicles. (C) Outer hair cells (OHC) synapse onto branches of afferent nerve terminals (Na) which often display a characteristic invagination. Efferent nerve fibers (Ne) synapse directly onto the OHC body, and the OHC has a subsynaptic cistern (*arrow*) opposite the terminal specialization. (D) Type I ganglion cells from IHC afferent terminals are large, myelinated (*double arrows*), and bipolar. Each gives off a myelinated process (*arrow*) that enters the modiolus. Type II ganglion cells from OHC afferent terminals are smaller, pseudomonopolar, and unmyelinated or thinly myelinated. Each has an unmyelinated process entering the modiolus.

presence of efferent terminals in young animals may reflect the age of the animal. In mice, there are many efferent terminals on IHCs up to at least postnatal day 19 (Emmerling et al. 1990; Sobkowicz and Slapnick 1994).

Although most efferent fibers contain acetylcholine and stain positively for acetylcholinesterase, some of the efferents near the IHC are acetylcholinesterase-negative (Emmerling et al. 1990). Based on both branching pattern and immunocytochemical staining, there appear to be other populations of efferent fibers that run in the inner spiral bundle. Fibers that stain positively for GABA (Thompson, Cortez, and Igarashi 1986; Eybalin et al. 1988; Usami et al. 1988; Whitlon and Sobkowicz 1989; Eybalin and Altschuler 1990) run for long distances, and GABA-like immunoreactivity was found in one-half of the varicosities in the inner spiral bundle. These data suggest that the acetylcholinesterase/acetylcholine-staining fibers and the glutamic acid decarboxylase (GAD)/GABA-staining fibers are components of separate fiber systems. Fibers have also been identified that stain positively for tyrosine hydroxylase and dopamine (Jones, Fex, and Altschuler 1987; Usami et al. 1988; Eybalin 1993), enkephalins (Altschuler et al. 1894; Fex and Altschuler 1986; Eybalin 1993), dynorphins (Altschuler et al. 1985), and calcitonin gene-related peptide (Sliwinska-Kowalska et al. 1989; Kuriyama et al. 1990; Eybalin 1993).

6.1.3 Reciprocal Synapses on Inner Hair Cells

Reciprocal synapses on IHCs, where one nerve terminal contains both afferent (a postsynaptic density opposite an IHC synaptic bar) and efferent (an accumulation of presynaptic vesicles) synaptic specializations, have been found in cultures of developing organ of Corti (Sobkowicz, Slapnick, and August 1993) but have not been seen in adult cochleae.

6.2 Outer Hair Cell Innervation

6.2.1 Afferent Fibers and Type II Ganglion Cells

OHC afferent fibers are highly branched with small bouton endings, and each OHC synapses directly with terminals from many different afferent fibers (Smith 1975; Berglund and Ryugo 1987; Dannhof and Bruns 1993). Because they are highly branched, one spiral fiber can receive information from 6 to 100 OHCs, most often with OHCs within the same row. The branching pattern for afferent fibers contacting OHCs is not uniform throughout the cochlea—it differs by the row innervated and changes systematically from base to apex. For comparative purposes, the innervation density can be approximated by knowing the number of afferents crossing the tunnel (always at the base of the tunnel rather than at midlevel), the number of unmyelinated afferents in the osseous spiral lamina, or the number of type II ganglion cells.

The majority of the type II ganglion cells contact OHCs in the first row, and these OHCs are preferentially contacted by the largest number of terminals. The remaining type II ganglion cells are more highly branched and preferentially contact OHCs in the third row and OHCs at the apex. In bat and rat cochleae, there appear to be two morphologically distinct types of afferent endings in the apex (Pujol and Lenoir 1986; Vater, Lenoir, and Pujol 1992), and it has been suggested that these may be collaterals from type I ganglion cells (Gil-Loyzaga and Pujol 1990; Puel and Pujol 1992).

The synaptic specializations between the OHC and its afferent terminals differ from those seen at IHCs. In the active zone, there is an invagination of the OHC, with a corresponding evagination of the nerve terminal (Fig. 2.15). There are clear, irregular synaptic vesicles in this area, and a synaptic rod, although lacking in some species (Dunn and Morest 1975), is sometimes present (Saito 1980). In guinea pig cochleae fixed for electron microscopy, it has been noted that there are parallel and perpendicular synaptic bodies at the synaptic specialization. Circadian changes have been observed in the perpendicular synaptic bodies but not in the parallel synaptic bodies (Swetlitschkin and Vollrath 1988). In the periactive zone, coated vesicles are attached to the presynaptic membrane to form endocytotic profiles. The pre- and postsynaptic regions are separated by a synaptic cleft. Some irregularly sized vesicles and electron-dense material accumulate on the postsynaptic membrane, and the afferent bouton contains a few mitochondria.

The OHC afferents are rich in microtubules and spiral in bundles (outer spiral bundles) between the Deiters cells. To exit the organ of Corti, the OHC afferents spiral in the apical direction for varying distances (up to a quarter of a turn), cross the lower part of the tunnel, pass between the inner pillar cells, and go out through the habenulae perforatae with the radial afferents. While the radial IHC afferents become thicker and myelinated at this point, the OHC afferent fibers remain unmyelinated and gradually increase in diameter as they approach their somata in the spiral ganglion (Kiang et al. 1982; Romand and Romand 1984). Thus, the axon and the dendrite of each OHC ganglion cell are of equal diameter.

The type II ganglion cells (Fig. 2.15) are smaller than the type I cells. They are unmyelinated, are covered by a simple sheath of Schwann cells, and are surrounded by a basement membrane. They tend to be pseudomonopolar. These neurons contain a rather large nucleus, which exhibits numerous deep infoldings and contains a prominent nucleolus. The cytoplasm is characteristically rich in neurofilaments, and there are many ribosomes, mitochondria, rough endoplasmic reticulum not usually organized into Nissl bodies, Golgi apparatus, occasional dense-cored vesicles, dense bodies, multivesicular bodies, and SSCs. Unmyelinated nerve processes make contact with the somata and dendrites of the type II cells in the spiral ganglion (Kimura, Ota, and Takahashi 1979; Romand and Romand 1984; Kimura, Bongiorno, and Iverson 1987; Hozawa and Kimura 1990).

Type II ganglion cells make up only 5% of the ganglion cells of the cochlea, and unmyelinated fibers from these cells have been traced to the cochlear nucleus.

6.2.2 Efferent Innervation of Outer Hair Cells

The base of the OHC is the site of innervation of the large-diameter, myelinated efferent fibers originating from cell bodies located in subnuclei medial to the medial superior olivary nuclei. Projections may be either ipsilateral or contralateral, running to the peripheral cochlea in the intraganglionic spiral bundle. A small component of the OHC efferent fibers arises from neurons projecting to both cochleas (Aschoff and Ostwald 1987).

Efferent fibers destined for OHCs are myelinated by Schwann cells in the osseous spiral lamina. After they pass through the habenula, they are no longer myelinated and continue without branching under the IHCs. There they can spiral for a short distance in the inner spiral bundle, pass through the inner pillar cells, and spiral for a distance in the tunnel spiral bundle prior to crossing the tunnel radially (Spoendlin 1979) at a midlevel to reach the OHC region. They branch almost immediately in all directions and run in the outer spiral bundle; each of the branches ends with a vesicle-filled varicosity at the base of an OHC (Engström 1958; Spoendlin 1968). Some efferent fibers have been observed to synapse on the afferent fibers in an axodendritic arrangement. It is not yet clear if these are part of the medial or lateral efferent systems (Guinan, Warr, and Norris 1982).

There are differences between the innervation of the OHC rows, with more efferent terminals on the first row of OHCs at each place along the length of the BM. In most species, there is a base-to-apex gradation of efferent innervation, with a larger number of efferent terminals ending on each OHC at the base of the cochlea. While the efferent fibers are traditionally thought to synapse only at the base of the OHCs, up to 25% of the OHCs in the apex of the cochlea have terminals synapsing above the nucleus (Altschuler et al. 1984; Hashimoto and Kimura 1988; Liberman, Dodds, and Pierce 1990). The branching pattern of the efferent fibers in these regions also differs, based on Golgi staining, Maillet staining, and tracer studies using horseradish peroxidase or biocytin. At the base, the efferent fibers branch with projections spread only within a restricted area. In the apex of the cochlea, the efferent fibers spiral for a distance in the outer spiral bundles and project to OHCs spatially separated from each other.

In the nerve terminal cytoplasm (Fig. 2.15), many clear synaptic vesicles accumulate at the electron-dense projections of the presynaptic membrane, and a few dense-cored vesicles and mitochondria are observed (Kimura 1975; Saito 1980). Both pre- and postsynaptic membranes have increased electron densities and are separated by a synaptic cleft. In the hair cell cytoplasm, a flattened subsynaptic cistern, continuous with the subsurface cistern lining the lateral wall of the hair cell, is observed beneath the postsynaptic membrane (Kimura 1975). This differs from the efferent synapse on the afferent nerve fibers below the IHCs, which has no specialization on the postsynaptic membrane. Cholinergic efferents can be identified by acetylcholinesterase histochemistry and by immunocytochemistry for the enzyme choline acetyltransferase, which is used in its synthesis. Terminals can be further identified by immunocytochemical labeling for synaptophysin (Liberman, Dodds, and Pierce 1990), an integral membrane protein specific for synaptic vesicles. Stains for these substances display a close agreement with regard to the number of efferent endings labeled and the number of endings observed by semiserial electron microscopy.

Subsets of efferent fibers have been identified by their differing immunoreactivities to other neurotransmitter and neurotransmitter-related substances such as GAD and GABA (Thompson, Cortez, and Igarashi 1986; Eybalin et al. 1988; Usami et al. 1988; Whitlon and Sobkowicz 1989), enkephalins (Altschuler et al. 1984; Fex and Altschuler 1986; Eybalin 1993), dynorphins (Altschuler et al. 1985), and calcitonin gene-related peptide (Sliwinska-Kowalska et al. 1989; Kuriyama et al. 1990; Eybalin 1993). These classes of efferent endings at the base of the OHCs differ in their location and branching pattern. The fibers containing acetylcholine and GABA have a reciprocal distribution, with acetylcholine-containing fibers found in large numbers at the base of the cochlea and GABA-containing fibers preferentially distributed at the apex. The efferents containing acetylcholine have an overlapping pattern of distribution, and those containing GABA have a very restricted distribution, innervating patches and narrow columns of OHC in the radial direction. Terminals of the GABA-containing fibers make contact with OHCs above the nucleus near the reticular lamina, or make axodendritic synapses on OHC afferent fibers located beneath the OHCs.

6.2.3 Reciprocal Synapses at Outer Hair Cells

In some species, there is evidence for reciprocal synapses on OHCs, where a single nerve ending possesses two types of synaptic specializations, afferent and efferent (Nadol 1983). Within a bouton ending, the afferent synapse is located at the periphery. The corresponding active zone in the hair cell is characterized by an indentation of the membrane, and within the hair cell there is a presynaptic structure of synaptic bar with synaptic vesicles. Postsynaptically the nerve terminal membrane displays an accumulation of electron-dense material. The efferent-type synapse is found in part in the central region of the bouton ending. Presynaptically the nerve terminal has an area with synaptic vesicles near its plasma membrane, and pre- and postsynaptic membranes both have an electron-dense layer. Postsynaptically the hair cell has a subsynaptic cistern opposite the nerve terminal. Thus, at reciprocal synapses, both afferent and efferent information may be transferred between an OHC and one nerve terminal.

6.3 Sympathetic Innervation to the Cochlea

It has long been suspected that the vasculature of the cochlea, like that of other organs of the body, is innervated by the sympathetic nervous system. Recent evidence suggests that sympathetic innervation arises from both the stellate and the superior cervical ganglia (Spoendlin 1981; Laurikainen et al. 1993). With the introduction of techniques for histochemical and immunocytochemical identification of nerve fibers and electron microscopic analysis of nerve fiber terminals, small, unmyelinated sympathetic nerve fibers were observed. It is difficult to determine the cells on which these fibers exert their influence in the normal cochlea, for two reasons: (1) The unmyelinated sympathetic fibers are ultrastructurally similar to the unmyelinated fibers of the efferent olivocochlear bundle. (2) There are no traditional "synaptic specializations" between postganglionic sympathetic fibers and their postsynaptic target. Transmitter is liberated from synaptic vesicles localized in varicosities along the length of the thin fibers and reaches receptor molecules on the postsynaptic cells by diffusion.

At the light microscopic level, identification of the catecholaminergic fibers can be based on their localization by fluorescent histochemical methods (Spoendlin and Lichtensteiger 1966; Terayama, Holz, and Beck 1966; Spoendlin 1981). With an electron microscope, fibers and varicosities can be identified by the presence of dense-cored synaptic vesicles (Terayama, Yamamoto, and Sakamoto 1968; Densert and Flock 1974). However, since catecholamines (dopamine) are present in the olivocochlear efferent nerve fibers within the organ of Corti (Jones, Fex, and Altschuler 1987), and dense-cored vesicles are found in nerve terminals other than those of sympathetic fibers, these methods of identification are not sufficient.

More specific methods use uptake of radioactively labeled transmitters or antibodies to identify individual neurotransmitters and enzymes related to their synthesis. On the basis of these results, the vasculature-associated and vasculature-independent sympathetic fibers in the cochlea have been identified as noradrenergic (Eybalin, Calas, and Pujol 1983; Usami et al. 1988; Eybalin 1993). These techniques, coupled with selective lesions of the sympathetic chain ganglion and pretreatment of the cochlea with agents that enhance or reduce the amount of noradrenergic transmitter substances stored in synaptic vesicles, have allowed investigators to map the origin, distribution, and sites of synaptic interaction of fibers within the cochlea (Spoendlin and Lichtensteiger 1966; Terayama, Holz, and Beck 1966; Terayama, Yamamoto, and Sakamoto 1968; Ross 1973; Densert and Flock 1974; Spoendlin 1981; Eybalin, Calas, and Pujol 1983; Usami et al. 1988; Brechtelsbauer et al. 1990; Hozawa and Kimura 1990).

Sympathetic nerve fibers innervating the cochlea enter as a perivascular plexus along with the cochlear artery in the internal auditory meatus. The perivascular plexus continues along the spiral modiolar artery, along the larger branches of the spiral modiolar artery, and along the smaller branches of the spiral modiolar artery. It runs with the arterioles and small vessels exiting to the different turns of the cochlea, and fibers are found in the cochlear nerve, the spiral ganglion, Rosenthal's canal, and the osseous spiral lamina. The perivascular plexus can also be observed in the first part of the radiating arterioles and the collecting venules that supply the lateral wall. As the vessels exit from the bone to ramify in the capillary beds of the spiral ligament and stria vascularis, they lose their smooth muscle cells and their sympathetic innervation.

In addition to the perivascular plexus in the cochlea, there is a vessel-independent network of sympathetic fibers. These fibers enter into the osseous spiral lamina associated with the small blood vessels, but the fibers branch away from the vessels and associate with the auditory nerve afferent cell bodies in the spiral ganglion, as well as with afferent and efferent nerve fibers in Rosenthal's canal. In many cases they reach as far as the habenulae perforatae, where there are concentrations of varicosities filled with synaptic vesicles around the nonmyelinated portion of the afferent axons. In no case have they been found to enter the organ of Corti. Since noradrenaline may be released from the nerve fiber varicosities into the extracellular space, it is free to diffuse away from the site of release and may interact with adrenergic receptors on the surface of any one of a number of surrounding cell types.

Thus, it is possible that outflow from the sympathetic nervous system causes primary and secondary effects in the cochlea, in some way altering the metabolism of strial, supporting, and receptor cells (Ross et al. 1974; Densert 1975; Ross 1978) as well as influencing the blood vessels and nerve fibers with which they are in close contact. The existence of a parasympathetic innervation (Ross 1973) is still in question.

7. Structural and Biochemical Organization of the Cytoskeleton

7.1 Cytoskeleton of the Organ of Corti

Analysis of cells in the organ of Corti using immunocytochemistry has given us information on the structural and biochemical organization of the different cell types that permits us to better predict the functions that these cells perform. In mammalian cells, the two proteins responsible for providing the structural basis for cell shape and the substrates for changes in cell shape are actin and tubulin. They are present in both monomeric and polymerized forms, but act as structural or contractile components mainly when they are polymerized into microfilaments and microtubules, respectively. Actin-binding and microtubule-associated proteins regulate microfilament and microtubule number, length, organization, and location in cells. Calcium is thought to play a major role in cells as a signal that controls metabolism, cytoskeletal integrity, and changes in cell shape. Intermediate filaments, which are intermediate in size between microfilaments and microtubules, are thought to contribute to maintaining stable cell shape over long periods of time.

The localization and suggested functional importance of several proteins found in the cells of the organ of Corti are presented here (Figs. 2.16, 2.17, 2.18). Additional proteins intimately associated with the cytoskeleton on the inside and outside surfaces of the cells are not covered in detail. These include cadherins, transmembrane glycoproteins that mediate calcium-dependent cell-to-cell adhesion and make and retain close intercellular contact within the reticular lamina (Raphael et al. 1988; Whitlon 1993). They also include enzymes that are responsible for maintaining ionic gradients, such as Na^+, K^+-ATPase (Schulte 1993), Ca^{2+}-ATPase (Schulte and Adams 1989a), and neurotransmitter receptors (Canlon, Cartaud, and Changeux 1989; Zenner et al. 1989; Plinkert et al. 1990).

7.2 Sensory Hair Cells

7.2.1 Actin and Actin-Binding Proteins in Hair Cells

Immunocytochemical studies have shown that actin is present predominantly as filaments in the hair cells of the mammalian cochlea in at least five different regions (Flock and Cheung 1977; Slepecky and Chamberlain 1982; Arima, Uemura, and Yamamoto 1987; Thorne et al. 1987; Slepecky 1989a; Weaver, Hoffpauir, and Schweitzer 1993). In the stereocilia, it is present in bundles of parallel filaments in the shaft of each stereocilium (Fig. 2.16), some of which continue down into the cuticular plate as the rootlet. Apically, it is present in the cuticular plate (Fig. 2.16) as a meshwork of filaments and as a circumferential ring of filaments associated with the adherens junctions at the reticular lamina. It is present as a cortical lattice just inside the cell membrane along the lateral wall, and it is present in the hair cell cytoplasm.

Each of these regions has different actin-binding proteins (Drenckhahn et al. 1982; Flock, Bretscher, and Weber 1982; Slepecky and Chamberlain 1985b; Holley and Ashmore 1990; Ylikoski et al. 1990; Slepecky and Ulfendahl 1992; Nishida et al. 1993; Slepecky and Savage 1994). It is these proteins that determine the state of microfilament assembly and disassembly, as well as the structural and mechanical properties of the individual microfilaments and of the microfilament bundles. Profilin is an actin-binding protein that sequesters actin monomers and regulates polymerization into filaments. Tropomyosin wraps around actin filaments and main-

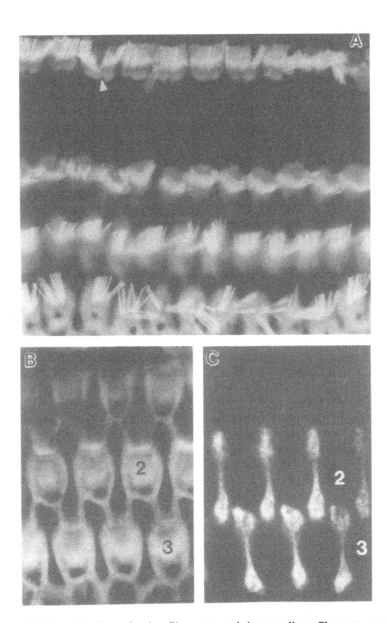

FIGURE 2.16 Localization of microfilaments and intermediate filaments at two different levels in whole mounts of the organ of Corti. (A) Actin (labeled with fluorescent phalloidin) is present in the stereocilia and the cuticular plates of the inner (top row) and outer (bottom three rows) hair cells. *Arrowhead* indicates the actin-free region of the cuticular plate where the basal body is situated. (B) When labeled with fluorescent phalloidin, actin is present in the cuticular plates of the outer hair cells. The actin at the junctional complex along the circumference of both the outer hair cells (in row 2 and 3), and the Deiters cells is prominent. (C) When labeled with antibodies to cytokeratin, intermediate filaments are seen in the Deiters cell processes surrounding the outer hair cells but not in the hair cells (2 and 3). (Confocal scanning micrographs from the guinea pig provided by Dr. Yoash Raphael, University of Michigan, Ann Arbor, MI. From Raphael and Altschuler © 1991, reprinted by permission of John Wiley and Sons, Inc.)

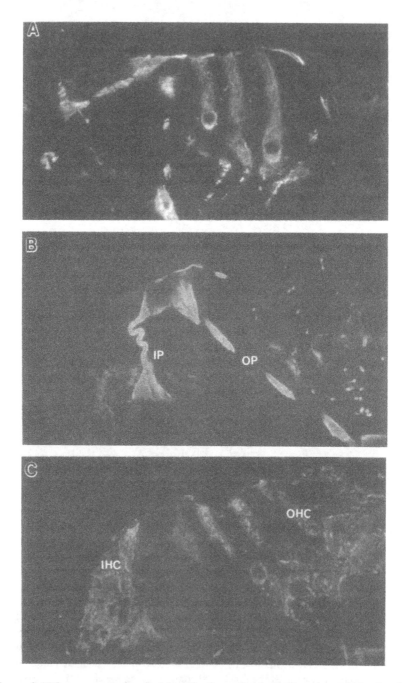

FIGURE 2.17 Immunocytochemical localization of tubulin in microtubules in the sensory and supporting cells of the guinea pig cochlea. (A) Antibodies to all forms of tubulin show microtubules in the sensory and supporting cells. (B) Stable microtubules, as indicated by staining with antibodies to acetylated tubulin, are

(*continued*)

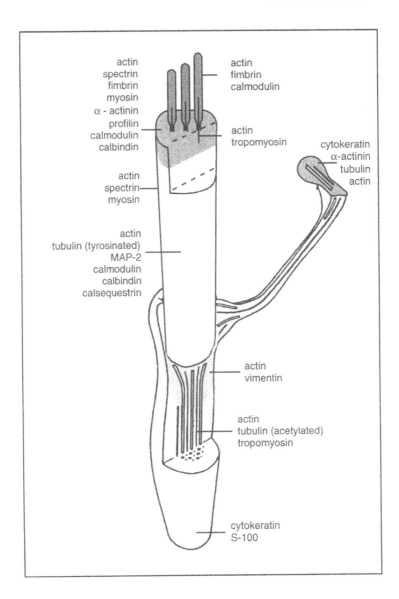

FIGURE 2.18 Diagram of outer hair cell and Deiters cell showing the location of many of the proteins associated with the cytoskeleton that have been localized by immunocytochemistry.

FIGURE 2.17 (*continued*) found in the inner (IP) and outer (OP) pillar cells, as well as in the processes of the Deiters cells (small bright dots to the right of the OP). (C) Newly synthesized tubulin, indicating more dynamic and less stable microtubules, is localized by staining with antibodies to tyrosinated tubulin. Tyrosinated tubulin is present in networks of microtubules found predominantly in the inner (IHC) and outer (OHC) hair cells, and less abundantly in the Deiters cell bodies below the OHCs.

tains them in a rigid conformation. Fimbrin cross-links filaments of the same orientation into rigid bundles, spectrin cross-links filaments into meshworks, and α-actinin anchors the ends of filaments. A summary diagram (Fig. 2.18) shows the location of many of the proteins that have been found.

The co-localization studies show that the actin filaments in the stereocilia are cross-linked by fimbrin, while the actin filaments in the rootlet are coated by tropomyosin. Both of these proteins are thought to contribute to the stiffness of the stereocilia. Within the body of the cuticular plate, actin is present with the cross-linking proteins spectrin, fimbrin, and α-actinin and the actin monomer–sequestering protein profilin. Myosin is also present in the cuticular plate, and if it interacts with actin filaments of opposite polarity as it does in other cell types, it might provide tension in the cuticular plate that would modulate the stiffness or position of the stereocilia, ultimately affecting transduction. Within the circumferential ring, actin and α-actinin are prominent, and vinculin and spectrin are present. A constriction in this region of the reticular lamina could increase the tension of the reticular lamina as well as altering the position or stiffness of the stereocilia.

The actin filaments in the lateral wall forming the cortical lattice are cross-linked by spectrin, and the linkage (similar to that in the cytoskeleton of red blood cells) is thought either to be involved in fast OHC motility or to be flexible enough to follow the cell shape changes caused by both fast and slow types of OHC motility. The presence of myosin in this region also supports some type of active movement. The distribution of the proteins and cortical lattice is discussed further in Holley, Chapter 7. In guinea pigs, an actin-containing structure called the infracuticular network runs from the bottom surface of the cuticular plate toward the nucleus of the OHCs. Its function is not known, but it does not appear to be involved in any of the OHC specific motile responses, since only the long apical hair cells in the guinea pig contain it, and it is not obviously present in the OHCs of any other species (Thorne et al. 1987; Slepecky 1989b).

7.2.2 Tubulin and Microtubule-Associated Proteins in Hair Cells

Tubulin is present as microtubules (Fig. 2.17), found in networks throughout the cell bodies of both the IHCs and the OHCs (Steyger et al. 1989; Furness, Hackney, and Steyger 1990; Slepecky and Ulfendahl 1992). When fluorescently labeled antibodies are used to localize tubulin, labeling of microtubules is more intense in IHCs than in OHCs. In both cell types, microtubules are found in arrays below the cuticular plate. On the side of the hair cell toward the lateral wall of the cochlea, in the cuticular plate–free region where the basal body is found (and the original site of the kinocilium during development), a large number of microtubules are found. Labeling indicates many microtubules running parallel to the long axis of the IHCs and OHCs, between the subcuticular and synaptic regions. In OHCs, intense labeling and many microtubules are seen in the subnuclear region.

Recent staining with antibodies to different tubulin isoforms indicates that the microtubules in the sensory hair cells are composed predominantly of the tyrosinated form of tubulin (Slepecky and Ulfendahl 1992 and Fig. 2.17), suggesting that these are dynamic microtubules, undergoing rapid cycles of polymerization and depolymerization. MAP-2, a protein thought to reversibly cross-link microtubules with microfilaments, is restricted to OHCs (Slepecky and Ulfendahl 1992). These findings provide biochemical evidence for remodeling of the OHC cytoskeleton, which may be correlated with, or even cause, chemically induced OHC shape changes.

7.2.3 Calcium-Binding Proteins and Calcium-Dependent Regulatory Proteins in Hair Cells

Calcium is thought to play a major role in cells, and the release of intracellular free calcium ions signals changes to the cell that can result in modulation of its structure (through the cytoskeleton) and metabolism (through enzyme regulation). For calcium to be effective as a signaling mechanism, the concentration of free calcium ions must be maintained at low levels in resting cells so that focal increases can trigger specific events. Calcium-binding proteins have been suggested to play two functionally different roles in cells. Some are primarily calcium buffers (such as calbindin and calsequestrin), responsible for maintaining low levels of intracellular free calcium. Others are calcium sensors and triggers (such as calmodulin) that detect changes in the level of intracellular free calcium and activate proteins and enzymes. The role of calcium in the organ of Corti is discussed further by Wangemann and Schacht, Chapter 3.

Calcium-binding proteins and calcium-dependent regulatory proteins have been well studied in the adult mammalian cochlea (Dechesne and Thomasset 1988; Legrand et al. 1988; Slepecky, Ulfendahl, and Flock 1988; Eybalin and Ripoll 1990; Dechesne et al. 1991; Slepecky and Ulfendahl 1993). It is clear from these studies that both IHCs and OHCs contain significant amounts of these proteins (Fig. 2.18) but that there are quantitative and qualitative differences between species. There are differences in the intensity of staining along the length of the sensory epithelium from apex to base, suggesting that at least calbindin and parvalbumin display longitudinal gradients (Pack and Slepecky 1995).

The stereocilia and cuticular plate of both IHCs and OHCs contain calmodulin (Slepecky and Ulfendahl 1993). The cell bodies of IHCs have been shown to contain calmodulin, calbindin, calretinin, and parvalbumin (Legrand et al. 1988; Eybalin and Ripoll 1990; Dechesne et al. 1991; Slepecky and Ulfendahl 1993). OHCs also contain calmodulin and calbindin, but in addition they contain the protein calsequestrin (Slepecky and Ulfendahl 1993). On the basis of its location in muscle (where it is specifically located in the sarcoplasmic reticulum) and in nonmuscle cells (where it is found in small membrane vesicles called calciosomes), the presence of calsequestrin in OHCs suggests the presence of an inositol trisphosphate-sensitive storage organelle. Unexpectedly, the various calcium-binding proteins and precipitable calcium appear to be diffusely distributed throughout the cytoplasm (Slepecky and Ulfendahl 1993) and not localized to the SSCs.

7.3 Supporting Cells

Actin, tubulin, and intermediate filaments comprise the cytoskeleton of the supporting cells within the organ of Corti (Fig. 2.18). As in the hair cells, there is a network of filaments at the apical portion of the tunnel pillar cells and the Deiters cells. Unlike the hair cells, the supporting cells contain organized bundles of filaments and tubules that span the distance between the reticular lamina and the BM. These are thought to play a major role in mechanical support, holding the sensory cells up off the BM and transmitting movement of the BM to the stereocilia at the reticular lamina.

7.3.1 Actin and Actin-Binding Proteins in Supporting Cells

Actin is a major component of the filament network at the apical and basal parts of both types of supporting cells (Flock, Bretscher, and Weber 1982; Slepecky and Chamberlain 1985b; Arima, Uemura, and Yamamoto 1986), with tiny cross-links interconnecting the randomly oriented filaments. A similar network of actin-containing filaments is present in the midregion of the Deiters cells, just below the cup on which the OHCs rest (Slepecky and Savage 1994; Pack and Slepecky 1995). Actin is also a component of the bundles that span the length of the supporting cells. The actin filaments in these bundles (Fig. 2.17) co-localize with tropomyosin (Slepecky and Chamberlain 1983), and interdigitate with and are cross-linked to microtubules (Flock, Bretscher, and Weber 1982; Slepecky and Chamberlain 1985b).

7.3.2 Microtubules and Microtubule-Associated Proteins in Supporting Cells

The microtubules in the supporting cells (Fig. 2.17), which are a component of the bundles (Slepecky and Chamberlain 1985b; Steyger et al. 1989), differ from the microtubules in the sensory hair cells in that they contain 15 rather than 13 protofibrils (Saito and Hama 1982; Kikuchi et al. 1991). Recent immunocytochemical studies indicate that the tubulin present in both pillar cells and Deiters cells no longer contains tyrosine (a marker for newly synthesized tubulin) and has been posttranslationally modified (Fig. 2.17). The presence of acetylated and glutamylated tubulin predominantly in the supporting cells (Slepecky, Henderson, and Saha 1995) suggests that the microtubules are long-lived and more stable and further supports a role for the pillar cells and Deiters cells in structural support.

7.3.3 Intermediate Filaments in Supporting Cells

Intermediate filaments are so cell type–specific that their presence has been used to classify cells according to their origin. Cytokeratins are characteristic of epithelial cells with desmosomes; vimentin is characteristic of cells of mesenchymal origin (fibroblasts); desmin filaments occur in striated and smooth muscle cells; glial fibrillary acid protein (GFAP) is present in glial (nonneuronal) cells; and neurofilaments are found in neurons. Several studies have investigated the presence of intermediate filaments in the adult cochlea of various species (Raphael et al. 1987; Schrott, Egg, and Spoendlin 1988; Schulte and Adams 1989b; Anniko and Arnold 1990; Bauwens et al. 1991; Kuijpers et al. 1991; Shi et al. 1993). Although localized in supporting cells of the organ of Corti, there is general agreement that intermediate filaments are not present in the sensory hair cells (Raphael et al. 1987; Schulte and Adams 1989b; Bauwens et al. 1991; Kuijpers et al. 1991).

Cytokeratins are present in all cells lining the endolymphatic space of the cochlear duct except the sensory hair cells (Raphael et al. 1987; Bauwens et al. 1991; Kuijpers et al. 1991). A subset of these cells, including the inner and outer pillar cells, Deiters cells, inner border cells, inner phalangeal cells; outer sulcus cells, and spiral prominence cells, also contains vimentin (Schulte and Adams 1989b; Oesterle, Sarthy, and Rubel 1990; Bauwens et al. 1991). In the pillar and Deiters cells, the location of the two intermediate filaments is complementary. Cytokeratin is present at both ends of the cells in the apical domain at the reticular lamina (Fig. 2.16) and in the basal domain near the BM. Vimentin is found in the middle regions of these cells. In the Deiters cells, large amounts of vimentin fill the region of the cells just under the base of the OHCs. The finding that cytokeratin and vimentin are both present in some of the supporting cells is surprising, since it is rare to find two types of intermediate filament proteins coexpressed in the same cell, and it is rare to find vimentin in epithelial cells.

7.3.4 Calcium-Binding Proteins in Supporting Cells

The major calcium-binding protein found in supporting cells in the gerbil and guinea pig organ of Corti appears to be S-100 (Foster et al. 1994), and its localization is specific to supporting cells (Pack and Slepecky 1995). S-100 is abundant in the inner border and inner phalangeal cells surrounding the IHCs and in the pillars, Deiters, Hensen, and Claudius cells. Outside the organ of Corti, it has been observed in the glial cells surrounding nerve fibers and in cells of the spiral ligament and stria vascularis at the lateral wall (Shi et al. 1992; Foster et al. 1994).

Calretinin, a calcium-binding protein that is not specific to supporting cells, is present in the Deiters cells (but not the Hensen cells) in the guinea pig and gerbil cochleae (Dechesne et al. 1991; Pack and Slepecky 1995), but may be absent from supporting cells in the rat cochlea (Dechesne et al. 1991). Calbindin has been shown to be present at low levels in the pillar cells of the rat cochlea (Legrand et al. 1988; Dechesne et al. 1991), but not in the guinea pig or gerbil cochlea (Slepecky and Ulfendahl 1993; Pack and Slepecky 1995).

The role of calcium-binding proteins in supporting cells is not known, but the presence of creatine kinase in inner phalangeal cells and Deiters cells (Spicer and Schulte 1992) and the presence of transient calcium fluxes in Deiters cells (Dulon, Moataz, and Mollard 1993) suggest that the different cell types included under the general heading of "supporting cells" have more functions than purely structural support.

8. Blood Supply and Vasculature

Extensive studies (Smith 1972; Axelsson and Ryan 1988) have found widespread similarities in the vascular anatomy of cochleae among various mammalian species, including humans. Not surprisingly, the arrangement of the vessels has both a radial and a spiral component. Large arteries and veins spiral in the modiolus. Smaller vessels radiate to supply the cells in the osseous spiral lamina and the lateral wall, where capillaries ramify and provide nutrients and oxygen to the cochlea. Consistent differences appear along the length of the cochlea. The apex is characterized by fewer radiating arterioles and a thin stria vascularis. In basal turns, the stria is thicker, well-organized spiral vessels run along the stria's inferior edge, and more arteriovenous shunts within the spiral ligament allow blood to bypass the stria vascularis.

Of functional importance are the relatively large diameters of the capillaries of the lateral wall. In the guinea pig, they range from 7 to 9 μm in the spiral ligament and are approximately 12 μm in diameter in the stria (Smith 1972; Miles and Nuttall 1988). In cochleae from young normal animals, most capillaries are filled with moving red blood cells. In cochleae from aging animals and in pathological specimens, it is possible to find intravascular strands where capillaries are closed down to a thick strand, and avascular channels where capillaries narrow and lack red blood cells.

8.1 Large Blood Vessels of the Modiolus

The spiral modiolar artery and vein are prominent structures that can be seen easily in an open cochlea, filled with red blood cells and following a tortuous course around the bone of the modiolus. In the arteries, a layer of endothelial cells is surrounded by one to three layers of smooth muscle cells, adventitial cells, and elastic laminae. Initially the vessels are surrounded by thin smooth muscle cells, but as they become smaller they are first covered entirely and then covered only partially by pericytes. Arterioles branch off the spiral modiolar artery and subsequently branch to form capillary beds in the modiolus, spiral ganglion, osseous spiral lamina, and tympanic lip. Radiating arterioles pass out through the bony partition of the coils, over the scala vestibuli of each turn, into the lateral wall (Figs. 2.19, 2.20).

In the lateral wall, branches from the radiating arterioles form distinct capillary beds in the spiral ligament, stria vascularis, and spiral prominence. The walls of the capillaries are formed by flattened endothelial cells, which have a basement membrane on their outer surface and face the lumen on their inner surface. The capillaries can be fenestrated, with spaces between the adjacent endothelial cells through which substances may pass through into connective tissue. Or the capillaries may contain endothelial cells that are connected by tight junctions preventing diffusion of substances through the intercellular spaces. The capillary beds are drained by collecting venules that pass through the bony partition on the scala tympani side and empty into one or two veins running spirally around the modiolus. The large veins are closely enveloped by a net of syncyticum-like cells.

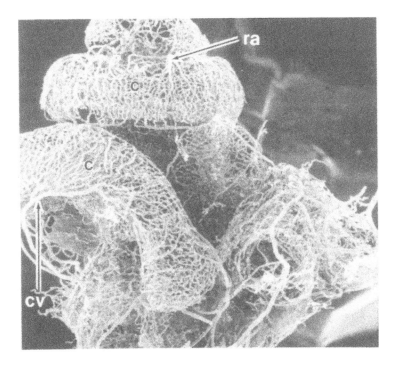

FIGURE 2.19 The radiating arterioles (ra), capillary beds (c) in the stria vascularis and spiral ligament, and collecting venules (cv) are easily seen in a corrosion cast of the blood vessels. (Scanning electron micrograph of the gerbil cochlea provided by Dr. R.A. Tange, AMC Amsterdam, The Netherlands, from Tange and Hodde 1984.)

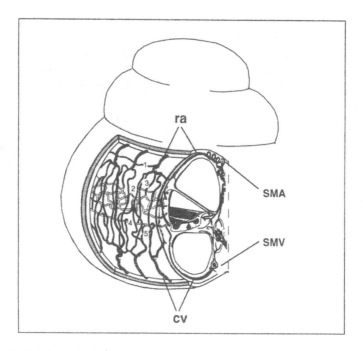

FIGURE 2.20 Schematic diagram of the cochlear vasculature. Radiating arterioles (ra) branch off the spiral modiolar artery (SMA) and run toward the lateral wall in the bone over the scala vestibuli. Branches of the radiating arterioles supply (1) capillaries that run in the suprastrial region, (2) capillaries that form a network in the stria vascularis, (3) arteriovenous arcades, (4) capillaries in the spiral prominence, and (5) capillaries in the lower spiral ligament. The capillaries are drained by the collecting venules (cv), which exit the lateral wall through the bone under the scala tympani and empty into the spiral modiolar vein (SMV). (Diagram modified from Smith 1972.)

The vasculature throughout the cochlea thus appears segmentally arranged, and the capillary beds in the modiolar regions are completely separated from those in the lateral wall by avascular structures such as Reissner's membrane, the TM, and the peripheral portions of the BM. However, because of the anastomoses within the capillary networks, there are multiple possibilities for shunting blood in the spiral direction. Moreover, since the spiral capillary systems run at right angles to the arterioles and venules, it is thought that the blood flow in the spiral vessels is much slower than in the radiating arterioles and collecting venules.

8.2 Capillary Beds in the Modiolus and Osseous Spiral Lamina

The capillaries in the modiolus are fenestrated for rapid transfer of fluids and solutes. The cells surrounding these fenestrated vessels show an unusual protein metabolism, and they have been compared to the cells of the choroid plexus that produce cerebrospinal fluid (Balogh and Koburg 1965). Thus, they have been named the cochlear plexus. However, the modiolar connective tissue cells are fibroblast-like and are not like the cells in the choroid plexus, which are epithelial and rest on a basement membrane.

The capillaries in the spiral ganglion and the osseous spiral lamina form a loose spiral network within the dense connective tissue. They continue into the spiral limbus, where they supply the interdental cells, which are thought to secrete components of the TM, and the specialized vasculoepithelial areas near Reissner's membrane. Capillaries also continue out to the tympanic lip portion of the spiral limbus, near the BM. Here they loop to form a network, which in most species is the closest blood supply to the sensory cells of the organ of Corti. In a few species, there is an additional vessel that runs under the BM in the region of the tunnel called the vas spirale. These vessels are surrounded by a prominent basement membrane and pericytes. The capillaries are not fenestrated, but the endothelial cells are coupled by tight junctions that contribute to the blood–perilymph barrier (Jahnke 1980).

8.3 Capillary Beds in the Lateral Wall

Along the entire length of the cochlea in the lateral wall adjacent to the otic capsule, capillaries are found in the spiral ligament and the stria vascularis. Within the superior portion of the spiral ligament near Reissner's membrane, the radiating arterioles branch (Fig. 2.20). Separate arterioles supply capillary nets found in the spiral ligament (the suprastrial region and the spiral vessel at Reissner's membrane), the stria vascularis, the peripheral region of the spiral ligament, and the inferior portion of the spiral ligament (the spiral prominence).

In the spiral vessel at Reissner's membrane, its endothelial cells contain bundles of intracellular filaments, suggesting that blood flow is very fast and that the filaments, organized into stress fibers, may be contractile. Around the spiral vessel is a wide perivascular fluid space, which may provide an area for plasma filtrate to accumulate.

In the portion of the spiral ligament behind the stria vascularis, the capillaries are lined with a continuous layer of endothelial cells connected by tight junctions and containing microvesicles associated with transcellular transport (Takahashi and Kimura 1970). The capillaries have a prominent basement membrane and are surrounded by pericytes. The capillaries and the arteriovenous shunts intermingle with loosely packed fibrocytes and extracellular matrix material.

In the inferior portion of the spiral ligament, in the subepithelial connective tissue of the spiral prominence, capillaries run mainly free in the extracellular space. The capillaries are larger in diameter than strial capillaries and have a thinner basement membrane. The vessels are closely associated with root processes from external sulcus cells and type II fibrocytes. Microsphere studies have been used to study blood flow, and the results indicate that the flow to the spiral prominence is the highest in the lateral wall (Angelborg et al. 1987).

The endothelial cells of the capillary wall prevent free diffusion of solutes from the blood into the spiral ligament tissue spaces containing perilymph (Juhn 1988). The blood-perilymph barrier is localized to the vascular epithelium and is characterized anatomically by tight junctions between adjacent endothelial cells and a lack of intercellular spaces (fenestrations). Substances may move out from the blood through the endothelial cells mediated by binding to specific receptors, uptake, vesicular transport through the cell, and release. The presence of both pinocytotic vesicles and the GLUT1 glucose transporter protein (Ito, Spicer, and Schulte 1993) in capillary cells suggests active transport of molecules across the blood-perilymph barrier.

The capillaries in the stria vascularis form the densest network in the lateral wall. This intraepithelial network is completely surrounded by tight junctions between adjacent marginal cells and adjacent basal cells, which together form a separate intrastrial compartments (Figs. 2.3, 2.21). The capillaries of the stria vascularis are ultrastructurally very similar to those in the spiral ligament, although their relations to adjacent cells may differ. Strial capillary endothelial cells form a continuous sheet with no fenestrations.

Since the adjacent cells are also connected by tight junctions, any substances in the blood must pass through the endothelial cells into the intrastrial compartment by vesicular transport and not by diffusion between cells. Tracer molecules have been used to map the permeability of strial capillaries (Yamamoto and Nakai 1964; Duvall, Quick, and Sutherland 1971; Winther 1971; Gorgas and Jahnke 1974; Santos-Sacchi and Marovitz 1980; Juhn 1988), and it is suggested that a blood-strial barrier exists that is even more restrictive than the blood-brain barrier, since the density of pinocytotic vesicles in strial vessels is low (Yamamoto and Nakai 1964). As in the spiral ligament, GLUT1 is believed to transport glucose from the strial capillaries into the intrastrial space (Ito, Spicer, and Schulte 1993).

Within the stria, the endothelial cells of the capillaries are surrounded by pericytes and thick basement membranes. They, in turn, are almost completely surrounded by marginal cell processes and, less frequently, intermediate cell processes. Projections of the basement membrane are found between marginal and intermediate cells and may provide a pathway for the distribution of metabolites. This differs from the arrangement of capillaries in the peripheral region of the spiral ligament behind the stria vascularis, where capillaries run through the loosely packed extracellular matrix material and are not in close contact with the fibrocytes.

All of these capillary beds are drained by comparatively few, but large, collecting venules that run back toward the spiral modiolar vein, by transit of the bone that separates the cochlear turns on the scala tympani side of the cochlear duct.

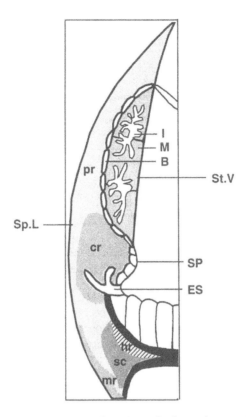

FIGURE 2.21 Diagram of a cross section through the stria vascularis and spiral ligament. The stria vascularis (St.V) is composed of three cell types. Marginal (M) cells line the surface facing endolymph. Intermediate (I) cells have processes that interdigitate with the marginal cells and have extensive contact with the strial capillaries. Basal (B) cells face the spiral ligament. The spiral ligament (Sp.L) can be divided into several basic regions based on connective tissue cell shape, orientation, and relationship to extracellular matrix material. The peripheral region (pr) is located behind the stria vascularis. The central region (cr) includes the root processes of the external sulcus cells (ES). The subcentral region (sc) contains radially oriented fibroblasts. The marginal region (mr) is filled with fibroblasts that are specialized for anchoring and creating tension in the basilar membrane. The hyaline region (hr) is within the basilar crest. (Diagram based on results in spiral ligament from Henson and Henson 1988.)

9. Vascularized Regions Near Reissner's Membrane Specialized for Ion Transport

Several areas in the cochlea appear to be specialized for active fluid transport by cells. Two of these areas are located in the perilymphatic space at opposite ends of Reissner's membrane (Fig. 2.22) — one at its point of insertion into the spiral limbus and another at the point of insertion into

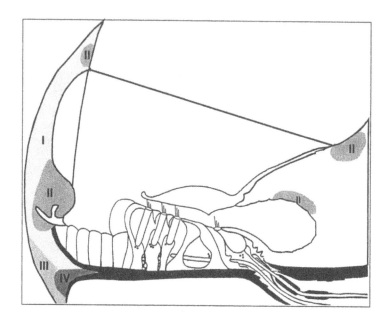

FIGURE 2.22 Diagram showing the areas of the cochlea with fibrocytes specialized for ion transport activity based on the presence of the enzymes Na^+, K^+-ATPase, carbonic anhydrase, and creatine kinase (based on results from Spicer and Schulte 1991). Type I fibrocytes contain carbonic anhydrase and creatine kinase. Type II fibrocytes contain Na^+, K^+-ATPase. Type III fibrocytes contain carbonic anhydrase and creatine kinase. Type IV fibrocytes contain variable amounts of all three enzymes.

the spiral ligament. Both are highly vascularized. Originally these regions were named vasculoepithelial zones (Borghesan 1957; Ishiyama, Keels, and Weibel 1970), because the cells adjacent to capillaries were found to have organelles common to ion-transporting epithelial cells. It is now known that this name is incorrect because the cells are not epithelial. There are no tight junctions between them, just as the cells lining the bony surfaces of the scala tympani and scala vestibuli lack tight junctions.

Rather, the "dark" cells adjacent to the capillaries in these regions are fibrocytes. These fibrocytes, specialized for ion transport, have a dense cytoplasm with abundant ribosomes, endoplasmic reticulum, and Golgi apparatus. Long cytoplasmic processes contain abundant mitochondria with longitudinal cristae, and the cells have high levels of Na^+, K^+-ATPase activity (Schulte and Adams 1989a). Thus, the supralimbal and suprastrial fibrocytes appear to be morphologically and functionally similar to each other. Their presence on both sides of Reissner's membrane suggests that they might maintain and/or produce perilymph or regulate the electrolyte balance between blood and perilymph. They are also morphologically similar to the type II fibrocytes in the spiral prominence.

10. Spiral Ligament and Spiral Prominence

10.1 Arrangement of Cells

The spiral ligament consists of connective tissue cells, epithelial cells, blood vessels, and extracellular matrix material (Fig. 2.21). It lies adjacent to the otic capsule and provides support for the highly vascularized stria, as well as being adjacent to the BM, with which it is functionally linked. In the upper and lower regions, facing the scale vestibuli and the scala tympani, respectively, the fibrocytes of the spiral ligament are in direct contact with the perilymph. In the connective tissue region behind the stria vascularis, the fibrocytes contact the basal cells and are loosely scattered in the extracellular matrix of collagen fibers. In the spiral prominence and external sulcus, epithelial cells joined by tight junctions are in contact with endolymph and form a barrier to prevent mixing of the perilymph in the spiral ligament and scala tympani with the endolymph in the scala media. The basilar crest consists of extracellular matrix material and few cells.

In the cellular regions of the spiral ligament, the cells are spindle-shaped and grouped together. At points where they are in contact, the cell membranes are highly convoluted, increasing their surface area. The capillaries are highly branched and anastomose as they wind tortuously among the cells and the connective extracellular matrix. The arteriovenous shunts, which run directly between the superior and inferior portions of the spiral ligament, are thicker in diameter, straight, and unbranched.

The spiral ligament can be divided into five regions based on connective tissue cell shape, orientation, and relationship to extracellular matrix material (Fig. 2.21), and these divisions have held up across many species (Henson and Henson 1988). The peripheral region is located behind the stria vascularis. The central region is in the area of the spiral prominence and appears dark in tissue sections stained with basic dyes. The subcentral region contains radially oriented fibrocytes and is near the basilar crest, adjacent to the point of insertion of the BM. The hyaline region is within the basilar crest, adjacent to the subcentral region. The marginal region is filled with fibrocytes that, based on their constituent proteins, have contractile abilities and are thought to both anchor and create tension in the BM.

The spiral prominence is a highly vascularized region of the spiral ligament that bulges into the scala media just above the external sulcus cells and below the stria vascularis (Figs. 2.21, 2.24). Depending on the species, a single layer of flat to cuboidal cells lines its endolymphatic surface, with a basement membrane beneath the epithelium.

Between the spiral prominence and the Claudius cells on the BM, external sulcus cells have their apical surfaces in endolymph. There are two types of external sulcus cells, one containing vimentin and the other not containing vimentin (Schulte and Adams 1989b). The degree to which the apical surfaces of the external sulcus cells come into contact with endolymph varies, and often they are covered almost entirely by Claudius cells. Their basal surface sends long projections into the connective tissue of the spiral prominence, and these peglike structures are intimately associated with the blood vessels and the highly infolded cell membranes of the type II fibrocytes displaying Na^+, K^+-ATPase activity. Thus, cells in the region of the spiral prominence and external sulcus are probably actively involved in the regulation of fluid and ion composition in the cochlear duct.

10.2 Fibrocytes Specialized for Ion Transport Activity

The stromal cells of the spiral ligament and spiral prominence are fibrocytes; in early analyses based on ultrastructure they were classified into two different populations (Takahashi and Kimura 1970; Morera, del Sasso, and Iurato 1990; Santi 1988). Type I fibrocytes contained few cellular organelles, whereas type II fibrocytes possessed abundant mitochondria. More recent histochemical and immunocytochemical analyses suggest that the fibrocytes play a role in fluid transport and maintaining ionic gradients (Lim, Karaabinas, and Trune 1983; Watanabe and Ogawa 1984; Schulte and Adams 1989a), and that the fibrocytes can be divided into four types (Fig. 2.22) based on the number and isoforms of enzymes present (Spicer and Schulte 1991). Their distribution corresponds roughly with the four anatomically defined regions of the spiral ligament (Henson and Henson 1988).

Type I fibrocytes contain carbonic anhydrase and creatine kinase, and appear moderately electron-dense with slight infoldings of their cell membranes. They are spread diffusely in the spiral ligament lateral to the stria vascularis. Type II fibrocytes are smaller cells than type I fibrocytes. They have a highly convoluted shape with numerous cytoplasmic exten-

---

FIGURE 2.23 Localization of the different isoforms of Na^+, K^+-ATPase in the lateral wall. (A) Hematoxylin and eosin (H&E)-stained section showing Reissner's membrane (RM), the spiral ligament (Sp.L), the stria vascularis (St.V), the spiral prominence (SP), and the basilar membrane (BM). (B,D) The α_1 and β_1 subunit isoforms are colocalized and predominantly distributed in cells bordering the scale media, such as fibrocytes in the supralimbal region near Reissner's membrane and in the spiral prominence. This subunit combination is commonly found in epithelial cell types that help to maintain ion gradients. (B,E) The α_1 and β_2 subunit isoforms are colocalized in the strial marginal cells. The β_2 subunit has a fairly limited distribution, found primarily in the brain, pineal, and choroid plexus. (C) The localization of the α_2 subunit isoform is restricted to a small percentage of the cells in the suprastrial and supralimbal regions. (Micrographs of serial paraffin sections through the gerbil cochlea provided by Ms. Joan McGuirt and Dr. Bradley Schulte, Medical University of South Carolina, Charleston, SC.)

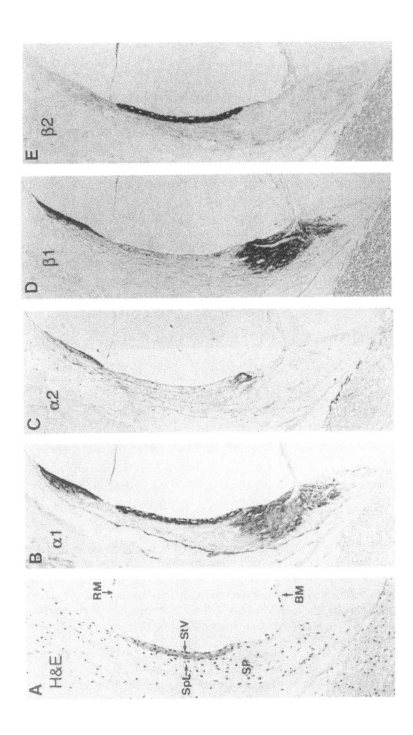

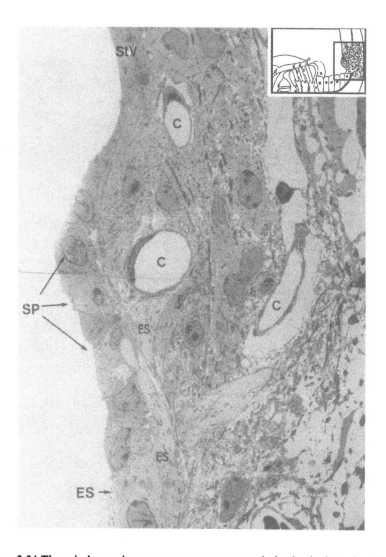

FIGURE 2.24 The spiral prominence appears as a convex bulge in the lateral wall of the cochlea, just below the stria vascularis (StV). On the surface, the spiral prominence epithelial cells (SP) face the scala media. Below them are the external sulcus cells (ES), which also face the scale media. Behind the spiral prominence cells are capillaries (c) which are in contact with the root processes from external sulcus cells (ES) and type II fibrocytes. (Fig. 28.19 from Slepecky 1993. Reprinted by permission of Churchill Livingstone.)

sions, yet are nonpolarized with a uniform distribution of the ion-transporting enzyme Na^+, K^+-ATPase. These type II fibrocytes are similar to the type II cells characterized earlier by ultrastructure alone (Takahashi and Kimura 1970). Their electron-dense cytoplasm is rich in mitochondria, and they have numerous cytoplasmic extensions with highly infolded cell membranes to greatly increase their surface. They are found in the spiral prominence close to root processes of the external sulcus cells. In the suprastrial and supralimbal areas near the insertion of Reissner's membrane, fibrocytes with similar ultrastructural characteristics and Na^+, K^+-ATPase content are found. Type III fibrocytes lie adjacent to bone in the inferior region of the spiral ligament and contain the enzymes carbonic anhydrase and creatine kinase. Type IV fibrocytes are located more superficially in the inferior part of the spiral ligament, and contain carbonic anhydrase and Na^+, K^+-ATPase in variable amounts (Spicer and Schulte 1991).

The fibrocytes also can be identified by their cytoskeletal components. Type II fibrocytes can be distinguished from type I fibrocytes because the type II cells lack vimentin (Schulte and Adams 1989a). In bats and mice, type III fibrocytes have been characterized as anchoring cells, and may be specialized for developing or reacting to tension in the BM – spiral ligament complex. They are distinguished by the presence of actin, abundant cytoplasmic filaments, and other proteins necessary to provide a biochemical basis for tension development (Henson et al. 1985). Little information is available on these cells in other species, and it is not known if type III fibrocytes actually have contractile properties.

Morphological observations support the idea that the fibrocytes in the spiral ligament are involved in ion transport. The fibrocytes are coupled to each other by numerous gap junctions (Takahashi and Kimura 1970; Reale et al. 1975), which imply electrical or ionic coupling between cells. The fibrocytes are also coupled by gap junctions to the basal cells of the stria (Forge 1984; Kikuchi et al. 1995). Thus, type II fibrocytes in the superior and inferior portions of the spiral ligament, through their Na^+, K^+-ATPase activity, can effectively take up endolymphatic potassium that has leaked into the perilymphatic space. This potassium could be passed sequentially through gap junctions to type I fibrocytes in the peripheral region of the spiral ligament, on to basal cells of the stria vascularis, and on to intermediate cells. Thus, potassium can enter into the intrastrial space and be recycled back to endolymph by the marginal cells (see Wangemann and Schacht, Chapter 3). Therefore it is of interest that the greatest density of the type II fibrocyte processes is in the spiral prominence, and that they are more abundant at the base of the cochlea (Spicer and Schulte 1991), where they should be located if their presence is involved with generating and maintaining the larger endocochlear potential recorded there.

11. The Stria Vascularis

The stria vascularis is a vascularized epithelial tissue at the lateral wall of the cochlear duct (Figs. 2.2, 2.19, 2.20, 2.21, 2.25). It is the only known epithelium where blood vessels lie between two continuous cell layers each joined by tight junctions. It is thought that the stria secretes potassium into the cochlear endolymph and that this contributes directly to the endolymphatic potential. Cells within the stria are rich in Na^+, K^+-ATPase (Kerr, Ross, and Ernst 1982; Iwano et al. 1989; Schulte and Adams 1989a). The presence of carbonic anhydrase in strial cells has been suggested histologically (Lim, Karaabinas, and Trune 1983; Watanabe and Ogawa 1984), but the presence of the enzyme has not yet been confirmed by immunocytochemistry. Since this region of the lateral wall is thought to play such an important role in cochlear function, quantitative analysis of the stria vascularis tissue has been performed to measure the relative volume of strial cells and capillaries, the width of the stria, and the number of marginal cells (Santi, Lakhani, and Bingham 1983; Carlisle and Forge 1989; Lohuis, Patterson, and Rarey 1990).

Three cell types are present in the stria vascularis (Smith 1957; Rodriguez-Echandia and Burgos 1965): marginal cells, intermediate cells, and basal cells. The intrastrial space is protected from endolymph by the tight junctions of the marginal cells (Jahnke 1975; Reale et al. 1975), and it is similarly protected from perilymph diffusion by the tight junctions of the basal cells. However, gap junctions couple basal cells to adjacent basal cells, to fibrocytes in the spiral ligament (Forge 1984; Kikuchi et al. 1995), and to intermediate cells (Forge 1984; Kikuchi et al. 1995). Thus, glucose and potassium can be actively transported from perilymph through the basal cells to the marginal cells of the stria vascularis. This selective access of ions and metabolites in perilymph to the intrastrial space, coupled with the limited permeability of the strial capillaries (Gorgas and Jahnke 1974; Santos-Sacchi and Marovitz 1980), suggests that endolymph is derived from perilymph rather than from blood.

11.1 Cell Types in the Stria Vascularis

11.1.1 Marginal Cells

The marginal cells are dark-appearing cells (Figs. 2.21, 2.25) with sparse microvilli and a moderately thick layer of cell coat material on their apical surfaces, facing the endolymph. The marginal cells contain abundant mitochondria, microtubules (Santos-Sacchi 1978, 1892), and cytokeratin (Schrott, Egg, and Spoendlin 1988; Bauwens et al. 1991; Kuijpers et al. 1991) but lack vimentin (Schulte and Adams 1989b). By and large, there is a linear relationship between the number of marginal cells and the overall size of the stria (Santi, Lakhani, and Bingham 1983; Carlisle and Forge 1989; Lohuis, Patterson, and Rarey 1990). Cytoplasmic extensions from the marginal cells completely surround most of the capillaries in the stria.

Marginal cells are joined at their lateral surfaces by tight and adherens junctions. Below the junctional complex, each cell's basolateral membrane is extensively folded, and the projections are tightly packed with mitochon-

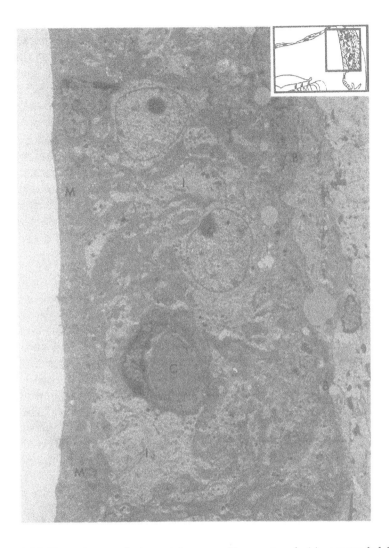

FIGURE 2.25 The stria vascularis contains a capillary network (c) surrounded by dark-staining marginal cells (M) bordering the scala media, light-staining intermediate cells (I), and flattened basal cells (B) bordering the spiral ligament. Tight junctions between adjacent marginal cells facing endolymph and between adjacent basal cells facing the spiral ligament form the intrastrial space. (Fig. 28.20 from Slepecky 1993. Reprinted by permission of Churchill Livingstone.)

dria. The lateral surfaces of adjoining cells have elaborate interdigitations, and the cells have only a narrow intercellular space. The cells exhibit Ca^{2+}-ATPase activity on their apical membranes (Yoshihara and Igarashi 1987) and Na^+, K^+-ATPase activity on their basolateral membranes (Kerr, Ross, and Ernst 1982; Iwano et al. 1989; Schulte and Adams 1989a). The cells contain a dense network of tubulocisternal endoplasmic reticulum (Forge

1982), coated vesicles, vacuoles, and glycogen and stain positively for creatine kinase (Spicer and Schulte 1992). These specializations are characteristic of cells with a high rate of metabolism and a transport or secretory function.

Although the marginal cells were originally suggested to be coupled to each other (Reale et al. 1975) and to intermediate cells (Jahnke 1975) by gap junctions, this finding has not been confirmed by more recent work (Forge 1984; Kikuchi et al. 1995). The finding that marginal cells are coupled to basal cells (Forge 1984) has also not been confirmed (Kikuchi et al. 1995). The reasons for the controversy stem from the problems inherent in the study of these junctions in cells with such highly infolded membranes. For freeze-fracture analysis, it is necessary to have fracture planes that reveal the membranes on both types of cells, and to be able to positively identify the cells types. In transmission electron microscopy, the junctions must be oriented precisely perpendicular to the plane of section. These events rarely occur in strial specimens.

11.1.2 Intermediate Cells

The intermediate cells form a discontinuous layer beneath the marginal cells. They do not contact the endolymph. They are relatively clear cells, have a highly convoluted surface with cytoplasmic extensions, and contain vimentin. Their intermediate distribution between the marginal and the basal cells allows them to bridge the other two cell types, and they are thought to play a pivotal role in generating the positive component of the endocochlear potential. Whereas the intermediate cells are termed the "light-staining" cells of the stria, these cells have been shown to be melanocytes capable of synthesizing melanin pigment granules (Hilding and Ginzburg 1977), and they contain a significant number of the granules. Albino animals have a normal distribution of melanocytes, but because of one of several biochemical defects, they are unable to synthesize melanin pigment. In the albino guinea pig cochlea, where the melanocytes are amelanotic, intermediate cells occupy significantly less volume than usual, with a compensatory increase in marginal cell volume (Conlee et al. 1994). White spotting mutants, on the other hand, lack the intermediate cells, and the endocochlear potential in them is close to zero. This suggests that intermediate cells are necessary for the development or activity of marginal cells or for providing potassium for movement into the endolymph (Schulte and Steel 1994).

11.1.3 Basal Cells

The basal cells are flat, overlapping cells joined by tight junctions and forming a continuous cell layer. They separate the marginal cells, intermediate cells, and capillaries of the stria vascularis from the spiral ligament and its perilymph. The layer extends beyond the marginal and intermediate cells to the endolymphatic surface at the upper and lower ends of the stria.

The layer forms a sleeve around each capillary entering and leaving the stria. Basal cells, like intermediate cells, also contain vimentin (Schrott, Egg, and Spoendlin 1988; Schulte and Adams 1989b; Bauwens et al. 1991) and lack Na^+, K^+-ATPase.

The tight junctions between adjacent basal cells have more fused regions and connecting strands than do tight junctions between any other cell types in the cochlea. Also, gap junctions occupy a large proportion of basal cell membrane and are found between adjacent basal cells, between basal cells and intermediate cells, and between basal cells and spiral ligament cells (Forge 1984; Bagger-Sjöbäck et al. 1987; Kikuchi et al. 1995). No continuous basement membrane separates the basal cells or stria from the underlying spiral ligament. It is thought that glucose is transported across the basal cells from the interstitial fluid of the spiral ligament by GLUT1, a glucose transporter protein. In the gerbil, GLUT1 appears to be an exclusive marker for basal cells among all epithelial cells of the adult cochlea (Ito, Spicer, and Schulte 1993).

12. Summary

Anatomical studies on cells of the mammalian inner ear have provided us with many clues as to their different roles in the perception of sound. However, cells do not act in isolation, and hearing depends critically on interactions between cells—some structurally related, others spatially separate but functionally related. Many of the most interesting problems in cochlear research today are focused on the nature of these interactions. What is the relationship between the stereocilia and the TM, and can it be modified during life or repaired following trauma? What is it that enables cells to maintain their position within an epithelium; how are cells replaced either by cells of similar type during regeneration or by cells of differing type during repair? How are changes in the structure of cells correlated with changes in their function following mechanical trauma, treatment with ototoxic drugs, and aging? How is morphology correlated with physiological properties of nerve fibers connecting the auditory periphery to the central nervous system? How are changes in the lateral wall (vasculature and cell type) related to changes in hearing? It is hoped that the reader, now knowledgeable about the morphology and biochemistry of the individual cells in the cochlear duct, will be motivated to begin or continue study of the cochlea and the basic mechanisms related to hearing.

Acknowledgments

There are many giants on whose shoulders I have been standing in the process of writing this chapter. My past and present colleagues, through their expertise and persistence, have contributed to the elucidation of structure–function relationships in this small and difficult-to-access organ. Many have provided me with data, micrographs, and critical comments that have been incorporated into this review. It is unfortunate that all of the excellent research on cochlear structures cannot be adequately described or cited. My students, by continuing to ask questions to which I do not know the answers, have also contributed significantly to the scope of the manuscript. Many others, too numerous to mention, have provided thoughtful and constructive criticism as well as excellent technical assistance. Their help and support is most gratefully acknowledged. The research from my laboratory related to the structure and cell biology of the organ of Corti has been supported by the Deafness Research Foundation and grants from the National Institutes of Health.

Abbreviations Used

B	basal cells
BM	basilar membrane
c	capillary
C	Claudius cells
cn	cover net of tectorial membrane
cp	cuticular plate
cv	collecting venule
D	Deiters cell
ES	external sulcus
H	Hensen cell
I	intermediate cell
IB	inner border cell
IHC	inner hair cell
IP	inner pillar cell
IPh	inner phalangeal cell
M	marginal cell
mn	marginal net of tectorial membrane
Mo	modiolus
mz	middle zone of tectorial membrane
n	nucleus
N	nerve
Na	afferent nerve
Ne	efferent nerve
OHC	outer hair cell
OP	outer pillar cell
OSB	outer spiral bundle
OSL	osseous spiral lamina
OT	outer tunnel
R	rootlet

ra	radiating arteriole
RM	Reissner's membrane
S	stereocilium
SL	spiral limbus
SM	scala media
SMA	spiral modiolar artery
SMV	spiral modiolar vein
SP	spiral prominence
Sp. L	spiral ligament
SSC	subsurface cistern
ST	scala tympani
St. V	stria vascularis
T	tunnel
TM	tectorial membrane

References

Altschuler RA, Parakkal MH, Rubio JA, Hoffman DW, Fex J (1984) Enkephalin-like immunoreactivity in the guinea pig organ of Corti: ultrastructural and lesion studies. Hear Res 16:17-31.

Altschuler RA, Hoffman DW, Reeks KA, Fex J (1985) Localization of dynorphin B-like and α-neo-endorphin-like immunoreactivities in the guinea pig organ of Corti. Hear Res 17:249-258.

Angelborg C, Engström H (1972) Supporting elements in the organ of Corti. Fibrillar structures in the supporting cells of the organ of Corti of mammals. Acta Otolaryngol Suppl 301:49-60.

Angelborg CA, Engström H (1973) The normal organ of Corti. In: Møler AR (ed) Basic Mechanisms in Hearing. New York: Academic Press, pp. 125-182.

Angelborg C, Engström B (1974) The tympanic covering layer. An electron microscopic study in the guinea pig. Acta Otolaryngol Suppl 319:43-56.

Angelborg C, Slepecky N, Larsen HC, Soderberg L (1987) Colored microspheres for blood flow determinations twice in the same animal. Hear Res 27:265-269.

Anniko M, Arnesen AR (1988) Cochlear nerve topography and fiber spectrum in the pigmented mouse. Arch Otorhinolaryngol 245:155-159.

Arima T, Uemura T, Yamamoto T (1985) Structural features of the basal lamina in Reissner's membrane of the guinea pig. Acta Otolaryngol 100:194-200.

Arima T, Uemura T, Yamamoto T (1986) Cytoskeletal organization in the supporting cell of the guinea pig organ of Corti. Hear Res 24:169-175.

Arima T, Uemura T, Yamamoto T (1987) Three-dimensional visualizations of the inner ear hair cell of the guinea pig. A rapid-freeze, deep-etch study of filamentous and membranous organelles. Hear Res 25:61-68.

Arima T, Kuraoka A, Toriyama R, Shibata Y, Uemura T (1991) Quick-freeze, deep-etch visualisation of the "cytoskeletal spring" of cochlear outer hair cells. Cell Tiss Res 263:91-97.

Arnesen AR, Osen KK (1978) The cochlear nerve in the cat: topography, tonotopy and fiber spectrum. J Comp Neurol 178:661-678.

Arnold W, Anniko M (1990) Structurally based new functional interpretations of the subsurface cisternal network in human outer hair cells. Acta Otolaryngol

109:213-220.

Aschoff A, Ostwald J (1987) Different origins of cochlear efferents in some bat species, rats, and guinea pigs. J Comp Neurol 264:56-72.

Assad JA, Corey DP (1992) An active motor mediates adaptation by vertebrate hair cells. J Neurosci 12:3291-3309.

Axelsson A (1968) The vascular anatomy of the cochlea in the guinea pig and man. Acta Otolaryngol 243:1-134.

Axelsson A, Ryan A (1988) Comparative study of the vascular anatomy in the mammalian cochlea. In: Jahn AF, Santos-Sacchi JR (eds) Physiology of the Ear. New York: Raven Press, pp. 295-316.

Bagger-Sjöbäck D, Engström B, Steinholtz L, Hillerdahl M (1987) Freeze-fracturing of the human stria vascularis. Acta Otolaryngol 103:64-72.

Balogh K, Koburg E (1965) Der Plexus Cochlearis. Arch Ohr Nas Kehkopfhel 185:638-690.

Bannister LH, Dodson HC, Astbury AR, Douek EE (1988) The cortical lattice: a highly ordered system of subsurface filaments in the guinea pig cochlear outer hair cells. Prog Brain Res 74:213-219.

Bauwens LJM, Veldman JE, Ramaekers FCS, Bouman H, Huizing EH (1991) Expression of intermediate filament proteins in the adult human cochlea. Ann Otol Rhinol Laryngol 100:211-218.

Berglund AM, Ryugo DK (1986) A monoclonal antibody labels type II neurons of the spiral ganglion. Brain Res 382:327-332.

Berglund AM, Ryugo DK (1987) Hair cell innervation by spiral ganglion neurones in the mouse. J Comp Neurol 255:560-571.

Bodian D (1983) Electron microscopic atlas of the simian cochlea. Hear Res 9:201-246.

Bohne BA, Carr DC (1985) Morphometric analysis of hair cells in the chinchilla cochlea. J Acoust Soc Am 77:153-158.

Borghesan E (1957) Modality of the cochlear humoral circulation. Laryngoscope 67:1266-1285.

Brechtelsbauer PB, Prazma J, Garret CG, Carrasco VN, Pillsbury HC (1990) Catecholaminergic innervation of the inner ear. Arch Otolaryngol 116:566-574.

Bredberg G (1968) Cellular pattern and nerve supply of the human organ of Corti. Acta Otolaryngol Suppl. 236:85.

Bredberg G, Lindeman HH, Ades H, West R (1970) Scanning electron microscopy of the organ of Corti. Science 170:861-863.

Brown MC (1987) Morphology of labeled afferent fibers in the guinea pig cochea. J Comp Neurol 260:591-604.

Bruns V (1980) Basilar membrane and its anchoring system in the cochlea of the greater horseshoe bat. Anat Embryol 161:29-50.

Bruns V, Schmieszek E (1980) Cochlear innervation of the greater horseshoe bat: demonstration of an acoustic fovea. Hear Res 3:27-43.

Cabezudo LM (1978) The ultrastructure of the basilar membrane in the cat. Acta Otolaryngol 86:160-175.

Canlon B, Cartaud J, Changeux JP (1989) Localization of alpha-bungarotoxin binding sites on outer hair cells from the guinea pig cochlea. Acta Physiol Scand 137:549-550.

Carlisle L, Forge A (1989) The vessels of the stria vascularis: quantitative comparison of three rodent species. Hear Res 38:111-118.

Carlisle L, Zajic G, Altschuler RA, Schacht J, Thorne PR (1988) Species differences

in the distribution of infracuticular F-actin in outer hair cells of the cochlea. Hear Res 33:201-206.
Comis SD, Pickles JO, Osborne MP (1985) Osmium tetroxide postfixation in relation to the cross-linkage and spatial organization of stereocilia in the guinea pig cochlea. J Neurocytol 14:113-130.
Conlee JW, Gerity LC, Westenberg IS, Creel DJ (1994) Pigment-dependent differences in the stria vascularis of albino and pigmented guinea pigs and rats. Hear Res 72:108-124.
Corey DP, Hudspeth AJ (1983) Kinetics of the receptor current in bullfrog saccular hair cells. J Neurosci 3:962-976.
Dannhof BJ, Bruns V (1993) The innervation of the organ of Corti in the rat. Hear Res 66:8-22.
Dechesne CJ, Thomasset M (1988) Calbindin (CaBP 28 kDa) appearance and distribution during development of the mouse inner ear. Dev Brain Res 40:233-242.
Dechesne CJ, Winsky L, Kim HN, Goping G, Vu TD, Wenthold RJ, Jacobowitz DM (1991) Identification and ultrastructural localization of a calretinin-like calcium-binding protein (protein 10) in the guinea pig and rat inner ear. Brain Res 560:139-148.
Densert O (1975) The effect of 6-hydroxydopamine on the rabbit cochlea. Acta Otolaryngol 79:339-351.
Densert O, Flock A (1974) An electron microscopic study of adrenergic innervation in the cochlea. Acta Otolaryngol 77:185-197.
Dieler R, Shehata-Dieler WE, Brownell WE (1991) Concomitant salicylate-induced alterations of outer hair cell subsurface cisternae and electromotility. J Neurocytol 20:637-653.
Drenckhahn D, Kellner J, Mannherz HG, Groschel-Stewart U, Kendrick-Jones J, Scholey J (1982) Absence of myosin-like immunoreactivity in stereocilia of cochlear hair cells. Nature 300:531-532.
Dulon D, Moataz R, Mollard P (1993) Characterization of calcium signals generated by extracellular nucleotides in supporting cells of the organ of Corti. Cell Calcium 14:245-254.
Dunn RA, Morest DK (1975) Receptor synapses without synaptic ribbons in the cochlea of the cat. Proc Natl Acad Sci USA 72:3599-3603.
Duvall AJ, Rhodes VT (1967) Reissner's membrane: an ultrastructural study. Arch Otolaryngol 80:143-151.
Duvall AJ, Sutherland CR (1970) The ultrastructure of the extrasensory cells in the cochlear duct. In: Paparella MM (ed) Biochemical Mechanisms in Hearing and Deafness. Springfield, IL: Charles C. Thomas, pp. 149-170.
Duvall AJ, Quick C, Sutherland C (1971) Horseradish peroxidase in the lateral cochlear wall: an electron microscopic study of transport. Arch Otolaryngol 93:304-316.
Echteler SM, Fay RR, Popper AN (1994) Structure of the mammalian cochlea. In: Fay RR, Popper AN (eds) Comparating Hearing: Mammals. New York: Springer Verlag, pp. 134-171.
Ekström von Lubitz DKJ (1981) Subsurface tubular system in the outer sensory cells of the rat cochlea. Cell Tiss Res 220:787-795.
Emmerling MR, Sobkowicz HM, Levenick CV, Scott GL, Slapnick SM, Rose JE (1990) Biochemical and morphological differentiation of acetylcholinesterase-positive efferent fibers in the mouse cochlea. J Elect Microscop Tech 15:123-143.

Engström H (1958) Structure and innervation of the inner ear sensory epithelia. Int Rev Cytol 7:535–585.

Evans BN (1990) Fatal contractions: ultrastructural and electromechanical changes in outer hair cells following transmembraneous electrical stimulation. Hear Res 45:265–282.

Eybalin M (1993) Neurotransmitters and neuromodulators of the mammalian cochlea. Physiol Rev 73:309–373.

Eybalin M, Altschuler RA (1990) Immunoelectron microscopic localization of neurotransmitters in the cochlea. J Elect Microscop Tech :209–224.

Eybalin M, Ripoll C (1990) Immunolocalisation de la parvalbumine dans deux types de cellules glutamatergiques de la cochlée du cobaye; les cellules ciliées internes et les neurones du ganglion spiral. Compt Rend Acad Sci Paris 310:639–644.

Fex J, Altschuler RA (1986) Neurotransmitter-related immunocytochemistry of the organ of Corti. Hear Res 22:249–263.

Flock Å, Cheung HC (1977) Actin filaments in sensory hairs of inner ear receptor cells. J Cell Biol 75:339–343.

Flock Å, Kimura R, Lundquist PG, Wersall J (1962) Morphological basis of directional sensitivity of the outer hair cells in the organ of Corti. J Acoust Soc Am 34:1351–1355.

Flock Å, Flock B, Murray E (1977) Studies on the sensory hairs of receptor cells in the inner ear. Acta Otolaryngol 83:85–91.

Flock Å, Cheung HC, Flock B, Utter G (1981) Three sets of actin filaments in sensory cells of the inner ear. Identification and functional orientation determined by gel electrophoresis, immunofluorescence and electron microscopy. J Neurocytol 10:133–147.

Flock Å, Bretscher A, Weber K (1982) Immunohistochemical localization of several cytoskeletal proteins in inner ear sensory and supporting cells. Hair Res 6:75–89.

Flock Å, Flock B, Ulfendahl M (1986) Mechanisms of movement in outer hair cells and a possible structural basis. Arch Otorhinolaryngol 243:83–90.

Forge A (1982) A tubulo-cisternal endoplasmic reticulum system in the potassium transporting marginal cells of the stria vascularis and the effects of the ototoxic diuretic ethacrynic acid. Cell Tiss Res 226:375–387.

Forge A (1984) Gap junctions in the stria vascularis and effects of ethacrynic acid. Hear Res 13:189–200.

Forge A (1987) Specializations of the lateral membrane of inner hair cells. Hear Res 31:99–110.

Forge A (1991) Structural features of the lateral walls in mammalian cochlear outer hair cells. Cell Tiss Res 265:473–483.

Forge A, Davis S, Zajic G (1991) Assessment of ultrastructure in isolated cochlear hair cells using a procedure for rapid freezing before freeze-fracture and deep-etching. J Neurocytol 20:471–484.

Forge A, Zajic G, Li L, Nevill G, Schacht J (1993) Structural variability of the sub-surface cisternae in intact, isolated outer hair cells shown by fluorescent labelling of intracellular membranes and freeze-fracture. Hear Res 64:175–183.

Foster JD, Drescher MJ, Hatfield JS, Drescher DG (1994) Immunohistochemical localization of S-100 protein in auditory and vestibular end organs of the mouse and hamster. Hear Res 74:67–76.

Friede RL (1984) Cochlear axon calibers are adjusted to characteristic frequencies. J Neurol Sci 66:193–200.

Furness DN, Hackney CM (1985) Cross-links between stereocilia in the guinea pig

cochlea. Hear Res 18:177-188.
Furness DN, Hackney CM (1990) Comparative ultrastructure of subsurface cisternae in inner and outer hair cells of the guinea pig cochlea. Eur Arch Otorhinolaryngol 247:12-15.
Furness DN, Hackney CM, Steyger PS (1990) Organization of microtubules in cochlear hair cells. J Elect Microscop Tech 15:261-279.
Gil-Loyzaga P, Pujol R (1990) Neurotoxicity of kainic acid in the rat cochlea during early developmental stages. Eur Arch Otorhinolaryngol 248:40-48.
Gil-Loyzaga PG, Gabrion J, Uziel A (1985) Lectins demonstrate the presence of carbohydrates in the tectorial membrane of mammalian cochlea. Hear Res 20:1-8.
Ginzberg RD, Morest DK (1983) A study of cochlear innervation in the young cat with the Golgi method. Hear Res 10:227-246.
Gorgas K, Jahnke K (1974) The permeability of blood vessels in the guinea pig cochlea. II Vessels in the spiral ligament and the stria vascularis. Anat Embryol 146:33-42.
Guinan JJ, Warr WB, Norris BE (1982) Differential olivocochlear projections from lateral versus medial zones of the superior olivary complex. J Comp Neurol 221:358-370.
Gulley RL, Reese TS (1976) Intercellular junctions in the reticular lamina of the organ of Corti. J Neurocytol 5:479-507.
Gulley RL, Reese TS (1977) Regional specializations of the hair cell plasmalemma in the organ of Corti. Anat Rec 189:109-124.
Hackney CM, Furness DN, Benos DJ (1991) Localization of putative mechano-electrical transducer channels in cochlear hair cells by immunoelectron microscopy. Scan Microscop 5:741-746.
Harding GW, Baggot PJ, Bohne BA (1992) Height changes in the organ of Corti after noise exposure. Hear Res 63:26-36.
Hashimoto S, Kimura RS (1988) Computer-aided three-dimensional reconstruction and morphometry of the outer hair cells of the guinea pig cochlea. Acta Otolaryngol 105:64-74.
Hasko JA, Richardson GP (1988) The ultrastructural organization and properties of the mouse tectorial membrane matrix. Hear Res 35:21-28.
Henson MM, Henson OW (1988) Tension fibroblasts and the connective tissue matrix of the spiral ligament. Hear Res 35:237-258.
Henson MM, Henson OW (1991) Specializations for sharp tuning in the mustached bat: the tectorial membrane and spiral limbus. Hear Res 56:122-132.
Henson MM, Jenkins DB, Henson OW (1982) The cells of Boettcher in the bat. Hear Res 7:91-103.
Henson MM, Jenkins DB, Henson OW (1983) Sustentacular cells of the organ of Corti—the tectal cells of the outer tunnel. Hear Res 10:153-166.
Henson M, Henson OW, Jenkins DB (1984) The attachment of the spiral ligament to the cochlear wall: anchoring cells and the creation of tension. Hear Res 16:231-242.
Henson MM, Burridge K, Fitzpatrick D, Jenkins DB, Pillsbury HC, Henson OW (1985) Immunocytochemical localization of contractile and contraction associated proteins in the spiral ligament of the cochlea. Hear Res 20:207-214.
Hilding DA, Ginzburg RD (1977) Pigmentation of the stria vascularis. Arch Otolaryngol 84:24-37.
Hirokawa N, Tilney LG (1982) Interactions between actin filaments and between actin filaments and membranes in quick-frozen and deeply etched hair cells of the

chick ear. J Cell Biol 95:249–261.
Holley MC, Ashmore JF (1988) A cytoskeletal spring in cochlear outer hair cells. Nature 335:635–637.
Holley MC, Ashmore JF (1990) Spectrin, actin and the structure of the cortical lattice in mammalian cochlear outer hair cells. J Cell Sci 96:283–291.
Holley MC, Kalinec F, Kachar B (1992) Structure of the cortical cytoskeleton in mammalian outer hair cells. J Cell Sci 102:569–580.
Hoshino T (1977) Contact between the tectorial membrane and the cochlear sensory hairs in the human and the monkey. Arch Otorhinolaryngol 217:53–60.
Howard J, Hudspeth AJ (1988) Compliance of the hair bundle associated with gating of mechanoelectrical transduction channels in the bullfrog's saccular hair cell. Neuron 1:189–199.
Hozawa K, Kimura RS (1990) Cholinergic and noradrenergic nervous systems in the cynomolgus monkey cochlea. Acta Otolaryngol 110:46–55.
Hudspeth AJ (1982) Extracellular current flow and the site of transduction by vertebrate hair cells. J Neurosci 2:1–10.
Hudspeth AJ, Corey DP (1977) Sensitivity, polarity, and conductance change in the response of vertebrate hair cells to controlled mechanical stimuli. Proc Natl Acad Sci USA 74:2407–2411.
Hunter-Duvar I (1978) Electron microscope assessment of the cochlea 2. Reissner's membrane and endocytosis of cell debris. Acta Otolaryngol Suppl 24:24–32.
Ishiyama E, Cutt RA, Keels WW (1970) Distribution and ultrastructure of the Boettcher's cells in mammals. Ann Otol Rhinol Laryngol 79:54–69.
Ishiyama E, Keels EW, Weibel J (1970) New anatomical aspects of the vasculoepithelial zone of the spiral limbus in mammals. An electron microscope study. Acta Otolaryngol 70:319–328.
Ito M, Spicer SS, Schulte BA (1993) Immunohistochemical localization of brain type glucose transporter in mammalian inner ears: comparison of developmental and adult stages. Hear Res 71:230–238.
Itoh M (1982) Preservation and visualization of actin-containing filaments in the apical zone of cochlear sensory cells. Hear Res 6:277–289.
Iurato S (1960) Submicroscopic structure of the membranous labyrinth. I. The tectorial membrane. Z Zellforsch 52:105–128.
Iurato S (1961) Submicroscopic structure of the membranous labyrinth. II. The epithelium of Corti's organ. Z Zellforsch 53:259–298.
Iurato S (1962) Submicroscopic structures of the membranous labyrinth. III. The supporting structures of Corti's organ (basilar membrane, limbus spiralis and spiral ligament). Z Zellforsch 56:40–96.
Iurato S (1967) Submicroscopic Structure of the Inner Ear. Oxford: Pergamon Press.
Iurato S, Franke K, Luciano L, Wermbter G, Pannese E, Reale E (1976a) Fracture faces of the junctional complexes in the reticular membrane of the organ of Corti. Acta Otolaryngol 81:36–47.
Iurato S, Franke K, Luciano L, Wermbter G, Pannese E, Reale E (1976b) Intercellular junctions in the organ of Corti as revealed by freeze fracturing. Acta Otolaryngol 82:57–69.
Iwano T, Yamamoto A, Omori K, Akayama M, Kumazawa T, Tashiro Y (1989) Quantitative immunocytochemical localization of Na^+K^+ ATPase-α subunit in the lateral wall of rat cochlear duct. J Histochem Cytochem 37:353–363.
Jahnke K (1975) The fine structure of freeze-fracture intercellular junctions in the guinea pig inner ear. Acta Otolaryngol Suppl 336:1–40.

Jahnke K (1980) The blood-perilymph barrier. Arch Oto-Rhino-Laryngol 228:29–34.

Jaramillo F, Hudspeth AJ (1990) Localization of the hair cell's transduction channels at the hair bundle's top by iontophoretic application of a channel blocker. Neuron 7:409–420.

Jones N, Fex J, Altschuler RA (1987) Tyrosine hydroxylase immunoreactivity identifies possible catecholaminergic fibers in the organ of Corti. Hear Res 30:33–38.

Jorgensen JM, Flock Å (1976) Non-innervated sense organs of the lateral line. J Neurocytol 5:33–41.

Juhn SK (1988) Barrier systems in the inner ear. Acta Otolaryngol Suppl 458:79–83.

Keithley EM, Ryan AF, Woolf NK (1993) Fibronectin-like immunoreactivity of the basilar membrane of young and aged rats. J Comp Neurol 327:612–617.

Kellerhals B, Engström H, Ades HW (1967) Die Morphologie des Ganglion Spiral Cochleae. Acta Otolaryngol Suppl 226:1–78.

Kerr TP, Ross MD, Ernst SA (1982) Cellular localization of Na^+, K^+-ATPase in the mammalian cochlear duct: significance for cochlear fluid balance. Am J Otolaryngol 3:332–338.

Khalkhali-Ellis Z, Hemming FW, Steel KP (1987) Glycoconjugates of the tectorial membrane. Hear Res 1:81–94.

Kiang NYS, Rho JM, Northrop CC, Liberman MC, Ryugo DK (1982) Hair cell innervation by spiral ganglion cells in adult cats. Science 217:175–177.

Kikuchi T, Takasaka T, Tonosaki A, Kator Y, Shinkawa H (1991) Microtubules of guinea pig cochlear epithelial cells. Acta Otolaryngol 111:286–290.

Kikuchi T, Kimura RS, Paul DL, Adams JC (1995) Gap junction systems in the rat cochlea: immunohistochemical and ultrastructural analysis. Anat Embryol 191:101–118.

Kimura RS (1966) Hairs of the cochlear sensory cells and their attachment to the tectorial membrane. Acta Otolaryngol 61:55–72.

Kimura RS (1975) The ultrastructure of the organ of Corti. Int Rev Cytol 42:173–222.

Kimura RS (1986) An electron microscopic study of cochlear nerve fibers followed serially from spiral ganglion to organ of Corti. Ear Res Jpn 17:4–7.

Kimura RS, Ota CY, Takahasi T (1979) Nerve fiber synapses on spiral ganglion cells in the human cochlea. Ann Oto Rhinol Laryngol 88:1–17.

Kimura RS, Bongiorno CL, Iverson NA (1987) Synapses and ephapses in the spiral ganglion. Acta Otolaryngol Suppl 438:3–18.

Kronester-Frei A (1978) Ultrastructure of the different zones of the tectorial membrane. Cell Tiss Res 193:11–23.

Kuijpers W, Tonnaer ELGM, Peters TA, Ramaekers FCS (1991) Expression of intermediate filament proteins in the mature inner ear of the rat and guinea pig. Hear Res 52:133–146.

Kuriyama H, Shioska S, Sekitani M, Tohyama Y, Kitajiri M, Yamashita T, Kumazawa T, Tohyama M (1990) Electron microscopic observation of calcitonin gene-related peptide-like immunoreactivity in the organ of Corti of the rat. Brain Res 517:76–80.

Laurikainen EA, Kim D, Didier A, Ren T, Miller JM, Quirk WS, Nuttall AL (1993) Stellate ganglion drives sympathetic regulation of cochlear blood flow. Hear Res 64:199–204.

Legrand C, Brehier A, Clavel MC, Thomasset M, Rabie A (1988) Cholecalcin (28kDa CaBP) in the rat cochlea. Dev Brain Res 38:121–129.

Lenoir M, Puel JL, Pujol R (1987) Stereocilia and tectorial membrane development in the rat cochlea. A scanning electron microscopy study. Anat Embryol 175:477-487.
Liberman MC (1980a) Efferent synapses in the inner hair cell area of the cat cochlea: an electron microscopic study of serial sections. Hear Res 3:189-204.
Liberman MC (1980b) Morphologial differences among radial afferent fibers in the cat cochlea. Hear Res 3:45-63.
Liberman MC, Oliver ME (1984) Morphometry of intracellularly labelled neurons of the auditory nerve. J Comp Neurol 223:163-176.
Liberman MC, Dodds LW, Pierce S (1990) Afferent and efferent innervations of the cat cochlea: quantitative analysis with light and electron microscopy. J Comp Neurol 301:443-460.
Lim DJ (1972) Fine morphology of the tectorial membrane. Arch Otolaryngol 96:199-215.
Lim DJ (1986) Functional structure of the organ of Corti: a review. Hear Res 22:117-146.
Lim DJ, Karaabinas C, Trune DR (1983) Histochemical localization of carbonic anhydrase in the inner ear. Am J Otolaryngol 4:33-42.
Lohuis PJFM, Patterson K, Rarey KE (1990) Quantitative assessment of the rat stria vascularis. Hear Res 47:95-102.
Meyer zum Gottesberge AM, Tsujikawa S (1993) Glycerol effect on the guinea pig tectorial membrane. Eur Arch Otorhinolaryngol 250:88-91.
Miles FP, Nuttall AL (1988) In vivo capillary diameters in the stria vascularis and spiral ligament of the guinea pig cochlea. Hear Res 33:191-200.
Morera D, del Sasso A, Iurato S (1980) Submicroscopic structure of the spiral ligament in man. Rev Laryngol 101:73-85.
Morrison D, Schindler RA, Wersäll J (1975) A quantitative analysis of the afferent innervation of the organ of Corti in the guinea pig. Acta Otolaryngol 79:1-23.
Munyer PD, Schulte BA (1991) Immunohistochemical identification of proteoglycans in gelatinous membranes of cat and gerbil inner ear. Hear Res 52:369-378.
Munyer PD, Schulte BA (1994) Immunohistochemical localization of keratan sulfate and chondroitin 4- and 6-sulfate proteoglycans in subregions of the tectorial and basilar membranes. Hear Res 79:83-93.
Nadol JB (1978) Intercellular junctions in the organ of Corti. Ann Oto Rhino Laryngol 87:70-80.
Nadol JB (1983) Serial section reconstruction of the neural poles of hair cells in the human organ of Corti. II. Outer hair cells. Laryngoscope 93:780-791.
Nadol JB (1988) Comparative anatomy of the cochlea and auditory nerve in mammals. Hear Res 34:253-266.
Nadol JF (1989) Morphometric analysis of normal human spiral ganglion cells. Abstract ARO 12:76.
Nadol JB, Mulroy MJ, Goodenough DA, Weiss TF (1976) Tight and gap junctions in a vertebrate inner ear. Am J Anat 147:281-302.
Neugebauer DC, Thurm U (1986) Surface charges influence the distances between vestibular stereovilli. Naturwissenschaft 73:508-509.
Nielsen DW, Slepecky N (1986) Stereocilia: In: Altschuler RA, Hoffman DW, Bobbin RP (eds) Neurobiology of Hearing: The Cochlea. New York: Raven Press, pp. 23-46.
Nishida Y, Fujimoto T, Takagi A, Honjo I, Ogawa K (1993) Fodrin is a constituent of the cortical lattice in outer hair cells of the guinea pig cochlea: immunocyto-

chemical evidence. Hear Res 65:274-280.
Nomura Y (1970) Histochemistry of the cochlea. Seitai Nokagak 21:68-80.
Oesterle EC, Sarthy PV, Rubel EW (1990) Intermediate filaments in the inner ear of normal and experimentally damaged guinea pigs. Hear Res 47:1-16.
Oesterle EC, Dallos P (1989) Intracellular recordings from supporting cells in the guinea pig cochlea: AC potentials. J Acoust Soc Am 86:1013-1032.
Osborne MP, Comis SD (1990) High resolution scanning electron microscopy of stereocilia in the cochlea of normal, postmortem and drug-treated guinea pigs. J Elect Microscop Tech 15:245-260.
Osborne MP, Comis SC, Pickles JO (1988) Further observations on the fine structure of tip links between stereocilia of the guinea pig cochlea. Hear Res 35:99-108.
Pack AK, Slepecky NB (1995) Cytoskeletal and calcium-binding proteins in the mammalian organ of Corti: cell type-specific proteins displaying longitudinal and radial gradients. Hear Res 92:119-135.
Pickles JO, Comis SD, Osborne MP (1984) Cross-links between stereocilia in the guinea pig organ of Corti, and their possible relation to sensory transduction. Hear Res 15:103-112.
Plinkert PK, Gitter AH, Zimmerman U, Kirchner T, Tzartos S, Zenner HP (1990) Visualization and functional testing of acetylcholine receptor-like molecules in cochlear outer hair cells. Hear Res 44:25-34.
Preston RE, Wright CG (1974) Pinocytosis in the pillar cells of the organ of Corti. Acta Otolaryngol 78:333-340.
Prieto JJ, Merchan JA (1987) Regional specialization of the cell coat in the hair cells of the organ of Corti. Hear Res 31:223-228.
Prieto JJ, Rubio ME, Merchan JA (1990) Localization of anionic sulfate groups in the tectorial membrane. Hear Res 45:283-294.
Puel J, Pujol R (1992) Selective glutamate antagonists block the excitotoxicity caused either by glutamate agonists or by ischemia; In: Cazals Y, Demany L, Horner K (eds) Advances in Biosciences: Auditory Physiology and Perception. Oxford: Pergamon Press, pp. 589-598.
Pujol R, Carlier E (1981) Cochlear synaptogenesis after section of the efferent bundle. Dev Brain Res 3:151-154.
Pujol R, Lenoir M (1986) The four types of synapses in the organ of Corti. In: Altschuler RA, Hoffman DW, Bobbin RP (eds) Neurobiology of Hearing: The Cochlea. New York: Raven Press, pp. 161-172.
Qvortrup K, Rostgaard J (1990a) Mesothelium of Reissner's membrane in guinea pigs: an electron microscopic study. Eur Arch Otorhinolaryngol 248:57-62.
Qvortrup K, Rostgaard J (1990b) Three-dimensional organization of a transcellular tubulocisternal endoplasmic reticulum in epithelial cells of Reissner's membrane in the guinea pig. Cell Tiss Res 261:287-299.
Raphael Y, Altschuler RA (1991) Reorganization of cytoskeletal and junctional proteins during cochlear hair cell degeneration. Cell Motil Cytoskel 18:215-227.
Raphael Y, Altschuler RA (1992) Early microfilament reorganization in injured auditory epithelia. Exp Neurol 115:32-36.
Raphael Y, Wroblewski R (1968) Linkage of sub-membrane cisterns with the cytoskeleton and the plasma membrane in cochlear outer hair cells. J Submicrosc Cytol 18:730-733.
Raphael Y, Marshak GH, Barash A, Geiger B (1987) Modulation of intermediate-filament expression in developing cochlear epithelium. Differentiation 35:151-162.

Raphael Y, Volk T, Crossin KL, Edelman GM, Geiger B (1988) The modulation of cell adhesion expression and intercellular junction formation in the developing avian inner ear. Dev Biol 128:222-235.

Raphael Y, Lenoir M, Wroblewski R, Pujol R (1991) The sensory epithelium and its innervation in the mole rat cochlea. J Comp Neurol 314:367-382.

Rasmussen G (1940) Studies of the VIIIth cranial nerve of man. Laryngoscope 50:67-83.

Rasmussen GL (1953) Further observations of the efferent cochlear bundle. J Comp Neurol 99:61-74.

Reale E, Luciano L, Franke K, Pannese E, Wermbter G, Iurato S (1975) Intercellular junctions in the vascular stria and spiral ligament. J Ultrastruct Res 53:284-297.

Reiss G, Raphael Y (1992) Atypical cells in the normal guinea pig organ of Corti as revealed by scanning electron microscopy. Microscop Res Tech 20:288-297.

Richardson GP, Russell IJ, Duance VC, Bailey AJ (1987) Polypeptide composition of the mammalian tectorial membrane. Hear Res 25:45-60.

Rodriguez-Echandia EL, Burgos MH (1965) The fine structure of the stria vascularis of the guinea pig inner ear. Z Zellforsch 67:600-619.

Romand R, Romand MR (1984) The ontogenesis of pseudomonopolar cells in spiral ganglion of cat and rat. Acta Otolaryngol 97:239-249.

Romand R, Romand MR, Marty R (1981) Regional differences in fiber size in the cochlear nerve. J Comp Neurol 198:1-5.

Romand R, Hafidi A, Despres G (1988) Immunocytochemical localization of neurofilament protein subunits in the spiral ganglion of the adult rat. Brain Res 462:167-173.

Romand R, Sobkowicz H, Emmerling M, Whitlon D, Dahl D (1990) Patterns of neurofilament stain in the spiral ganglion of the developing and adult mouse. Hear Res 49:119-126.

Rosenbluth J (1962) The fine structure of the acoustic ganglion in the rat. J Cell Biol 12:329-359.

Ross MD (1973) Autonomic components of the VIII nerve. Adv Otorhinolaryngol 20:316-336.

Ross MD (1978) Glycogen accumulation in Reissner's membrane following chemical sympathectomy with 6-hydroxydopamine. Acta Otolaryngol 86:313-330.

Ross MD, Burkel W (1973) Multipolar neurons in the spiral ganglion of the rat. Acta Otolaryngol 76:381-394.

Ross MD, Liu R, Preston RE, Wright CG (1974) Changes in conformation in hair cell stereocilia of the rat spiral organ of Corti after 6-hydroxydopamine as revealed by scanning electron microscopy. Audiology 13:290-301.

Roth B, Bruns V (1992) Postnatal development of the rat organ of Corti. Anat Embryol 185:571-581.

Russell IJ (1983) Origin of the receptor potential in inner hair cells of the mammalian cochlea—evidence for Davis' theory. Nature 301:334-446.

Russell IJ, Richardson GP, Cody AR (1986) Mechanosensitivity of mammalian auditory hair cells in vitro. Nature 321:517-519.

Saito K (1980) Fine structure of the sensory epithelium of guinea pig organ of Corti: afferent and efferent synapses of hair cells. J Ultrastruct Res 71:222-232.

Saito K (1983) Fine structure of the sensory epithelium of guinea-pig organ of Corti: subsurface cisternae and lamellar bodies in the outer hair cells. Cell Tiss Res 229:467-481.

Saito K, Hama K (1982) Structural diversity of microtubules in the supporting cells of the sensory epithelium of guinea pig organ of Corti. J Electron Microsc 311:278–281.

Santi PA (1988) Cochlear microanatomy and ultrastructure. In: Jahn AF, Santos-Sacchi J (eds) Physiology of the Ear. New York: Raven Press, pp. 173–199.

Santi PA, Anderson CB (1987) A newly identified surface coat on cochlear hair cells. Hear Res 27:47–65.

Santi PA, Lakhani BN, Bingham C (1983) The volume density of cells and capillaries of the normal stria vascularis. Hear Res 11:7–22.

Santi PA, Larson JT, Furcht LT, Economou TS (1989) Immunohistochemical localization of fibronectin in the chinchilla cochlea. Hear Res 39:91–102.

Santi PA, Lease MK, Harrison RP, Wicker EM (1990) Ultrastructure of proteoglycans in the tectorial membrane. J Electron Microsc Tech 15:293–300.

Santos-Sacchi J (1978) Cytoplasmic microtubules in strial marginal cells. Arch Otorhinolaryngol 218:297–300.

Santos-Sacchi J (1982) An electron microscopic study of microtubules in the development of marginal cells of the mouse stria vascularis. Hear Res 6:7–13.

Santos-Sacchi J (1986) Dye coupling in the organ of Corti. Cell Tiss Res 245:525–529.

Santos-Sacchi J, Marovitz WF (1980) An evaluation of normal strial capillary transport using the electron-opaque tracers ferritin and iron dextran. Acta Otolaryngol 89:12–26.

Schrott A, Egg G, Spoendlin H (1988) Intermediate filaments in the cochleas of normal and mutant (w/wv, sl/sld) mice. Arch Otorhinolaryngol 245:250–254.

Schulte BA (1993) Immunohistochemical localization of intracellular Ca^{++}-ATPase in outer hair cells, neurons and fibrocytes in the adult and developing inner ear. Hear Res 65:262–273.

Schulte BA, Adams JC (1989a) Distribution of immunoreactive Na^+, K^+-ATPase in gerbil cochlea. J Histochem Cytochem 37:127–134.

Schulte BA, Adams JC (1989b) Immunohistochemical localizations of vimentin in the gerbil inner ear. J Histochem Cytochem 37:1787–1797.

Schulte BA, Steel KP (1994) Expression of α and β subunit isoforms of Na, K-ATPase in the mouse inner ear and changes with mutations at the Wv or Sld loci. Hear Res 78:259–260.

Schwartz A (1986) Auditory nerve and spiral ganglion cells. In: Altschuler RA, Hoffman DW, Bobbin RP (eds) Neurobiology of Hearing: The Cochlea. New York: Raven Press, pp. 271–282.

Shepherd GM, Barres BA, Corey DP (1989) "Bundle blot" purification and initial protein characterization of hair cell stereocilia. Proc Natl Acad Sci USA 86:4973–4977.

Shi SR, Tandon AK, Cote C, Kalra KL (1992) S-100 protein in human inner ear: use of a novel immunohistochemical technique on routinely processed, celloidin-embedded human temporal bone sections. Laryngoscope 102:734–738.

Shi SR, Tandon AK, Hausmann RRM, Kalra KL, Taylor CR (1993) Immunohistochemical study of interemediate filament proteins on routinely processed, celloidin-embedded human temporal bone sections using a new technique for antigen-retrieval. Acta Otolaryngol 113:48–54.

Simmons DD, Liberman MC (1988) Afferent innervation of outer hair cells in adult cats. J Comp Neurol 270:132–144.

Slepecky NB (1989a) Cytoplasmic actin and cochlear outer hair cell motility. Cell

Tiss Res 257:69-75.

Slepecky NB (1989b) An infracuticular network is not required for outer hair cell shortening. Hear Res 38:135-140.

Slepecky NB (1993) Ultrastructure of the inner ear. In Friedmann I, Arnold W (eds) Pathology of the Ear. Edinburgh: Churchill Livingstone.

Slepecky NB (1995) Sensory and supporting cells of the organ of Corti—cytoskeletal organization related to cellular function. In: Flock Å, Ottoson D, Ulfendahl M (eds) Active Hearing. Amsterdam: Elsevier Scientific 87:104.

Slepecky NB, Chamberlain SC (1982) Distribution and polarity of actin in the sensory hair cells of the chinchilla cochlea. Cell Tiss Res 224:15-24.

Slepecky NB, Chamberlain SC (1983) Distribution and polarity of actin in inner ear supporting cells. Hear Res 10:359-370.

Slepecky NB, Chamberlain SC (1985a) The cell coat of inner ear sensory and supporting cells as demonstrated by ruthenium red. Hear Res 17:281-288.

Slepecky NB, Chamberlain SC (1985b) Immunoelectron microscopic and immunofluorescent localization of cytoskeletal and muscle-like contractile proteins in inner ear sensory hair cells. Hear Res 20:245-260.

Slepecky NB, Chamberlain SC (1987) Tropomyosin co-localizes with actin microfilaments and microtubules within supporting cells of the inner ear. Cell Tiss Res 248:63-66.

Slepecky NB, Hamernik R, Handerson D, Coling D (1982) Correlation of audiometric data with changes in cochlear hair cell stereocilia resulting from impulse noise trauma. Acta Otolaryngol 93:329-340.

Slepecky NB, Henderson CG, Saha S (1995) Post-translational modifications of tubulin suggest that dynamic microtubules are present in sensory cells and stable microtubules are present in supporting cells of the mammalian cochlea. Hear Res 91:136-147.

Slepecky NB, Ligotti P (1992) Characterization of inner ear sensory hair cells after rapid-freezing and free-substitution. J Neurocytol 21:374-381.

Slepecky NB, Savage J (1994) Expression of actin isoforms in the guinea pig organ of Corti: muscle isoforms are not detected. Hear Res 73:16-26.

Slepecky NB, Ulfendahl M (1992) Actin-binding and microtubule associated proteins in the organ of Corti. Hear Res 57:201-215.

Slepecky NB, Ulfendahl M (1993) Evidence for calcium-binding proteins and calcium-dependent regulatory proteins in sensory cells of the organ of Corti. Hear Res 70:73-84.

Slepecky NB, Hamernik RP, Henderson D (1981) The consistent occurrence of a striated organelle (Friedmann body) in the inner hair cells of the normal chinchilla. Acta Otolaryngol 91:189-198.

Slepecky NB, Ulfendahl M, Flock Å (1988) Effects of caffeine and tetracine on outer hair cell shortening suggest intracellular calcium involvement. Hear Res 32:11-22.

Slepecky NB, Cefaratti LK, Yoo TJ (1992) Type II and type IX collagen form heterotypic fibers in the tectorial membrane of the inner ear. Matrix 12:80-96.

Slepecky NB, Savage JE, Cefaratti LK, Yoo TJ (1992) Electron microscopic localization of type II, IX and V collagen in the organ of Corti. Cell Tiss Res 267:413-418.

Sliwinska-Kowalska M, Parakkal M, Schneider ME, Fex J (1989) CGRP-like immunoreactivity in the guinea pig organ of Corti: a light and electron microscopic study. Hear Res 42:83-96.

Smith C (1957) Structure of the stria vascularis and spiral prominence. Ann Otol

Rhinol Laryngol 66:521-536.
Smith CA (1968) Ultrastructure of the organ of Corti. Adv Sci 24:419-433.
Smith CA (1972) Vascular patterns of the membranous labyrinth. In: de Lorenzo AJD (ed) Vascular Disorders and Hearing Defects. Baltimore: University Park Press, pp. 1-18.
Smith CA (1975) Innervation of the cochlea of the guinea pig by use of the Golgi stain. Ann Otol Rhinol Laryngol 84:443-458.
Smith CA, Sjostrand FS (1961a) Structure of the nerve endings on the external hair cells of the guinea pig cochlea as studied by serial sections. J Ultrastruct Res 5:523-556.
Smith CA, Sjostrand FS (1961b) A synaptic structure in the hair cells of the guinea pig cochlea. J Ultrastruct Res 5:184-192.
Sobkowicz HM, Slapnick SM (1994) The efferents interconnecting auditory inner hair cells. Hear Res 75:81-92.
Sobkowicz HM, Rose JE, Scott GL, Levenick CV (1986) Distribution of synaptic ribbons in the developing organ of Corti. J Neurocytol 15:693-714.
Sobkowicz HM, Slapnick SM, August BK (1993) Presynaptic fibers of spiral neurons and reciprocal synapses in the organ of Corti in culture. J Neurocytol 22:979-993.
Spicer SS, Schulte BA (1991) Differentiation of inner ear fibrocytes according to their ion transport related activity. Hear Res 56:53-64.
Spicer SS, Schulte BA (1992) Creatine kinase in epithelium of the inner ear. J Histochem Cytochem 40:185-192.
Spicer SS, Schulte BA (1993) Cytologic structures unique to Deiters cells of the cochlea. Anat Rec 237:421-430.
Spicer SS, Schulte BA (1994a) Differences along the place-frequency map in the structure of supporting cells in the gerbil cochlea. Hear Res 79:161-177.
Spicer SS, Schulte BA (1994b) Ultrastructural differentiation of the first Hensen cell as a distinct cell type. Anat Rec 240:149-156.
Spoendlin H (1966) The organization of the cochlear receptor. Adv Otorhinolaryngol 13:1-227.
Spoendlin H (1968) Ultrastructure and peripheral innervation pattern of the receptor in relation to the first coding of the acoustic message. In: DeReuck AVS, Knight J (eds) Hearing Mechanisms in Vertebrates. London: Churchill, pp. 89-119.
Spoendlin H (1973) Innervation of the cochlear receptor. In: Møller A (ed) Basic Mechanisms in Hearing. New York: Academic Press, pp. 185-230.
Spoendlin H (1979) Neural connections of the outer hair cell system. Acta Otolaryngol 87:381-387.
Spoendlin H (1981) Autonomic innervation of the inner ear. Adv Otorhinolaryngol 27:1-13.
Spoendlin H, Lichtensteiger W (1966) The adrenergic innervation of the labyrinth. Acta Otolaryngol 61:423-434.
Spoendlin H, Schrott A (1988) The spiral ganglion and the innervation of the human organ of Corti. Acta Otolaryngol 105:403-410.
Steel KP (1986) Tectorial membrane. In: Altschuler RA, Hoffman DW, Bobbin RP (eds) Neurobiology of Hearing: The Cochlea. New York: Raven Press, pp. 139-148.
Steyger PS, Furness DN, Hackney CM, Richardson GP (1989) Tubulin and microtubules in cochlear hair cells: comparative immunocytochemistry and ultrastructure. Hear Res 42:1-16.

Strelioff D, Flock A (1984) Stiffness of sensory cell hair bundles in the isolated guinea pig cochlea. Hear Res 15:19-28.

Sugiyama S, Spicer SS, Munyer PD, Schulte BA (1991) Histochemical analysis of glycoconjugates in gelatinous membranes of the gerbil's inner ear. Hear Rees 55:263-272.

Sugiyama S, Spicer SS, Munyer PD, Schulte BA (1992) Ultrastructural localization and semiquantitative analysis of glycoconjugates in the tectorial membrane. Hear Res 58:35-46.

Suzuki H, Lee YC, Tachibana M, Hozawa K, Watayna H, Takasaka T (1992) Quantitative carbohydrate analyses of the tectorial and otoconial membranes of the guinea pig. Hear Res 60:45-52.

Swetlitschkin R, Vollrath L (1988) Synaptic bodies in the different rows of outer hair cells in the guinea pig cochlea. Ann Oto Rhinol Laryngol 97:308-312.

Takahashi T, Kimura RS (1970) The ultrastructure of the spiral ligament in the rhesus monkey. Acta Otolaryngol 69:46-60.

Takasaka T, Shinkawa H, Hashimoto H, Watanuki K, Kawamoto K (1983) High-voltage electron microscopic study of the inner ear. Technique and preliminary results. Ann Otol Rhinol Laryngol Suppl 101:1-12.

Tange RA, Hodde KC (1984) The microvasculature of the cochlea and the vestibular system as seen in scanning electron microscopy. Clin Otolaryngol 9:306.

Terayama Y, Holz E, Beck C (1966) Adrenergic innervation of the cochlea. Ann Otol Rhinol Laryngol 75:69-86.

Terayama Y, Yamamoto K, Sakamoto T (1968) Electron microscopic observations on the postganglionic sympathetic fibers in the guinea pig cochlea. Ann Otol Rhinol Laryngol 72:1152-1171.

Thalmann I (1993) Collagen of accessory structures of organ of Corti. Conn Tiss Res 29:199-201.

Thalmann I, Thallinger G, Comegys TH, Thalmann R (1985) Collagen II the predominate protein of the tectorial membrane. Otorhinolaryngology 48:116-123.

Thalmann I, Thallinger G, Comegys TH, Crouch EC, Barrett N, Thalmann R (1987) Composition and supramolecular organization of the tectorial membrane. Laryngoscope 97:357-367.

Thomasen E (1966) The ultrastructure of the spiral ganglion in the guinea pig. Acta Otolaryngol Suppl 224:442-448.

Thompson GC, Cortez AM, Igarashi M (1986) GABA-like immunoreactivity in the squirrel organ of Corti. Brain Res 372:72-79.

Thorne PR, Carlisle L, Zajic G, Schacht J, Altschuler RA (1987) Differences in the distribution of F-actin in outer hair cells along the organ of Corti. Hear Res 30:253-266.

Tilney LG, DeRosier DJ (1985) The organization of actin filaments in the stereocilia of the hair cells of the cochlea. In: Drescher DG (ed) Auditory Biochemistry. Springfield, IL: Charles C. Thomas, pp. 281-309.

Tilney LG, DeRosier DJ, Mulroy MJ (1980) The organization of actin filaments in the stereocilia of cochlear hair cells. J Cell Biol 86:244-259.

Tilney LG, Tilney MS, DeRosier DJ (1992) Actin filaments, stereocilia, and hair cells. How cells count and measure. Annu Rev Cell Biol 8:257-274.

Torihara K, Morimitsu T, Suganuma T (1995) Anionic sites of Reissner's membrane, stria vascularis, and spiral prominence. J Histochem Cytochem 43:299-305.

Tucker JB, Paton CC, Richardson GP, Mogensen MM, Russell IJ (1992) A cell surface-associated centrosomal layer of microtubule-organizing material in the inner pillar cell of the mouse cochlea. J Cell Sci 102:215-226.
Usami S-I, Hozawa J, Tazawa M, Yoshihara T, Igarashi M, Thompson GC (1988) Immunocytochemical study of catecholaminergic innervation in the guinea pig cochlea. Acta Otolaryngol Suppl 447:36-45.
Vater MM, Lenoir M (1992) Ultrastructure of the horseshoe bat's organ of Corti. I Scanning electron microscopy. J Comp Neurol 318:367-379.
Vater M, Lenoir M, Pujol R (1992) Ultrastructure of the horseshoe bat's organ of Corti. II Transmission electron microscopy. J Comp Neurol 318:380-391.
Warr WB (1992) Organization of olivocochlear efferent systems in mammals. In: Webster DB, Popper A, Fay RR (eds) Mammalian Auditory Pathways: Neuroanatomy. New York: Springer-Verlag, pp. 410-448.
Warr WB, Guinan JJ (1979) Efferent innervation of the organ of Corti: two separate systems. Brain Res 173:152-155.
Watanabe K, Ogawa A (1984) Carbonic anhydrase activity in stria vascularis and dark cells in vestibular labyrinth. Ann Otol Rhinol Laryngol 93:262-266.
Watanuki K (1968) Some morphological observations of Reissner's membrane. Acta Otolaryngol 66:40-48.
Weaver SP, Hoffpauir J, Schweitzer L (1993) Actin distribution along the lateral wall of gerbil outer hair cells. Brain Res Bull 31:225-228.
Whitlon DS (1993) E-cadherin in the mature and developing organ of Corti of the mouse. J Neurocytol 22:1030-1038.
Whitlon DS, Sobkowicz HM (1989) GABA-like immunoreactivity in the cochlea of the developing mouse. J Neurocytol 18:505-518.
Winther F (1971) The permeability of the guinea pig cochlear capillaries to horseradish peroxidase. Z Zellforsch 114:193-202.
Woolf NK, Koern FJ, Ryan AF (1992) Immunohistochemical localization of fibronectin-like protein in the inner ear of the developing gerbil and rat. Dev Brain Res 65:21-33.
Wright A (1984) Dimensions of the cochlear stereocilia in man and the guinea pig. Hear Res 13:89-98.
Xie DH, Henson MM, Bishop AL, Henson OW (1993) Efferent terminals in the cochlea of the mustached bat: quantitative data. Hear Res 66:81-90.
Yamamoto K, Nakai Y (1964) Electron microscopic studies on the functions of the stria vascularis and the spiral ligament in the inner ear. Ann Otorhinolaryngol 73:332-342.
Ylikoski J, Pirvora U, Narvanen O, Virtanen I (1990) Nonerythroid spectrin (fodrin) is a prominent component of the cochlear hair cells. Hear Res 43: 199-204.
Yoshihara T, Igarashi M (1987) Cytochemical localization of calcium ATPase activity in the lateral cochlear wall of the guinea pig. Arch Otorhinolaryngol 243: 395-400.
Zenner HP, Reuter G, Plinkert PK, Zimmermann U, Gitter AH (1989) Outer hair cells possess acetylcholine receptors and produce motile responses in the organ of Corti. In: Wilson JP, Kemp DT (eds) Cochlear Mechanisms. New York: Plenum Publishing Corp., pp. 93-98.
Zwislocki JJ, Slepecky NB, Cefaratti LK, Smith RL (1992) Ionic coupling among cells in the organ of Corti. Hear Res 57: 175-194.

3
Homeostatic Mechanisms in the Cochlea

PHILINE WANGEMANN AND JOCHEN SCHACHT

1. Introduction

1.1 The Concept of Homeostasis

The challenge of every cell is to maintain intracellular conditions that may vastly differ from the external environment, yet still communicate with this environment. Walter Cannon (1929) first applied the term "homeostasis" to the concept, originally formulated by Claude Bernard (1878), of the constancy of the *milieu interne* as essential for the existence of free-living organisms. Broadly defined, homeostasis represents the sum of the physiological processes in an organism, a multicellular system, or a cell that maintain the relative stability of its internal environment and thus provide the basis for its survival and function. The inner ear, as suggested by Hawkins (1973), possesses a variety of microhomeostatic mechanisms that sustain the integrity, sensitivity, and dynamic range of the organ of Corti. They make possible its function as a transducer, although they do not include the transduction process itself.

Maintenance of the internal milieu is an active process requiring the intricate interaction of numerous biochemical and biophysical processes, utilizing most of the metabolic energy of the cell. Homeostatic mechanisms respond to specific intra- and extracellular signals that report on the physiological balance or imbalance of a cell or an organ: changes in the demand for energy, pH shifts, ionic imbalances, and trophic signals of hormones or local mediators. In contrast to sensory transduction, which is based on short-term processing of information, homeostatic mechanisms may respond along more varied time scales. An ionic imbalance may be restored within milliseconds, whereas other responses operate over minutes or even hours if synthesis of proteins or the transcription of genetic information is involved. Homeostasis is so pervasive that its complexity tends to be overlooked until the consequence of failed homeostasis is manifested as cellular dysfunction and pathophysiology.

Homeostasis is controlled on two different levels. One is the maintenance of the individual intracellular milieu. The other is the maintenance of the extracellular milieu within a multicellular organization. Membrane transport, ion channels, and both synthetic and catabolic pathways all operate on well-conserved principles. An individual cell, however, may employ a unique combination of mechanisms geared to its specific needs in a particular environment. This ability of a cell to sustain its intracellular and extracellular milieu depends, in turn, on external factors: an appropriate supply of energy substrates, essential nutrients, and oxygen, and the means for removal of CO_2 and metabolic waste products.

Homeostasis is not based on an unvarying set of reactions. As the environment changes, by necessity the mechanisms to sustain the integrity of the system also change. Stress and insult trigger biochemical defenses that are an integral part of homeostasis and that are essential to protect the cell from external noxious influences. They operate in conjunction with and in addition to protective mechanisms that defend the cell against hazardous products of its own metabolism. Finally, in addition to the homeostasis of the normally functioning adult system, unique adaptive mechanisms are induced in development, repair, and regeneration, processes that lie beyond the scope of this chapter.

1.2 Cochlear Homeostasis

Hallowell Davis's model of auditory transduction (Davis 1965) assigned distinctly different and essential functions to the basilar membrane and the lateral wall tissues. His "battery theory" placed the site of transduction proper in the hair cells, which function as modulators of a DC current supplied via the endolymph from the stria vascularis acting as a battery. This basic distinction between the major functions of the two tissues still holds today.

In this chapter, the discussion of homeostasis in the organ of Corti will emphasize the mechanisms that individual cells employ to maintain their intracellular environment and viability. It is understood that such individual cell homeostasis is as applicable to cells in the lateral wall tissues as it is to sensory or supporting cells in the organ of Corti. However, the stria vascularis sustains the endolymph's unique electrochemical composition and its potential through a combination of processes specific to the inner ear. Thus, the discussion of the stria will focus on mechanisms that combine to maintain the extracellular milieu and the driving forces upon which auditory transduction depends.

2. Energy Metabolism

2.1 Basic Metabolism

Energy metabolism is central to all aspects of the physiology of a cell. Enzymes and substrates of glycolysis and the citric acid cycle were the first biochemical constituents to be quantitated in individual cochlear tissues (Thalmann 1971). These studies provided indirect support for Davis's "battery theory" that the stria vascularis provided the energy for the cochlear transducer. The stria vascularis consumes oxygen and energy substrates at a much higher rate than the organ of Corti (reviewed by Thalmann 1976; Schacht and Canlon 1988). Lower oxygen consumption in the organ of Corti does not, however, indicate a predominance of the glycolytic nonoxidative pathway in this tissue. The absence of high levels of lactate, the end product of anaerobic glycolysis, in inner ear tissues and fluids (Scheibe et al. 1981) argues against a major contribution of this pathway.

Although glucose, fatty acids, and amino acids are all potential sources of energy, glucose is the major energy substrate for the inner ear (Kambayashi et al. 1982b). Glucose is supplied through the vascular system and crosses the blood–perilymph barrier (see Section 6.4.1) by a facilitated transport mechanism (Ferrary et al. 1987) in which the brain-type glucose transporter GLUT1 may participate (Ito, Spicer, and Schulte 1993). The GLUT1 transporter is also present in strial basal cells where it may facilitate additional movement of glucose from the perilymph into the intrastrial compartment for utilization by strial marginal cells.

The perilymph can thus serve as a reservoir for the entry of glucose into inner ear tissues. This becomes evident when the vascular system is perfused with substrate-free solutions. The endocochlear and cochlear microphonic potentials are maintained under such conditions, whereas simultaneous vascular and perilymphatic perfusions cause an immediate decline of the potentials (Kambayashi et al. 1982a). Although perilymph also supplies oxygen (Lawrence and Nuttall 1972), it does not represent a large reservoir. Anoxia abolishes the endocochlear potential within seconds (Konishi, Butler, and Fernández 1961), probably because of the high oxygen requirement of the stria vascularis.

Glucose uptake from the extracellular fluids into cells of the organ of Corti is also mediated by facilitated transport, evidenced by the presence of the GLUT5 transporter in outer hair cells (Nakazawa, Spicer, and Schulte 1995). Insulin-stimulated glucose uptake is absent from the cochlea, based on biochemical studies (Wang and Schacht 1990) and the lack of immunocytochemical evidence for the GLUT4 transporter (Nakazawa, Spicer, and Schulte 1994).

Adenosine triphosphate (ATP) is the ultimate chemical form of biological energy. Homeostatic mechanisms operating at the levels of oxidative phosphorylation, substrate availability, and enzyme activities assure constant cellular ATP levels. As more ATP is hydrolyzed in energy-consuming reactions, the pathways of its formation are activated. Thus, the concentration of ATP will remain constant under physiological conditions, and any change will indicate a pathophysiological rather than a physiological state. The energy metabolism of both the stria vascularis and the organ of

Corti is indeed sensitive to pathological stress such as prolonged ischemia (Thalmann, Marcus, and Thalmann 1978), which breaks down regulatory processes. In contrast, ATP and related biochemical parameters remain unaltered under various conditions of noise exposure, including those leading to a temporary threshold shift (Thalmann 1976).

This constancy of ATP levels does not mean, however, that ATP utilization and energy metabolism remain constant under acoustic stimulation. To the contrary, processing of physiological information in general increases ATP consumption, and a higher rate of ATP synthesis is necessary to maintain homeostasis. Thus, although the *levels* of energy metabolites may not change, their *turnover* (i.e., the *rate of their formation and utilization*) will. Measurements of metabolic rates in the inner ear confirm changes associated with auditory stimulation.

2.2 Response to Acoustic Stimulation

The fact that energy metabolism rises when a cell engages in physiological activity has most directly been demonstrated with the deoxyglucose method. Deoxyglucose, a poorly metabolized analog of the natural substrate glucose, can function as a tracer of glucose utilization at the cellular level (Sokoloff 1977).

Deoxyglucose uptake in the cochlea of an unstimulated animal is high in the stria vascularis and low in the organ of Corti (Ryan et al. 1985), providing a good correlate to the rate of oxygen consumption in these tissues. Moderate acoustic stimulation (55–85 dbA) significantly increases deoxyglucose uptake in auditory structures of mice, including the organ of Corti, the lateral wall tissues, the 8th nerve, and the inferior colliculus (Canlon and Schacht 1983). Studies in the gerbil further localized enhanced metabolism in all three cell layers of the stria vascularis and in the inner hair cells and nerve endings underneath the outer hair cells of the organ of Corti (Ryan et al. 1985). While an increased energy flux in sensory cells can be explained as a correlate of transduction, the increase in the stria vascularis points to the unique tissue interactions in the cochlea serving the extracellular milieu. During auditory transduction, increased K^+ fluxes and transient fluctuations of the endolymphatic potential apparently are appropriate feedback mechanisms to maintain strial homeostasis.

2.3 Cochlear Blood Flow

The cellular requirement for glucose is met by its extraction from the circulation via facilitated transport. The activity of transporters is not directly regulated by metabolic feedback. Rather, glucose homeostasis is achieved by a coupling of cellular metabolism to enhanced delivery of nutrients and oxygen through local blood flow. Since one of the results of glucose metabolism is the production of protons, increased metabolic activity is associated with acidification. Acidification, in turn, causes vasodilation so that enhanced metabolism triggers a greater supply of glucose through increased blood flow, as demonstrated in the central nervous system (Reivich 1974). Therefore, acoustic stimulation leading to greater energy utilization should affect cochlear blood flow.

The blood supply to the mammalian cochlea is largely concentrated in the lateral wall tissues. Increased blood flow with any level of sound exposure has rarely been observed (Prazma, Rodgers, and Pillsbury 1983). Recent experiments, however, have confirmed small changes in microcirculation in response to moderate sound stimulation, establishing a potential link to the deoxyglucose studies. Oxygenation of perilymph rises by about 20% with sound intensities of 85-90 dB (Scheibe, Haupt, and Ludwig 1992), and red blood cell velocity increases in the vessels of the lateral wall tissues (Quirk et al. 1992). In contrast, intense noise exposure consistently reduces cochlear blood flow (Thorne and Nuttall 1987). This is in apparent contradiction to expected increased energy demands but in agreement with capillary vasoconstriction (Hawkins 1971), decreased oxygenation of endolymph (Thorne and Nuttall 1989), and decreased deoxyglucose uptake (Canlon and Schacht 1983), which occur under such conditions. It must be assumed that traumatic sound levels trigger pathophysiological responses of the cochlea.

Cochlear blood flow and metabolism thus respond to acoustic stimulation, suggesting that homeostatic mechanisms control blood flow and thereby oxygen and nutrient supply. Anatomical evidence suggests the sympathetic noradrenergic system as a potential modulator of cochlear blood flow. Its fibers extend to the spiral modiolar artery and the arterioles in the modiolus and osseous spiral lamina (Spoendlin and Lichtensteiger 1966), possibly even to locations in or near the lateral wall (Brechtelsbauer et al. 1990). Functionally, electrical stimulation of either the superior cervical ganglion or the stellate ganglion reduces cochlear blood flow, a phenomenon likely mediated by β-adrenergic mechanisms (Laurikainen et al. 1993). In addition, a variety of receptors for vasoactive effectors are associated with the labyrinthine and cochlear arteries. Agents such as adrenergic agonists (Ohlsén et al. 1991) or vasoactive peptides (angiotensin II; Quirk et al. 1988; substance P: McLaren et al. 1993) influence cochlear blood flow. Although this evidence points to elaborate regulatory and autoregulatory mechanisms (Brown and Nuttall 1994), cochlear feedback systems responsive to sound stimulation and metabolism remain to be explored.

3. Intracellular Regulation by Second Messengers and Calcium

The response of cochlear energy metabolism and blood flow to acoustic stimulation demonstrates that the homeostasis of a system is a dynamic process. These system responses are the net effect of individual cellular reactions adapting to changing conditions. Enzymes, transport processes, and ion channels all operate at rates that vary with physiological requirements. An acute adjustment of biochemical processes relies on feedback mechanisms in which enzyme activities are allosterically modulated by the concentrations of intermediates or end products. A second mode of regulation is the reversible posttranslational modification of proteins, primarily by phosphorylation. Most cellular proteins may undergo phosphorylation reactions that modify their properties. Specific phosphatases restore the original state of the proteins by hydrolyzing the phosphate groups. For adaptation on a longer time scale, enzyme levels are regulated by gene induction through the action of steroid hormones, insulin, and growth factors.

3.1 Protein Phosphorylation

Protein kinases include several families of proteins subject to regulation by Ca^{2+} and second messengers, which in turn are activated by hormones, neurotransmitters, or modulators (Nairn, Hemmings, and Greengard 1985). Thus, protein kinases are an important link in transducing extracellular signals into intracellular physiological responses. The sequence consisting of agonist stimulation of a receptor, generation of second messengers, activation of protein kinases, and phosphorylation of target proteins is one of the fundamental processes in cellular biochemistry and physiology (Fig. 3.1).

Protein kinases can be distinguished by their activating second messengers: protein kinase A (PKA), activated by cyclic AMP; protein kinase G (PKG), activated by cyclic GMP; protein kinase C (PKC), activated by diacylglycerol or lipids; and calmodulin-dependent kinase (CaM kinase), activated by Ca^{2+} and calmodulin. These kinases phosphorylate the amino acids serine or threonine in their target proteins. A family of kinases distinct in their activating mechanism and their substrates, but of no less importance, is the protein-tyrosine kinases. Trophic factors active in the inner ear (Avila et al. 1993; Ylikoski et al. 1993) are known stimulators of these enzymes, but information on the cochlear activity of tyrosine kinases is not yet available.

In contrast, the activity of serine or threonine kinases is well documented in the inner ear. Proteins in different inner ear tissues are phosphorylated in reactions catalyzed by PKA, PKG, PKC, and CaM kinase (Coling and Schacht 1991). CaM kinases seem to play a major role in the organ of Corti, since protein phosphorylation in this tissue is most responsive to Ca^{2+} and calmodulin and less so to other activators. Such a role may be associated with the modulation of auditory-specific events. In actomyosin-like motile systems, one of the participating enzymes is the Ca^{2+}- and calmodulin-dependent myosin light chain kinase, which has been identified in the organ

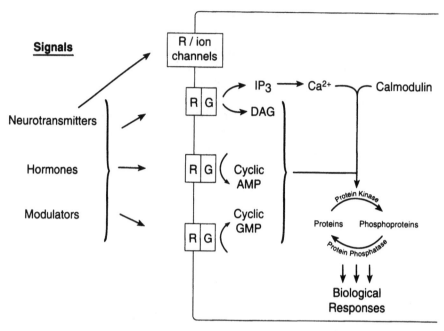

FIGURE 3.1 Schematic representation of second messenger actions. Neurotransmitters, neuromodulators, and hormones bind to specific receptors (R) at the plasma membrane that are coupled to ion channels (ionotropic receptors) or the formation of second messengers (metabotropic receptors). G proteins (G) couple metabotropic receptor stimulation to the activation of enzymes that produce the second messengers inositol trisphosphate (IP_3) plus diacylglycerol (DAG), cyclic AMP, or cyclic GMP. Specific protein kinases are activated either directly by the second messengers or by calmodulin combining with Ca^{2+}, liberated from intracellular storage sites by IP_3. Phosphorylation and dephosphorylation of target proteins is the molecular switch for biological responses.

of Corti (Sziklai, Kiss, and Ribari 1986). Furthermore, calmodulin antagonists block the Ca^{2+}-dependent "slow motility" in isolated outer hair cells (Dulon, Zajic, and Schacht 1990), suggesting the involvement of protein kinase reactions in this process.

3.2 Second Messengers

With few exceptions (e.g., steroid hormones triggering genomic effects), extracellular signals do not enter the cell but exert their action at specific receptors at the plasma membrane. These may either directly activate ion channels (ionotropic receptors) or trigger the generation of second messengers (metabotropic receptors). At these metabotropic receptors, conversion to intracellular signals is accomplished by the activation of membrane-bound mediators called G proteins, which couple receptors to specific enzymes that produce intracellular second messengers. Enzymes amplify the signal by catalyzing second messenger formation.

Second messengers and protein kinase activators are widely distributed in the cochlea. High levels of calmodulin are found in hair cells (Slepecky and Ulfendahl 1993; see Section 3.3.3). Cyclic GMP (Guth and Stockwell 1977) follows a cochlear gradient correlating with the gradient in cholinergic innervation (Thalmann, Paloheimo, and Thalmann 1979). Cyclic AMP and its synthesizing enzyme, adenylyl cyclase, are found in all inner ear tissues (Thalmann, Paloheimo, and Thalmann 1979; Bagger-Sjöbäck, Filipek, and Schacht 1980). High concentrations of adenylyl cyclase are present in strial marginal cells (Zajic, Anniko, and Schacht 1983) where cyclic AMP is involved in the regulation of K^+ secretion (Julien et al. 1994; Sunose et al. 1995).

Phosphoinositides were reported in the inner ear before their function as precursors of second messengers was known (Schacht 1974). The components of this second messenger system are present in all cochlear tissues, including hair cells (Schacht and Zenner 1987; Williams, Zenner, and Schacht 1987), supporting cells of the organ of Corti (Dulon, Zajic, and Schacht 1993), the stria vascularis, and the spiral ligament (Orsulakova, Stockhorst, and Schacht 1976). Their role in cochlear transmembrane signaling was confirmed when it was shown that phosphoinositide metabolism and inositol trisphosphate (IP_3) formation is stimulated by acetylcholine as well as by ATP (Ono and Schacht 1987; Guiramand et al. 1990; Niedzielski, Ono, and Schacht 1992; Ogawa and Schacht 1993).

At metabotropic receptors, the transmembrane coupling between the external receptors and the cytoplasmic effector enzymes is accomplished through specific members of a family of G proteins. Consistent with the existence of different second messengers in tissues of the cochlea, several G proteins are present: those that are generally associated with the stimulation or inhibition of adenylyl cyclase (G_s and G_i, respectively; Canlon, Homburger, and Bockaert 1991; Tachibana et al. 1992), the regulation of ion channels (G_o; Canlon, Homburger, and Bockaert 1991), and the stimulation of phospholipase C (G_p; Ogawa and Schacht 1994). Only G_p has been experimentally linked to its presumed function in the inner ear. This G protein was characterized in the cochlear neuroepithelium by its differential sensitivity to cholera and pertussis toxins and is associated with the muscarinic activation of inositol phosphate release (Ogawa and Schacht 1994).

The pharmacological profile of inositol release and the participation of G_p make the biochemical response to acetylcholine a classical muscarinic system, probably of the M_3 subtype (Guiramand et al. 1990; Niedzielski, Ono, and Schacht 1992). This characterization is supported by the expression of m_3 (as well as of the related m_1 and m_5) receptors in the cochlea (Drescher et al. 1992). It differs, however, from cholinergic actions determined electrophysiologically in recordings from the perfused cochlea or from isolated outer hair cells. For example, a nicotinic-like receptor mediates the effects of acetylcholine on otoacoustic emissions (Kujawa et al. 1992) and of cholinergic antagonists on the action of medial efferents (Kujawa et al. 1994). Likewise, the pharmacology of the acetylcholine-induced K^+ current in outer hair cells in vitro points to a nonclassical receptor of the nicotinic family (Housley and Ashmore, 1991; Erostegui, Norris, and Bobbin 1994), as do the pharmacological characteristics observed in hair cells of the chick (Fuchs and Morrow 1992a,b; Martin and Fuchs 1992). The discrepancy with biochemical studies can best be reconciled with the assumption that different receptor types are associated with cholinergic synapses in the cochlea. Biochemical studies in complex tissue preparations may detect the sum of responses from a multitude of receptors.

ATP affects the phosphoinositide system in the cochlear neuroepithelium via P_{2y} receptors (Niedzielski and Schacht 1992; Ogawa and Schacht 1993). A different P_2 receptor subtype may be located in the apex of the outer hair cells in contact with endolymph. Its presence was deduced from measurements of response latencies and amplitudes by whole-cell patch clamping (Housley, Greenwood, and Ashmore 1992). The current characterized in these experiments may represent an ATP-gated ion channel, leaving unresolved the localization of the metabotropic receptor (see Kros, Chapter 6).

Purinoceptors are also present in the lateral wall tissues of the cochlea where the P_{2y} subtype mediates IP_3 release (Ogawa and Schacht 1995). P_{2y} receptors localized in the basolateral membrane of strial marginal cells and P_{2U} receptors in both the apical and the basolateral membrane participate in the regulation of K^+ secretion (Liu, Kozakura, and Marcus 1995).

3.3 Calcium Homeostasis

Most biochemical pathways, transduction mechanisms, or cell-to-cell signaling systems are regulated in one form or another by the intracellular concentration of free Ca^{2+}. Small and rapid fluctuations from resting levels differentially modulate a variety of cellular processes. An effective regulation of cellular processes by Ca^{2+} requires in turn that its concentration be strictly controlled. Moreover, excess Ca^{2+} can overstimulate enzymatic activities, leading to, among other things, generation of free radicals (for a discussion of free radicals, see Section 7), membrane damage, and DNA fragmentation. The end result may be Ca^{2+}-mediated cell death (Farber 1990). Therefore, cells have developed highly efficient ways to regulate intracellular Ca^{2+} levels. One level of regulation resides at the plasma membrane where influx is controlled by Ca^{2+} channels and export is accomplished by Na^+–Ca^{2+} exchangers and Ca^{2+} pumps. A second level of control is associated with intracellular Ca^{2+}-binding proteins and Ca^{2+}-storing organelles.

3.3.1 Calcium Transport

Ca^{2+} channels regulate the influx of Ca^{2+} into cells along its electrochemical gradient (Fig. 3.2). Voltage-grated channels are selectively permeable, whereas stretch-activated channels primarily provide a nonspecific entry pathway. Ligand-gated channels that are also nonspecific Ca^{2+} entry pathways are associated with the action of neurotransmitters, particularly glutamate acting on N-methyl-D-aspartate (NMDA)-receptors and acetylcholine acting on nicotinic receptors. These channels are part of the processes that elevate intracellular Ca^{2+} in response to external stimuli and are discussed in Kros, Chapter 6, and Sewell, Chapter 9. Two homeostatic mechanisms operate at the plasma membrane of most cells to restore or maintain low Ca^{2+} levels. These are a Ca^{2+}-ATPase and an electrogenic Na^+-Ca^{2+} exchanger. Both of these mechanisms are documented in cochlear outer hair cells (Ikeda et al. 1992b; Schulte 1993).

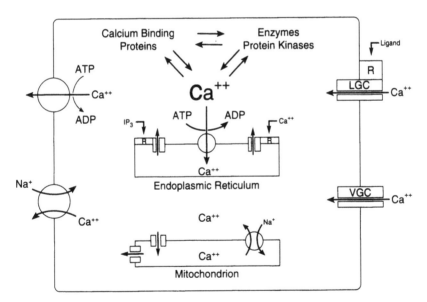

FIGURE 3.2 At the plasma membrane, ligand-gated channels (LGC; R = receptors) and voltage-gated channels (VGC) are entry pathways for Ca^{2+}, while ATPases (pump) and antiporters are exit mechanisms. Free Ca^{2+} inside the cell is in equilibrium with Ca^{2+} bound to calcium-binding proteins or sequestered in mitochondria or specific storage organelles associated with the endoplasmic reticulum. Sequestration into these storage sites is promoted by an ATPase and release is promoted by IP_3 acting on IP_3 receptors or by Ca^{2+} itself acting on ryanodine receptors.

3.3.2 Calcium-Binding Proteins

Although physiological regulation is mediated by intracellular free Ca^{2+}, the majority of intracellular Ca^{2+} exists in bound form. Only about 0.01% of total Ca^{2+} may be available in free ionic form in most cell types. The concentration of intracellular free Ca^{2+} is about 60–200 nM in isolated outer hair cells from the guinea pig cochlea (Ashmore and Ohmori 1990; Dulon, Zajic, and Schacht 1990; Ikeda et al. 1992b). In contrast, total Ca^{2+} in most mammalian cells is in the millimolar range. Sequestration in Ca^{2+}-storing organelles and binding to specific Ca^{2+}-binding proteins accounts for this difference.

Ca^{2+}-binding proteins participate in Ca^{2+} buffering, transport, and modulation of enzyme activities. Binding proteins in mammalian hair cells include the Ca^{2+}-buffering proteins calbindin (Legrand et al. 1988; Slepecky and Ulfendahl 1993), calretinin (Dechesne et al. 1993), calsequestrin (Slepecky and Ulfendahl 1993), and parvalbumin (Eybalin and Ripoll 1990). Of the Ca^{2+}-sensing proteins that may activate enzymes and Ca^{2+}-dependent processes, calmodulin and S-100, but not caldesmon, are present in the cochlea (Slepecky and Ulfendahl 1993; Foster et al. 1994). The sensory cells appear to be major sites of localization of Ca^{2+}-binding proteins, although the supporting cells of the organ of Corti may also contain some of them (e.g., calretinin: Dechesne et al. 1993; S-100: Foster et al. 1994). Whereas calretinin and parvalbumin are preferentially localized to inner hair cells, calsequestrin is found in outer hair cells only. This is interesting in view of the presumed function of calsequestrin as a major Ca^{2+} storage protein in skeletal and smooth muscle and its general absence from neuronal cell populations.

3.3.3 Calcium Stores (see Fig. 3.2)

Ca^{2+} stores are considered morphologically and functionally distinct entities from cytosolic Ca^{2+}-binding proteins. Ca^{2+}-storing organelles have a much higher Ca^{2+} buffering capacity and actively compartmentalize Ca^{2+} by an energy-dependent mechanism catalyzed by an intracellular Ca^{2+}-ATPase. The release of Ca^{2+} from these stores is receptor-mediated and permits fast and controlled regulation of intracellular free Ca^{2+}. Channels in their membranes allow efflux of Ca^{2+} along its concentration gradient when opened by the action of second messengers such as inositol trisphosphate (IP_3-sensitive stores) or Ca^{2+} itself (Ca^{2+}-dependent Ca^{2+} release from ryanodine-sensitive stores) (Berridge 1993). IP_3-sensitive stores can be responsible for the mediation of signals from neurotransmitters, neuromodulators, or hormones; the ryanodine receptor may serve as an amplification mechanism of voltage-gated or ligand-gated Ca^{2+} entry by promoting Ca^{2+}-induced Ca^{2+} release.

The existence of ryanodine receptors in outer hair cells was implied by the ability of caffeine to raise intracellular Ca^{2+} levels (Ashmore and Ohmori 1990). Other pharmacological evidence, however, suggests that caffeine solutions may affect outer hair cells osmotically and thus raise intracellular Ca^{2+} by a receptor-independent mechanism (Skellet et al. 1995). No further morphological or functional information is currently available on ryanodine-sensitive Ca^{2+} stores in the cochlea.

IP_3-sensitive Ca^{2+} stores are present in several cell types of the cochlea. Outer hair cells (Ashmore and Ohmori 1990; Ikeda et al. 1991), inner hair cells (Dulon, Mollard, and Aran 1991), and supporting cells (Ashmore and Ohmori 1990; Dulon, Zajic, and Schacht 1993) respond to the purinergic agonist ATP with an elevation of intracellular Ca^{2+} independent of the presence of extracellular Ca^{2+}. Since ATP is coupled to metabotropic receptors and IP_3 formation in the cochlea, this intracellular Ca^{2+} release argues for the existence of IP_3-sensitive Ca^{2+} stores. The evidence is equivocal whether acetylcholine can elevate intracellular free Ca^{2+} by a similar metabotropic action in the cochlea (Shigemoto and Ohmori 1990). At least one study suggested that elevated intracellular free Ca^{2+} is associated with a nicotinic-type acetylcholine receptor channel (Doi and Ohmori 1993). Another study showed that free Ca^{2+} increases after exposure to carbachol but not similar concentrations of acetylcholine (Nilles et al. 1994), and yet another found no changes in intracellular Ca^{2+} in response to cholinergic agonists (Ikeda et al. 1991). A confounding factor possibly accounting for some of the discrepancies is that both ATP (via P_{2x} receptors) and acetylcholine (via α_7-receptors) may elevate intracellular Ca^{2+} by gating of nonspecific ion channels.

The subcellular localization of Ca^{2+} stores in cells of the cochlea is uncertain. Ca^{2+} stores are characterized by the presence of the IP_3 (or ryanodine) receptor, Ca^{2+}-ATPase, and calsequestrin or a functionally related Ca^{2+}-binding protein. Calciosomes, Ca^{2+} stores in nonmuscle cells, contain calsequestrin-like proteins and have been proposed to serve as IP_3-sensitive Ca^{2+} stores (Volpe et al. 1988; Henson et al. 1989). In outer hair cells, the localization of IP_3-sensitive Ca^{2+} stores is of particular interest because of the Ca^{2+}-dependent slow motility of these cells (see Holley, Chapter 7). There has been much speculation that the subsurface cisternae represent reservoirs of Ca^{2+}, since they morphologically resemble the sarcoplasmic reticulum of muscle cells and are localized in proximity to the actin network of outer hair cells (Flock, Flock, and Ulfendahl 1986). The concept of muscle-like, actin-mediated outer hair cell contraction may not, however, be strictly applicable. Detailed morphological comparisons show significant differences between the potential contractile elements of these two cell types (see Slepecky, Chapter 2), and muscle-like isoforms of actin appear to be absent in outer hair cells (Slepecky and Savage 1994). Also, although the levels of free Ca^{2+} (Yamashita et al. 1990) and bound Ca^{2+} (Ikeda and Takasaka 1993) and the activity of Ca^{2+}-ATPase (Schulte 1993) are high along the lateral wall of the outer hair cells, the light microscopic resolution achieved in these studies cannot rule out localization to the cisternae or the plasma membrane.

In contrast, calsequestrin appears to be distributed throughout the cytoplasm in outer hair cells in a staining pattern suggestive of its presence in calciosomes (Slepecky and Ulfendahl 1993). If calsequestrin is accepted as an indicator of IP_3-sensitive Ca^{2+} stores, this rules out the subsurface cisternae as major Ca^{2+} storage sites. However, the association of calsequestrin with calciosomes in non-muscle cells is controversial (Rossier and Putney 1991), and a calsequestrin-like protein, calreticulin, has been identified as another major Ca^{2+}-binding protein of the endoplasmic lumen (Milner et al. 1991). Its presence in the cochlea has not yet been reported. Another argument against localized Ca^{2+} stores is the diffuse distribution of total Ca^{2+} in outer hair cells, which correlates with calcium-binding proteins but not with the subsurface cisternae (Slepecky and Ulfendahl 1993). Thus, although Ca^{2+} stores are undoubtedly present in hair cells, their precise location remains uncertain.

Mitochondria have long been known for their capacity for Ca^{2+} retention, particularly under pathological conditions (Somlyo 1984). Direct measurements of mitochondrial free Ca^{2+} recently demonstrated that this pool was able to follow rapid cytosolic Ca^{2+} oscillations triggered by release from IP_3-sensitive stores (Rizzuto et al. 1993). Thus, it is possible that the numerous mitochondria along the lateral wall of the outer hair cells contribute to local Ca^{2+} micro-homeostasis.

3.4 The Physiological Role of Calcium

Second messengers and Ca^{2+} mediate both visual and olfactory transduction. However, reactions involving second messengers have relatively long time courses and are not suited for auditory transduction in which stimuli as fast as 20 kHz or more are processed. Although acoustic stimulation does affect the metabolism of second messengers in the cochlea (Ono and Schacht 1989), the targets of second messenger action may be homeostatic mechanisms and the modulation of the transduction process rather than mechanoelectric transduction proper. The discussion of free Ca^{2+} in outer hair cells has, in recent years, centered around its putative role in the slow shape changes, or "slow motility," of these cells (see Holley, Chapter 7). Although such a role seems accepted, the source of this Ca^{2+} has been established. Furthermore, regulation of slow motility cannot be the only role of Ca^{2+} in hair cells. Regulation by Ca^{2+} is multifaceted, and protein phosphorylation, pH, and volume regulation are influenced by Ca^{2+} in hair cells, as discussed below. Differential activation of these processes is accomplished by temporal and spatial restrictions of the Ca^{2+} signal and its amplitude. First, enzymes and binding proteins have different affinities for Ca^{2+} and are therefore activated at different free Ca^{2+} concentrations.

Second, the effective free diffusion range of Ca^{2+} before it is buffered by binding proteins is 0.1 μm (Allbritton, Meyer, and Stryer 1992). Its effectiveness is therefore highly local. Thus, it must be expected that Ca^{2+} originating from different sources (e.g., Ca^{2+} channels or intracellular-stores) acts on different targets. In this vein, it has been postulated that Ca^{2+} entering the hair cell via voltage-gated channels constitutes a pool distinct from Ca^{2+} released from intracellular stores (Dulon and Schacht 1992). Neurotransmitters and second messengers may thus regulate cytoplasmic enzyme activities (energy metabolism, protein phosphorylation, channel activities), while Ca^{2+} influx through voltage-gated channels could trigger motile responses of the cortical skeleton (Dulon and Schacht 1992; Niedzielski, Ono, and Schacht 1992).

4. Intracellular Ion and Fluid Balance in the Organ of Corti

4.1 Sodium and Potassium Balance

The intracellular ion composition and pH maintain optimal conditions for cellular metabolism, stabilization of proteins and DNA, maintenance of membrane potential, transport processes, maintenance of cell volume and cell shape, and ability to respond to environmental stimuli. Within certain margins, deviations from normal conditions can be tolerated by cells and balanced by homeostatic mechanisms. In fact, small fluctuations around cellular resting conditions constitute the basis of many biological signaling mechanisms. Transduction, action potentials, neurotransmission, and regulatory processes are accompanied by rapid changes in intracellular ions, mediated through fluxes down their electrochemical gradients regulated by specific ion channels (see Kros, Chapter 6).

A cell may expend a major portion of its metabolic energy to maintain transmembrane gradients of the electrolytes K^+, Na^+, Ca^{2+}, and Cl^-. In general, the cytosolic K^+ concentration is above its electrochemical equilibrium (145 mM cytosolic vs 5 mM extracellular), whereas cytosolic Na^+ is below electrochemical equilibrium (5 mM vs 150 mM). Cytosolic Cl^- concentrations are more variable and may be at equilibrium in some cell types (e.g., muscle) and above the electrochemical equilibrium in others (e.g., vestibular dark cells; >100 mM cytosolic vs 150 mM extracellular). A steeper gradient exists for free Ca^{2+}. Its intracellular concentration (nanomolar) is 10^4 to 10^5 times lower than its extracellular concentration (2-5 mM). The exact concentrations of ions vary by type of cell and extracellular fluid. Furthermore, a cell is not a homogeneous entity: cellular organelles may maintain their own homeostasis and create a milieu sometimes strikingly different in ionic composition and pH from the cytoplasm. For example, calcium-storing organelles actively sequester and concentrate Ca^{2+}, and lysosomes enclose an acidic environment approximately three pH units lower than the cytoplasm.

Ions are transported either by ATPases energized by hydrolysis of ATP ("ion pumps" for Na^+-K^+, H^+, Ca^{2+}) or by symporters or antiporters (also known as cotransporters or exchangers), which derive their energy for uphill transport of one ion from the electrochemical gradient of another (Na^+-H^+, K^+-H^+, Cl^--HCO_3^-, Na^+-$2Cl^-$-K^+). Carrier-mediated transport coupled to the Na^+ gradient is also common for amino acids, neurotransmitters, and, in some tissues, glucose. Since the establishment of an electrochemical gradient requires energy, ATP is ultimately involved in all forms of ion transport. The primary active transport system in most animal cells is a Na^+,K^+-ATPase that removes Na^+ from the cell and imports K^+, usually at a Na^+/K^+ ratio of 3:2. In the cochlea, Na^+,K^+-ATPase is acutely regulated by the intracellular Na^+ and the extracellular K^+ concentrations and by the cytosolic pH (Kuijpers and Bonting 1969). Its long-term (from minutes to days) modulation is under hormonal control, most notably by adrenal corticosteroids (Rossier, Geering, and Kraehenbuhl 1987).

Although active ion transport is necessary in every cell, its magnitude may vary. For example, an argument can be made that hair cells require only little Na^+,K^+-ATPase activity by virtue of their unique exposure to extracellular fluids of vastly different composition (endolymph and perilymph; see Section 6.2). In cells surrounded by interstitial fluid, homeostasis is challenged by the outward leakage of K^+ and the inward flux of Na^+. In hair cells, however, the outward flux of K^+ across the lateral membrane may be balanced, at least in part, by the inward flux of K^+ across the apical surface, which is exposed to the K^+-rich endolymph. Thus, the "standing current" (see discussion in Section 6) in the cochlea may not impose large energy requirements on hair cells, since it is carried by K^+ ions that are replenished by entry from the endolymph. The energy is contributed by the stria vascularis, which maintains the endolymph composition and the endocochlear potential. In fact, the existence of Na^+,K^+-ATPase in hair cells had long been debated. Nevertheless, hair cells are exposed to the driving force of a Na^+ gradient (Sellick and Johnstone 1975; Ikeda and Morizono 1990) and these ions must be removed from the cell by a Na^+,K^+-ATPase. Measurements in isolated outer hair cells of membrane potentials and intracellular Na^+ and K^+ concentrations have recently established the presence of Na^+,K^+-ATPase (Sunose et al. 1992).

4.2 pH Regulation

Cell metabolism is highly pH-dependent. Without adequate pH control, severe metabolic disturbances may arise. Intracellular and extracellular pH usually do not differ greatly in multicellular organisms, but a cell may be subjected to pH fluctuations both by its own metabolism (production and utilization of protons) and by the extracellular environment (for example, metabolic or respiratory acidosis or alkalosis). Intrinsic passive buffering capacities of cells are limited by the concentration of physiological buffers such as bicarbonate ($HCO_3 \leftrightarrow OH^- + CO_2$) and phosphate ($H_2PO_4^- \leftrightarrow H^+ + HPO_4^{2-}$) and vary from one cell type to another. Therefore, active transport of ions is required in the maintenance of intracellular pH.

A cell has at least two mechanisms to balance its pH: the export of protons (H^+) to lower its acidity (increase pH) and the export of hydroxyl ions (OH^-) to increase its acidity (decrease pH). This is accomplished by transporter-mediated processes. Protons are carried by a Na^+-H^+ exchanger or H^+-monocarboxylate cotransporters and alkaline equivalents are removed by a Cl^--HCO_3^- exchanger. The Na^+-H^+ antiporter is sensitive to intracellular pH: it is activated by an increase in protons, thus promoting their export. In the Cl^--HCO_3^- antiporter, HCO_3^- is the carrier of the hydroxyl ion, since HCO_3^- is in equilibrium with CO_2 + OH^-. This equilibrium is facilitated by carbonic anhydrase, a ubiquitous cellular enzyme that is present in high concentrations in lateral wall tissues (Erulkar and Maren 1961) and hair cells (Lim, Karabinas, and Trune 1983).

In outer hair cells, acid–base regulation has been studied in detail with the use of a fluorescent pH indicator (Ikeda et al. 1992a). The dynamic changes following manipulation of the internal pH and the effect of pharmacological blockers of transporters and enzymes were consistent with the generally established features of cellular pH regulation. CO_2 entering the hair cell or produced by oxidative metabolism is in rapid equilibrium with HCO_3^-. The extrusion of HCO_3^- and thereby of OH^- is accomplished by a HCO_3-Cl^- exchanger. Conversely, intracellular H^+ can be regulated and exported by the Na^+-H^+ exchanger. The combination of these reactions balances the intracellular pH in outer hair cells.

4.3 Volume Regulation

Cell volume is linked to ion transport, since small inorganic ions are the main osmolytes in the cytosol. Acute osmotic stress is relieved by ion movements through specific channels and transporters (Eveloff and Warnock 1987) such that extracellular osmolarity will regulate ion transport via cell volume as a second messenger. Cell swelling during hypo-osmotic challenges can be counteracted by a regulatory volume decrease through extrusion of KCl via parallel K^+ and Cl^- channels or a KCl symporter. Conversely, cell shrinkage during hyperosmotic challenges is voided by regulatory volume increases through uptake of NaCl via Na^+-Cl^--K^+

cotransporter, parallel $Cl^--HCO_3^-$ and Na^+-H^+ exchangers, or Na^+-Cl^- symporter. In vestibular dark cells, regulatory volume decreases are mediated through apical K^+ and basolateral Cl^- channels (Shiga and Wangemann 1995), whereas regulatory volume increases involve the $Na^+-Cl^--K^+$ cotransporter (Wangemann and Shiga 1994a).

The regulation of cell volume has received special attention in outer hair cells. Their function as modulators of the transduction process is linked to their mechanical characteristics and their influence on the micromechanics of the basilar membrane. Outer hair cells can respond to hypotonic media with a regulatory volume decrease, but the channels involved have not been precisely delineated. K^+ and Cl^- channels were postulated to contribute to volume regulation in outer hair cells subjected to tetanic stimulation (Ohnishi et al. 1992). This has yet to be reconciled with the fact that under isosmotic conditions there is no evidence for Cl^- conductance in the plasma membrane of the outer hair cell (Sunose et al. 1992). On the other hand, intracellular Ca^{2+} seems to be involved in osmotic regulation, as it is in other cell types (Eveloff and Warnock 1987). During cell swelling in a hyposmotic solution, intracellular Ca^{2+} is elevated (Harada, Ernst, and Zenner 1993), presumably by entry through stretch-activated Ca^{2+} channels. An increase in the internal concentration of Ca^{2+} reduces the cell volume in response to hypotonic shock (Dulon, Zajic, and Schacht 1990). This is consistent with the facts that stretch-activated Ca^{2+} channels exist in outer hair cells (Ding, Salvi, and Sachs 1991) and that gadolinium ions (Gd^{3+}), which inhibit stretch-activated channels, block the regulatory response to hypotonic challenge (Crist, Fallon, and Bobbin 1993; Harada, Ernst, and Zenner 1993).

5. Intercellular Communication in the Organ of Corti

Both homeostasis and stimulus processing in a sensory system depend on integrated cell activities. Homeostatic interactions between different cell types and tissues in the generation of endolymph are discussed in Section 6. Cell communication during stimulus processing from mechanoelectrical transduction to neurotransmission and efferent feedback are covered in Kros, Chapter 6; Holley, Chapter 7; Guinan, Chapter 8; and Sewell, Chapter 9 (see also Eybalin 1993 for a review). In addition, both structural and functional interactions between supporting cells and between supporting and sensory cells contribute to the microhomeostasis of the organ of Corti.

5.1 Cell-Cell Interactions

A variety of specialized proteins participate in the establishment of connections of cells to one another and to the surrounding extracellular matrix.

Among them are collagens, which constitute the major matrix proteins, and cell-adhesion molecules (CAMs) such as integrins or cadherins. The glycocalix, the surface coat of cells consisting of glycolipids, glycoproteins, and proteoglycans, has frequently been probed by staining reactions at the light microscopic and electron microscopic level in the organ of Corti (see discussions by Santi and Anderson 1987; Plinkert, Plinkert, and Zenner 1992). The glycoprotein laminin may promote the attachment of epithelial cells to basement membranes by preferentially binding to type IV collagen. Fibronectin, a cell-coat glycoprotein with high-affinity binding sites for all types of collagen, is also present in the inner ear (Santi et al. 1989). Cadherins mediate Ca^{2+}-dependent cell adhesion during development and in adult tissues. Epithelial cadherin (E-cadherin) is present in the adherens-type junctional complexes between supporting cells in the organ of Corti of the mouse (Whitlon 1993). Another cadherin, ACAM, participates in the formation of junctions both between adjacent supporting cells and between supporting and sensory cells in the chick (Raphael et al. 1988). Adherens junctions in the reticular lamina may provide major support for its rigidity, and together with "tight" junctions they may function as a barrier between the endolymphatic and perilymphatic spaces (see Slepecky, Chapter 2).

Tight junctions, however, do not provide for information exchange. It is gap junctions that afford such intimate contact between cells by acting as channels for the exchange of small molecules between the cytoplasm of adjacent cells (see Slepecky, Chapter 2). These membrane structures are a significant pathway for the exchange of nutrients, transmission of electrical signals, and coordination of physiological responses between cells (Loewenstein 1981).

The existence of gap junctions has been documented in the cochlea based on their characteristic morphology (Jahnke 1975; Gulley and Reese 1976; Slepecky, Chapter 2) and protein content (Kikuchi et al. 1995). Their presence between adjacent supporting cells, radially and longitudinally, would promote extensive communication. Transfer of dyes, a method to visualize cell coupling, indicates functional gap junctions between the Hensen and Deiters cells in isolated organ of Corti preparations from the guinea pig (Santos-Sacchi 1986). Low resistance between supporting cells corroborates the concept of electrical coupling (Oesterle and Dallos 1989, 1990). Additionally, coupling between outer hair cells and supporting cells has been proposed for the gerbil cochlea based on dye transfer and electrical recordings in situ (Zwislocki et al. 1992). This issue, however, is controversial, since morphological studies in different species have consistently shown that there are no gap junctions associated with hair cells in the inner ear.

The importance of gap junctions for the homeostasis of the organ of Corti remains speculative. Buffering of K^+, metabolite exchange, or control of supporting cell tonus are all feasible but remain to be experimentally supported.

5.2 Neurotransmitter and Potassium Buffering

The participation of supporting cells in excitatory transductional events between neurons and glia is well documented. The flux of ions and transmitter molecules during cell depolarization and neurotransmission generates a transient homeostatic imbalance not only in the sensory cells proper but also in their immediate environment. This imbalance is primarily created by the extracellular presence of the released neurotransmitter and of an increased concentration of extracellular K^+, which carries the current in excitable cells. Both need to be efficiently removed to terminate synaptic events, prevent a lateral spread of excitation, and restore excitability. Neuronal and glial cells cooperate in these mechanisms.

In the cochlea, the neurotransmitters or neurotransmitter candidates are glutamate (afferent, excitatory), α-aminobutyric acid (GABA; efferent, inhibitory), acetylcholine (efferent), and ATP (afferent excitatory or efferent). Although other neuroactive substances have also been identified, little is known about their physiology and biochemistry (Eybalin 1993; see Sewell, Chapter 9). Neurotransmitters, with the exception of acetylcholine, are removed from the synaptic cleft by high-affinity uptake into the releasing cells as well as the surrounding cells. High-affinity uptake of both GABA and glutamate was confirmed in the cochlea by radioautography after application of radioactively labeled compounds (Gulley, Fex, and Wenthold 1979; Schwartz and Ryan 1983). GABA uptake mechanisms appear to be widespread and not restricted to fibers expressing GABA and its synthesizing enzyme, glutamate decarboxylase (Eybalin et al. 1988). However, this widespread distribution would assure effective clearance of GABA from extracellular spaces.

Glutamate removal is more complex, as it may involve a glutamate–glutamine cycle between neurons and glia that serves both to terminate the action of the transmitter and to replenish it, as suggested for the central nervous system (Fonnum 1991). Radiotracer studies in the cochlea are consistent with uptake into the efferents and a glutamate–glutamine cycle operating between the inner hair cells and the glial cells in the osseous spiral lamina (Eybalin and Pujol 1983; Ryan and Schwartz 1984). In this cycle, glutamate would be removed by glial cells, where it is metabolized to glutamine. Glutamine, in turn, is released and then taken up by inner hair cells, where it is converted back to glutamate. However, it has been argued (Eybalin 1993) that the distance between hair cells and glial cells in the cochlea makes this cycle an unlikely mechanism for rapid glutamate removal from the synapse. Rather, glial cells may be involved in a glutamate–glutamine cycle removing glutamate from perilymph and thus protecting the cochlea from the excitotoxicity of excess extracellular glutamate.

Elevations of extracellular K^+ occur in the vicinity of excitable cells upon depolarization, and maintenance of a constant extracellular K^+ is essential to prevent deleterious effects on neuronal function. The spread of an increased K^+ concentration is prevented by "K^+ buffering," a combination of diffusion, active uptake, and transport into glial cells. Uptake into glial cells includes activation of a glial Na^+,K^+-ATPase by extracellular K^+. Spatial K^+ buffering can be accomplished by uptake at sites of high concentration and discharge at distant regions of lower K^+ concentration (Katzman 1976). In the cochlea, both the standing and the transduction currents are primarily carried by K^+. In fact, K^+ concentrations are increased in the organ of Corti following exposure to intense sound (Johnstone et al. 1989). They can double in the vicinity of the inner hair cells, whereas increases are less pronounced in the tunnel of Corti, perhaps reflecting diffusion into the surrounding fluids. K^+ concentrations quickly return to normal following the cessation of the stimulus, indicating the presence of uptake systems of K^+ into surrounding cells. It remains to be established whether a Na^+,K^+-ATPase participates in analogy to glial mechanisms and where it would be located. A pathway for spatial K^+ buffering may also exist, since elevation of perilymphatic K^+ stimulates K^+ secretion by the stria vascularis (Salt and Ohyama 1993).

6 Inner Ear Fluids and Their Role in Transduction

6.1 The Standing Current

When von Békésy (1950) first measured the DC potentials within the cochlea, he described a positive potential in the endolymphatic space and a negative potential inside the organ of Corti. The presence of such potentials would drive a circulating current, an idea that essentially laid the basis for Davis's mechanoelectrical "battery theory" of cochlear transduction (Davis 1957). This theory postulates that the acoustic stimulus itself does not need to generate the energy for transduction. Rather, the energy is provided by a standing current flowing through the hair cells. Transduction will only change the resistance of the hair cells and thereby the current flow (Johnstone and Sellick 1972; Dallos 1973). This places the source of metabolic energy for transduction in the tissues that maintain the current and establishes a unique relationship between the sensory cells and the supporting tissues in the cochlea.

The stria vascularis is believed to be solely responsible for the generation of the standing current, since it secretes K^+ into the endolymph and generates the large potential across the epithelia lining the scala media (see Section 6.5). This transepithelial potential of about +80 mV (endolymphatic side positive) is the endocochlear potential, and it drives the standing current in conjunction with the negative membrane potential of the hair cells and the steep gradient of K^+ between endolymph and perilymph.

The standing current generated by the stria vascularis flows radially through the scala media toward two current sinks (Fig. 3.3). One part of the

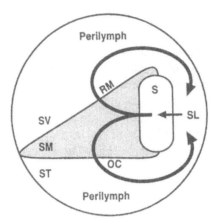

FIGURE 3.3 The standing current is generated by the stria vascularis (S). One part flows from the scala media (SM) through the organ of Corti (OC) into the scala tympani (ST). The other part flows through Reissner's membrane (RM) into the scala vestibuli (SV). Both branches of the current return radially via the spiral ligament (SL) to the stria vascularis.

current flows through the organ of Corti containing the sensory hair cells, where it enters the scala tympani. The other part flows through Reissner's membrane where the current crosses into the scala vestibuli. Both branches of the current return radially via the spiral ligament to the stria vascularis (Zidanic and Brownell 1990). The spatial separation of the current source in the stria vascularis from the sensory cells by almost half a millimeter may be a prerequisite for the high sensitivity of the auditory system. This arrangement attenuates the noise originating from blood flow in the highly vascularized stria.

Between the current source and the two sinks, the standing current flows through the inner ear fluids. The composition of the inner ear fluids is crucial, since they provide the charge carrier and the ionic milieu for the transduction process. Furthermore, they serve as a reservoir of metabolic substrates for the surrounding tissues (Section 3).

6.2 Composition of Inner Ear Fluids

Three distinct extracellular fluids have been identified in the inner ear: endolymph, perilymph, and intrastrial fluid (Fig. 3.3; see also Slepecky, Chapter 2). In the cochlea, endolymph is in contact with the tectorial membrane and the apical portion of the sensory cells, and provides the environment of the transduction channel. Perilymph surrounds the basolateral membrane of the sensory cells and the synapses and nerve terminals of both the afferent and the efferent innervation. The salient feature of inner ear fluid composition is the high K^+, low Na^+, and low Ca^{2+} content of endolymph (Table 3.1). Endolymph is in fact the only extracellular body fluid of this composition. In contrast, the ionic contents of perilymph and intrastrial fluid largely conform to those of other extracellular fluids (Tables 3.2, 3.3, 3.4).

Endolymph is also hyperosmotic to perilymph by 11–40 mOsmol/kg H_2O (Konishi, Hamrick, and Mori 1984; Sterkers, Ferrary, and Amiel 1984; Sziklai et al. 1992). Furthermore, the protein and amino acid content differs quantitatively between endolymph and perilymph. The amino acid profiles have been important in determining the relationship of these fluids to one another and to plasma and cerebrospinal fluid, as will be discussed below.

Perilymph itself is not uniform throughout the cochlea. This seems remarkable, since the two perilymphatic scalae are connected via the helicotrema at the apex and via the extracellular spaces in the spiral ligament over the length of the cochlea. Measurements of radial communication between the scala vestibuli and the scala tympani indeed suggest that the two fluid spaces are essentially a single compartment, at least in the

TABLE 3.1 Composition of cochlear and vesticular endolymph.

Component	Cochlear endolymph	Vestibular endolymph
Na^+	1.3	9
K^+	157	149
Ca^{2+}	0.023	0.197
Cl^-	132	n.a.
HCO^{3-}	31	n.a.
Aspartate	0.015	0.014
Glutamate	0.053	0.055
Glutamine	0.054	0.036
Glycine	0.071	0.027
Alanine	0.040	0.028
Glucose	0.6	n.a.
Urea	4.9	n.a.
Protein (mg/100 ml)	38	n.a.
pH	7.4	n.a.

Values in Tables 3.1–5 are averages of data obtained in guinea pigs and rodents, assembled from various references. No major differences are assumed to exist among mammalian species. The tables are not exhaustive: not all constituents have been listed. The selected amino acids are the most abundant in either endolymph or perilymph. According to convention, values in the tables are "activities" for Ca^{2+}, Mg^{2+}, and pH, but concentrations for other constituents. The activity of an ion indicates its unbound, dissociated ionic form, i.e., the form that is generally available for biological reactions and thus determines its physiological activity. In physiological solutions, only a small discrepancy exists between the activity and the total concentration of Na^+, K^+, or Cl^-. However, the activity of divalent cations is frequently much smaller than their concentration, due to chelation by organic substances as discussed for free vs bound Ca^{2+} in Section 3.3. Values (except for protein) are millimolar (mM).

Total volume of cochlear endolymph is approximately 2 µl (guinea pig: Salt and Thalmann 1988b).

n.a. = data not available.

TABLE 3.2 Composition of the perilymphs of scala vestibuli and scala tympani. For general comments, see Table 3.1.

Component	Perilymph of scala vestibuli	Perilymph of scala tympani
Na^+	141	148
K^+	6.0	4.2
Ca^{2+}	0.6	1.3
Cl^-	121	119
HCO_3^-	18	21
Aspartate	0.003	0.002
Glutamate	0.008	0.007
Glutamine	0.645	0.625
Glycine	0.289	0.322
Alanine	0.367	0.321
Glucose	3.8	3.6
Urea	5.2	5.0
Protein (mg/100 ml)	242	178
pH	7.3	7.3

Total volume of cochlear perilymph is approximately 16 μl (guinea pig: Salt and Thalmann 1988b).

TABLE 3.3 Ionic composition of the intrastrial fluid. For general comments, see Table 3.1.

Na^+	85
K^+	13
Ca^{2+}	0.8
Cl^-	55
HCO_3^-	n.a.
pH	n.a.

Data for the intrastrial fluid should be considered preliminary because of analytical difficulties related to the small volume of this compartment.

third turn of the cochlea (Salt, Ohyama, and Thalmann 1991a,b). Nevertheless, the concentrations of Na^+ and K^+ differ significantly between the perilymph of the scala tympani and the perilymph of the scala vestibuli (Table 3.2). The higher K^+ concentration in the scala vestibuli may be caused by the standing current, one part of which passes across Reissner's membrane into the scala vestibuli while the other part passes through the basilar membrane into the scala tympani (Zidanic and Brownell 1990; Salt and Ohyama 1993; Fig. 3.3). Morphologically, the return pathway for both currents via the lateral wall tissues is narrower between the scala vestibuli and the stria vascularis than between the scala tympani and the stria vascularis. This may lead to an accumulation of K^+ in the scala vestibuli (Salt and Ohyama 1993). An asymmetry of access to the lateral wall from the two perilymphatic scalae is also supported by the kinetics of K^+ diffusion. The current calculated from the flux of K^+ is quantitatively similar to the standing current through Reissner's membrane (Zidanic and

TABLE 3.4 Composition of cerebrospinal fluid and plasma. For general comments, see Table 3.1.

Component	Cerebrospinal fluid	Plasma
Na^+	149	145
K^+	3.1	5.0
Ca^{2+}	1.2	2.6
Cl^-	129	106
HCO_3^-	19	18
Aspartate	0.003	n.a.
Glutamate	0.003	n.a.
Glutamine	0.450	n.a.
Glycine	0.010	n.a.
Alanine	0.036	n.a.
Glucose	4.81	8.3
Urea	5.2	6.5
Protein (mg/100 ml)	24	4238
pH	7.3	7.3

Brownell 1990; Salt and Ohyama 1993). The origin of the difference in Na^+ concentration between the two perilymphatic scalae is not clear, but may lie in the requirement for osmotic balance.

Likewise, small gradients are present within endolymph. The concentrations of K^+, Na^+, Cl^-, and Ca^{2+} and the endocochlear potential vary longitudinally from base to apex of the cochlea (Sterkers, Ferrary, and Amiel 1984; Sterkers et al. 1984; Salt et al. 1989; Sziklai et al. 1992). One function of the ionic gradients could be to promote iontophoretic transport, for example, of Ca^{2+} from the apex of the cochlea into the vestibular labyrinth and toward the endolymphatic sac (Thalmann, Salt, and DeMott 1988; Salt et al. 1989).

The electrochemical gradients in the endolymph suggest the presence of parallel gradients in ion transport or permeability. Na^+, K^+-ATPase activity is indeed highest in the base of the cochlea (Kuijpers and Bonting 1969). It is interesting in this context that base-to-apex gradients seem rather ubiquitous in the inner ear. For example, gradations exist in the length of the sensory cells and their stereocilia, or the size of the basilar and tectorial membranes (see Slepecky, Chapter 2, Section 3.1); the innervation pattern of hair cells and the distribution of neurotransmitters change continually from base to apex (see Holley, Chapter 7); glycogen, lipid metabolism, second messengers, and other small constituents of cells vary quantitatively (Thalmann, Paloheimo, and Thalmann 1979; Niedzielski and Schacht 1991). This multitude of gradients suggests important functional consequences, which have yet to be established.

6.3 Physiologic Significance of Fluid Composition

The specific ion concentrations of endolymph and perilymph maintain structure and function in the cochlea. Variations in endolymphatic K^+,

Na^+, or Ca^{2+} affect the conformation of the tectorial membrane (Kronester-Frei 1979). Outer hair cell shape, and thereby the micromechanics of the basilar membrane, are sensitive to extracellular K^+ (Zenner, Zimmermann, and Schmitt 1985), osmolarity (Dulon, Aran, and Schacht 1987), and Ca^{2+} (Pou et al. 1991). The tip links connecting individual cilia of hair cells and believed to be associated with the transduction channel are lost in the absence of Ca^{2+} (see Kros, Chapter 6).

The function of the cochlea depends strongly on the K^+ concentration of the fluids, since the transduction current through the hair cells is mostly carried by this ion. This is due to the high K^+ and low Na^+ concentrations of the endolymph, the K^+ selectivity of ion channels in the basolateral membrane of the hair cells, and the low K^+ and high Na^+ concentrations of the perilymph. It is probable that the exquisite sensitivity of the cochlea could not be reached with, for example, Na^+ as the main charge carrier. K^+ has several distinct advantages. Entry of K^+ via the apical transduction channel and exit via the basolateral K^+-selective channels occur along its electrochemical potential. The energy requirements of the hair cells for the transduction process are therefore relatively low, and the blood supply does not need to be immediately adjacent. Furthermore, the influx of K^+ disturbs the cytosolic homeostasis very little, since the steady-state K^+ concentration in the hair cells is high (Ikeda and Morizono 1990). Finally, K^+ channels in the basolateral membrane not only serve as exit mechanisms for the charge carrier of the transduction current; they also generate a membrane potential of about -60 mV in hair cells based upon the K^+ concentration difference between cytosol and perilymph. In series with the endocochlear potential of about $+80$ mV, this large potential gradient of 140 mV drives the transduction current across the apical transduction channel.

Electrophysiological measurements confirm the critical role of ion concentrations in the inner ear fluids. Lowered perilymphatic K^+ concentrations reduce the endocochlear potential, the cochlear microphonic, and the compound action potential and increase the resistance of the cochlear duct (Konishi and Kelsey 1973; Marcus, Marcus, and Thalmann 1981). Elevated perilymphatic K^+ concentrations transiently increase the endocochlear potential and lower the compound action potential (Salt and Stopp 1979). Cochlear potentials are also affected by experimental manipulations of the endolymphatic Ca^{2+} (Marcus et al. 1983; Tanaka and Salt 1994) and the perilymphatic Ca^{2+} or Mg^{2+} concentrations (Bobbin, Fallon, and Kujawa 1991).

In spite of the importance of Ca^{2+} homeostasis for the transduction process, buffering systems for Ca^{2+} like those operating intracellularly (see Section 3.3) may not be present in inner ear fluids (Juhn and Youngs 1976; Bosher and Warren 1978; but see Ferrary et al. 1988). Conceivably, the cells bounding these compartments regulate the Ca^{2+} transport pathways to endolymph and perilymph.

6.4 Homeostasis of Perilymph

The perilymph-containing spaces of the inner ear are in contact not only with each other but also with other fluid- and air-filled compartments. The perilymph of the scala tympani is separated from the air-filled middle ear by the round window membrane and is in connection with the cerebrospinal fluid via the cochlear aqueduct. The perilymph of the scala vestibuli is separated by the stapes footplate in the oval window from the middle ear and connected to the vestibular perilymph. The complexity of the perilymph-filled spaces and the fact that the perilymph and cerebrospinal fluid are under pressure complicate the investigation of perilymph homeostasis.

Two contrasting mechanisms can be conceived for the site of generation of the cochlear fluids. One is a continuum of ion-transporting mechanisms controlling homeostasis locally along the length of the cochlear duct. The other is the generation of the fluid at a distinct site, its distribution throughout the cochlea by bulk flow, and its resorption at another distinct location. The ductlike morphology and the interconnectivity of the scalae suggest the possibility of longitudinal flow of perilymph. However, the absence of significant bulk flow of perilymph of both the scala tympani and the scala vestibuli (Ohyama, Salt, and Thalmann 1988) suggests that perilymph homeostasis is primarily regulated locally. It does not preclude that certain constituents may be secreted at one site and diffuse to another where they may be resorbed.

Four cell types or tissues of the cochlea are presently being considered as potential participants in the generation of perilymph: the stromal cells under the spiral prominence (also called the inferior portion of the spiral ligament), the stromal cells under the spiral limbus, the supralimbal zone (also called the vasculoepithelial zone), and the suprastrial region (Ishiyama, Keels, and Weibel 1970; Takahashi and Kimura 1970; Hawkins, Johnsson, and Preston 1972; Schulte and Adams 1989; Kimura, Nye, and Southward 1990). Each of these areas is characterized by the presence of blood vessels and specialized fibrocytes containing a high activity of Na^+, K^+-ATPase (Jones-Mumby and Axelsson 1984; Kerr, Ross, and Ernst 1982; Schulte and Adams 1989; Iwano et al. 1989). These features by themselves are, however, not sufficient to establish involvement in perilymph homeostasis, and further functional evidence is needed.

6.4.1 The Barrier Between Blood and Perilymph

Perilymph originates, at least in part, by transcellular transport of solutes from blood rather than by serum ultrafiltration. The kinetics of tracers entering perilymph from blood (Juhn, Rybak, and Prado 1981; Sterkers et al. 1982; Ferrary et al. 1987) and the quantitative differences between the composition of perilymph and plasma (Thalmann et al. 1992) suggest the presence of a selective barrier between these two compartments. Furthermore, characteristics of transport-mediated processes, namely specificity and competitive inhibition, have been shown for the transfer of glucose, Cl^-, and organic anions from blood into perilymph (Rybak et al. 1984; Sterkers et al. 1984; Ferrary et al. 1987). Thus, a functional "blood-labyrinth barrier" or "blood-perilymph barrier" must exist (Juhn, Rybak, and Prado 1981; Sterkers et al. 1982).

Some, but not all, properties of this barrier are similar to those of the blood-brain barrier: (1) Both barriers consist of endothelial cells lining the capillaries lacking fenestration and linked to each other by tight junctions (see Slepecky, Chapter 2). (2) The ratio of the apparent permeability to Na^+ and urea is similar (Sterkers et al. 1987). (3) Both barriers contain a facilitated uptake mechanism for D-glucose (Crone 1965; Ferrary et al. 1987). (4) Cl^- entry across both barriers is sensitive to inhibition of acetazolamide (Vogh and Maren 1975; Sterkers et al. 1984). (5) The changes in permeability of both barriers induced by several hormones are similar (Gross et al. 1981; Sokrab et al. 1988; Inamura and Salt 1992; Schmidley et al. 1992). Among the contrasts between the two barriers is the lack of astrocytes in the cochlea that contact the endothelial cells of the brain in a characteristic fashion. Some cochlear endothelial cells are, however, surrounded by one or two layers of pericytes, which may play an analogous role (Takahashi and Kimura 1970). Furthermore, the anion transport inhibitor probenecid reduces the permeability of the blood-labyrinth barrier to furosemide but has no effect on the blood-cerebrospinal fluid barrier (Rybak et al. 1984).

6.4.2 The Connection to the Cerebrospinal Fluid

The cochlear aqueduct is an anatomical connection of the scala tympani to the cerebrospinal fluid-filled subarachnoid space of the cisterna magna. It may, however, not be a functional connection, because cerebrospinal fluid does not contribute significantly to the perilymph of the scala tympani. The fact that there is virtually no flow of perilymph argues against influx of cerebrospinal fluid (Ohyama, Salt, and Thalmann 1988). Furthermore, perilymph composition differs distinctly from that of cerebrospinal fluid (Tables 3.2, 3.4). This has now firmly been established through the analysis of sufficiently small (200 nl) perilymph samples. Earlier contradictory evidence can be explained by contamination of large samples by cerebrospinal fluid (Hara, Salt, and Thalmann 1989). The low K^+ concentration in the scala tympani also does not originate from a contribution by cerebrospinal fluid. Rather, it is the result of a higher clearance rate for K^+ out of the scala tympani than out of the scala vestibuli (see Section 6.2).

6.5 Homeostasis of Endolymph

Like perilymph homeostasis, endolymph homeostasis seems to be controlled locally. It is estimated that the minimal flow measured in the scala media cannot contribute significantly to the turnover of endolymph (Salt et al. 1986; Salt and Thalmann 1988a). However, movements of solutes can be driven via diffusion along a chemical gradient or, in the case of charged substances, via electrophoresis along an electrical gradient, as has been suggested for endolymphatic Ca^{2+} in the cochlea (Salt et al. 1989).

The ionic composition of endolymph is a product of active transport processes, since, in the presence of an endocochlear potential of about +80 mV, none of the major ions are in equilibrium. The equilibrium potentials (Table 3.5) of K^+, Ca^{2+}, and H^+ are smaller than the endocochlear potential, indicating that these ions must be actively transported into the endolymph. Conversely, the equilibrium potentials for Na^+ and for Cl^- and HCO_3^- suggest that these ions must be actively transported out of the endolymph. This is corroborated by the fact that the large differences between the K^+, Na^+, and Ca^{2+} concentrations in endolymph and perilymph are reduced during anoxia, when active processes are inhibited (Salt and Konishi 1979; Konishi and Mori 1984; Ikeda et al. 1987; Komune et al. 1993).

6.5.1 The Barrier Between Endolymph and Perilymph

The anatomical barrier between endolymph and perilymph consists of the epithelia lining the scala media, the saccule, the utricle, the three semicircular canals, the endolymphatic duct, and the endolymphatic sac. Physiological evidence suggests a functional barrier for ions and organic compounds but not necessarily for water.

One indication of a functional barrier is a difference in ionic composition that requires active transport for maintenance. Additionally, glucose, proteins, and most, but not all, amino acids are present in endolymph in much lower concentrations than in perilymph (see Table 3.1), a fact that is compatible with selective transport processes operating between these two

TABLE 3.5 Equilibrium potentials between cochlea endolymph and perilymph.

Ion	E (mV)
Na^+	+123
K^+	−89
Ca^{2+}	+48
Cl^-	+2
HCO_3^-	+12
H^+	+6

Equilibrium potentials were derived from data given in Tables 3.1 and 3.2 and calculated according to the Nernst equation:

$$E = (RT/zF)\ln[i]_p/[i]_e$$

where R is the gas constant, T is the temperature in kelvin, z is the valence of the ion i, F is the Faraday constant, and $[i]_p$ and $[i]_e$ are the concentrations or activities of the ion i in perilymph and endolymph, respective.

fluids. The kinetics of the distribution of tracers between blood, cerebrospinal fluid, perilymph, and endolymph support this concept (Konishi and Hamrick 1978; Konishi, Hamrick, and Walsh 1978; Sterkers et al. 1982). In contrast, the apparent water permeability of the endolymph–perilymph barrier remains unresolved, since several different approaches did not yield unambiguous and quantitative results (Bosher and Warren 1971; Sterkers et al. 1982; Konishi, Hamrick, and Mori 1984; Thalmann, Salt, and DeMott 1988).

The existence of both a blood–perilymph and a perilymph–endolymph barrier raises the question whether the precursor of endolymph is perilymph or blood. Current evidence suggests perilymph as the source of endolymph. First, the compartmental distribution of radioactive tracers is consistent with a K^+ and Cl^- exchange occurring first between blood and perilymph and subsequently between perilymph and endolymph, rather than directly between blood and endolymph (Konishi and Hamrick 1978; Konishi, Hamrick, and Walsh 1978; Sterkers et al., 1982). Second, K^+-free perfusion of both perilymphatic scalae causes a rapid decline of the endocochlear potential, whereas K^+-free perfusion of the vasculature is ineffective (Konishi and Kesley 1973; Wada et al. 1979; Marcus, Marcus, and Thalmann 1981). Since K^+ secretion and generation of the endocochlear potential are linked, these findings also suggest that perilymph rather than blood supplies the endolymphatic K^+.

6.5.2 Reissner's Membrane

About half of the standing current generated by the stria vascularis passes through Reissner's membrane (Zidanic and Brownell 1990; Salt and Ohyama 1993). Regulation of such a large ion flux would strongly influence cochlear homeostasis. However, although several ion transporters have been demonstrated in Reissner's membrane, very little is known about the physiological aspects of ion transport in this tissue.

Reissner's membrane is an avascular structure consisting of an epithelial monolayer facing endolymph at its apical side and perilymph at its basolateral side. The perilymphatic side is covered with a layer of mesothelial cells of unknown function (Duvall and Rhodes 1967). The epithelial cells contain a Na^+, K^+-ATPase in their basolateral membrane and a Ca^{2+}-ATPase in their apical membrane (Yoshihara et al. 1987; Iwano et al. 1989). Cl^- permeability seems to be absent, at least under unstimulated conditions (Marcus 1984).

The second messenger, cyclic AMP, may be a regulator of ion transport across Reissner's membrane (Thalmann and Thalmann 1978; Zajic, Anniko, and Schacht 1983). In other epithelia, intracellular cyclic AMP can modulate ion transport, including several pathways specific for anions (Petersen and Reuss 1983; Reuss 1987; Sunose et al. 1993). Perfusion of the scala vestibuli with stimulators of cyclic AMP affects endolymphatic Cl^-

and endocochlear potential in a fashion compatible with the activation of Cl^- transport mechanisms (Doi, Mori, and Matsunaga 1990). However, the precise location and identity of the transport pathway(s) modified by cyclic AMP remain to be determined.

6.5.3 Stria Vascularis

The stria vascularis is a multilayered structure formed by three different cell types: marginal, intermediate, and basal cells (see Dallos, Chapter 1). Marginal cells are epithelial cells bordering endolymph and linked to each other by apically located "intermediate to tight" junctional complexes (Jahnke 1975). Intermediate cells stem from the neural crest, contain melanin granules in pigmented animals, and form a discontinuous layer between the marginal and the basal cells. Intermediate cells are in contact neither with endolymph nor with perilymph, but are connected via gap junctions to basal cells (Kikuchi et al. 1995). Basal cells, considered to be of mesenchymal origin, are linked to each other by gap junctions and "very tight" junctional complexes (Jahnke 1975). Furthermore, basal cells are connected via gap junctions to fibrocytes of the spiral ligament. The endothelial cells of the intrastrial vasculature are joined by tight junctions (zonulae occludentes) and devoid of fenestrae (Jahnke 1980). The tight junctions connecting marginal, basal, and endothelial cells make the intrastrial space a separate fluid compartment (Table 3.3).

The stria vascularis is considered the source of the endocochlear potential, the major driving force for the current through the sensory cells (see Section 6.1). This widely accepted hypothesis is primarily based on indirect evidence and an elegant but simple experiment. Tasaki and Spyropoulos (1959) opened the cochlear duct near the apex, drained most of the endolymph, and touched the surface of different tissues with the tip of a blunt microelectrode. The surface of the stria vascularis was the only area that showed a positive potential with respect to a remote reference electrode, suggesting that the stria vascularis was the source of the endocochlear potential.

The stria vascularis is also assumed to be the source of the high endolymphatic K^+ concentration. The kinetics of K^+ entry into endolymph are similar when the scala vestibuli or the scala tympani is perfused, suggesting the lateral wall as the route of K^+ entry, since it is equidistant from both scalae (Konishi, Hamrick, and Walsh 1978; but see Salt and Ohyama 1993). This concept was recently supported by the demonstration of K^+ secretion in strial marginal cell epithelium (Wangemann, Liu, and Marcus 1995).

Epithelial cells homologous to those in the stria vascularis exist in the vestibular system. Morphological similarities between strial marginal and vestibular dark cells have long been recognized (Kimura 1969), and these cells also share functional features (Wangemann et al. 1995a). Vestibular and cochlear endolymph, which are connected to each other via the ductus reuniens, are similar in composition (Table 3.1). Similar ion transport systems appear to be operating in both cell types, including transporters of identical molecular composition. For example, the same slowly voltage-activated K^+ channel (I_{sk} or minK) is responsible for K^+ secretion across the apical membrane in both strial marginal and vestibular dark cells (Marcus and Shen 1994; Sunose et al. 1994), and the same Cl^- channel is responsible for the Cl^- conductance in the basolateral membrane (Marcus, Takeuchi, and Wangemann 1993 Takeuchi et al. 1995). Pharmacological evidence suggests the involvement of the Na^+-Cl^--K^- cotransporter in K^+ secretion in both cell types (Marcus and Shipley 1994; Wangemann, Liu, and Marcus 1995). Furthermore, only in these two cell types is Na^+,K^+-ATPase composed of α_1-β_2 subunits, whereas in most other inner ear cells it is α_1-β_2 or α_2-β_1 (McGuirt and Schulte 1994; Schulte and Steel 1994). Thus, consideration of results from vestibular dark cells may well complement models of ion transport in strial marginal cells.

6.6 Models of Ion Transport in the Stria Vascularis

Several models for the generation of the endocochlear potential and the secretion of K^+ have been proposed, most notably the "two-pump model," the "single-pump model," and the "two-cell model." Common to all three models is the participation of the Na^+,K^+-ATPase in the basolateral membrane of strial marginal cells, based on strong experimental evidence. Anoxia rapidly depletes ATP in parallel with the decline of the endocochlear potential (Thalmann, Kusakari, and Miyoshi 1973). The specific blocker ouabain inhibits both the generation of the endocochlear potential and the secretion of K^+, providing an additional link between these processes and Na^+,K^+-ATPase activity (Konishi and Mendelsohn 1970; Konishi, Hamrick, and Walsh 1978). In vitro, generation of a transepithelial current indicative of K^+ secretion is inhibited by ouabain in strial cells (Marcus, Marcus, and Greger 1987; Marcus and Shipley 1994). This is consistent with the observation that Na^+,K^+-ATPase is present in the basolateral membrane of strial marginal and vestibular dark cells at a similar high density (Kerr, Ross, and Ernst 1982; Yoshihara et al. 1987; Iwano et al. 1989).

Models for the generation of the endocochlear potential and the secretion of K^+ differ in the cellular location of the generation of the endocochlear potential. The two-pump model and the single-pump model assume that the marginal cells are sufficient to secrete K^+ and to generate the endocochlear potential. In the two-pump model, the endocochlear potential is generated across the apical membrane of marginal cells, whereas in the single-pump model, the endocochlear potential is generated across the basolateral membrane of marginal cells. In contrast, the two-cell model assumes a cooperation between marginal cells and basal cells within the lateral wall.

The two-pump model (Fig. 3.4A) assumes that marginal cells take up K^+ and extrude Na^+ via the Na^+,K^+-ATPase located in the basolateral membrane. The model is based on the idea that the membrane potential of marginal cells is negative with respect to endolymph. This requires that K^+ be actively transported across the apical membrane. Therefore, the model postulates an electrogenic K^+ transport mechanism in the apical membrane that generates the positive endocochlear potential (Sellick and Johnstone 1975). However, this model is rather unlikely, since such a K^+-transporting ATPase has never been established for the stria vascularis (Marcus et al. 1994; but see Ferrary et al. 1993) and has essentially been ruled out for vestibular dark cells (Marcus and Shen 1994). Furthermore, the resting potential of the marginal cells is positive with respect to the endolymph (see below).

The single-pump model (Fig. 3.4B), like the two-pump model, assumes that marginal cells take up K^+ and extrude Na^+ by a Na^+,K^+-ATPase located in the basolateral membrane (Offner, Dallos, and Cheatham 1987). In contrast to the two-pump model, it assumes an Na^+ conductance in the basolateral membrane as the source of the positive endocochlear potential (Offner 1991). Also in contrast to the two-pump model, both the single-pump model and the two-cell model assume that K^+ diffuses across the apical membrane along its electrochemical gradient (Kuijpers and Bonting 1970; Johnstone and Sellick 1972). Such a diffusional pathway for K^+ has been demonstrated in the apical membrane of strial marginal cells (Sakagami et al. 1991; Sunose et al. 1994; Wangemann, Liu, and Marcus 1995) and in vestibular dark cells (Marcus, Liu, and Wangemann 1994; Marcus and Shen 1994).

K^+ secretion via this apical K^+ channel is driven by the apical membrane potential, which is up to 10 mV more positive than the endocochlear

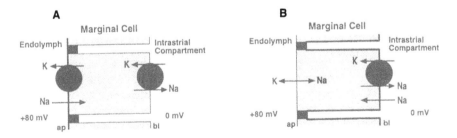

FIGURE 3.4 Single-cell models of endocochlear potential generation. (A) The two-pump model. The endocochlear potential is generated by an electrogenic K^+-ATPase across the apical membrane (ap) of marginal cells. A Na^+ conductance is present in the apical membrane, and the Na^+,K^+-ATPase is localized in the basolateral membrane (bl) of the cell. (B) The single-pump model. The endocochlear potential is generated by a Na^+ conductance in the basolateral membrane (bl) of marginal cells where the Na^+,K^+-ATPase is also present. The apical membrane (ap) possesses channels for Na^+ and K^+.

potential (Melichar and Syka 1987; Offner, Dallos, and Cheatham 1987; Ikeda and Morizono 1989). The source of the positive potential is a Na^+ conductance in the basolateral membrane inferred from the dependence of the endocochlear potential on the presence of Na^+ in the intrastrial space (Offner, Dallos, and Cheatham 1987).

The two-cell model (Fig. 3.5), like the two preceding models, assumes that marginal cells take up K^+ and extrude Na^+ by an Na^+,K^+-ATPase located in the basolateral membrane of marginal cells. Additional uptake of K^+ occurs via the Na^+-Cl^--K^+ cotransporter driven by the Na^+ gradient generated by the Na^+,K^+-ATPase. Cl^- recycles via Cl^- channels in the basolateral membrane, whereas K^+ is secreted via K^+ channels in the apical

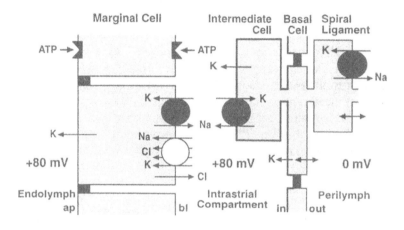

FIGURE 3.5 The two-cell model of endocochlear potential generation. The stria vascularis has two barriers, the marginal cells and the basal cells, which separate endolymph from the intrastrial compartment and the intrastrial compartment from perilymph, respectively. The two cell types are connected to each other by tight junctions. The endocochlear potential is generated by a K^+ conductance in the inner membrane of basal cells (or in intermediate cells or in both cell types) in conjunction with a high cytosolic K^+ concentration and a low K^+ concentration in the intrastrial compartment. The inner membrane of basal cells (in) is electrically coupled via gap junctions to intermediate cells, and the outer membrane of basal cells (out) is electrically coupled via gap junctions to fibrocytes of the spiral ligament. Na^+,K^+-ATPase maintains the high cytosolic K^+ concentration in fibrocytes, intermediate cells, and basal cells. K^+ uptake via the Na^+,K^+-ATPase and the Na^+-Cl^--K^+ cotransporter in the basolateral membrane (bl) of marginal cells maintains the low K^+ concentration in the intrastrial space. K^+ is secreted into endolymph via K^+ channels in the apical membrane (ap) of marginal cells. Even though there is no voltage across marginal cells, they participate indirectly in the generation of the endocochlear potential due to their involvement in the maintenance of the low K^+ concentration in the intrastrial compartment. This model is well supported in regard to ion transport in strial marginal cells due to the homology between strial marginal cells and vestibular dark cells. The model is more speculative as far as ion transport in intermediate cells, basal cells, and fibrocytes is concerned.

membrane. In contrast to the other models, the two-cell model does not assume that the endocochlear potential is generated across the membrane of marginal cells. Rather, it postulates that the endocochlear potential is generated by a K^+ conductance in the inner membrane of basal cells (Salt, Melichar, and Thalmann 1987).

The participation of the Na^+-Cl^--K^+ cotransporter in the secretion of K^+ and the generation of the endocochlear potential is suggested by several lines of evidence. The loop diuretics furosemide, piretanide, and bumetanide, which are blockers of the cotransporter, rapidly decrease the positive endocochlear potential in vivo (Kusakari et al. 1987a,b; Rybak and Whitworth 1986). In vitro, these drugs inhibit K^+ secretion and the transepithelial potential generated by both strial marginal cells (Wangemann, Liu, and Marcus 1995) and vestibular dark cells (Marcus, Marcus, and Greger 1987; Ferrary et al. 1993; Marcus and Shipley 1994), and prevent K^+ uptake into vestibular dark cells (Wangemann and Marcus 1990). Furthermore, the endocochlear potential depends on the presence of Na^+ and Cl^- in the intrastrial compartment (Shindo et al. 1992). The combination of the Na^+,K^+-ATPase and the Na^+-Cl^--K^+ cotransporter is theoretically very energy-efficient. Assuming their normal stoichiometries, 5 K^+ are taken up across the basolateral membrane for each ATPase hydrolyzed.

The Na^+-Cl^--K^+ cotransporter may be the only significant Na^+ entry pathway of marginal cells. The loop diuretic furosemide protects the ATP content of the stria vascularis during anoxia without inhibiting strial Na^+,K^+-ATPase directly (Kusakari et al. 1978a), similar to mechanisms in the thick ascending limb of the kidney (Wangemann and Greger 1990). Furosemide directly inhibits the Na^+-Cl^--K^+ cotransporter, whereas the ensuing depletion of intracellular Na^+ blocks Na^+,K^+-ATPase activity and thereby ATP consumption. This depletion of Na^+ is possible only if no other significant Na^+ entry pathway is present. Additional compelling evidence against Na^+-conductive basolateral membrane of strial marginal cells comes from recent findings that this membrane is Cl^--conductive rather than Na^+-conductive (Takeuchi et al. 1995; Wangemann, Liu, and Marcus 1995).

The absence of a major Na^+ conductance renders the one-cell model unlikely, and a different source of the endocochlear potential is required. The two-cell model assumes that the endocochlear potential is generated by a K^+ conductance in the outer membranes of basal cells (Salt, Melicher, and Thalmann 1987), and furthermore that the K^+ concentration is high in basal cells and low in the intrastrial space. This is supported by the finding of a space in the stria vascularis of low K^+ concentration but high voltage (in the range of the endocochlear potential), which may represent the intrastrial compartment (Salt, Melichar, and Thalmann 1987; Ikeda and Morizono 1989). Additional support for the generation of the endocochlear potential by a K^+ conductance comes from the sensitivity of the endococh lear potential to the K^+ channel blocker Ba^{2+} (Marcus, Rokugo, and Thalmann 1985). Even though basal cells themselves do not contain Na^+, K^+-ATPase, they are coupled by gap junctions to fibrocytes of the spiral ligament and intermediate cells high in Na^+, K^+-ATPase (Iwano et al. 1989; Schulte and Adams 1989; McGuirt and Schulte 1994; Schulte and Steel 1994; Kikuchi et al. 1995). Morphological considerations support the idea that fibrocytes in the spiral ligament are involved in ion transport (Slepecky, Chapter 2).

7. Current Developments

This discussion can only touch upon some of the topics that deserve our attention and is by no means intended to be exhaustive. Instead, it is meant as a discussion of recent developments that merit attention but that do not seem to be mature enough to be integrated into our conception of cochlear homeostatic mechanisms.

7.1 Unique Features of the Sensory Epithelium

Two major proteins specific to the organ of Corti have been identified: the organ of Corti-specific proteins 1 and 2 (OCP1 and OCP2; Thalmann et al. 1980). They are acidic proteins of molecular weight 37 and 22.5 kDa, respectively, and are present in high concentrations in the organ of Corti but essentially absent from tissues outside the ear. Their partial amino acid sequences exhibit only limited homology with known target proteins (Thalmann et al. 1993), rendering them even more likely to be unique to the inner ear. However, their significance remains elusive.

Identification of other molecules specific to the cochlea has continued with the use of refined techniques for the production of antibodies. Monoclonal antibodies have successfully been raised against specific antigens associated with stereocilia of avian and mammalian hair cells (Richardson, Bartolami, and Russell 1990; Ptok et al. 1991), with supporting cells (Zajic et al. 1991), and with membrane structures of outer hair cells (Holley and Richardson 1994). The steps from demonstrating an antigenic site to its characterization and delineation of function, however, have yet to be made.

7.2 Energy Metabolism and Blood Flow

Blood flow and the supply of oxygen and nutrients are central points in inner ear homeostasis. Although blood flow responds to vasoactive effectors (Quirk et al. 1988; Ohlsén et al. 1991; McLaren et al. 1993), the details of this regulation at the cellular level remain to be established. For example, nitric oxide, a vasodilator and modulator, dilates cochlear vessels (Brechtelsbauer, Nuttall, and Miller 1994). The mechanisms of its generation, the cells of origin, and the signals that trigger its release need to be determined in this (and other) cases to understand the physiological regulation of blood flow. Novel approaches, such as isolation of individual cell types from the cochlear vasculature (Lamm, Zajic, and Schacht 1994), will aid in the characterization of receptors, messenger systems, and effector mechanisms.

A related issue is the interaction between energy demand in the cochlea and control of blood supply. Regulation by a neural loop through sensory afferents is a possibility (Schacht and Canlon 1985). Although there is no direct evidence for such a feedback mechanism, a vasodilatory neural process indeed appears to be operating in the cochlea (Sillman et al. 1989). Vasodilation by sound-induced biochemical changes would constitute a direct local effector system, but, as detailed before (Section 2.3), documentation of changes in cochlear microcirculation induced by moderate sound exposure is scant. Thus, although there is clear evidence of mechanisms for regulation of cochlear blood flow (Brown and Nuttall 1994), it remains to be established how these mechanisms are mediated and how they are linked to acoustically evoked stimulation.

7.3 The Physiological Role of Second Messengers

Since IP_3 is involved in the regulation of intracellular free Ca^{2+}, it seems likely that some of the actions of Ca^{2+} (described in Section 3) are mediated by this second messenger. However, it remains to be established what these actions are and where they are located in the cochlea. Two sites for metabotropic (IP_3-mediated) cholinergic transmission are possible: the medial efferent synapses at the outer hair cells, and the lateral efferent axodendritic synapses on the afferent fibers innervating the inner hair cells. Biochemical evidence based on base–apex gradients of stimulated IP_3 release and selective destruction of hair cells does not rule out either site (Niedzielski, Ono, and Schacht 1992; Bartolami et al. 1993).

Speculation as to an involvement of second messengers in the slow motility of outer hair cells has come from the exposure of cells to IP_3, to which they responded with shortening (Schacht and Zenner 1987). However, several caveats apply. The experiments were conducted with detergent-permeabilized cells to allow the impermeate IP_3 intracellular access. By the same token, Ca^{2+} liberated from storage sites is no longer compartmentalized in detergent-treated cells, and may reach targets from which it is normally excluded. Thus, responses in vitro may differ significantly from the in vivo situation. Furthermore, it is unresolved whether acetylcholine can trigger contractile events in outer hair cells (see discussions in Bobbin et al. 1990; Dulon and Schacht 1992; Gitter 1992).

Likewise, there is little basis for a hypothesis of ATP action via P_{2y} receptors. A modulation of active cochlear mechanical responses through P_{2y}-like receptors has been suggested from antagonist actions on distortion product acoustic emissions (Kujawa, Fallon, and Bobbin 1994). This could involve outer hair cells or supporting cells, since P_{2y}-mediated increases of Ca^{2+} from intracellular stores have been observed in both cell types (Ashmore and Ohmori 1990; Niedzielski and Schacht 1992; Dulon, Zajic, and Schacht 1993). Furthermore, supporting cells may participate in regulating the micromechanics of the basilar membrane (Dulon, Blanchet, and Laffon 1994). Firm evidence, however, for the participation of second messengers in these processes is lacking.

What potential physiological roles remain then for second messengers in the cochlea? Obviously, many of the homeostatic functions documented in other tissues may be applicable to the inner ear, and some of them could impact on transduction mechanisms. IP_3 and other messengers may regulate voltage-dependent Ca^{2+} channels, plasma membrane-bound ion exchangers, neurotransmitter receptors, or motile responses. Finally, we should note that second messenger pathways do not operate in isolation from one another. Although for the skew of clarity we have treated these homeostatic mechanisms as individual processes, it should be borne in mind that subtleties of regulation are accomplished by "cross-talk" in which pathways influence each other's activity (Nishizuka 1992). This aspect of signal processing has yet to be addressed in the inner ear.

7.3.1 Nitric Oxide

The intercellular messenger nitric oxide (NO) has been implicated as a neurotransmitter, vasodilator, calcium regulator, and cell toxin (Moncada, Palmer, and Higgs 1991). The enzyme that produces NO, nitric oxide synthase, is present in the peripheral auditory system (Fessenden, Coling, and Schacht 1994). Enzyme activity in the lateral wall, neuroepithelium, and auditory nerve arises mainly from the neuronal type I isoform. The observation that neuronal NO synthase predominates in the peripheral auditory system is supported by NADPH-diaphorase staining of nerve endings and efferent fibers innervating the outer hair cells. Since NO is a diffusible gaseous messenger, it may link receptor activation by neurotransmitters to changes in surrounding cell populations. NO may also mediate glutamate excitotoxicity, and may thus be involved in traumatic events triggered by acoustic overstimulation or pharmacological overactivation of cochlear glutamate receptors (Pujol et al. 1993). The finding of NO synthase in the major structures of the auditory periphery implicates NO as an important signaling molecule in cochlear physiology and pathophysiology.

7.4 *Regulation of Intercellular Communication*

The physiological role of gap junctions in the organ of Corti remains speculative, although their existence is firmly established (see Section 5). The regulatory aspects of gap junctions, equally uncharted for the organ of

Corti, are interesting in view of some of the unknown second messenger functions in the cochlea. In other cell systems, the permeability of gap junctions may be under the control of Ca^{2+} (Loewenstein, Nakas, and Socolar 1967) or protein kinase C, and thus of the phosphoinositide second messenger system (Murray et al. 1993). This signaling system is active in supporting cells of the organ of Corti (see Section 3). Modification of gap junctions and regulation of cell coupling are possible functions of second messengers and of phosphoinositide signaling in particular.

Another intriguing facet of intercellular communication in the organ of Corti is the response of supporting cells to chemical stimuli. Increases in intracellular free Ca^{2+} are triggered by the putative neuromodulator ATP in Deiters and Hensen cells (Dulon, Moataz, and Mollard 1993; Dulon, Zajic, and Schacht 1993). Since Ca^{2+}-dependent motile responses are also seen in supporting cells (Dulon, Blanchet, and Laffon 1994), it is possible that modulation of cochlear micromechanics depends not only on outer hair cells but also on an integrated action of different structures on the basilar membrane.

7.5 Inner Ear Fluids

Intense acoustic stimulation transiently increases the perilymphatic K^+ concentration around the basolateral membrane of hair cells (Johnstone et al. 1989). Independently of the possibility of K^+ buffering (Section 5.2), an elevation of perilymphatic K^+ should ultimately increase K^+ secretion into endolymph. Indeed, elevation of the K^+ concentration in the scala tympani causes a large increase in the K^+ concentration of the scala vestibuli. This occurs most likely via stimulation of K^+ secretion in the stria vascularis and an increase in K^+ flux from the endolymph across Reissner's membrane (Salt and Ohyama 1993). This hypothesis is supported by the finding that K^+ secretion by strial marginal cells is stimulated by an elevation of the basolateral K^+ concentration (Wangemann et al. 1995a). It is presently not known, however, whether the K^+ concentration in the intrastrial compartment of the cochlea follows increases in the perilymphatic K^+ concentration surrounding the hair cells.

A decrease in perilymphatic osmolarity can be expected to cause water influx into the endolymph (Thalmann, Salt, and Demott 1988) due to the water permeability of the perilymph–endolymph barrier (Bosher and Warren 1971; Juhn, Ryback, and Prado 1981; Sterkers et al. 1982). The influx of water will swell the endolymphatic compartment and dilute its constituents. This poses the question of which cells and which mechanisms maintain endolymph volume. For example, a hypo-osmotic challenge on the basolateral side of the vestibular dark cells stimulates the apical K^+ channel. Depending on the ionic gradient, K^+ flux across this channel can be secretory or resorptive (Wangemann et al. 1995b). The question of ion and volume regulation of endolymph is particular interest in view of the etiology of endolymphatic hydrops, a pathological increase in endolymph volume.

7.5.1 Ion Transport Mechanisms

A number of questions arise with regard to the models of strial ion transport and the generation of the endocochlear potential. Two cellular barriers for ion transport across the stria vascularis have to be considered, since ions in the endolymph such as K^+ and Cl^- originate from perilymph rather than from blood (Sterkers et al. 1982). Both barriers consist of cells joined by tight junctions. One barrier is the marginal cell epithelium, the other the basal cell layer. So far no direct evidence exists for specific ion transport mechanisms in basal cells. Indeed, the presence of K^+ conductance postulated as a source for the endocochlear potential in the two-cell model (Section 6.6) has yet to be demonstrated. The basal cells may achieve a high cytosolic K^+ concentration due to communication via gap junctions with fibrocytes of the spiral ligament. An analogous cell-to-cell transport of K^+ has been proposed in the vestibular labyrinth based on preliminary evidence for gap junctions between fibrocytes (Kikuchi et al. 1995). However, in both the vestibule and the cochlea, physiological evidence for such a mechanism remains to be established.

The assumption that strial marginal cells or their vestibular counterparts alone maintain the composition of endolymph may need to be refined. It is likely that vestibular transitional cells are involved in Na^+ reabsorption, since they contain several Na^+ transporters, including Na^+, K^+-ATPase (Wangemann and Marcus 1989; Spicer, Schulte, and Adams 1990), Na^+, H^+ exchanger (Wangemann, Shiga, and Marcus 1993), and a Na^+ or nonselective cation conductance (Wangemann and Shiga 1994b). Furthermore, pharmacological evidence obtained in the vestibular labyrinth suggests the involvement of an amiloride-sensitive Na^+ channel in Na^+ reabsorption (Ferrary et al. 1989). In the cochlea, such a Na^+ channel may be present in the apical membrane of the strial marginal cells and Reissner's membrane (Iwasa et al. 1994). Other cochlear epithelia, such as the spiral prominence and the interdental cells of the spiral limbus, could be involved in ion transport and maintenance of endolymph, in addition to the stria vascularis and Reissner's membrane. Both lie close to blood vessels and contain Na^+, K^+-ATPase in their basolateral membranes (Jones-Mumby and Axelsson 1984; Iwano et al. 1989).

7.5.2 Regulation of Ion Transport

The rate of ion transport is acutely regulated by the intra- and extracellular concentrations of the transported ions themselves. In addition, intracellular signals such as pH, Ca^{2+}, cyclic AMP, cell volume, or other second messengers may influence transport processes. A small cytosolic acidification activates the apical K^+ secretion, whereas a larger cytosolic acidification inhibits the basolateral Na^+, K^+-ATPase and thereby K^+ secretion (Wangemann, Liu, and Shiga 1995). Elevation of the cytosolic Ca^{2+} opens a nonselective cation channel in the apical membranes of strial marginal cells and vestibular dark cells, providing an additional pathway for K^+ secretion and Na^+ reabsorption (Marcus, Takeuchi, and Wangemann 1992; Takeuchi, Marcus, and Wangemann 1992; Sunose et al. 1993). Cell swelling stimulates transepithelial K^+ secretion and activates the apical I_{sK} channel (Marcus and Shen 1994; Wangemann et al. 1995b). Elevation of the cytosolic cyclic AMP concentration stimulates transepithelial K^+ secretion in strial marginal and vestibular dark cells by activating the apical I_{sK} channel (Sunose, Liu, and Marcus 1995).

Ion transport processes can also be affected by hormones. Frequent targets for hormonal regulation include the Na^+, K^+-ATPase, the Na^+-H^+ exchanger, and various ion channels. The I_{sK} channel in strial marginal and vestibular dark cells is inhibited via apical and basolateral P_{2U} receptors and activated via basolateral P_{2y} receptors (Liu, Kozakura, and Marcus 1995). Receptors for mineralocorticoids and glucocorticoids are present in the cochlea and in the vestibular labyrinth (Rare and Luttge 1989; Rarey, Curtis, and Ten Cate 1993; Pitovski, Drescher, and Drescher 1993). Alteration of circulating levels of these hormones alters the morphology of the stria vascularis and vestibular dark cells (ten Cate and Rarey 1991; Ten Cate, Curtis, and Rarey 1992) and the activity of Na^+, K^+-ATPase (Pitovski et al. 1993). Other receptors potentially related to fluid regulation in the inner ear include β_2-adrenergic receptors (Schacht 1985; Koch and Zenner 1988) and receptors for atrial natriuretic hormone (Meyer zum Gottesberge and Lamprecht 1989; Koch, Gloddek, and Gutzke 1992). It will be interesting to establish the links between these observations, the expression and regulation of ion transporters, and the physiology and pathophysiology of inner ear fluids.

7.6 Homeostasis Under Stress

Every cell possesses homeostatic mechanisms with which it can defend its internal milieu against attacks upon its integrity. Noise overstimulation or trauma and drugs or environmental toxins are among the challenges to the inner ear. Furthermore, constant "oxidative stress" comes from cellular metabolism itself in the form of production of free radicals.

7.6.1 Defense Mechanisms Related to Free Radicals

Free radicals are highly reactive and extremely detrimental to the integrity of the cell, their uncontrolled action eventually leading to cell death. Predictably, hair cells are sensitive to free radicals. The membrane integrity of isolated outer hair cells exposed to free radicals is compromised within seconds, leading to swelling of the cells and their eventual collapse (Dulon,

Zajic, and Schacht 1989). This may be accompanied by the uncontrolled entry of Ca^{2+} (Ikeda, Sunose, and Takasaka 1993), which also contributes to the swelling and eventual collapse of the cells.

A cell has several mechanisms for its defense. Glutathione is the most important physiological radical scavenger. It is ubiquitous, is present in high (millimolar) concentrations, and may function in enzymatic and nonenzymatic scavenging processes and detoxifying reactions. Additionally, free radicals can be eliminated or their formation prevented by the enzymes superoxide dismutase and catalase. A remarkable protective capacity of the inner ear can be inferred from the relatively high activities of glutathione transferase, glutathione reductase, and glutathione peroxidase, enzymes of glutathione homeostasis. The activity of glutathione reductase in the organ of Corti is comparable to that of the liver, the major organ for body self-defense. Likewise, the activities of superoxide dismutase, catalase, and glutathione peroxidase are higher in the organ of Corti and lateral wall tissues of the guinea pig than in other sensory or neural tissue (Pierson and Gray 1982; ElBarbary, Altschuler, and Schacht 1993).

7.6.2 Induced Stress Responses

Changes in gene expression and synthesis of heat-shock proteins (HSPs) are common stress responses in prokaryotic and eukaryotic cells (Morimotor 1993). HSPs (also termed "stress-shock proteins") represent a family of related proteins, some of which are inducible (e.g., HSP 72) and some of which are constitutive; that is, present in cells under normal conditions (e.g., HSP 70). HSPs seem to protect other proteins from denaturation or cellular organelles from disassembly.

HSPs in the mammalian cochlea increase with noxious stimuli, such as transient ischemia (Myers et al. 1992) or acoustic overstimulation (Lim et al. 1993). Under either condition, synthesis of HSP 72 is primarily induced in outer hair cells. Another member of the family, HSP 27, is expressed constitutively in stereocilia, where it may protect actin against depolymerization. Its concentration increases in outer hair cell somata but not inner hair cells in a time course compatible with recovery from temporary threshold shift (Altschuler et al. 1996). HSP 90, which is possibly associated with steroid receptor activity, is constitutive in inner and outer hair cells, and noise modulates its levels in both cell types (Altschuler et al. 1996). The presence of HSPs, however, is not restricted to sensory cells. Supporting cells also express some of the isoforms (Thompson and Neely 1992), and stress responses have been observed in the stria vascularis. The function of this important class of proteins in the cochlea remains to be established.

8. Conclusion

This brief discussion of gaps in our knowledge only points to specific questions that are already beginning to be addressed. The paramount problem, however, is understanding the complexity of a sensory system that is optimally geared toward the processing of complex signals. The basis for the exquisite capabilities of the auditory system lies in the morphological substrate and functional cooperation of sensory and supporting cells, neuroepithelium, and lateral wall tissues. Understanding these complexities that give us insight into subtleties of auditory transduction, be it the interaction of ion movements, second messengers or other homeostatic mechanisms. Therefore, research needs to focus on individual components of the inner ear to provide a basis for an understanding of this complex and integrated sensory system.

Acknowledgments

The writing of this chapter was concluded in spring 1994. Thereafter, references were updated but no new material was included. The authors wish to thank their colleagues Don Coling, Sherry Crann, Evelyne Ferrary, James Fessenden, Andrew Forge, Joseph E. Hawkins, Jr., Hakim Hiel, Matthew Holley, Kathryn Kimmel, Corné Kros, Daniel C. Marcus, Alec Salt, and Norma Slepecky for valuable comments and suggestions. The authors' research related to cochlear mechanisms is supported by grants DC-01098 and DC-00212 (to P.W.) and DC-00078 and DC-00124 (to J.S.) from the National Institutes of Health.

References

Allbritton NL, Meyer T, Stryer L (1992) Range of messenger action of calcium ion and inositol 1,4,5-trisphosphate. Science 258:1812–1815.

Altschuler RA, Lim HH, Ditto J, Dolan D, Raphael Y (1996) Protective mechanisms in the cochlea: heat shock proteins. In: Salvi RJ, Henderson D, Fiorini F, Colletti V (eds) Auditory Plasticity and Regeneration. New York: Tieman Medical Publications (in press).

Ashmore JF, Ohmori H (1990) Control of intracellular calcium by ATP in isolated outer hair cells of the guinea-pig cochlea. J Physiol 428:109–131.

Avila MA, Varela-Nieto I, Romero G, Mato JM, Giraldez F, Van de Water TR, Represa J (1993) Brain-derived neurotrophic factor and neurotrophin-3 support the survival and neuritogenesis response of developing cochleovestibular ganglion neurons. Dev Biol 159:266–275.

Bagger-Sjöbäck D, Filipek CS, Schacht J (1980) Characteristics and drug responses of cochlear and vestibular adenylate cyclase. Arch Otorhinolaryngol 228:217–222.

Bartolami S, Ripoll C, Planche M, Pujol R (1993) Localisation of functional muscarinic receptors in the rat cochlea: evidence for efferent presynaptic autoreceptors. Brain Res 626:200–209.

Bernard C (1878) Leçons sur les phénomènes de la vie. Paris: Baillière.

Berridge MJ (1993) Inositol trisphosphate and calcium signalling. Nature 361:315–325.

Bobbin RP, Fallon M, Puel J-L, Bryant G, Bledsoe SC, Zajic G, Schacht J (1990) Acetylcholine, carbachol, and GABA induce no detectable change in the length of isolated outer hair cells. Hear Res 47:39–52.

Bobbin RP, Fallon M, Kujawa SG (1991) Magnitude of the negative summating potential varies with perilymph calcium leevls. Hear Res 56:101–110.

Bosher SK, Warren RL (1971) A study of the electrochemistry and osmotic relationships of the cochlear fluids in the neonatal rat at the time of the development of the endocochlear potential. J Physiol (Lond) 212:739–761.

Bosher SK, Warren RL (1978) Very low calcium content of cochlear endolymph, an extracellular fluid. Nature 273:377–378.

Brechtelsbauer PB, Prazma J, Garrett CG, Carrasco VN, Pillsbury HC 3d (1990) Catecholaminergic innervation of the inner ear. Otolaryngol Head Neck Surg 103:566–574.

Brechtelsbauer PB, Nuttall AL, Miller JM (1994) Basal nitric oxide production in regulation of cochlear blood flow. Hear Res 77:38–42.

Brown JN, Nuttall AL (1994) Autoregulation of cochlear blood flow in the guinea pig. Am J Physiol 266:458–467.

Canlon B, Schacht J (1983) Acoustic stimulation alters deoxyglucose uptake in the mouse cochlea and inferior colliculus. Hear Res 10:217–226.

Canlon B, Homburger V, Bockaert J (1991) The identification and localization of the guanine nucleotide protein G_o in the auditory system. Eur J Neurosci 3:1338–1342.

Cannon WB (1992) Organization for physiological homeostasis. Physiol Rev 9:399–431.

Coling DE, Schacht J (1991) Protein phosphorylation in the organ of Corti: differential regulation by second messengers between base and apex. Hear Res 57:113–120.

Crist JR, Fallon M, Bobbin RP (1993) Volume regulation in cochlear outer hair cells. Hear Res 69:194–198.

Crone C (1965) Facilitated transfer of glucose from blood into brain tissue. J Physiol (Lond) 181:103–113.

Dallos P (1973) Cochlear potentials. In: Dallos P (ed) The Auditory Periphery: Biophysics and Physiology. New York: Academic Press, pp. 218–390.

Davis H (1957) Biophysics and physiology of the inner ear. Physiol Rev 37:1–49.

Davis H (1965) A model for transducer action in the cochlea. Cold Spring Harbor Symp Quant Biol 30:181–190.

Dechesne CJ, Winsky L, Moniot B, Raymond J (1993) Localization of calretinin mRNA in rat and guinea pig inner ear by in situ hydridisation using radioactive and non-radioactive probes. Hear Res 69:91–97.

Ding JP, Salvi RJ, Sach F (1991) Stretch-activated ion channels in guinea pig outer hair cells. Hear Res 56:19–28.

Doi K, Mori N, Matsunaga T (1990) Effects of forskolin and 1,9-dideoxy-forskolin on cochlear potentials. Hear Res 45:157–163.

Doi T, Ohmori H (1993) Acetylcholine increases intracellular Ca^{2+} concentration and hyperpolarizes the guinea-pig outer hair cell. Hear Res 67:179–188.

Drescher DG, Upadhyay S, Wilcox E, Fex J (1992) Analysis of muscarinic receptor subtypes in the mouse cochlea by means of the polymerase chain reaction. J Neurochem 59:765–767.

Dulon D, Schacht J (1992) Motility of cochlear outer hair cells. Am J Otol 13:108–112.

Dulon D, Aran J-M, Schacht J (1987) Osmotically induced motility of outer hair cells: implications for Meniere's disease. Arch Otorhinolaryngol 244:104–107.

Dulon D, Zajic G, Schacht J (1989) Photo-induced irreversible shortening and swelling of isolated cochlear outer hair cells. Int J Radiat Biol 55:1007–1014.

Dulon D, Zajic G, Schacht J (1990) Increasing intracellular free calcium induces circumferential contractions in isolated outer hair cells. J Neurosci 10:1388–1397.

Dulon D, Mollard P, Aran J-M (1991) Extracellular ATP elevates cytosolic Ca^{2+} in cochlear inner hair cells. NeuroReport 2:69–72.

Dulon D, Moataz R, Mollard P (1993) Characterization of Ca^{2+} signals generated by extracellular nucleotides in supporting cells of the organ of Corti. Cell Calcium 14:245–254.

Dulon D, Zajic G, Schacht J (1993) $InsP_3$ releases intracellular stored calcium in Deiters' cells of the organ of Corti. Abs Assoc Res Otolaryngol 16:117.

Dulon D, Blanchet C, Laffon E (1994) Photo-released intracellular Ca^{2+} evokes reversible mechanical responses in supporting cells of the guinea-pig organ of Corti. Biochem Biophys res Commun 201:1263–1269.

Duvall AJ, Rhodes VT (1967) Reissner's membrane. An ultrastructural study. Arch Ototlaryngol 86:143–151.

ElBarbary A, Altschuler R, Schacht J (1993) Glutathione S-transferase in the organ of Corti of the rat: enzymatic activity, subunit composition and immunohistochemical localization. Hear Res 71:80–90.

Erostegui C, Norris CH, Bobbin RP (1994) In vitro pharmacologic characterization of a cholinegic receptor on outer hair cells. Hear Res 74:135–147.

Erulkar SD, Maren TH (1961) Carbonic anhydrase in the inner ear. Nature 189:459–460.

Eveloff JL, Warnock DG (1987) Activation of ion transport systems during cell volume regulation. Am J Physiol 252:F1–F10.

Eybalin M (1993) Neurotransmitters and neuromodulators of the mammalian cochlea. Physiol Rev 73:309–373.

Eybalin M, Pujol R (1983) A radioautographic stdy of [^{3}H]L-glutamate and [^{3}H]L-glutamine uptake in the guinea pig cochlea. Neuroscience 9:863–871.

Eybalin M, Ripoll C (1990) Immunolocalisation de la parvalbumine dans deux types de cellules glutamatergiques de la cochlée du cobaye: les cellules ciliées internes et les neurones du ganglion spiral. CR Acad Sci Paris 310:639–644.

Eybalin M, Parnaud C, Geffard M, Pujol R (1988) Immunoelectron microscopy identifies several types of GABA-containing efferent synapses in the guinea-pig organ of Corti. Neuroscience 24:29–38.

Farber JL (1990) The role of calcium in lethal cell injury. Chem Res Toxicol 3:503–508.

Ferrary E, Sterkers O, Saumon G, Tran Ba Huy P, Amiel C (1987) Facilitated transfer of glucose from blood into perilymph in the rat cochlea. Am J Physiol 253:F59–F65.

Ferrary E, Tran Ba Huy P, Roinel N, Bernard C, Amiel C (1988) Calcium and the inner ear fluids. Acta Otolaryngol Suppl (Stockh) 460:13–17.

Ferrary E, Bernard C, Oudar O, Sterkers O, Amiel C (1989) Sodium transfer from endolymph through a luminal amiloride-sensitive channel. Am J Physiol 257:F182–F189.

Ferrary E, Barnard C, Oudar O, Loiseau A, Sterkers O, Amiel C (1993) N-Ethylmaleimide-inhibited electrogenic K^+ secretion in the ampulla of the frog semicircular canal. J Physiol (Lond) 461:451–465.

Fessenden JD, Coling DE, Schacht J (1994) Detection and characterization of nitric oxide synthase in the mammalian cochlea. Brain Res 668:9-15.

Flock A, Flock B, Ulfendahl M (1986) Mechanisms of movement in OHCs and a possible structural basis. Arch Otorhinolaryngol 243:83-90.

Fonnum F (1991) Neurochemical studies on glutamate-mediated neurotransmission. In: Meldrum BS, Moroni F, Woods JH (eds) Excitatory Amino Acids. New York: Raven Press, pp. 15-25.

Foster JD, Drescher MJ, Hatfield JS, Drescher DG (1994) Immunohistochemical localization of S-100 protein in auditory and vestibular end organs of the mouse and hamster. Hear Res 74:67-76.

Fuchs PA, Murrow BW (1992a) Cholinergic inhibition of short (outer) hair cells of the chick's cochlea. J Neurosci 12:800-809.

Fuchs PA, Morrow BW (1992b) A novel cholinergic receptor mediates inhibition of chick cochlear hair cells. Proc R Soc Lond B 248:35-40.

Gitter AH (1992) The length of isolated outer hair cells is temperature dependent. ORL 54:121-123.

Gross PM, Teasdale GM, Anerson WJ, Harper AM (1981) H^2-receptors mediate increases in permeability of the blood-brain barrier during arterial histamine infusion. Brain Res 210:396-400.

Guiramand J, Mayat E, Bartolami S, Lenoir M, Rumigny J-F, Pujol R, Récasens M (1990) A M3 muscarinic receptor coupled to inositol phosphate formation in the rat cochlea? Biochem Pharmacol 39:1913-1919.

Gulley RL, Reese TS (1976) Intercellular junctions in the reticular lamina of the organ of Corti. J Neurocytol 5:479-507.

Gulley RL, Fex J, Wenthold RJ (1979) Uptake of putative transmitters in the organ of Corti. Acta Otolaryngol 88:177-182.

Guth PS, Stockwell M (1977) Guanylate cyclase and cyclic guanosine monophosphate in the guinea-pig cochlea. J Neurochem 28:263-265.

Hara A, Salt AN, Thalmann R (1989) Perilymph composition in scala tympani of the cochlea: influence of cerebrospinal fluid. Hear Res 42:265-271.

Harada N, Ernst A, Zenner HP (1993) Hyposmotic activation hyperpolarizes outer hair cells of guinea pig cochlea. Brain Res 614:205-211.

Hawkins JE Jr (1971) The role of vasoconstriction in noise-induced hearing loss. Ann Otol Rhinol Laryngol 80:903-913.

Hawkins JE (1973) Comparative otopathology: aging, noise, and ototoxic drugs. Adv Otorhinolaryngol 20:125-141.

Hawkins JE Jr, Johnsson L-G, Preston RE (1972) Cochlear microvasculature in normal and damaged ears. Laryngoscope 82:1091-1104.

Henson JH, Begg DA, Beaulieu SM, Fishkind DJ, Bonder EM, Terasaki M, Lebeche D, Kaminer B (1989) A calsequestrin-like protein in the endoplasmic reticulum of the sea urchin: localization and dynamics in the egg first cell cycle embryo. J Cell Biol 109:149-161.

Holley MC, Richardson GP (1994) Monoclonal antibodies specific for endoplasmic membranes of mammalian cochlear outer hair cells. J Neurocytol 23:87-96.

Housley GD, Ashmore JF (1991) Direct measurement of the action of acetylcholine on isolated outer hair cells of the guinea pig cochlea. Proc R Soc Lond B 244:161-167.

Housley GD, Greenwood D, Ashmore JF (1992) Localization of cholinergic and purinergic receptors on outer hair cells isolated from the guinea-pig cochlea. Proc R Soc Lond B 249:265-273.

Ikeda K, Morizono T (1989) Electrochemical profiles for monovalent ions in the stria vascularis: cellular model of ion transport mechanisms. Hear Res 39:279–286.

Ikeda K, Morizono T (1990) Electrochemical aspects of cations in the cochlear hair cell of the chinchilla: a cellular model of the ion movement. Eur Arch Otorhinolaryngol 247:43–47.

Ikeda K, Takasaka T (1993) Confocal laser microscopical images of calcium distribution and intracellular organelles in the outer hair cell isolated from the guinea pig cochlea. Hear Res 66:169–176.

Ikeda K, Kusakari J, Takasaka T, Saito Y (1987) The Ca^{2+} activity of cochlear endolymph of the guinea pig and the effect of inhibitors. Hear Res 26:117–125.

Ikeda K, Saito Y, Sunose H, Nishiyama A, Takasaka T (1991) Effects of neuroregulators on the intracellular calcium level in the outer hair cell isolated from the guinea pig. ORL 53:78–81.

Ikeda K, Saito Y, Nishiyama A, Takasaka T (1992a) Intracellular pH regulation in isolated cochlear outer hair cells of the guinea pig. J Physiol 447:627–648.

Ikeda K, Saito Y, Nishiyama A, Takasaka T (1992b) Sodium-calcium exchange in the isolated cochlear outer hair cells of the guinea pig studied by fluorescence image microscopy. Pflügers Arch 420:493–499.

Ikeda K, Sunose H, Takasaka T (1993) Effects of free radicals on the intracellular calcium concentration in the isolatd outer hair cell of the guinea pig cochlea. Acta Otolaryngol (Stockh) 113:137–141.

Inamura N, Salt AN (1992) Permeability changes of the blood-labyrinth barrier measured in vivo during experimental treatments. Hear Res 61:12–18.

Ishiyama E, Keels EW, Weibel J (1970) New anatomical aspects of the vasculoepithelial zone of the spiral limbus in mammals. An electron microscopic study. Acta Otolaryngol (Stockh) 70:319–328.

Ito M, Spicer SS, Schulte BA (1993) Immunohistochemical localization of brain type glucose transporter in mammalian inner ears: comparison of developmental and adult stages. Hear Res 71:230–238.

Iwano T, Yamamoto A, Omori K, Akayama M, Kumazawa T, Tashiro Y (1989) Quantitative immunocytochemical localization of Na^+, K^+-ATPase α-subunit in the lateral wall of rat cochlear duct. J Histochem Cytochem 37:353–363.

Iwasa KH, Mizuta K, Lin DJ, Benos DJ, Tachibana M (1994) Amiloride-sensitive channels in marginal cells in the stria vascularis of the guinea pig cochlea. Neurosci Lett 172:163–166.

Jahnke K (1975) The fine structure of free-fractured intercellular junctions in the guinea pig inner ear. Acta Otolaryngol Suppl (Stoch) 336:1–40.

Jahnke K (1980) The blood-perilymph barrier. Arch Otorhinolaryngol 228:29–34.

Johnstone BM, Sellick PM (1972) The peripheral auditory apparatus. Q Rev Biophys 5:1–57.

Johnstone BM, Patuzzi R, Syka J, Sykova E (1989) Stimulus-related potassium changes in the organ of Corti of guinea-pig. J Physiol (Lond) 408:77–92.

Jones-Mumby CJ, Axelsson A (1984) The vascular anatomy of the gerbil cochlea. Am J Otolaryngol 5:127–137.

Juhn SK, Youngs JN (1976) The effect on perilymph of the alteration of serum glucose or calcium concentration. Laryngoscope 86:273–279.

Juhn SK, Rybak LP, Prado S (1981) Nature of blood-labyrinth barrier in experimental conditions. Ann Otol Rhinol Laryngol 90:135–141.

Julien N, Loiseau A, Sterkers O, Amiel C, Ferrary E (1994) Antidiuretic hormone restore the endolymphatic longitudinal K^+ gradient in the Brattleboro rat cochlea. Pflügers Arch 426:446-452.

Kambayashi J, Kobayashi T, Demott JE, Marcus NY, Thalmann I, Thalmann R (1982a) Effect of substrate-free vascular perfusion upon cochlear potentials and glycogen of the stria vascularis. Hear Res 6:223-240.

Kambayashi J, Kobayashi T, Demott JE, Marcus NY, Thalmann I, Thalmann R (1982b) Minimal concentrations of metabolic substrates capable of supporting cochlear potentials. Hear Res 7:105-114.

Katzman R (1976) Maintenance of a constant brain extracellular potassium. Fed Proc 35:1244-1247.

Kerr TP, Ross MD, Ernst SA (1982) Cellular localization of Na^+, K^+-ATPase in the mammalian cochlear duct: significance for cochlear fluid balance. Am J Otolaryngol 3:332-338.

Kikuchi T, Kimura RS, Paul DL, Adams JC (1995) Gap junctions in the rat cochlea: immunohistochemical and ultrastructural analysis. Anat Embryol (Berlin) 191:101-118.

Kimura RS (1969) Distribution, structure, and function of dark cells in the vestibular labyrinth. Ann Otol Rhinol Laryngol 78:542-561.

Kimura RS, Nye CL, Southard RE (1990) Normal and pathologic features of the limbus spiralis and its functional significance. Am J Otolaryngol 11:99-111.

Koch T, Zenner HP (1988) Adenylate cyclase and G-proteins as a signal transfer system in the guinea pig inner ear. Arch Otorhinolaryngol 245:82-87.

Koch T, Gloddek B, Gutzke S (1992) Binding sites of atrial natriuretic peptide (ANP) in the mammalian cochlea and stimulation of cyclic GMP synthesis. Hear Res 63:197-202.

Komune S, Nakagawa T, Hisashi K, Kimituki T, Uemura T (1993) Movement of monovalent ions across the membranes of marginal cells of the stria vascularis in the guinea pig cochlea. ORL J Otorhinolaryngol Relat Spec 55:61-67.

Konishi T, Butler RA, Fernández C (1961) Effect of anoxia on cochlear potentials. J Acoust Soc Am 33:349-356.

Konishi T, Hamrick PE (1978) Ion transport in the cochlea of guinea pig. II. Chloride transport. Acta Otolaryngol (Stockh) 86:176-184.

Konishi T, Kelsey E (1973) Effect of potassium deficiency on cochlear potentials and cation contents of the endolymph. Acta Otolaryngol (Stockh) 76:410-418.

Konishi T, Mendelsohn M (1970) Effect of oubain on cochlear potentials and endolymph composition in guinea pigs. Acta Otolaryngol (Stockh) 69:192-199.

Konishi T, Mori H (1984) Permeability to sodium ions of the endolymph-perilymph barrier. Hear Res 15:143-149.

Konishi T, Hamrick PE, Walsh PJ (1978) Ion transport in guinea pig cochlea. I. Potassium and sodium transport. Acta Otolaryngol (Stockh) 86:22-34.

Konishi T, Hamrick PE, Mori H (1984) Water permeability of the endolymph-perilymph barrier in the guinea pig cochlea. Hear Res 15:51-58.

Kronester-Frei A (1979) The effect of changes in endolymphatic ion concentrations on the tectorial membrane. Hear Res 1:81-94.

Kuijpers W, Bonting SL (1969) Studies on (Na^+-K^+)-ATPase. XXIV. Localization and properties of ATPase in the inner ear of the guinea pig. Biochim Biophys Acta 173:477-485.

Kuijpers W, Bonting SL (1970) The cochlear potentials. II. The nature of the cochlear endolymphatic resting potential. Pflügers Arch 320:359-372.

Kujawa SG, Glattke TJ, Fallon M, Bobbin RP (1992) Intracochlear application of acetylcholine alters sound-induced mechanical events within the cochlear partition. Hear Res 61:106–116.

Kujawa SG, Fallon M, Bobbin RP (1994) ATP antagonists cibaron blue, basilen blue and suramin alter sound-evoked responses of the cochlea and eighth nerve. Hear Res 78:181–188.

Kujawa SG, Glattke TJ, Fallon M, Bobbin RP (1994) A nicotinic-like receptor mediates suppression of distortion product otoacoustic emissions by contralateral sound. Hear Res 74:122–134.

Kusakari J, Ise I, Comegys TH, Thalmann I, Thalmann R (1978a) Effect of ethacrynic acid, furosemide, and ouabain upon the endolymphatic potential and upon high energy phosphates of the stria vascularis. Laryngoscope 88:12–37.

Kusakari J, Ise I, Comegys TH, Thalmann I, Thalmann R (1978b) Reduction of the endocochlear potential by the new "loop" diuretic, bumetanide. Acta Otolaryngol (Stockh) 86:336–341.

Lamm K, Zajic G, Schacht J (1994) Living isolated smooth muscle cells, pericytes and endothelial cells from inner ear vessels: a new approach to study the regulation of cochlear microcirculation and permeability. Hear Res 81:83–90.

Laurikainen EA, Kim D, Didier A, Ren T, Miller JM, Quirk WS, Nuttall AL (1993) Stellate ganglion drives sympathetic regulation of cochlear blood flow. Hear Res 64:199–204.

Lawrence M, Nuttall AL (1972) Oxygen availability in tunnel of Corti measured by microelectrode. J Acoust Soc Am 52:566–573.

Legrand C, Brehier A, Clavel MC, Thomasset M, Rabie A (1988) Cholecalcin (28-kDa CaBP) in the rat cochlea. Development in normal and hypothyroid animals. An immunocytochemical study. Brain Res 466:121–129.

Lim D, Karabinas C, Trune DR (1983) Histochemical localization of carbonic anhydrase in the inner ear. Am J Otolaryngol 4:33–42.

Lim HH, Jenkins OH, Myers MW, Miller JM, Altschuler RA (1993) Detection of HSP 72 synthesis after acoustic overstimulation in rat cochlea. Hear Res 69:146–150.

Liu J, Kozakura K, Marcus DC (1995) Evidence for purinergic receptors in vestibular dark cell and strial marginal cell epithelia of the gerbil. Audit Neurosci 1:331–340.

Loewenstein WR (1981) Junctional intercellular communication: the cell-to-cell membrane channel. Physiol Rev 61:829–913.

Loewenstein WR, Nakas M, Socolar SJ (1967) Junctional membrane uncoupling. Permeability transformation at a cell membrane junction. J Gen Physiol 50:1865–1891.

Marcus DC (1984) Characterization of potassium permeability of cochlear duct by perilymphatic perfusion of barium. Am J Physiol 247:C240–C246.

Marcus DC, Shen Z (1994) Slowly activating, voltage-dependent K^+ conductance is apical pathway for K^+ secretion in vestibular dark cells. Am J Physiol Cell Physiol 267:C857–864.

Marcus DC, Shipley A (1994) Potassium secretion by vestibular dark cell epithelium demonstrated by vibrating probe. Biophys J 66:1939–1942.

Marcus DC, Marcus NY, Thalmann R (1981) Changes in cation contents of stria vascularis with ouabain and potassium-free perfusion. Hear Res 4:149–160.

Marcus DC, Rokugo M, Ge XX, Thalmann R (1983) Response of cochlear potentials to presumed alterations of ionic conductance: endolymphatic perfusion of barium, valinomycin and nystatin. Hear Res 12:17–30.

Marcus DC, Rokugo M, Thalmann R (1985) Effects of barium and ion substitutions in artificial blood on endocochlear potential. Hear Res 17:79–86.

Marcus DC, Marcus NY, Greger R (1987) Sidedness of action of loop diuretics and ouabain on nonsensory cells of utricle: a micro-Ussing chamber for inner ear tissues. Hear Res 30:55–64.

Marcus DC, Takeuchi S, Wangemann P (1992) Ca^{2+}-activated nonselective cation channel in apical membrane of vestibular dark cells. Am J Physiol 262:C1423–C1429.

Marcus DC, Takueuchi S, Wangemann P (1993) Two types of chloride channel in the basolateral membrane of vestibular dark cell epithelium. Hear Res 69:124–132.

Marcus DC, Liu J, Shiga N, Wangemann P (1994) N-Ethylmaleimide stimulates and inhibits ion transport in vestibular dark cells of gerbil. Audit Neurosci 1:101–109.

Marcus DC, Liu J, Wangemann P (1994) Transepithelial voltage and resistance of vestibular dark cell epithelium from the gerbil ampulla. Hear Res 73:101–308.

Martin AR, Fuchs PA (1992) The dependence of calcium-activated potassium currents on membrane potential. Proc R Soc Lond B 250:71–76.

McGuirt JP, Schulte BA (1994) Distribution of immunoreactive α and β subunit isoforms of Na, K-ATPase in the gerbil inner ear. J Histochem Cytochem 42:843–853.

McLaren GM, Quirk WS, Laurikainen E, Coleman JKM, Seidman MD, Dengerink HA, Nuttall AL, Miller JM, Wright JW (1993) Substance P increases cochlear blood flow without changing cochlear electrophysiology in rats. Hear Res 71:183–189.

Melichar I, Syka J (1987) Electrophysiological measurements of the stria vascularis potentials in vivo. Hear Res 25:35–43.

Meyer zum Gottesberg A, Lamprecht J (1989) Localization of the atrial natriuretic peptide binding sites in the inner ear tissue—possibly an additional regulating system. Acta Otolaryngol (Stockh) Suppl 468:53–57.

Milner RE, Baksh S, Shemanko C, Carpenter MR, Smillie L, Vance JE, Opas M, Michalak M (1991) Calretinin, and not calsequestrin, is the major calcium binding protein of smooth muscle sarcoplasmic reticulum and liver endoplasmic reticulum. J Biol Chem 266:7155–7165.

Moncada S, Palmer RM, Higgs EA (1991) Nitric oxide: physiology, pathophysiology, and pharmacology. Pharm Rev 43:109-42.

Morimoto RI (1993) Cells in strses: transcriptional activation of heat shock genes. Science 259:1409–1410.

Murray SA, Plummer HK III, Leonard EE Jr, Deshmukh P (1993) Regulation of the 12-O-tetradecanoyl-phorbol-13-acetate-induced inhibition of intercellular communication. Anat Rec 235:1–11.

Myers MW, Quirk WS, Rizk SS, Miller JM, Altschuler RA (1992) Expression of the major mammalian stress protein in the rat cochlea following transient ischemia. Laryngoscope 102:981–987.

Nairn AC, Hemmings HC, Greengard P (1985) Protein kinases in the brain. Annu Rev Biochem 54:931–976.

Nakazawa K, Spicer SS, Schulte BA (1995) Postnatal expression of the facilitated glucose transporter, GLUT5, in gerbil outer hair cells. Hear Res 82:93–99.

Niedzielski AS, Schacht J (1991) Phospholipid metabolism in the cochlea: differences between base and apex. Hear Res 57:107–112.

Niedzielski AS, Schacht J (1992) P$_2$ purinoceptors stimulate inositol phosphate release in the organ of Corti. Neuroreport 3:273-275.

Niedzielski AS, Ono T, Schacht J (1992) Cholinergic regulation of the phosphoinositide second messenger system in the organ of Corti. Hear Res 59:250-254.

Nilles R, Järlebark L, Zenner HP, Heilbronn E (1994) ATP-induced cytoplasmic [Ca^{2+}] increases in isolated cochlear outer hair cells. Involved receptor and channel mechanisms. Hear Res 73:27-34.

Nishizuka Y (1992) Signal transduction: crosstalk. Trends Biochem Sci 17:367.

Oesterle EC, Dallos P (1989) Intracellular recordings from supporting cells in the guinea-pig cochlea: AC potentials. J Acoust Soc Am 86:1013-1032.

Oesterle EC, Dallos P (1990) Intracellular recordings from supporting cells in the guinea-pig cochlea: DC potentials. J Neurophysiol 64:617-636.

Offner FF (1991) Ion flow through membranes and the resting potential of cells. Membr Biol 123:171-182.

Offner FF, Dallos P, Cheatham MA (1987) Positive endocochlear potential: mechanism of production by marginal cells of stria vascularis. Hear Res 29:117-124.

Ogawa K, Schacht J (1993) Receptor-mediated release of inositol phosphates in the cochlear and vestibular sensory epithelia of the rat. Hear Res 69:207-214.

Ogawa K, Schacht J (1994) G-proteins coupled to phosphoinositide hydrolysis in the cochlear and vestibular epithelia of the rat are insensitive to cholera and pertussis toxins. Hear Res 74:197-203.

Ogawa K, Schacht J (1995) P$_{2y}$ purinergic receptors coupled to phosphoinositide hydrolysis in tissues of the cochlear lateral wall. Neuroreport 6:1538-1540.

Ohlsén KA, Baldwin DL, Nuttall AL, Miller JM (1991) Influence of topically applied adrenergic agents on cochlear blood flow. Circ Res 69:509-518.

Ohnishi S, Hara M, Inoue M, Yamashita T, Kumazawa T, Minato A, Inagaki C (1992) Delayed shortening and shrinkage of cochlear outer hair cells. Am J Physiol 263:C1088-1095.

Ohyama K, Salt AN, Thalmann R (1988) Volume flow rate of perilymph in the guinea-pig cochlea. Hear Res 35:119-129.

Ono T, Schacht J (1987) Effect of cholinergic agents on phospholipid metabolism in the guinea pig cochlea. Audiol Jpn 30:607-608.

Ono T, Schacht J (1989) Acoustic stimulation increases phosphoinositide breakdown in the guinea pig cochlea. Neurochem Int 14:327-330.

Orsulakova A, Stockhorst E, Schacht J (1976) Effect of neomycin on phosphoinositide labeling and calcium binding in guinea pig inner ear tissues in vivo and in vitro. J Neurochem 26:285-290.

Petersen KU, Reuss L (1983) Cyclic AMP-induced chloride permeability in the apical membrane of *Necturus* gallbladder epithelium. J Gen Physiol 81:705-729.

Pierson MG, Gray BH (1982) Superoxide dismutase activity in the cochlea. Hear Res 6:141-152.

Pitovski DZ, Drescher MJ, Drescher DG (1993) High affinity aldosterone binding sites (type I receptors) in the mammalian inner ear. Hear Res 69:10-14.

Pitovski DZ, Drescher MJ, Kerr Tp, Drescher DG (1993) Aldosterone mediates an increase in [^{3}H]ouabain binding at Na$^+$,K$^+$-ATPase sites in the mammalian inner ear. Brain Res 601:273-278.

Plinkert PK, Plinkert B, Zenner HP (1992) Carbohydrates in the cell surface of hair cells from the guinea pig cochlea. Eur Arch Otorhinolaryngol 249:67-73.

Pou AM, Fallon M, Winbery S, Bobbin RP (1991) Lowering extracellular calcium decreases the length of isolated outer hair cells. Hear Res 52:305-311.

Prazma J, Rodgers GK, Pillsbury HC (1983) Cochlear blood flow: effect of noise. Arch Otolaryngol 109:611-615.

Ptok M, Nair TS, Altschuler RA, Schacht J, Carey TE (1991) Monoclonal antibodies to inner ear antigens: II. Antigens expressed in sensory cell stereocilia. Hear Res 57:79-90.

Pujol R, Puel J-L, Gervais d'Aldin C, Eybalin M (1993) Pathophysiology of the glutamatergic synapses in the cochlea. Acta Otolaryngol 113:330-334.

Quirk WS, Wright JW, Dengerink HA, Miller JM (1988) Angiotensin II-induced changes in cochlear blood flow and blood pressure in normotensive and spontaneously hypertensive rats. Hear Res 32:129-136.

Quirk WS, Avinash G, Nuttall AL, Miller JM (1992) The influence of loud sound on red blood cell velocity and blood vessel diameter in the cochlea. Hear Res 63:102-107.

Raphael Y, Volk T, Crossin KL, Edelman GM, Geiger B (1988) The modulation of cell adhesion molecule expression and intercellular junction formation in the developing avia inner ear. Dev Biol 128:222-235.

Rarey KE, Luttge WG (1989) Presence of type I and type II/IB receptors for adrenocorticosteroid hormones in the inner ear. Hear Res 41:217-222.

Rarey KE, Curtis LM, ten Cate WJ-F (1993) Tissue specific levels of glucocorticoid receptor within the rat inner ear. Hear Res 64:205-210.

Reivich M (1974) Blood flow metabolism couple in brain. Res Publ Assoc Res Nerv Ment Disorders 53:125-140.

Reuss L (1987) Cyclic AMP inhibits Cl^-/HCO_3^- exchange at the apical membrane of *Necturus* gallbladder epithelium. J Gen Physiol 90:173-196.

Richardson GP, Bartolami S, Russell IJ (1990) Identification of a 275kD protein associated with the apical surfaces of sensory hair cells in the avian inner ear. J Cell Biol 110:1055-1066.

Rizzuto R, Brini M, Murgia M, Pozzan T (1993) Microdomains with high Ca^{2+} close to IP_3-sensitive channels that are sensed by neighboring mitochondria. Science 262:744-747.

Rossier BC, Geering K, Kraehenbuhl JP (1987) Regulation of the sodium pump: how and why? Trends Biol Sci 12:483-487.

Rossier MF, Putney JW (1991) The identity of the calcium-storing, inositol 1,4,5-trisphosphate-sensitive organelle in non-muscle cells: calciosome, endoplasmic reticulum . . . or both? Trends Neurosci 14:310-314.

Ryan AF, Schwartz IR (1984) Preferential glutamine uptake by cochlear hair cells: implications for the afferent transmitter. Brain Res 290:376-379.

Ryan AF, Woolf NK, Catanzaro A, Braverman S, Sharp FR (1985) Deoxyglucose uptake patterns in the auditory system: metabolic response to sound stimulation in the adult and neonate. In: Drescher D (ed) Auditory Biochemistry. Springfield, IL: Charles C. Thomas, pp. 401-421.

Rybak LP, Whitworth C (1986) Comparative ototoxicity of foresemide and piretanide. Acta Otolaryngol (Stockh) 101:59-65.

Rybak LP, Green TP, Juhn SK, Morizono T (1984) Probenecid reduces cochlear effects and perilymph penetration of furosemide in chinchilla. J Pharmacol Exp Ther 230:706-709.

Sakagami M, Fukazawa K, Matsunaga T, Fujita H, Mori N, Takumi T, Ohkubo H, Nakanishi S (1991) Cellular localization of rat Isk protein in the stria vascularis

by immunohistochemical observation. Hear Res 56:168-172.
Salt AN, Konishi T (1979) Effects of noise on cochlear potentials and endolymph potassium concentration recorded with potassium-selective electrodes. Hear Res 1:343-363.
Salt AN, Ohyama K (1993) Accumulation of potassium in scala vestibuli perilymph of the mammalian cochlea. Ann Otol Rhinol Laryngol 102:64-70.
Salt AN, Stopp PE (1979) The effect of raising the scala tympani potassium concentration on the tone-induced cochlear responses of the guinea pig. Exp Brain Res 36:87-98.
Salt AN, Thalmann R (1988a) Rate of longitudinal flow of cochlear endolymph. In: Nadol JB Jr (ed) Second International Symposium on Meniere's Disease. Amsterdam: Kugler & Ghedini, pp. 69-73.
Salt AN, Thalmann R (1988b) Cochlear fluid dynamics. In: Jahn AF, Santos-Sacchi J (eds) Physiology of the Ear. New York: Raven Press, pp. 341-357.
Salt AN, Thalmann R, Marcus DC, Bohne BA (1986) Direct measurement of longitudinal endolymph flow rate in the guinea pig ochlea. Hear Res 23:141-151.
Salt AN, Melichar I, Thalmann R (1978) Mechanisms of endocochlear potential generation by stria vascularis. Laryngoscope 97:984-991.
Salt AN, Inamura N, Thalmann R, Vora A (1989) Calcium gradients in inner ear endolymph. Am J Otolaryngol 10:371-375.
Salt AN, Ohyama K, Thalmann R (1991a) Radial communication between the perilymphatic scalae of the cochlea. I: Estimation by tracer perfusion. Hear Res 56:29-36.
Salt AN, Ohyama K, Thalmann R (1991b) Radial communication between the perilymphatic scalae of the cochlea. II: Estimation by bolus injection of tracer into the sealed cochlea. Hear Res 56:37-43.
Santi PA, Anderson CB (1987) A newly identified surface coat on cochlear hair cells. Hear Res 27:47-65.
Santi PA, Larson JT, Furcht LT, Economu TS (1989) Immunohistochemical localization of fibronectin in the chinchilla cochlea. Hear Res 39:91-102.
Santos-Sacchi J (1986) Dye coupling in the organ of Corti. Cell Tiss Res 245:525-529.
Schacht J (1974) Interaction of neomycin with phosphoinositide metabolism in guinea pig inner ear and brain tissues. Ann Otol 83:613-618.
Schacht J (1985) Hormonal regulation of adenylate cyclase in the stria vascularis of the mouse. Hear Res 20:9-13.
Schacht J, Canlon B (1985) Noise-ineduced changes of cochlear energy metabolism. In: Drescher D (ed) Auditory Biochemistry. Springfield, IL: Charles C. Thomas, pp. 389-400.
Schacht J, Canlon B (1988) Biochemistry of the inner ear. In: Alberti PW, Ruben RJ (eds) Otologic Medicine and Surgery. New York: Churchill Livingstone, pp. 151-178.
Schacht J, Zenner HP (1987) Evidence that phosphoinositides mediate motility in cochlear outer hair cells. Hear Res 31:155-159.
Scheibe F, Haupt H, Rothe E, Hache U (1981) Zur Glukose-, Pyruvat-und Laktatkonzentration von Perilymphe, Blut und Liquor cerebrospinalis unbelasteter und schallbelasteter Meerschweinchen in Äthylurethannarkose. Arch Otorhinolaryngol 233:89-97.
Scheibe F, Haupt H, Ludwig C (1992) Intensity-dependent changes in oxygenation of cochlear perilymph during acoustic exposure. Hear Res 63:19-25.

Schmidley JW, Dadson J, Iyer RS, Salomon RG (1992) Brain tissue injury and blood-brain barrier opening induced by injection of LGE2 or PGE2. Prostagl Leukot Essent Fatty Acids 47:105-110.

Schulte BA (1993) Immunohistochemical localization of intracellular Ca-ATPase in outer hair cells, neurons and fibrocytes in the adult and developing inner ear. Hear Res 65:262-273.

Schulte BA, Adams JC (1989) Distribution of immunoreactive Na^+, K^+-ATPase in gerbil cochlea. J Histochem Cytochem 37:127-134.

Schulte BA, Steel KP (1994) Expression of α and β subunit isoforms of Na,K-ATPase in the mouse inner ear and changes with mutations at the W^v or Sl^d loci. Hear Res 78:65-76.

Schwartz IR, Ryan AF (1983) Differential labeling of sensory cell and neuronal populations in the guinea pig organ of Corti following amino acid incubations. Hear Res 9:185-200.

Sellick PM, Johnstone BM (1975) Production and role of inner ear fluid. Prog Neurobiol 5:337-362.

Shiga N, Wangemann P (1995) Ion selectivity of volume regulatory mechanisms presnt during a hyposmotic challenge in vestibular dark cells. Biochim Biophys Acta 1240:48-54.

Shigemoto T, Ohmori H (1990) Muscarinic agonists and ATP increase the intracellular Ca^{2+} concentration in chick cochlear hair cells. Hear Res 61:35-46.

Shindo M, Miyamoto M, Abe N, Shida S, Murakami Y, Imai Y (1992) Dependence of endocochlear potential on basolateral Na^+ and Cl^- concentration: a study using vascular and perilymph perfusion. Jpn J Physiol 42:617-630.

Sillman JS, Masta RI, LaRuere MJ, Nuttall AL, Miller JM (1989) Electrically stimulated increases in cochlear blood flow: II. Evidence for neural mediation. Otolaryngol Head Neck Surg 101:362-374.

Skellett RA, Crist JR, Fallon M, Bobbin RP (1995) Caffeine-induced shortening of isolated outer hair cells: an osmotic mechanism of action. Hear Res 87:41-48.

Slepecky NB, Savage JE (1994) Expression of actin isoforms in the guinea pig organ of Corti: muscle isoforms are not detected. Hear Res 73:16-26.

Slepecky NB, Ulfendahl M (1993) Evidence for calcium-binding proteins and calcium-dependent regulatory proteins in sensory cells of the organ of Corti. Hear Res 70:73-84.

Sokoloff L (1977) Relation between physiological function and energy metabolism in the central nervous system. J Neurochem 29:13-26.

Sokrab TE, Johansson BB, Tengvar C, Kalimo H, Olsson Y (1988) Adrenaline-induced hypertension: morphological consequences of the blood-brain barrier disturbance. Acta Neurol Scand 77:387-396.

Somlyo AP (1984) Cellular site of calcium regulation. Nature 309:516-517.

Spicer SS, Schulte BA, Adams JC (1990) Immunolocalization of Na^+, K^+-ATPase and carbonic anhydrase in the gerbil's vestibular system. Hear Res 43:205-217.

Spoendlin H, Lichtensteiger W (1966) The adrenergic innervation of the labyrinth. Acta Otolaryngol 61:423-434.

Sterkers O, Saumon G, Tran Ba Huy P, Amiel C (1982) K, Cl, and H_2O entry in endolymph, perilymph, and cerebrospinal fluid of the rat. Am J Physiol 243:F173-F180.

Sterkers O, Ferrary E, Amiel C (1984) Inter- and intracompartmental osmotic gradients within the rat cochlea. Am J Physiol 247:F602-F606.

Sterkers O, Saumon G, Tran Ba Huy P, Ferrary E, Amiel C (1984) Electrochemical

heterogeneity of the cochlear endolymph: effect of acetazolamide. Am J Physiol 246:F47–F53.
Sterkers O, Ferrary E, Saumon G, Amiel C (1987) Na and nonelectrolyte entry into inner ear fluids of the rat. Am J Physiol 253:F50–F58.
Sunose H, Ikeda K, Saito Y, Nishiyama A, Takasaka T (1992) Membrane potential measurements in isolated outer hair cells of the guinea pig cochlea using conventional microelectrodes. Hear Res 62:237–244.
Sunose H, Ikeda K, Saito Y, Nishiyama A, Takasaka T (1993) Nonselective cation and Cl channels in luminal membrane of the marginal cell. Am J Physiol Cell Physiol 265:C72–C78.
Sunose H, Ikeda K, Suzuki M, Takasaka T (1994) Voltage-dependent small K channel in luminal membrane of marginal cells of stria vascularis dissected from guinea pig cochlea. Assoc Res Otolaryngol 17:133.
Sunose H, Liu J, Marcus DC (1995) Elevated intracellular cAMP activates transepithelial potassium secretion and apical slowly activating potassium channel in strial marginal cells and in vestibular dark cells. Proc Sendai Symp 5 (in press).
Sziklai I, Kiss JG, Ribari O (1986) Inhibition of myosin light-chain kinase activity in the organ of Corti by 0.3–5 kilodalton substances of the otosclerotic perilymph. Arch Otorhinolaryngol 243:229–232.
Sziklai I, Ferrary E, Horner KC, Sterkers O, Amiel C (1992) Time-related alteration of endolymph composition in an experimental model of endolymphatic hydrops. Laryngoscope 102:431–438.
Tachibana M, Wilcox E, Yokotani N, Schneider M, Fex J (1992) Selective amplification and partial sequencing of cDNAs encoding G protein α subunits from cochlear tissues. Hear Res 62:82–88.
Takahashi T, Kimura RS (1970) The ultrastructure of the spiral ligament in the rhesus monkey. Acta Otolaryngol 69:46–60.
Takeuchi S, Marcus DC, Wangemann P (1992) Ca^{2+}-activated nonselective cation, maxi K^+ and Cl^- channels in apical membrane of marginal cells of stria vascularis. Hear Res 61:86–96.
Takeuchi S, Ando M, Kozakura K, Saito H, Irimajiri A (1995) Ion channels in basolateral membrane of marginal cells dissociated from gerbil stria vascularis. Hear Res 83:89–100.
Tanaka M, Salt AN (1994) Cochlear function is disturbed by micromolar increases of endolymph calcium. Assoc Res Otolaryngol 17:90.
Tasaki I, Spyropoulos CS (1959) Stria vascularis as source of endocochlear potential. J Neurophysiol 22:149–155.
ten Cate WJ, Rarey KE (1991) Plasma membrane modulation of ampullar dark cells by corticosteroids. Arch Otolaryngol Head Neck Surg 117:96–99.
ten Cate WJ, Curtis LM, Rarey KE (1992) Immunochemical detection of glucocorticoid receptors within rat cochlear and vestibular tissues. Hear Res 60:199–204.
Thalmann I, Thalmann R (1978) Reevaluation of adenylate cyclase in Reissner's membrane. Arch Otorhinolaryngol 221:311–312.
Thalmann I, Rosenthal Hl, Moore BW, Thalmann R (1980) Organ of Corti-specific polypeptides: OCP-I and OCP-II. Arch Otolaryngol 226:123–128.
Thalmann I, Comegys TH, Liu SZ, Ito Z, Thalmann R (1992) Protein profiles of perilymph and endolymph of the guinea pig. Hear Res 63:37–42.
Thalmann I, Suzuki H, McCourt DW, Comegys TH, Thalmann R (1993) Partial amino acid sequences of organ of Corti proteins OCP1 and OCP2: a progress

report. Hear Res 64:191-298.

Thalmann R (1971) Metabolic features of auditory and vestibular systems. Laryngoscope 81:1245-1260.

Thalmann R, Kusakari J, Miyoshi T (1973) Dysfunctions of energy releasing and consuming processes of the cochlea. Laryngoscope 83:1690-1712.

Thalmann RR (1976) Quantitative biochemical techniques for studying normal and noise-damaged ears. In: Henderson D, Hamernik RP, Dosanjh DS, Miller JH (eds) Effects of Noise on Hearing. New York: Raven Press, pp. 129-154.

Thalmann R, Marcus NY, Thalmann I (1978) Adenylate energy charge, energy status, and phosphorylation state of stria vascularis under metabolic stress. Laryngoscope 88:1985-1998.

Thalmann R, Paloheimo S, Thalmann I (1979) Distribution of cyclic nucleotides in the organ of Corti. Acta Otolaryngol 87:375-380.

Thalmann R, Salt AN, DeMott J (1988) Endolymph volume regulation. Possible mechanisms. In: Nadol JB Jr (ed) Second Symposium on Meniere's Disease. Amsterdam: Kugler & Ghedini, pp. 55-60.

Thompson AM, Neely JG (1992) Induction of heat shock protein in interdental cells by hyperthermia. Otolaryngol Head Neck Surg 107:769-774.

Thorne PR, Nuttall AL (1987) Laser Doppler measurements of cochlear blood flow during loud sound exposure in the guinea pig. Hear Res 27:1-10.

Thorne PR, Nuttall AL (1989) Alterations in oxygenation of cochlear endolymph during loud sound exposure. Acta Otolaryngol 107:71-79.

Vogh BP, Maren TH (1975) Sodium, chloride, and bicarbonate movement from plasma to cerebrospinal fluid in cats. Am J Physiol 228:673-683.

Volpe P, Krause KH, Hashimoto S, Zorzato F, Pozzan T, Meldolesi J, Lew DP (1988) "Calciosome," a cytoplasmic organelle: the inositol 1,4,5-triphosphate-sensitive calcium store of nonmuscle cells? Proc Natl Acad Sci USA 85:1091-1095.

von Békésy G (1950) DC potentials and energy balance of the cochlear partition. J Acoust Soc Am 22:576-582.

Wada J, Kambayashi J, Marus DC, Thalmann R (1979) Vascular perfusion of the cochlea: effect of potassium-free and rubidium-substituted media. Arch Otorhinolaryngol 225:79-81.

Wang S, Schacht T (1990) Insulin stimulates protein synthesis and phospholipid signaling systems but does not regulate glucose uptake in the inner ear. Hear Res 47:53-62.

Wangemann P, Greger R (1990) Piretanide inhibits the $Na^+2Cl^-K^+$ carrier in the thick ascending limb of the loop of Henle and reduces the metabolic fuel requirements of this nephron segment. In: Puschett JB, Greenberg A (eds) Diuretics III: Chemistry, Pharmacology, and Clinical Applications. New York: Elsevier, pp. 220-224.

Wangemann P, Marcus DC (1989) Membrane potential measurements of transitional cells from the crista ampullaris of the gerbil. Effects of barium, quinidine, quinine, tetraethylammonium, cesium, ammonium, thallium and ouabain. Pflügers Arch 414:656-662.

Wangemann P, Marcus DC (1990) K^+-induced swelling of vestibular dark cells is dependent on Na^+ and Cl^- and inhibited by piretanide. Pflügers Arch 416:262-269.

Wangemann P, Shiga N (1994a) Cell volume control in vestibular dark cells during and after a hyposmotic challenge. Am J Physiol Cell Physiol 226:C1046-C1060.

Wangemann P, Shiga N (1994b) Ba^{2+} and amiloride uncover or induce a pH-sensitive and a Na^+ or non-selective cation conductance in transitional cells of the inner ear. Pflügers Arch 426:258-266.

Wangemann P, Shiga N, Marcus DC (1993) The Na^+/H^+ exchanger in transitional cells of the inner ear. Hear Res 69:107-114.

Wangemann P, Liu J, Marcus DC (1995) Ion transport mechanisms responsible for K^+ secretion and the transepithelial voltage across marginal cells of stria vascularis in vitro. Hear Res 84:19-29.

Wangemann P, Liu J, Shiga N (1995) The pH-sensitivity of transepithelial K^+ transport in vestibular dark cells. J Membr Biol 147 255-262.

Wangemann P (1995) Comparison of ion transport mechanisms between vestibular dark cells and strial marginal cells. Hear Res 90:149-157.

Wangemann P, Liu J, Shen Z, Shipley A, Marcus DC (1995) Hypo-osmotic challenge stimulates transepithelial K^+ secretion and activates apical I_{sk} channel in vestibular dark cells. J Membr Biol 147:263-273.

Whitlon DC (1993) E-cadherin in the mature and develping organ of Corti of the mouse. J Neurocytol 22:1030-1038.

Williams SE, Zenner HP, Schacht J (1987) Three molecular steps of aminoglycoside ototoxicity demonstrated in outer hair cells. Hear Res 30:11-18.

Yamashita T, Amano H, Harada N, Su ZL, Kumazawa T, Tsunoda Y, Tashiro Y (1990) Calcium distribution and mobilization in single cochlear hair cells. Acta Otolaryngol 109:256-262.

Ylikoski J, Pirvola U, Moshnyakov M, Palgi J, Arumae U, Saarma M (1993) Expression patterns of neurotrophin and their receptor mRNAs in the rat inner ear. Hear Res 65:69-78.

Yoshihara T, Igarashi M, Usami S, Kanda T (1987) Cytochemical studies of Ca^{++}-ATPase activity in the vestibular epithelia of the guinea pig. Arch Otorhinolaryngol 243:417-423.

Zajic G, Anniko M, Schacht J (1983) Cellular localization of adenylate cyclase in the developing and mature inner ear of the mouse. Hear Res 10:249-261.

Zajic G, Nair TS, Ptok M, Van Waes C, Altschuler RA, Schacht J, Carey TE (1991) Monoclonal antibodies to inner ear antigens: I Antigens expressed by supporting cells of the guinea pig cochlea. Hear Res 52:59-71.

Zenner HP, Zimmerman U, Schmitt U (1985) Reversible contraction of isolated mammalian cochlear hair cells. Hear Res 22:83-90.

Zidanic M, Brownell WE (1990) Fine structure of the intracochlear potential field. I. The silent current. Biophys J 57:1253-1268.

Zwislocki JJ, Slepecky NB, Cefaratti LK, Smith RL (1992) Ionic coupling among cells in the organ of Corti. Hear Res 57:175-194.

4
Cochlear Micromechanics and Macromechanics

ROBERT PATUZZI

1 Introduction

1.1 Overview

The main task of the mammalian cochlea is to analyze sound in terms of its intensity, timing, and frequency content. To achieve this, sound in air is "matched" acoustically to the denser water medium within the cochlea by the middle ear, efficiently producing pressure fluctuations in the cochlear fluids and vibration of the ribbonlike *basilar membrane* (BM) and the sensory *organ of Corti* (OC) perched upon it. The vibration of the cochlear partition (BM and OC) normally takes the form of waves or ripples that travel away from the stapes and toward the cochlear apex (See Fig. 1.5, Dallos, Chapter 1). For a *pure-tone* stimulus, the wave grows as it travels, reaching a maximum at a position known as the wave's *characteristic place*, and then collapses abruptly so that no vibration exists beyond a cochlear position known as the wave's *cutoff region*. The characteristic place and cutoff region are positioned along the cochlear length according to a *place–frequency map*: high frequencies toward the cochlear base, closer to the stapes, and low frequencies closer to the cochlear apex. The vibration of the BM and OC relative to the surrounding bony structures is known as *macromechanical vibration*, while the complex vibration of the parts of the OC relative to one another is known as *micromechanical vibration*. It is worth stressing here that both these vibrations are normally of atomic or at most molecular dimensions, comparable with the width of a cell membrane (less than 10 nm), and they are not independent: not only does macromechanical vibration of the BM "drive" the micromechanical vibration of the OC, but the micromechanics greatly affects the macromechanics.

Once the OC is set in motion by the BM traveling wave, the mechanical arrangement of its parts transforms the vibration further, matching the acoustical properties of the BM and cochlear fluids to those of the hair bundles at the apex of the sensory hair cells. This leads to deflection of their hair bundles, which opens and closes *mechanoelectrical transduction*

(MET) *channels* near the tip of each stereocilium, producing a sound-induced change in the current through each hair cell known as its *receptor current,* and a change in the voltage across its membrane known as its *receptor potential* (Kros, Chapter 6). While the receptor potential within both types of hair cell, *inner hair cells* (IHCs) and *outer hair cells* (OHCs), acts to release neurochemicals and stimulate neurons, the receptor potential or current of OHCs also seems to control a molecular force generation mechanism within the OC (and probably within the OHCs themselves) known as the *active process* (de Boer, Chapter 5; Holley, Chapter 7).

When all goes well, these electrical and mechanical processes act synergistically to increase the vibration of the BM and OC 1000-fold or more. When this occurs, the cochlea can detect fluctuations in air pressure (sound) as low as 1/2,000,000,000 of normal atmospheric pressure (20 μPa rms; see Fay 1988), corresponding to vibrations of the BM and OC of only molecular dimensions (less than 1 nm). If the OHCs, the active process, or both are somehow disrupted, mechanical sensitivity is reduced and a sensorineural deafness results.

Despite its extreme sensitivity, for short periods the cochlea can withstand sound pressures that are about one million times greater in amplitude (and therefore with 10^{12} more energy) than sounds at the threshold of detection. It survives because the mechanical sensitivity is progressively reduced as the sound level increases. In essence, the cochlea possesses an "automatic gain control," and vibration grows with sound level at a much slower rate than would be expected from simple proportionality or *linearity.* This *compressive* or *nonlinear* growth of vibration ultimately produces a *dynamic range* of about 120 dB (1:1,000,000) from the softest to the loudest sounds the cochlea can detect without permanent damage. Nevertheless, although the ear's specifications are very impressive, they do not come cheaply. The cochlea has one of the highest metabolic rates of any organ in the body (Johnstone and Sellick 1972), and it can be disrupted by many physiological or pharmacological insults, producing sensorineural deafness. It is also prone to oscillate spontaneously, emitting sounds from the ear as *spontaneous otoacoustic emissions.* Once inside the cochlea, sound signals can also be very badly distorted. Whereas a mediocre stereo system may boast distortion that is less than 0.001%, the mechanical distortion within the cochlea can be as high as 3%! This distortion is so bad that when two pure tones are presented together, the vibration produced by one tone can completely suppress the vibration produced by the other. The distortion also produces additional vibrations that were not presented in the original stimulus. These *distortion tones* are perceived as additional sounds.

An understanding of these features of cochlear vibration requires an understanding of the structures within the ear and the complex mechanical interactions among them. Figure 4.1 presents a summary of the main stages of mechanical processing described in this chapter. In essence, cochlear vibration can be considered at many levels of complexity. Most simply

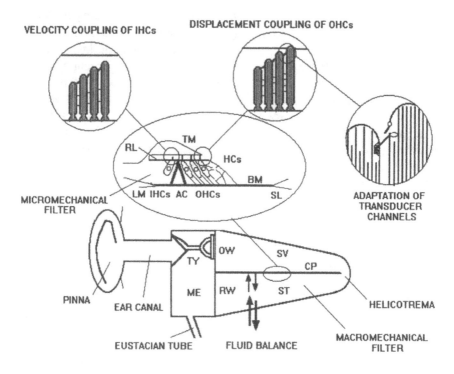

FIGURE 4.1. Sound stimuli entering the ear are modified in a frequency-dependent manner by a variety of elements (see text). Middle ear (ME), cochlear partition (CP), reticular lamina (RL), tectorial membrane (TM), outer hair cells (OHCs), tympanum (TY), oval window (OW), round window (RW), scala tympani (ST), scala vestibuli (SV), spiral lamina (LM), spiral ligament (SL), inner hair cells (IHCs), arch of Corti (AC), Hensen cells (HCs).

(Helmholtz 1863), different regions along the cochlear length are thought to vibrate independently, acting as tuning forks driven by pressure fluctuations in the fluids. The resonance in this case may be transverse, with the mass of the partition bouncing on the radially suspended BM, or the mass of the tectorial membrane (TM) bouncing radially against the hair bundles. The second level of complexity assumes that both resonances are important, and that they are independent and connected in cascade. In this case, the frequency filtering produced by the first filter (transverse BM resonance) would be modified by a second filter (radial TM resonance). A third level of complexity assumes that these resonators are not independent but are intimately coupled to form a complex second-order resonator. In this case, changing one resonator affects the other. A fourth level of complexity acknowledges that no micromechanical filter in one region can vibrate independently of all others, since they share the cochlear fluids and sit on a common BM platform: in other words, the micromechanical view must take into account macromechanical coupling. A fifth level of complexity considers "active" elements that utilize cochlear energy reserves to add energy to the incoming sound wave. Finally, a sixth level of complexity considers the possibility that mechanical elements may be nonlinear, producing acoustic distortion.

It should be stressed here that some of the "cochlear technology" used by mammals is different from that used by nonmammals, and that the exclusion of much of the research on nonmammalian species from this chapter does not indicate a lack of literature in this area: on the contrary, more is known about some aspects of cochlear function in nonmammals than in mammals (Lewis et al. 1985).

1.2 A Brief History of Cochlear Mechanics

In 1760, Cotugno announced that the cochlea of humans was filled with liquid (see historical review in Tonndorf 1981). Since it was generally thought to be filled with air, this statement was met with some disbelief. It required Meckel in 1777 to convince others that Cotugno was correct. He did so by freezing a fresh cadaver ear in the snow, producing the world's first cochlear ice-block. In 1851, Alfonse Corti described the structures of the sensory organ now named after him. Not surprisingly, it was unclear what each of its components was supposed to do, since the theory of sound propagation and its reception was still in its infancy. As for the mechanism of hearing and the analysis of sound, very little was known. Although Ohm had suggested in 1843 that it must occur by some form of Fourier or spectral analysis, it took Helmholtz (1863) to prove that this was the case. He described the cochlea in detail and proposed that structures within it resonated in sympathy with sound, much like a tuning fork or a piano string, with the various components of a complex sound distributed along the cochlea according to frequency. Even so, Helmholtz was unsure how the vibration took place, especially with the parts bathed in water. The fundamental mathematics of the vibration of membranes and fluids was not understood until Rayleigh first published *The Theory of Sound* in 1877. Nevertheless, based on the ear's inability to detect individual tones played in a musical "shaking" or "trill," Helmholtz reasoned that our frequency discrimination must be due to the vibration of a series of independent resonators within the cochlea. He also calculated that the vibration of these resonators must take a long time to die away between successive stimuli (almost ⅓ of a second at 110 Hz; Helmholtz, 1863, pp. 142–144). Later work by Gold (1948) suggested that the frequency analysis was indeed the result of resonators, and that the ringing time was too long to be consistent with the friction expected from the cochlea's watery contents. Gold concluded that if this were true, the resonators must be assisted to overcome friction in some manner.

Soon after, Georg von Békésy (see von Békésy 1960) described the vibration in various cochlear models and in temporal bone specimens from cadavers. Observing this vibration using stroboscopic illumination and direct microscopic examination, he noted two points about cochlear vibration that were not consistent with Helmholtz's view. First, the vibration did not occur as a simple resonance of *independent* mechanical resonators: the different regions along the cochlea seemed mechanically coupled to each other. Second, he observed a broad vibration pattern that was not consistent with the high degree of discrimination between frequencies that was observed in psychophysical experiments. An additional frequency-selective process seemed necessary to account for the difference between mechanical and psychophysical discrimination, possibly analysis within the central nervous system. When Tasaki (1954) reported that the response of auditory nerve fibers had a frequency selectivity very similar to that observed psychophysically (see Moore 1986), it seemed clear that the additional "tuning" must lie within the cochlea.

While work continued on the response properties of auditory nerve fibers, sophisticated experimental techniques for measuring vibration were not available. In 1958, however, Mössbauer described a phenomenon wherein the frequency or color of photons of light leaving a radioactive crystal could be changed, or Doppler shifted, if the radiating atom moved. In 1967, Johnstone and Boyle used this "Mössbauer technique" to measure the vibration of the structures within the cochlea in living animals at sound levels close to the threshold of hearing. Although their measurements were more accurate than those of von Békésy, and slightly more frequency-selective, they still fell short of explaining the high frequency-selectivity of auditory nerve fibers.

At this time, two aspects of the response of primary auditory neurons that could not be explained by the vibration described by von Békésy were coming into sharper focus. Neural *lability* was evident in that the sensitivity of auditory nerve fibers could change throughout an experiment and could be manipulated by the experimenter by draining cochlear fluid (Robertson 1974), inducing hypoxia in the experimental animal (Evans 1972), or exposing it to loud sound (Cody and Johnstone 1980) or various ototoxic drugs (see Ruggero and Rich 1991b). Frequency-dependent growth of afferent firing rate, or *nonlinearity,* was also evident in that the action potential firing of a single auditory neuron produced by one pure tone could be abolished by presenting another tone at a carefully chosen combination of frequency and intensity (Kiang et al. 1965; Sachs and Kiang 1968; Arthur, Pfeiffer, and Suga 1971). This *two-tone suppression* phenomenon was most evident if the initial probe tone was at a low intensity and had a frequency corresponding to the most sensitive frequency or characteristic frequency (CF) of the nerve fiber. It was also noted that when such combinations of tones were presented to an ear, additional tones could be heard at frequencies mathematically related to the two presented. Known as *combination tones,* or *distortion tones,* these had been described much earlier by Sorge, Tartini (see Wever 1947) and, in much more detail, by

Helmholtz (1863, p. 152). If the frequencies of the two tones presented (the primary tones) were f_1 and f_2, then the frequencies of the distortion tones were $nf_1 \pm mf_2$, where n and m are integers. Importantly, the auditory nerve fibers were found to respond to these additional frequency components as if they were presented as a sound stimulus. The distortion tones also seemed to propagate mechanically within the cochlea like sound (Goldstein and Kiang 1967; Smoorenburg 1972). In fact, it seemed inescapable that they *were* additional sounds, produced within the cochlea by some distorting or nonlinear mechanical process. This result was disturbing, because there had been no direct evidence for such distortion in cochlear vibration. Certainly the middle ear seemed not to distort sound to any significant degree (Wever and Lawrence 1954; Guinan and Peake 1967; Wilson and Johnstone 1975).

An additional sign of cochlear nonlinearity or distortion became apparent as the responses of auditory nerve fibers were characterized in more and more detail (Kiang et al. 1965). The increase in neural firing rate with sound level was quite different when the sound was presented at a fiber's CF and when the stimulus frequency was much lower. In essence, increasing the sound level for frequencies much lower than the neuron's CF produced a relatively rapid increase in firing rate as sound level was increased above threshold, whereas at or near the CF, the firing rate grew much less rapidly with sound level (see Fig. 4.10c). It should be remembered that at this time there was very little direct evidence that the vibration within the cochlea demonstrated any of these properties. In the early 1970s, however, Rhode (1971, 1973, 1980) had measured vibration in the squirrel monkey cochlea using the Mössbauer technique. He did find some evidence for nonlinearity, in that the vibration showed signs of frequency-dependent nonlinear growth with sound level. He also observed some sign of lability, with the vibration becoming less sensitive as the experiment proceeded. At the time, these responses were seen by many to be idiosyncratic of a particular species.

Nevertheless, many of the observations on the responses of auditory nerve fibers seemed most simply explained by a nonlinear or distorting vibration within the cochlea (Kim 1980), but doubts still remained. What was thought at the time to be a definitive experiment had been carried out by Evans and Wilson (1975), who recorded from auditory nerve fibers and measured vibration in the same animal (a cat). When they observed different response properties in the firing of the auditory nerve and the vibration of the structures within the cochlea, at almost the same time, it seemed clear that some other process or processes must be operating between the transverse vibration observed and the initiation of neural firing. The term "second-filter" was coined (Evans 1972) to describe those unknown processes accounting for the discrepancy between mechanical and neural responses, and a plethora of schemes was put forward to explain the second filter, including complex neural, biochemical, electrical, and/or mechanical processing within the cochlea. One favored (and highly plausible) explanation was that the transverse (up-down) vibration measured by all researchers at that time was not the vibration stimulating the sensory hair cells. It seemed clear that they were stimulated by a side-to-side rocking of their hair bundles (Flock 1965p; Hudspeth and Corey 1977). Much of the theoretical work that followed (for example, Zwislocki and Kletsky 1979; Allen 1980) attempted to determine whether a complicated micromechanical vibration of the OC occurred and whether it could explain the differences between the mechanical and the neural measurements (i.e., the "second filter").

In 1977, Russell and Sellick (see also Russell and Sellick 1978) had succeeded in impaling one type of sensory hair cell within the mammalian cochlea (the IHCs). They showed that many of the properties observed in the neural responses were already present in the IHCs that drive most of the afferent neurons. This result was at least consistent with a mechanical explanation for the second filter. The next year, Kemp (1978) observed that a brief acoustic click elicited an echo from the cochlea that could not be explained by simple linear vibration. Soon after, LePage and Johnstone (1980) observed vibration in the guinea pig cochlea that was reminiscent of that observed by Rhode (1971) in the squirrel monkey and was thought to be peculiar to that species. The following year, a sensitive, highly tuned, nonlinear, and labile vibration was observed in the guinea pig by Sellick, Patuzzi, and Johnstone (1982b) using the Mössbauer technique and by Khanna and Leonard (1982) using laser interferometry. It appeared that in previous experiments the vibration had been altered by the experimental procedures used, even in the work of Evans and Wilson who had explicitly set out to monitor vibration in animals with normal neural sensitivity. Whatever the cause of their seemingly anomalous result, highly tuned and nonlinear vibration has been observed repeatedly since 1982 (Patuzzi and Sellick 1983; Robles, Ruggero, and Rich 1991; Ruggero and Rich 1991a,b; Cooper and Rhode 1992; Ruggero, Rich, and Recio 1992), and with each new wave of experimentation, complex neural behavior has been traced to vibration within the cochlea.

Although these recent observations have simplified cochlear physiology to some extent, they have left a more fundamental question to be answered: Why does the cochlear partition vibrate as it does? As will be argued in the following pages and elsewhere in this volume (Holley, Chapter 7), aspects of cochlear vibration can be traced to molecular processes within the OHCs of the OC.

As early as 1972, Dallos and co-workers (Dallos et al. 1972) had pointed out that OHCs were necessary for normal cochlear sensitivity in that neural thresholds were elevated when OHCs were selectively destroyed with the aminoglycoside antibiotic, kanamycin. It was not clear at the time why this should occur, since OHCs only received 5% of the afferent neurons (Spoendlin 1969; see Slepecky, Chapter 2), and it was widely thought at the time that cochlear vibration was passive, linear, and robust. In the early 1980s, however, it was becoming increasingly clear that this view of cochlear mechanics was untenable. For example, Mountain (1980) reported that the electrical stimulation of the cochlear efferent neurons that innervate OHCs altered cochlear mechanics in some way, producing a change in the distortion tones measured in the ear canal with sensitive microphones (Guinan, Chapter 8). This suggested that the OHCs must play at least an indirect role in determining vibration. Mountain also noted that the acoustic distortion tones measured in the ear canal were changed when the endocochlear potential (EP) was lowered, suggesting that the electrical events within the cochlea could modify cochlear vibration. It was already known that stimulation of efferent neurons caused electrical changes within the cochlea (see Guinan, Chapter 8).

Then, in 1981/82, sensitive, nonlinear, and labile vibration was observed, and soon after, Brownell and co-workers (Brownell 1983; Brownell et al. 1985) described the motile behavior of OHCs in vitro in response to extracellular current stimulation. In the same year, Patuzzi, Sellick, and Johnstone (1983) reported that the vibration of the BM at high frequencies was reduced when low-frequency tones pushed the OHC receptor currents into partial saturation. At this stage, it seemed clear that the OHCs at least generated the electrical currents that synchronized the active forces, and possibly generated the forces as well. At about this time, Neely and Kim (1983) published an "active" model of the vibration of the cochlear partition in which OHCs generated forces that partially canceled friction. Weiss (1982) had already coined the term "bidirectional transduction" to describe the process whereby movement caused electrical events (forward transduction), and these electrical events in turn produced movement (reverse transduction) within individual hair cells. Some years earlier, Zwicker (1979) had proposed a similar theoretical scheme involving positive feedback to explain many cochlear phenomena, and had developed an electronic model of his scheme to demonstrate a number of its salient features (see Zwicker 1986).

Most recently it has been shown that OHCs are capable of axial expansion and contraction at high frequencies (Holley, Chapter 7). It has also been shown that most cochlear disruptions known to reduce neural sensitivity also reduce OHC receptor current, and that the mathematical relationships between the drop in receptor current and the (resultant) threshold elevation are quantitatively very similar for most cochlear disruptions (Patuzzi, Yates, and Johnstone 1989c). In the few exceptions where neural threshold is elevated without a reduction in the OHC receptor current, either the IHCs or the neurons are known to be disrupted and the detection of vibration is impaired, or the particular cochlear disruption is known to impair the motility of OHCs in vitro. Cochlear deafness seems due to either a failure in the detection of vibration by hair cells and neurons or an impairment of vibration itself, caused by an impaired electrical drive to the OHC motors or by broken OHC motors per se.

What follows is a description of the basic issues of cochlear vibration and the role of the OHCs in that vibration (see also Ashmore 1990; Dallos 1988; Duifhuis et al. 1993).

2. Micromechanical Vibration

A discussion of the micromechanical vibration of the OC can begin by assuming that (1) the macromechanical vibration of the BM platform upon which the OC rests is not at all affected by the complicated micromechanical properties of the OC (the OC is like a flea on the back of the BM elephant), and (2) the micromechanical vibration of each segment of the OC along the cochlear length is independent of all other segments (Helmholtz's original assumption in likening sections of the cochlear partition to resonating piano strings). Although neither assumption is correct, as discussed later, both are useful when introducing the main concepts of cochlear micromechanics.

2.1 Impedance and Acoustic Transformers

One view of cochlear mechanics is that the middle ear and cochlear microstructures are simply devices that "convert" or "transform" vibration of the floppy air medium in the ear canal into vibration of the relatively hard MET channels at the tops of the hair cells. To emphasize this point, Figure 4.2 presents six simple *acoustic transformers,* referred to here as the lever, the clothesline, the shear transformer, the pressure transformer, the

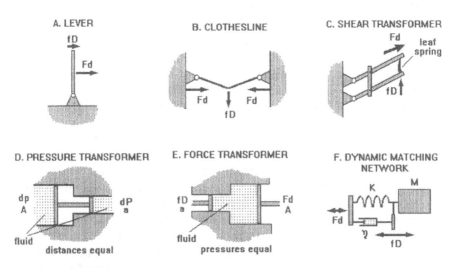

FIGURE 4.2 Six ideal acoustic transformers (see text). Each device transforms the combination of a small force (*f*) or pressure (*p*) and a large displacement (*D*) into the combination of large force (*F*) or pressure (*P*) and a small displacement (*d*). Large and small areas of pistons are shown as *A* and *a*, respectively. In each case the energy supplied to the device equals the energy supplied to an attached load. That is, the devices are passive, neither adding nor subtracting energy. (K, stiffness; M, mass; η, viscosity).

force transformer, and the dynamic matching network (see Section 2.2). In all cases depicted, a small force (f) or pressure (p) acting over a large distance (D) is converted to a large force (F) or pressure (P) acting over a small distance (d). The force transfer here is assumed to be passive, in that the energy supplied to the device is passed on to the acoustic load without loss or gain. On the other hand, the acoustic impedance at the site of force application *is* transformed. Clearly the impedance looking into the inputs (points fD of Fig. 4.2) of an acoustic transformer depends on the geometry of the device and on the nature of the acoustic load at its output (points Fd of Fig. 4.2). It can be simply shown that if a force at the input of a transformer is transformed into a force n times larger at its output, then the impedance felt at the input is a factor n^2 smaller than that of the acoustic load attached to the output. Such impedance transformation can be put to good use in coupling one sound-conducting medium to another with the least reflection, and therefore with the maximum efficiency or energy transfer. This process is known as *impedance matching* and is important in coupling the sound in air to the deformation of the MET channels of hair cells.

2.2 Micromechanics at Low Frequencies

Figure 4.3 presents one simple combination of acoustic transformers that serves as a first approximation to cochlear mechanics at low frequencies, where mass and viscosity are unimportant. Here, the middle ear is represented as a combination of a clothesline (Fig. 4.2B) and a lever (Fig. 4.2A). Once the stapes footplate is set in motion, the cochlea acts to a first approximation as a force transformer, with a small vibration of the relatively large stapes footplate producing a larger vibration of the BM at the characteristic place for the stimulus frequency. Transverse movement of the BM and OC is then transformed into a shearing motion between the reticular lamina and the TM (shear transformer 1 of Fig. 4.3), and between the reticular lamina and the BM (shear transformer 2 of Fig. 4.3). The deflection of the hair bundles of the hair cells, which pivot at their base, produces a shear motion between the individual stereocilia of the hair bundles (shear transformers 3 and 4), which finally applies forces to the MET channels, opening and closing them to produce the hair cell receptor current (Section 4.6). Remember that because it was assumed that each segment of the OC vibrates independently of every other segment, it is sufficient to consider only one location along the cochlea at a time.

2.3 Micromechanics and Morphology

Clearly, an understanding of the impedance transformations and modes of vibration of the OC in any particular cochlear region requires a detailed knowledge of its geometry (Rhode and Geisler 1967), the mass of each of its

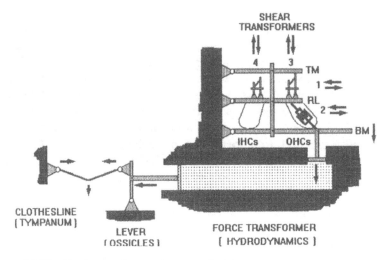

FIGURE 4.3 The ideal acoustic transformers of Figure 4.2 can be combined to give a simple micromechanical model of the OC at one location along the cochlear length. The four shear transformers shown are (1) between the tectorial membrane (TM) and reticular lamina (RL); (2) between the RL and basilar membrane (BM); (3) between adjacent stereocilia of the outer hair cells (OHCs) and (4) inner hair cells (IHCs). The force transducer shown between the BM and RL represents the action of the active OHCs. The relative elements of each transformer vary along the cochlea, and the positioning and existence of the hinges for some of the shear transformers are open to debate. Since these ideal transformers have no mass or damping, the model is only applicable for very low frequencies.

component parts, and the stiffness and friction of the coupling between them. The main variations in the geometry of the OC are discussed in Slepecky, Chapter 2. Essentially, in moving from base to apex, most cells, including the hair cells and their stereocilia, get longer and take up more of the width of the BM in absolute terms (see Slepecky, Chapter 2; Lim 1980). The BM itself also becomes wider and thinner in the apex, and its cross section changes from two distinct thickened regions separated by a thin "hinge" area in the base, to a simpler cross section with no marked grading in thickness across its width in the apex. The two distinct regions across the radial width of the BM are called the pectinate zone (the area that is furthest from the central axis of the cochlear spiral and devoid of hair cells) and the arcuate zone (the inner area supporting the OC). In some animals, notably some bats (Kössl and Vater 1995), there is an elaborate thickening of the BM, which presumably increases BM mass so that it forms a significant part of the mass of the whole cochlear partition. Normally, however, the BM mass is probably insignificant when compared to the mass of the cellular elements of the OC and that of the fluid trapped beneath the TM (the subtectorial space) and within the scala media.

The overall effect of this grading in geometry and morphology is to make the cochlear partition slightly larger, more massive, and less stiff in the apex than it is in the base. Although mass is estimated to increase very little in moving from base to apex, the stiffness of the BM is estimated to decrease by about 100:1 over the same range. The friction or damping associated with vibration of each of the components is very difficult to estimate, either theoretically or by direct measurement, but some estimates suggest that it may decrease in moving toward the apex, based on the fact that the distance between the TM and the reticular lamina is larger in the apex because of the lengthening of the hair cell stereocilia.

2.4 Micromechanics at High Frequencies

2.4.1 Simple Resonance: Mass, Stiffness, Damping, and an Active Process

Although the simple model of Figure 4.3 is useful as a starting point, it ignores mass and viscosity, and therefore frequency-dependent vibration. The simplest model that includes mass and viscosity is shown in Figure 4.4A. Here, the cross section of the cochlear partition at a particular location is represented as the combination of a discrete mass, restored to its rest position by a single spring, and damped by friction.[1] The resonator in this case could be the mass of the OC restored to its transverse rest position by the spring as a result of the transverse deflection of the BM, or the mass could be the TM restored to its radial rest position by the torsional stiffness of the hair bundles. The model of Figure 4.4A also includes a force generator that adds to the forces produced by the sound and can be called an *active process*. Moreover, because the active force is assumed to be controlled by a sensor that detects the resultant movement of the mass, it

---

[1] If the displacement (x) of the mass of Fig. 4.4 from its rest position is sinusoidal for an applied sinusoidal force [$x = X.\sin(\omega t)$], then the spring's restoring force is proportional to displacement such that $f_{spring} = -E.x$, where E is the spring's stiffness. The damping force is proportional to velocity t such that $f_{damp} = -\eta.dx/dt$, where η is the coefficient of friction for the damping element. The inertial reaction force of the mass, m, is proportional to its acceleration such that $f_{mass} = -m.d^2x/dt^2$. Since the applied force, f, and reaction forces must sum to zero, $f = E.x + \eta.dx/dt + m.d^2x/dt^2 = X.[(E - m.\omega^2).\sin(\omega t) + \eta.\omega.\cos(\omega t)]$. Since velocity is $v = dx/dt = X.\omega.\cos(\omega t) = V.\cos(\omega t)$, $f = V.[(E/\omega - m.\omega).\sin(\omega t) + \eta.\cos(\omega t)]$. For sinusoidal stimuli, the total reaction force is therefore the sum of two sinusoidal components: one in phase with the velocity [$\eta.\cos(\omega t)$] and the other 90° out of phase with it [$(E/\omega - m.\omega).\sin(\omega t)$]. This can be written $f = V.z.\sin(\omega t + \Phi)$, where $|z| = \sqrt{\eta^2 + (E/\omega - m.\omega)^2}$, and the phase angle $\Phi = \arctan[(E/\omega - m\omega)/\eta]$. The ratio of driving force at one point in a complex mechanical system divided by the velocity ($z = f/V$) at the same point is known as that point's *mechanical impedance*. The inverse ratio (velocity/force) is called the point's *mechanical admittance*. In the case of the resonator of Fig. 4.4, the magnitude of the admittance is $|y| = |V/f| = 1/|z| = 1/\sqrt{\eta^2 + (E/\omega - m.\omega)^2}$. The phase angle of the impedance and admittance are equal in magnitude but opposite in sign. Note that for a fixed force stimulus, the velocity is proportional to admittance.

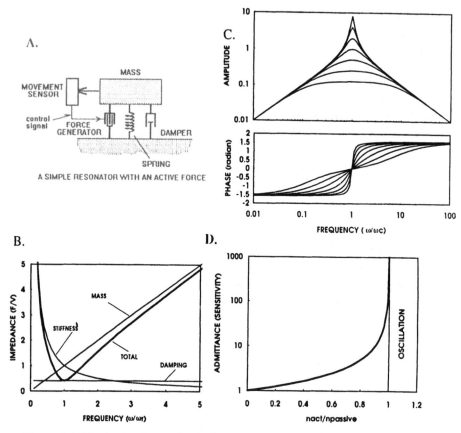

FIGURE 4.4 (A) A simple mechanical resonator with active feedback. (B) The magnitude of its acoustic impedance varies with frequency. The curve of "total" or summed impedance takes into account the phase of each component. (C) Amplitude of vibration is proportional to amplitude of admittance (1/impedance). Admittance reaches a maximum at the resonance frequency. The sharpness of tuning and the rate of phase change increase as damping is reduced (*topmost curve,* least damping; *bottom curve,* most damping). (D) Effective damping coefficient can be reduced by the active process which cancels the inherent friction ("negative damping"). As net damping coefficient ($\eta_{passive} - \eta_{act}$) decreases, vibration amplitude at resonance grows hyperbolically. If $\eta_{act}/\eta_{passive} > 1$, the system oscillates spontaneously.

represents *feedback* in the system. The force generator will ultimately represent the action of the OHCs.

It is worth summarizing the basic properties of such a simple mechanical resonator (Fig. 4.4). Not only is it the simplest frequency-dependent model of cochlear micromechanics, but it can represent many component parts of the OC when considering more complex micromechanics; it can represent a single element along a mass-loaded water surface, which is the simplest model for macromechanical vibration (Section 3.4); and it also forms the basis of the acoustic transformer referred to in this chapter as a "dynamic matching network" (Fig. 4.2F). The main properties of a simple resonator are as follows:

1. The reaction force of the resonator against the stimulus changes its nature with frequency (Fig. 4.4B). At low frequencies the spring (E) dominates, at high frequencies the mass (m) dominates, and near the resonance frequency [$\omega_R = \sqrt{E/m}$] these forces are almost equal in magnitude but opposite in direction, and the friction (η) dominates by default.

2. At the resonance frequency (ω_R), acoustic admittance (the inverse of impedance) and sensitivity are a maximum (Fig. 4.4C) and depend only on damping $|y| = 1/\eta$.

3. The sharpness of tuning depends on damping. Above or below ω_R, the admittance and sensitivity drop (Fig. 4.4C). At ω_{high} above and ω_{low} below ω_R, the resonator's response is decreased by a factor of 3dB (a drop in amplitude of $1\sqrt{2}$ and a drop of power of ½). The 3-dB bandwidth of the resonator is defined as the frequency difference $BW_{3dB} = (\omega_{high} - \omega_{low})$, and a useful parameter known as the quality factor or Q_{3dB} can be defined as $Q_{3dB} = \sqrt{(m.E)/\eta} = \omega_R/BW_{3dB}$. The admittance is then $|y| = \eta\sqrt{[1 + Q_{3dB}^2(\omega/\omega_R - \omega_R/\omega)^2]}$. In other words, reducing the damping η increases the quality factor Q_{3dB} and increases the sharpness of tuning (Fig. 4.4C). For cochlear vibration, an empirical Q_{10dB} figure is used instead of Q_{3dB} (Section 3.11).

4. The phase difference between the force stimulus and the velocity response varies with frequency (Fig. 4.4C) from $-90°$ at low frequencies, through zero at ω_R, to $+90°$ at very high frequencies. This 180° phase change is a signature of a simple resonator, which is why von Békésy's observations of larger phase changes in the cochlea indicated that Helmholtz was wrong in assuming simple resonance.

5. Damping determines the resonator's response time to transient stimuli. This was the point raised by Helmholtz and later by Gold: if the ringing time is very long, then the damping must be small.

6. If the force generator acts to cancel friction (rather than mass or stiffness), then near the resonance frequency, small changes in the active force can produce large changes in mechanical sensitivity (Fig. 4.4D). If the active force acts in a direction opposite to the passive damping, then the active force is given by $f_{active} = +\eta_{active}.dx/dt$. If the damping coefficient without the active element is called $\eta_{passive}$, then the total damping coefficient is $\eta = (\eta_{passive} - \eta_{active})$, and the force generator is said to produce negative damping. The admittance at resonance is then $1/\eta = 1/(\eta_{passive} - \eta_{active})$. If the force generator could progressively be "turned on" so that it increasingly canceled inherent damping, then the system's admittance and therefore sensitivity at resonance would increase hyperbolically as η_{active} increased (Fig. 4.4D). This is clearly a form of positive feedback. When $\eta_{passive} = \eta_{active}$, the total admittance is infinite (impedance is zero) and the system could move spontaneously without an applied force.

If $\eta_{active} > \eta_{passive}$, the resonator oscillates spontaneously. Figure 4.4D illustrates this point, plotting the relative admittance of the system at the resonance frequency versus the ratio ($\eta_{active}/\eta_{passive}$). The important point is that such a system with negative damping is critically dependent on the magnitude of the active feedback or negative damping (i.e., on η_{active}). Although it may confuse some readers that negative damping is a form of positive feedback, a moment's thought should resolve the problem: the term "negative damping" is a classic case of a double negative.

7. A simple resonator can act as a "dynamic matching network." When driven by an oscillating stimulus, a resonator converts the force and displacement applied to its input (F and d of Fig. 4.2F) into a smaller force and a larger displacement at its output (f and D of Fig. 4.2F). It differs from the first five acoustic transformers (Fig. 4.2A–2E) in that it only works for vibrations. Although it produces no impedance transformation for very slow displacements, at its resonance frequency it transforms impedance by the square of the resonator's quality factor: $(F/d)/(f/D) = Q_{3dB}^2$.

2.4.2 More Complicated Micromechanics

Although useful, the simple resonator is probably *too* simple to accurately represent the micromechanical vibration of the OC. If each component part of the OC has a significant mass, and is separated by a finite stiffness and damping, then all mass, stiffness, or damping elements cannot be lumped together. Nevertheless, it must be possible to approximate the more complicated OC motion by a combination of discrete elements, thereby allowing the estimation of the impedance transformations occurring within the OC. Unfortunately, there is a problem: there is little or no direct information allowing an estimate of the mass, stiffness, and damping of each element in the organ of Corti, and as a result, very few people can agree on the way that the OC vibrates at high frequencies. This is highlighted in Figure 4.5, in which six recent models of micromechanical vibration are shown, as discussed below.

2.4.2.1 What Can Most People Agree On?

There is *some* common ground in discussing OC micromechanics. Most workers agree that the movement that ultimately stimulates the hair cells to produce a receptor current is the deflection of the hair bundles in the radial direction (Flock 1965; Hudspeth and Corey 1977). There is also general agreement that the pillar cells that form the arch of Corti are relatively stiff. As a result, when vibration occurs, the BM must move up and down at the foot of the outer pillar cells, and the arch of Corti must pivot about the foot of the inner pillar cells, close to the edge of the spiral lamina. Clearly, in this case, the IHCs do not move as much as the OHCs in a transverse direction. A glance at a cross section of the OC should convince most people of this point: a large part of the cell bodies of the IHCs actually sit above the bony

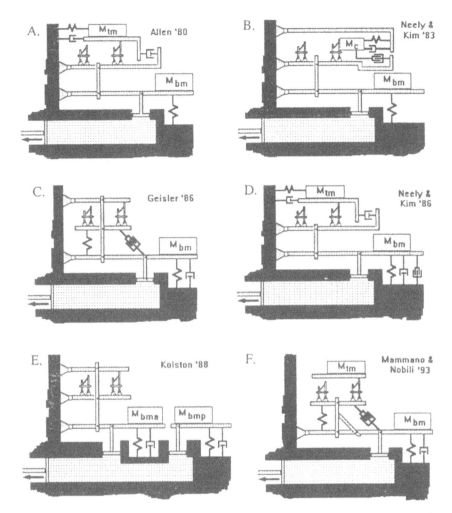

FIGURE 4.5 Using different assumptions about the relative stiffness and mass of various structures within the organ of Corti, modelers can produce a wide variety of micromechanical models. Six of them are shown here (see text).

spiral lamina, rather than on the moving BM. It is also well accepted that the reticular lamina, which forms a plate across the apex of the hair cells through which their hair bundles protrude, is relatively stiff and does not flex under normal stimulus conditions, but does rotate relative to the plane of the BM. In this case, it is unlikely to pivot at its extreme inner edge near the inner sulcus (Fig. 4.5A, for example), but is more likely to rock with the apex of the arch of Corti. In addition to rocking with the arch of Corti, the reticular lamina may also pivot about the apex of the arch of Corti (Fig. 4.5C,F). Clearly, both these actions would only occur if the cells surrounding the inner sulcus were compliant and the rotational stiffness about the apex of the arch of Corti were relatively small. As described in Holley (Chapter 7), recent observations suggest that the OHCs can change their length under some conditions. If this length change is to affect the vibration of the OC, the reticular lamina *must* pivot about the arch of Corti, otherwise the OHCs would be attempting to move against a very stiff load. If the reticular lamina refused to pivot, the OHCs could only affect vibration of the OC and BM by pushing on the pectinate region of the BM, via the Deiters cells, which does not appear to be an efficient way of maximizing the mechanical stimulus to the hair bundles of the IHCs.

2.4.2.2 Stiffness and Mass of the Basilar Membrane

Two parameters of obvious importance are the stiffness and mass of the BM. Direct measurement of stiffness have been performed in live animal preparations and in extirpated cochleae (Gummer, Johnstone, and Armstrong 1981; Olson and Mountain 1991, 1993). It is generally agreed that the stiffness measured using localized mechanical probes is not the stiffness of ultimate importance, since the normal stimulus is a distributed pressure across the BM width. The BM stiffness also appears to vary with amplitude of the probe displacement, and the stiffness obtained for the smallest displacements is assumed to most accurately reflect the stiffness of the BM–OC combination at normal sound levels. There is also some evidence that the arcuate and pectinate zones have different mechanical properties, although the significance of these differences is not clear. If these two regions of the BM vibrated independently, we would be faced with the problem of two "basilar membranes" moving in parallel. This raises the possibility of one region (the pectinate zone) moving freely, effectively "stealing" the pressure stimulus from the other region (the arcuate zone) without stimulating the hair cells (Kolston 1988). Although there is some evidence that there are differences in the mechanical properties and vibration across the BM width (Sellick, Yates, and Patuzzi 1983; Olson and Mountain 1993; Ulfendahl and Khanna 1993), the physiological significance of these differences is unclear.

As for the effective mass of the BM, most mathematical modelers include in their estimates the mass of all those structures and fluid compartments that are assumed to move in unison with the BM. These include the membrane itself, the BM cells, the OHCs, the supporting cells, and the fluids trapped within the OC. One structure that is not included in the effective mass of the BM is the TM, which is most often assumed to move independently in models incorporating micromechanical motion.

2.4.2.3 The Mass and Stiffness of the Tectorial Membrane

It is clearly important whether the TM moves in unison with the BM, or whether it forms a second resonator with a resonance frequency quite different from that of the BM. One extreme view is that the mass of the TM is insignificant, while its stiffness in the radial direction is very large (Fig. 4.5B,C,E). In this case, its resonance frequency would be very high, and far beyond the frequency range normally encountered by the ear. As a result, the TM would simply act as a tether, tying the tips of the hair bundles to the spiral limbus and the bending of the hair bundles would simply follow the motion of the BM and/or the reticular lamina. In another extreme view (Fig. 4.5F), the mass of the TM is thought to be very important, but its stiffness and damping are considered insignificant (Zwislocki, Chamberlain, and Slepecky 1988; Zwislocki and Ceferatti 1989). In this case, it would act simply as an inertial load, similar to the otoliths of the vestibular system or the sallets in some lizards (Wever 1978). Under these conditions it would offer very little impedance to movement at very low frequencies, where its inertia would be insignificant. This prediction seems to contradict many measurements of low-frequency microphonic potentials in extracellular fluid and macromechanical motion. Intermediate views hold that neither the TM's mass nor its stiffness in the radial direction can be ignored, and that they act together to form a resonant structure that moves optimally at its own resonance frequency, which is within the normal hearing range and different from the resonance frequency of the BM with its associated mass and stiffness. Zwislocki (1979) and Allen (1980) (Fig. 4.5A) have suggested that this relative motion between the TM and the BM–OC may account for some of the tuning in hair cells and primary afferent neurons. In particular, at a frequency just below the CF of a cochlear region, they have suggested that the TM resonator and the BM–OC resonator move with approximately the same amplitude and phase. As a result, the shear between them and the deflection of the hair bundles is minimized. Although there are clear examples of neural responses that suggest such a two-component resonance (Guinan and Gifford 1988), especially for neurons with relatively low CF (more apical regions), there is not a great deal of experimental data supporting this proposal in the high-frequency regions of the mammalian cochlea (Section 2.4.2.1).

In summary, many of the models presented in Figure 4.5 are similar. They differ only in whether the mass of the secondary resonator is dominated by the TM (Fig. 4.5A,F) or the hair bundles per se (Fig. 4.5B), and whether the stiffness is due to the hair bundles alone (Fig. 4.5F) or a combination of the hair bundles and the TM (Fig. 4.5A,B,D).

2.4.2.4 Coupling of the Hair Bundles to the Tectorial Membrane and Micromechanical Resonance

One important mechanism for reducing the ear's sensitivity to slow pressure changes is the viscous coupling of the hair bundles of the IHCs to the TM. In 1972, Dallos and colleagues reported that following the destruction of OHCs by kanamycin, the sound-evoked microphonic potential within the cochlear fluids was much smaller, and seemed to be synchronized with the velocity of the BM rather than with its displacement. This suggested that the two hair cell types may have different coupling to the TM. Later recordings of the receptor potentials within IHCs (Russell and Sellick 1978, 1983; Nuttall et al. 1981; Dallos and Santos-Sacchi 1983; Patuzzi and Yates 1987) and, to some extent, neural recordings of firing patterns of the primary afferent neurons (Sellick, Patuzzi, and Johnstone 1982b; Ruggero and Rich 1983; Ruggero, Robles, and Rich 1986a) supported the view that IHCs were stimulated in synchrony with the velocity of the BM. The simple interpretation of these results was that because the hair bundles of the IHCs were not embedded in the TM (see Slepecky, Chapter 2), they would not be stimulated by shearing movements between the TM and the reticular lamina. Instead, they should be viscously coupled to the TM and increase their sensitivity to sinusoidal stimulation by 6 dB/octave (doubling their sensitivity with each doubling of stimulus frequency). Clearly, this increase in sensitivity with frequency could not be without limit. In fact, Russell and Sellick (1983) suggested that this velocity or viscous coupling only occurred up to some relatively low frequency (perhaps 1000 Hz), above which the hair bundles became entrained with the displacement of the TM. Recordings of neural firing to low-frequency sinusoidal stimulation intended to resolve this issue produced confusing results, largely because the neurons could respond in synchrony with displacement of the BM down to very low frequencies (40 Hz and lower), and were stimulated by BM displacements in the "wrong" direction (toward the scala tympani). To explain this anomalous neural firing, Sellick, Patuzzi, and Johnstone (1982a) suggested an effect of the low-frequency extracellular microphonic potential on transmitter release at the base of the IHCs, while others suggested a complex micromechanical stimulation of the IHC hair bundles.

Subsequent hair cell recordings have suggested that the IHCs *are* maximally depolarized during movement of the BM toward the scala vestibular, but change to displacement coupling at about 500 Hz (see Dallos 1973b, 1985; Patuzzi and Yates 1987), probably due to the nature of their viscous coupling to the TM (Patuzzi and Yates 1987; Freeman and Weiss 1990a,b). This viscous coupling of IHCs has some other implications for cochlear micromechanics. First, if the hair bundles cannot move easily relative to the fluid in the subtectorial space for frequencies above 500 Hz, then the IHCs should be unable to act like high-frequency tuning forks, resonating *independently* of the motion of the TM and reticular lamina (as occurs in some reptiles). Second, just as the hair bundles of the IHCs cannot move independently of the overlying TM, the TM and reticular membrane would be unable to move independently of each other at high frequencies. Certainly the viscosity of the subtectorial space is judged by most modelers to dominate the damping of the motion of the partition.

2.4.2.5 Other Evidence for Complex Micromechanics

For some primary afferent neurons innervating the apex of the cochlea, the growth of neural firing rate with sound level can be quite unlike the growth expected in the transverse BM vibration and seen in most afferent neurons from basal regions of the cochlea. In particular, the firing rate can grow over a limited range of sound pressure, before dropping significantly at intermediate levels, and then rising again at higher sound levels. This nonmonotonic growth of neural firing rate suggests two stimulation mechanisms working in opposition (Guinan and Gifford 1988), or possibly a complicated micromechanical motion of the OC, as suggested by Allen (1980) and Zwislocki (1979). The response phase of neural firing certainly changes with sound level, and the high- and low-intensity components of the growth function are affected differently by a variety of cochlear manipulations, including loud sound, anoxia, and the loop diuretic furosemide, perhaps due to changes in cochlear micromechanics during these manipulations. Whether these responses are due to complex micromechanical motion or to other complex processing within the OC, such as electrical interactions, is difficult to say. There have been reports of complicated receptor potential waveforms in hair cells that have been attributed to complex micromechanical motion (Cody and Mountain 1989; Dallos and Cheatham 1989). A resolution of this issue must await direct observation of micromechanical vibration in these cochlear regions, or more elaborate recordings of indirect measures of vibration, such as hair cell receptor potentials, cochlear microphonic potential, or neural phase locking.

Finally, a simple comparison of the frequency dependence of hair cell and neural responses with that of macromechanical BM vibration should offer some evidence for complex micromechanical filtering by the OC. Most simply, if the tuning or frequency dependence of the hair cell and neural responses at a particular cochlear location is *identical* to that of the macromechanical (transverse) BM vibration at the same position (taking into account the viscous coupling of IHC hair bundles), then additional micromechanical filtering must be minimal and the radial shear of the hair bundles is simply a scaled version of the transverse displacement at the same position. Although there is some evidence that the tuning of hair cell and neural responses may be different from the macromechanical tuning under some conditions, it is more likely that these discrepancies are due to measurement artefacts due to the loading of the BM by the specifications of radioactive metal used for recording (Sellick, Yates, and Patuzzi 1983, but see Allen and Neely 1992).

3. Macromechanical Vibration

The assumptions made so far were that each segment of the OC along the length of the cochlea formed a separate micromechanical filter that vibrated independently of all others, and that this micromechanical vibration did not affect the macromechanical vibration of the whole cochlear partition. Both assumptions are incorrect. The first is incorrect because the individual micromechanical filters sit on a common BM trampoline and are bathed in a common cochlear fluid. The second is incorrect because the mass, stiffness, and damping of the OC form a significant part of the mass, stiffness, and damping of the whole cochlear partition, and cannot be ignored. In essence, to understand why Helmholtz was wrong in assuming independent vibration of the array of micromechanical filters, we need to understand the interactions between the pressure in the cochlear fluids, the long mobile cochlear partition, and the surrounding immobile cochlear walls.

3.1 The Helicotrema and the Rejection of Very Low Frequencies

The cochlea is routinely exposed to very intense low-frequency stimuli that would cause damage to hair cells and other structures if mechanisms did not exist to reject them. At the extreme apical end, the scala media and the partition taper prematurely to a blind end, leaving a small, perilymph-filled fluid junction between the scala tympani and the scala vestibuli (Slepecky, Chapter 2). Called the helicotrema, the canal ensures that slow or low-frequency pressure fluctuations introduced into either the scala vestibuli or the scala tympani are distributed almost equally to the other chamber, preventing a large differential pressure across the partition and damaging displacements of hair cells. For such low frequencies, an oscillatory movement of the stapes simply pumps perilymph through the helicotrema and displaces the round window. It should be emphasized, however, that the helicotrema is *only* effective for very low frequencies: above a particular cutoff frequency (normally a few hundred Hertz, depending on the size of the helicotrema and therefore the species; Dallos 1970; Ruggero, Robles, and Rich 1986a,b), the helicotrema is unable to pass pressure fluctuations because the combination of the fluid mass and viscosity within its small cross-sectional area forms an effective "acoustic plug." Many diagrams in elementary textbooks are incorrect, showing sound waves entering the cochlea at the stapes footplate, traveling along the length of the scala vestibuli before passing through the helicotrema, and returning to the basal end of the cochlea along the scala tympani. Most pressure fluctuations of audible frequency cannot pass through the helicotrema and do not make it as far as the helicotrema in any case. Rather, they take a "shortcut" through the cochlear partition near the cutoff region for the stimulus frequency.

It is also worth noting that the helicotrema is only one of at least four strategies used to minimize overstimulation of the hair cells by low-frequency pressure fluctuations, which are normally caused by atmospheric pressure changes, swallowing, postural changes, contraction of the middle ear muscles, intermittent closure of the external ear canal, or osmotic or hydrostatic imbalances in the cochlear fluids. The other mechanisms are depicted in Figure 4.1 and include the eustachian tube of the middle ear, which mainly rejects very slow changes in atmospheric pressure, the viscous coupling of the IHCs to the overlying TM, the adaptation of the MET process at the apex of the hair cells, and possibly the action of a neural reflex arc involving the efferent nerve fibers that synapse with the OHCs of the OC.

3.2 Coupling Along the Length of the Cochlea (Helmholtz's Mistake)

As already mentioned, Helmholtz assumed that the different regions of the cochlea resonated independently of each other, vibrating in sympathy with the frequency components of an incoming sound like the strings of a piano. Von Békésy's later observations of partition vibration emphasized that, unlike the strings of Helmholtz's piano, the cochlear partition is not bathed in air, which is relatively yielding when compared with the vibrating structures. Instead, it is suspended in an incompressible salt solution and encased in the cochlea's bony shell. As a result, the movement of any section of the partition has the potential to affect all other sections, either because they are coupled mechanically through the BM, or because they are coupled hydraulically through the fluid. A number of studies suggest that this *longitudinal coupling* is dominated by fluid coupling, with the forces along and through the membrane or OC being much less important. Von Békésy addressed this issue in his early studies, cutting the BM and prodding it with blunt probes. On the basis of the circular indentation produced around his probe, he reached the conclusion that there must be strong longitudinal coupling through the BM, in addition to any hydraulic coupling. However, he used cadaver specimens in which the physical properties of the BM were undoubtedly different from those in fresh or live specimens. Later experiments with fresher specimens by Voldrich (1978) showed little sign of such circular indentations. Rather, the BM was deflected only at the site of the probe, with adjacent sites unmoved. This indicated that the direct longitudinal coupling between adjacent regions of the BM was probably weak.

It is also important from a theoretical perspective whether the protein fibers of the BM are under resting tension, like the strings of Helmholtz's piano, or whether they act more like elastic beams without a resting tension along their length. In the first case, the BM would be a true acoustic "membrane," like a drum skin, with a resting tension and with its elastic restoring force generated by the small length increase produced by its deflection. In the second case, it would be more correctly classified as a "plate," with no resting tension and with its restoring force generated on deflection by a combination of stretch along one surface and compression on the other. Gummer and Johnstone (1983) have observed that there is very little separation of the cut edges of the BM when it is severed, suggesting that it is not under significant resting tension. More recent experiments by Olson and Mountain (1991, 1993) also suggest that the stiffening of the BM as it is deflected by a small probe is more consistent with the properties of a beam or plate, rather than that of a string or membrane. It seems that the "basilar membrane" should be named the "basilar plate," and it is more like a xylophone played under water than Helmholtz's piano played in air.

3.3 Straightening the Cochlear Spiral

The basilar "xylophone" is not only under water, it is spiral-shaped. This curvature of the cochlea is not of any great acoustic importance, especially at low frequencies where the cochlear dimensions are much smaller than the wavelength of sound in the cochlear fluids (Viergever 1978, Steel and Zais 1985). Assuming the velocity of sound (longitudinal pressure waves) in water to be 1500 m/s, we would require a frequency as high as 1.5 MHz before the wavelength of the longitudinal sound waves in the cochlear fluids approached a representative cochlear dimension of 1 mm. In other words, the cochlear pressure fluctuations at normal sound frequencies are normally so confined by the cochlea's bony walls that the intricacies of their shape and surface structure are of little concern.

3.4 Macromechanical Vibration as a Surface Acoustic Wave

This simple calculation also demonstrates that compression waves in the cochlear fluids should travel from the stapes to the cochlear apex within microseconds. How then is it that delays of milliseconds are detected in the responses of nerves at the cochlear apex (Salvi, Henderson, and Hamernik 1979; Ruggero 1980), or that Kemp's "cochlear echoes" can re-emerge from the cochlea milliseconds after the stimuli that produce them (Kemp 1978)? This apparent dilemma is resolved when it is realized that there are two different modes of wave propagation in a fluid, bulk compressional waves and surface acoustic waves, and that the vibration of the cochlear partition that stimulates nerves or elicits echoes is a surface acoustic wave,[2] and not a bulk compressional wave.

---

[2] That partition vibration is indeed a surface acoustic wave is emphasized if we consider a cochlea in which both the oval window and the round window possess a stapes, moving in opposite directions in a "push-pull" arrangement. Vibration is fed to this hypothetical cochlea as a *differential* pressure drive across the partition. If the fluid chambers above and below the partition are identical in size and shape, then the situation is symmetric and any change in pressure in the scala vestibuli will be matched by an equal and opposite pressure change in the scala tympani. In this case the forces acting on the partition will be the same as if the cochlea possessed

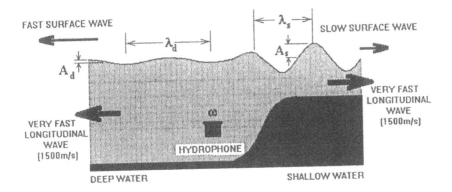

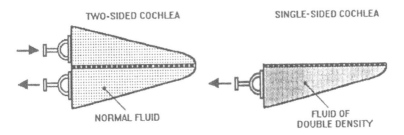

FIGURE 4.6 A sound source at the bottom of a water tank can produce waves that travel outward as fast longitudinal (compression/rarefaction) waves in bulk fluid and as slower transverse waves on the surface. The surface wave's speed depends on fluid density and depth, and on stimulus frequency ("dispersion"). In a hypothetical symmetric cochlea with two stapes driving the oval and round windows in antiphase (*lower left*), the forces on the partition would be equal to those in a cochlea with only one fluid chamber filled with fluid of double density (*lower right*). This transformation from the double-sided to the single-sided cochlea emphasizes that macromechanical vibration is the vibration of a water surface loaded by the cochlear partition.

These two forms of wave motion are illustrated in Figure 4.6, which depicts a marine hydrophone placed below the surface of an expanse of water of finite depth. The sound waves produced by the hydrophone can propagate outward either as horizontal particle oscillations parallel to the water surface (these are longitudinal compression waves and the normal mode of sound propagation in bulk fluid), or as vertical oscillations at the water surface that create a transverse wave or ripple that travels outward

---

only one chamber, filled with fluid twice as dense as that normally present, and the force applied to the single stapes were double that presented to either one of the two stapes (*lower right* of Fig. 4.6). When the two-sided cochlea is transformed into its single-sided equivalent, the vibration of the partition is clearly a surface or Rayleigh wave, subject to the well-known limitations on such waves.

from the hydrophone along the surface. The transverse surface motion is known as a Rayleigh wave or a surface acoustic wave (Ash and Paige 1985), and from everyday experience it is clear that the two perturbations do not travel at the same speed: longitudinal sound waves in water travel at about 1500 m/s, while surface waves, like the ripples on a pond surface, travel at a more leisurely speed that depends on the frequency of the sound, the density and surface tension of the water, and the depth of the water.

Vibration of the cochlear partition under some conditions resembles the vibration of a water surface with elastic surface tension (Fig. 4.7B), while under other conditions it resembles the vibration of a water surface loaded by a thin mass layer and restored to its rest position by a distributed elasticity, similar to gravity in the case of the water wave (Fig. 4.7C). The nature of the wave motion on such water surfaces can be likened to a von Békésy wave by using a mathematical relationship between the frequency (ω) of the stimulus and the wavelength (λ) known as the *dispersion relationship* for the surface. The dispersion relationship can be presented in a compact form if the *wave number* $k = 2\pi/\lambda$ is used instead of the wavelength λ directly. In the case of water with surface tension T, density ρ and depth h, restored to its rest position by a gravitational attraction g, the dispersion relationship can be shown (Main 1993) to be $\omega = \sqrt{[g.k + T.k^3 \rho]\tanh(k.h)}$.

In the case of a water surface loaded by a mass layer of density $m(\text{kg/m}^2)$ and restored to its rest position by a distributed elasticity $E(\text{N/m})$, the dispersion relationship can be shown to be $\omega = \sqrt{\dfrac{E.k.\sinh(k.h)}{[m.k.\sinh(k + \rho.\cosh(k.h)]}}$ (Yates 1976). The implications of these equations are illustrated in Figure 4.7 and described below:

1. The simple wave velocity (phase velocity) given by $v = f.\lambda = \omega.\lambda/2\pi = \omega/k$ is not constant in either case but is a function of the depth of the water: shallow water waves travel at a fixed velocity, whereas the velocity of deep water waves changes with frequency. As the frequency increases, the wavelength decreases relative to the depth of the fluid, shallow water waves become deep water waves, and the velocity changes from fixed to one that decreases with frequency. This difference in velocity with frequency explains the term "dispersion relationship": waves of different (high) frequency are dispersed along the surface as they travel, even though they may start at the same time and place (de Boer and Viergever 1984). This fact can be important in the cochlea because the presence of the cochlea's bony shell can alter the speed of a surface wave traveling along the partition if its frequency is low and its wavelength is long.

2. Although surface tension is unimportant at low frequencies where wavelengths are long, it begins to dominate wave propagation at high frequencies when the wavelengths become so short that significant shearing exists between adjacent regions on the water surface. In fact, at very high

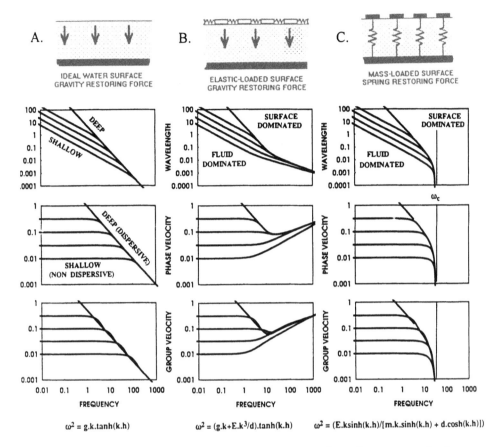

FIGURE 4.7 Water surface waves have different properties depending on water depth (see text). Curves show the variation in wavelength, phase velocity, and group velocity with stimulus frequency. Each curve in a panel is calculated from the dispersion relationship (given below panels) for a different depth of water. *Uppermost curves* (marked "deep") and *lowermost curves* (marked "shallow") represent the asymptotic curves obtained with water depth approaching infinity and zero, respectively. Successively lower curves are for shallower water. (A) An ideal water surface with gravity as the restoring force but without surface loading. (B) As for (A) but with surface tension (analogous to longitudinal coupling in the cochlea). (C) A mass-loaded water surface with stiffness as a restoring force rather than gravity (see text). Note that in the case of the mass-loaded surface (*far right*), the surface wave does not propagate for frequencies above a particular cutoff frequency (ω_c) which depends not on the depth but on the mass and stiffness of the surface layer. All values are in arbitrary units.

frequencies, the depth of the water becomes irrelevant and wave propagation is determined by the properties of the elastic surface layer. For water surfaces, the wavelength at which surface tension forces equal gravity forces is about 17 mm at 20°C. For shorter wavelengths, surface tension dominates. Since the magnitude of the longitudinal coupling along the cochlear partition is unknown, the analogous transition wavelength and frequency in the cochlea are unknown. And, since the mechanical properties of the cochlea vary along its length, these would vary from one cochlear region to another in any case. Interestingly, the speed of wave travel in this case actually increases with frequency at high frequencies (Fig. 4.7B), instead of remaining constant for very low frequencies or decreasing for moderate frequencies. This increase in speed with frequency at high frequencies is known as "anomalous dispersion."

3. For a surface loaded by a mass layer that is restored by a distributed elasticity (such as the BM loaded by OC), the surface wave's speed falls to zero as the frequency approaches the resonance frequency of the surface layer (i.e., the resonance frequency of the mass/stiffness combination). In other words, the wave "stalls" if the stimulus frequency is too high and *surface waves cannot propagate along a mass-loaded surface above the resonance frequency of the surface.* This stalling or cutoff phenomenon can be brought about either by increasing the stimulus frequency or by reducing the stiffness restoring the mass layer to its rest position, and both occur within the cochlea (Fig. 4.8).

As for the origin of the cutoff on the mass-loaded surface, it is most easily "explained" by noting that the wave along the mass-loaded water surface relies totally on the spring's restoring force: anything that abolishes this force eradicates the traveling wave. For a sinusoidal stimulus at any frequency, the spring's restoring force is partially canceled by the mass reaction force, which is in the opposite direction (from Section 2.4.1, the reaction force of a spring is synchronous with displacement, whereas that of a mass is synchronous with acceleration and therefore acts in the opposite direction at any instant). At the resonance frequency, the cancellation is complete, the spring restoring force is abolished, and the wave ceases to travel. Above the cutoff frequency, the inertial reaction force of the surface mass dominates and is in the wrong direction to support a traveling wave on the surface.

Finally, although the traveling waves along the BM are analogous to waves on a water surface under some conditions, there are important differences: in the case of a water surface, the restoring force is a mixture of gravity and surface tension, whereas the BM is restored to its rest position by a mixture of its transverse stiffness due to its plate nature ("gravity") and that due to the longitudinal coupling along its length ("surface tension").

4. Cochlear Micromechanics and Macromechanics   213

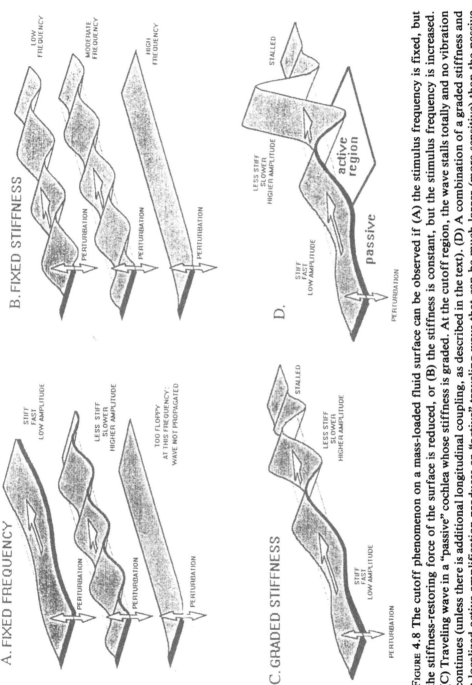

FIGURE 4.8 The cutoff phenomenon on a mass-loaded fluid surface can be observed if (A) the stimulus frequency is fixed, but the stiffness-restoring force of the surface is reduced, or (B) the stiffness is constant, but the stimulus frequency is increased. (C) Traveling wave in a "passive" cochlea whose stiffness is graded. At the cutoff region, the wave stalls totally and no vibration continues (unless there is additional longitudinal coupling, as described in the text). (D) A combination of a graded stiffness and a localized active amplification produces an "active" traveling wave that can be much larger (more sensitive) than the passive traveling wave and very localized along the vibrating surface.

3.5 Origin of the Place-Frequency Map in a Graded Cochlea

With this understanding of surface acoustic waves, it is possible to see how the typical macromechanical vibration pattern can occur in the cochlea with its grading of partition stiffness from base to apex (Fig. 4.8). As a sound stimulus of fixed frequency is presented to the outer ear, it produces vibration of the middle ear bones and stapes footplate, producing pressure fluctuations within the fluids of the scala vestibuli of the basal cochlear turn (like the hydrophone of Fig. 4.6). As a result, a traveling surface acoustic wave propagates along the BM toward the apex (Fig. 4.8C). As the wave moves apically, however, it encounters regions of the BM that are less stiff. As a result, the wave amplitude grows and the propagation velocity and therefore the wavelength decrease. Eventually, a cochlear region is reached where the stiffness is so low that the cochlear partition approaches resonance and the wave begins to stall. Near this cutoff region, the vibration amplitude and oscillatory pressure fall precipitously, because of the shunting of the fluid pressure drive by the highly mobile surface layer. It should be noted here that the peak of the traveling wave (the characteristic place) is not the site of resonance of the partition along the cochlea; rather, the resonance site corresponds to the cutoff region where vibration amplitude first falls to zero. Moreover, no traveling wave at a particular frequency can propagate beyond its cutoff region, unless there is direct mechanical coupling along the partition (similar to the "surface tension" case of Fig. 4.7B). Without such longitudinal coupling, apical regions beyond the cutoff region remain undisturbed by the traveling wave, unless, of course, they are affected indirectly by electrical, ionic, or biochemical perturbations that might spread from more basal regions where vibration is present.

3.6 Changes to the Passive Wave in an "Active" Cochlea

It has been assumed so far that the macromechanical vibration is purely "passive," or powered solely by the incoming sound. The *passive waves* propagating in such a passive system may or may not travel in the forward or reverse direction (from and to the stapes, respectively), depending on the site of the perturbation along the cochlea (the hydrophone of Fig. 4.6) and its frequency relative to the cutoff frequency of each cochlear region encountered in its travels.

In the last fifteen years, probably the most important realization in cochlear physiology has been that OHCs inject mechanical energy into a traveling wave as it passes them, creating a larger vibration than would be present otherwise. The resulting *active* wave (Fig. 4.8D) shares many of the features of the passive traveling wave (Fig. 4.8C), and it grows relative to the passive wave as long as the active hair cells are adding energy to it. With such active assistance, a much lower sound level is required to produce a particular amplitude of vibration or a particular hair cell or neural response. In addition, the active traveling wave is more localized along the cochlear length, apparently due to the decreased damping or improved "quality factor" for each incremental segment along the cochlear length, due to the partial cancellation of friction by the active hair cells. Finally, the active wave is labile, reverting to the passive traveling wave if the active process is disrupted, or at high sound levels when the active process appears to become relatively less important.

The progressive change from an active vibration pattern at low sound levels to a passive pattern at high sound levels (Johnstone, Patuzzi, and Yates 1986) is depicted in Figure 4.9, and is responsible for much of the nonlinearity in cochlear mechanics.[3] In Figure 4.9 the vibration profile (the traveling wave "envelope") produced by a pure tone of a specific frequency is depicted at low, medium, and high sound levels in a traumatized cochlea with no active process (left set of diagrams of Fig. 4.9) and in a normal cochlea with active assistance. As can be seen from Figure 4.9, the traumatized cochlea actually presents the simplest case, with the vibration profile simply scaling with the sound level, regardless of cochlear place or stimulus frequency. The more complex vibration pattern occurs in the normal active cochlea (*right* set of diagrams in Fig. 4.9), with a very sensitive and localized vibration pattern at very low sound levels, but a broad vibration pattern at high sound levels, similar to that present in the traumatized or passive cochlea (note the similarity of the vibration in the *topmost* diagrams). The reason for this convergence at high sound levels is most easily explained by stating that the normal cochlea at low sound levels has "something extra" that assists vibration in a very local region of the traveling wave, but this extra "something" is missing after trauma or at high sound levels. As already mentioned, and discussed in detail later, the extra "something" appears to be the assistance of the OHCs.

3.7 Nonlinear Growth of Vibration Amplitude with Sound Level

The gradual transformation of the vibration profile from the highly sensitive and localized active pattern at low stimulus levels to the less sensitive and less localized passive pattern at high stimulus levels (Fig. 4.9)

---

[3]This change in vibration pattern in a normal cochlea as sound level increase also explains the so-called half-octave shift observed with acoustic trauma, wherein maximum threshold elevation most often occurs at a frequency half an octave *above* the pure-tone exposure frequency (see Cody and Johnstone 1981; McFadden 1986; Patuzzi 1986). Put simply, as the sound level increases, the vibration profile changes so that the peak of the traveling wave at high sound levels is located more basally than the peak at low sound levels. As a result, the maximum trauma is produced in a more basal region: a region that normally responds to higher frequencies at low sound level (in fact, half an octave above the exposure frequency!).

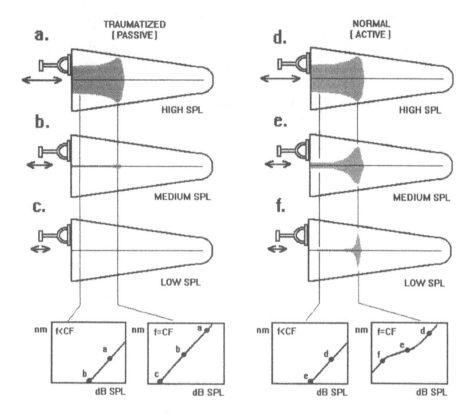

FIGURE 4.9 (A–C) In a traumatized cochlea with no active process, vibration is insensitive, poorly localized, and scales with stimulus sound pressure level (SPL). This produces linear input–output functions (*bottom left* diagrams) of vibration amplitude at all frequencies and at all regions along the cochlea. (D–F) With the active process, at low sound levels (F) the vibration at the peak is much more sensitive (× 1000), and the vibration profile is highly localized. However, as stimulus level increases (E), the vibration becomes more like the passive case (B). At very high sound levels (D), the vibration profile is very similar to the passive case (A versus D). The progressive reduction in active enhancement with sound level in a normal cochlea (D–F) produces input–output functions that are frequency-dependent (*bottom right* diagrams): when the recording site corresponds to the peak of the traveling wave (i.e., stimulus frequency is at or near the CF of the recording site), the input–output function is highly compressive or nonlinear, but when the recording site is substantially basal to the vibration peak (i.e., the stimulus frequency is much lower than the CF of the recording site), the input–output function is linear, as in the passive case.

determines how the amplitude of vibration grows as a function of sound level at any particular cochlear location. Whereas the vibration simply scales with sound level in a passive cochlea, regardless of cochlear position or stimulus frequency, in an active cochlea the vibration grows in a nonlinear or compressive fashion when the stimulus frequency is such that the peak of the traveling wave (the characteristic place) coincides with the observation site (*right* panels of Fig. 4.9). In such cases the vibration amplitude can grow with stimulus level as slowly as 0.2 dB/dB compared with 1 dB/dB for linear or proportional growth in a traumatized cochlea (Yates, Winter, and Robertson 1990), or in a normal cochlea with the stimulus frequency much lower than the CF of the observation site.

An example of the experimentally observed frequency-dependent growth of vibration with sound level is illustrated in Figure 4.10A, for a range of frequencies at, above, and below the CF of the observation site. In this case the data are from the basal turn of the guinea pig, where the CF was about 18 kHz. Note that when the stimulus frequency is much lower than the CF of the observation site, the growth of vibration with sound level is linear or proportional (growth functions with a slope of 1 dB/dB), whereas frequencies near the CF of the observation site produce a nonlinear or compressive growth of vibration with sound level. Figure 4.10B,C also illustrate the frequency-dependent nonlinear growth of the responses in IHCs and primary afferent neurons, respectively, in the same region of the guinea pig cochlea. For Figure 4.10B, the response is millivolts of depolarization of the IHC membrane potential (the DC receptor potential) as a function of sound level, whereas in Figure 4.10C the response is the action potential firing rate of the afferent neuron. It can be seen that the nonlinear growth or "saturation" of the IHC and afferent neuron responses at CF are even more marked than that observed in the macromechanical vibration, but this is to be expected: the receptor current generation process at the apex of an IHC saturates for large displacements of its hair bundle (Sellick and Russell

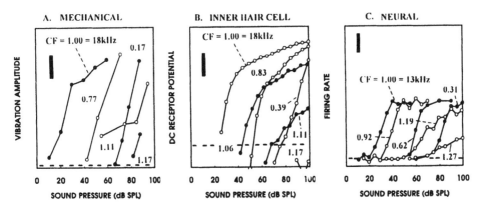

FIGURE 4.10 Frequency-dependent growth of (A) transverse vibration amplitude of BM with sound level (Sellick, Patuzzi, and Johnstone 1982b) is responsible for most of the frequency-dependent increase of (B) IHC depolarization (Patuzzi and Sellick 1983) and (C) primary afferent firing rate (Patuzzi and Robertson 1988). The more pronounced saturation of IHC depolarization at high sound levels that is not present in vibration is easily explained by saturation in the current generation mechanism at the apex of OHCs, whereas the even more pronounced saturation of neural firing rate at high sound levels is probably due to saturation of neurotransmission at sensory synapses. Frequency given as parameter normalized to CF.

1978; Goodman, Smith, and Chamberlain 1982; Patuzzi and Sellick 1983; Dallos 1985, 1986; Kros, Chapter 6), and the neurotransmitter release and reception processes at the primary afferent synapses also saturate for large depolarizations of the IHCs. This is clear from a comparison of the responses at low frequencies, where the vibration of the BM grows linearly with sound level (Fig. 4.10A at 0.17 CF or 3 kHz), but the IHC depolarization and afferent firing rate are still observed to saturate (Fig. 10B,C; Winter, Robertson, and Yates 1990; Yates, Winter, and Robertson 1990). Nevertheless, these data suggest that the compressive growth of the vibration at CF is mostly responsible for the large dynamic range of the IHCs and afferent neurons, and ultimately auditory perception in the whole animal. In particular, the absence of the compressive growth of vibration near the CF after cochlear trauma can produce an abnormally rapid growth in perceived loudness called "recruitment" (Yates 1990).

3.8 Panoramic Versus Local Views of Macromechanical Vibration

As already hinted at in describing the growth of vibration with sound level, the traveling wave can be viewed from two very different perspectives. On the one hand, a pure tone of fixed frequency and intensity can be presented to the cochlea and large tracts of the partition can be observed vibrating at the same time (the "panoramic view" of Fig. 4.11), or a small pinhole can be drilled in the cochlear wall to observe the vibration of a single point along the length of the partition (the "local view" of Fig. 4.12). In the local view, the changes in the amplitude and phase of the point's vibration can be observed as the frequency and intensity of a sinusoidal stimulus are altered. The relationship between the two views of vibration is illustrated in Figure 4.11. Although the more panoramic view is easier to grasp, it is extremely difficult to achieve technically. As a result, most experimental data are obtained and presented as *iso-input frequency tuning curves* (FTCs), where the vibration amplitude of a single location is plotted as a function of the frequency of a pure tone stimulus (Fig. 4.11). Such FTCs possess a large amplitude peak at the CF, a high-frequency region above the CF for which no response can be elicited at any sound level, and a low-frequency "tail" region where vibration is small but observable. Alternatively, *iso-output* or *iso-response* FTCs are presented, plotting the sound level required at each stimulus frequency to produce a predefined vibration amplitude at the observation site. Although the iso-output FTC normally resembles an upside-down version of the iso-input FTC (Fig. 4.11), it is not simply a mirror image of it, since the vibration amplitude does not necessarily scale with stimulus sound level (Section 3.7).

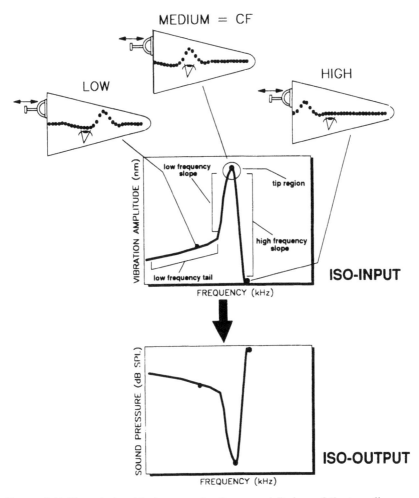

FIGURE 4.11 The relationship between the "panoramic" view of the traveling wave and the derived "local views" of iso-input or iso-output "frequency tuning curves" (FTCs). The small eye in each of the *upper panels* indicates the observation site for the recordings at which the iso-input and iso-output FTCs apply.

3.9 Mechanical Iso-input Frequency Tuning Curves

The extreme sensitivity of the cochlea referred to earlier is largely due to the fact that the vibration of the BM at the peak of the traveling wave is many times larger than the vibration of the stapes, as illustrated in Figure 4.12A. Here, iso-input FTCs are produced by plotting the amplitude of BM vibration as a function of stimulus frequency for a fixed amplitude of stapes vibration (similar to the iso-input FTC of Fig. 4.11). Three iso-input curves are presented: one from the 8-kHz region of the chinchilla (Robles, Ruggero, and Rich 1986) and two from the 18-kHz region of the guinea pig

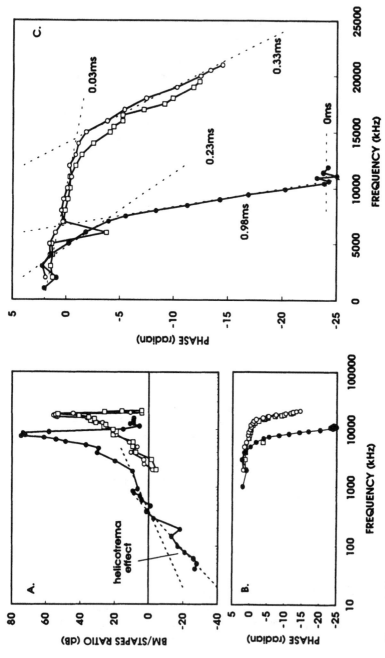

FIGURE 4.12 (A) Iso-input FTCs from transverse BM vibration at the 8-kHz region of the chinchilla cochlea (●, Robles, Ruggero, and Rich 1986) and the 18-kHz region of the guinea pig cochlea (o, □, Sellick, Yates, and Patuzzi 1983) derived on the basis of a fixed amplitude of stapes vibration. (B) The phase delay or lag between stapes and BM vibration, increasing monotonically with frequency, depends on species, stimulus frequency and intensity, position along the cochlea, and condition of the animal. (C) When the data of (B) are plotted on a linear frequency axis, the phase delays can be ascribed to equivalent time delays or "group delays" (shown next to each curve), which produce a linear relationship (*dotted lines*) between stimulus frequency and phase lag.

(Sellick, Yates, and Patuzzi 1983). It is clear that the amplitude of BM vibration is between 500 (55 dB) and 5000 (75 dB) times larger than the vibration of the stapes at the CF of each experimental preparation.

It may appear odd at first glance that the vibration of the BM should be so much larger than the vibration of the stapes: if the fluid is incompressible, where did the extra fluid displacement come from? This difference between the amplitude of stapes and BM vibration breaks no fundamental law of physics. A one-to-one relationship between the amplitude of stapes and BM vibration can be broken in a number of ways. First, the *area* of the stapes need not equal the *area* of the vibrating portion of the BM. As a result, the fluid *volumes* displaced by the stapes and by the partition can be equal without the *linear displacements* of the two structures being equal. Second, the fluid volume displaced by the stapes must only equal the *sum* of the fluid volumes displaced by the BM and that passing through the helicotrema. In fact, at low frequencies, the BM need not move at all, since all the fluid displaced by the stapes simply pumps through the helicotrema and displaces the round window. This shunting effect of the helicotrema is evident in Figure 4.12A at low frequencies (below 400 Hz), where the BM vibration can actually fall *below* that of the stapes. Finally, at high frequencies, where the helicotrema allows very little fluid to oscillate through it, the volume displacement of the stapes must still only equal the *summed* volume displacement along the whole length of the BM. In other words, for a small stapes displacement at one instant during pure tone stimulation, it is possible for the BM to have a very large displacement toward the scala tympani at one location along the membrane, as long as it is balanced by an approximately equal displacement toward the scala vestibuli at some other location. In other words, the 5000-fold larger vibration of the BM at one location is only possible because another region is vibrating almost as much in the opposite direction!

3.10 Phase Changes as a Function of Frequency

Phase changes as a function of frequency raise the issue of the relative timing of the vibration at various positions along the length of the BM. The vibration of the BM and the stapes are not only different in amplitude, but they occur at slightly different times (as with all waves). Figure 4.12B presents the phase difference in radians between the vibration of the BM and the stapes when a pure-tone stimulus is used. For relatively low frequencies (greater than the 400-Hz cutoff of the helicotrema), the phase of the sinusoidal vibration of the BM leads the vibration of the stapes by about a quarter of a cycle (1.6 radian). This is often explained by simple acoustics. At low frequencies, peak pressure in the cochlear fluids corresponds to peak velocity of the stapes footplate (Dallos 1973a), and at these frequencies the partition behaves as a simple spring-loaded piston with peak displacement corresponding to peak pressure in the cochlear fluids (Dancer and Franke 1980; Nedzelinsky 1980; Ruggero, Robles, and Rich 1986b). At higher frequencies, however, where the partition's acoustic admittance is no longer like that of a simple spring-loaded piston, the vibration of the BM begins to lag behind cochlear fluid pressure and therefore stapes vibration until, at frequencies much above the CF, the BM lags behind stapes vibration by 2.5 to 4 cycles of the sinusoidal stimulus (about 15 to 25 radians in Fig. 4.12B). Like sensitivity and tuning, these phase changes are intensity-dependent and also change with cochlear trauma (Sellick, Yates, and Patuzzi 1983).

Figure 4.12C presents the phase delay of the BM vibration (Fig. 4.12B) as a function of frequency on a linear rather than logarithmic frequency scale. This is useful because a fixed *time delay* between stapes vibration and BM vibration at a particular cochlear location would produce a linear relationship between phase and frequency. The time delay that would give the same rate of change of phase with increase in stimulus frequency is called the *group delay,* and is given by $T_g = d\phi/df$ with the frequency in cycles per second and the phase delay, ϕ, in cycles. The *dashed lines* on Figure 4.12C correspond to the lines of best fit to different regions of the phase versus frequency curves, and represent the equivalent group delays over the various frequency ranges. Clearly the group delay between vibration of the stapes and vibration of the BM depends on the species, the recording site, and the frequency of the stimulus. For example, for relatively low frequencies at the 18-kHz region in the guinea pig cochlea, the rate of change of phase of BM vibration with frequency suggests a time delay of only 0.03 ms, as might be expected, because the partition in the basal turn approximates a stiff plate at these frequencies. At higher frequencies, however, the rate of change of phase suggests an equivalent delay of about 0.3 ms. Clearly the traveling wave does not travel at the same speed for high- and low-frequency tones (dispersion), as discussed in Section 3.4. Moreover, although the 0.3-ms delay at 18 kHz is small, it is far longer than would be expected for a compressional sound wave traveling in the cochlear fluids. Assuming that the 18-kHz region is about 3 mm away from the stapes in the guinea pig (see Fig. 4.15B), the travel time of a compressional wave would be (3 mm/1500 m/s) = $2\mu s$. That is, the estimated group delay for the surface acoustic wave is about 150 times that expected for the compressional sound wave in the fluid. Similar changes can also be seen in the 8-kHz region of the chinchilla cochlea (Fig. 4.12C). At more apical regions of the cochlea, the group delay is even longer (up to 2 to 3 ms; Ruggero 1994). Clearly the traveling surface wave is much slower than a bulk compressional wave in the fluid, and cochlear vibration is not simple resonance. These relatively long delays may be crucial in analysis of complex signals by the central nervous system.

3.11 Mechanical Iso-output Frequency Tuning Curves

The data of Figure 4.12A also indicate that the cochlea's ability to discriminate frequency is remarkably good. In fact, the tuning of the vibration may explain most of the tuning of the hair cell and neural responses. This point is illustrated in Figure 4.13, which compares iso-response FTCs for a 1-nm vibration of the BM (Fig. 4.13A; Sellick, Patuzzi, and Johnstone 1982b), a 1-mV depolarization of an IHC (Fig. 4.13B; Sellick, Patuzzi, and Johnstone 1983), and the sound pressure required to just elevate the firing rate of a primary afferent neuron above its spontaneous firing rate (Fig. 4.13C; Sellick, Patuzzi, and Johnstone 1982b). All of these data are from the basal region of the guinea pig cochlea.

Figure 4.13 also highlights that the response of the cochlea at any particular location can be viewed as a highly tuned filter, most sensitive at the CF and progressively less sensitive at higher and lower frequencies. The frequency-discrimination power of such filters can be quantified in terms of the rate at which sensitivity drops as the stimulus frequency is increased above or decreased below the CF, that is, in terms of the high- and low-frequency slopes of the FTC (Fig. 4.13). In the case of the cochlea, sensitivity can rise at a rate between 10 and 200 dB/octave on the low-frequency site of the CF and can fall at the startling rate of 600 dB/octave on the high-frequency side (see Patuzzi and Robertson 1988). In other words, changing the stimulus frequency by only 1% can change vibration at a particular cochlear location almost 10-fold.

In Section 2.4 the quality factor for a simple resonant system was described. For a more complex mechanical system, such as the mammalian

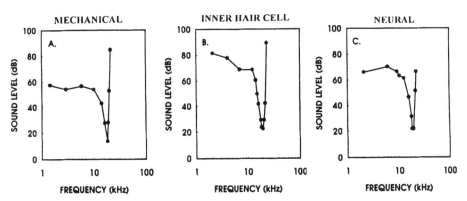

FIGURE 4.13 Iso-input frequency tuning curves for (A) transverse vibration of the BM (1-nm iso-displacement), (B) depolarization of the IHCs (1-mV iso-potential), and (C) neural firing threshold are very similar. The simplest explanation is that IHCs and afferent neurons are simple detectors of the vibration of the organ of Corti (see Patuzzi and Robertson 1988).

cochlea, the quality of tuning can be defined by a slightly different empirical factor called the Q_{10dB}. Here, the resonance frequency ω_R is replaced by the CF of the preparation (which is not the resonance frequency, as described in Section 2.4), and the 3-dB bandwidth is replaced by the 10-dB bandwidth, BW_{10dB}. This Q_{10dB} of the macromechanical vibration can vary between 1 and 300, depending on the stimulus level, the CF, the species, and the physiological condition of the cochlea, and explains much of the tuning of hair cell and neural responses. In the very basal region of the guinea pig cochlea, for example, the CF for the vibration of the OC is between 10 and 45 kHz, depending on the exact position, and the Q_{10dB} is about 10 in an animal in very good condition. If the physiological state of the cochlea is poor, the Q_{10dB} can drop to 1 (see Patuzzi and Robertson 1988). In some bats in good condition, the Q_{10dB} can be as high as 300 in specialized regions of the cochlea.

3.12 The Lability of Cochlear Vibration

Not only does the vibration of the BM and the OC explain much of the frequency selectivity and dynamic range of hair cells and primary afferent neurons, it also appears to explain much of the lability of their responses as well. This is illustrated in Figure 4.14. Figure 4.14A presents an example of the changes in the iso-response FTC for a primary afferent neuron before and after exposure to a traumatic pure tone (Cody and Johnstone 1980; see Schmiedt 1982). Before trauma, the afferent neuron was sensitive and sharply tuned to its CF. After an initial exposure of 7 minutes, it became less sensitive, with elevation of the FTC nearly the CF, but with little change in sensitivity on the low-frequency tail of the FTC. After an additional exposure to the traumatic tone, however, the FTC was elevated across the full frequency range tested (Fig. 4.14A, *uppermost curve*). The data of Figure 4.14B suggest that at least some portion of these changes in the neural FTC are due to changes in vibration within the cochlea. Figure 4.14B presents the mechanical FTC observed in the 18-kHz region of the guinea pig cochlea before and after degeneration of an experimental preparation with time. Like the changes in the neural FTC of Figure 4.14A, sensitivity is reduced first for frequencies near the CF (the tip of the FTC), with little change on the low-frequency tail (Rhode 1971; Sellick, Patuzzi, and Johnstone 1982b; Robles, Ruggero, and Rich 1986). Even after death, however, the vibration is altered little for frequencies on the low-frequency tail of the FTC (Fig. 4.14B). The reductions in mechanical and neural sensitivity in experiments such as those shown in Figure 4.14B are highly correlated (Fig. 4.14C), suggesting that at least under the conditions of these experiments, most if not all of the changes in the neural response are due to changes in the mechanical response of the cochlea.

4. Cochlear Micromechanics and Macromechanics   225

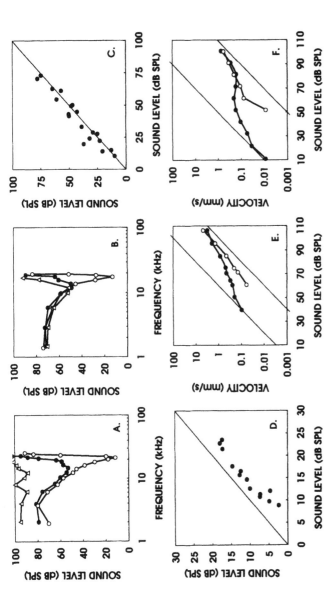

FIGURE 4.14 Much of the change in primary afferent activity after acoustic trauma can be attributed to changes in BM vibration. (A) Changes in the neural threshold FTC before (○) and after 7 min (●) or 11 min (△) of pure-tone overstimulation in a guinea pig (10 kHz at 110 dB SPL; Cody and Johnstone 1980). (B) Changes in the mechanical FTC observed before (○) and after (●) deterioration of an experimental preparation, or after death (△) (Sellick, Patuzzi, and Johnstone 1982). Note the similar changes to (A). (C) The correlation between mechanical sensitivity (dB SPL for 0.35-nm transverse vibration) and neural sensitivity (compound action potential or CAP visual detection threshold at CF) with deterioration of preparations similar to (B) (Sellick, Patuzzi, and Johnstone 1982b). (D) Correlation between neural (CAP threshold) and distortion tone sensitivity following acute acoustic trauma (Johnstone et al. 1990). (E,F) Two examples of the variation in the transverse velocity at CF versus sound level before (●) and after (○) exposure to a traumatic pure tone (Sellick, Patuzzi, and Johnstone 1982b; Patuzzi, Johnstone, and Sellick 1984).

Although the similar changes in Figure 4.14A,B suggest a mechanical origin for the reduced neural sensitivity after loud sound, there are other observations that support this view more directly. First, there is a good correlation between the reduction in neural sensitivity after loud sound and distortion tone generation, one of the indirect measures of cochlear vibration (Lonsbury-Martin et al. 1987; Johnstone et al. 1990; Martin et al. 1990). Figure 4.14D illustrates the correlation between the reduction in neural threshold produced by exposure to a traumatic pure tone and the increase in the sound level required to produce a fixed level of acoustic distortion in the ear canal. The origin of this acoustic distortion will be discussed later, but for the present it can be taken as a crude measure of the amplitude of vibration within the cochlea. Although this correlation between the neural and distortion tone sensitivity also suggests a mechanical origin for the reduction in neural sensitivity, direct observation of the vibration of the BM before and after exposure to a traumatic tone indicates even more clearly that the vibration of the BM can be altered by overstimulation. Figure 4.14E,F present two examples of the growth of vibration amplitude with sound level before and after acoustic overstimulation (Sellick, Patuzzi, and Johnstone 1982b; Patuzzi and Sellick 1983). Before trauma, the vibration was sensitive and grew nonlinearly with sound level, but after acoustic overstimulation, the vibration was less sensitive and grew almost proportionally or linearly with the sound level.

There is now direct and indirect evidence that other cochlear manipulations known to reduce neural sensitivity (creating deafness) do so by changing mechanical sensitivity (for example, a reduction of the EP with the loop diuretic furosemide; Evans and Klinke 1982; Ruggero and Rich 1991b). This clearly explains why it took so many years to observe normal vibration (Section 1.2): the animals were deafened by the surgery and cochlear manipulations required for mechanical measurement.

3.13 Normal and Abnormal Place-Frequency Maps

Cochlear trauma produces similar changes in neural sensitivity, tuning, and dynamic range, whether it is due to acoustic overstimulation, ototoxic drugs, or disease (see Patuzzi and Robertson 1988, and Section 3.12). After such trauma, the place-frequency map is also changed. This is illustrated in Figure 4.15, which presents neural data from the guinea pig basal turn (Robertson and Johnstone 1979). Figure 4.15A presents two FTCs from animals with normal cochlea, possessing normal sensitivity and tuning, and two additional FTCs from animals administered the ototoxic antibiotic kanamycin, some weeks before, which has destroyed the OHCs. After the destruction of the OHCs, the sensitivity and tuning of the neurons are altered, as indicated by the change in the FTCs of Figure 4.15A, and the mapping from stimulation frequency to innervation site or place has also changed (Fig. 4.15B). Just what such changes do to perception of frequency

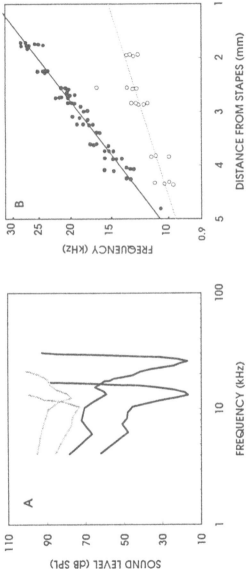

FIGURE 4.15 (A) Iso-output FTCs of primary afferent neurons of the auditory nerve. Each curve represents the sound level (dB SPL) required to initiate action potentials in an individual afferent neuron as a function of frequency. After selective destruction of OHCs with kanamycin (*lighter curves*), the afferent neurons become less sensitive and less sharply tuned, and their CFs change. (B) The relationship between neural innervation site and neural CF is given by a place-frequency map, which varies from species to species and with the condition of the cochlea. In this case, the map is for the basal turn of the guinea pig cochlea when the OHCs are normal (●) or after they have been destroyed with kanamycin (○) (Robertson and Johnstone 1979).

or pitch in the short and long term is not totally clear, but at least in the short term pitch perception is distorted in a process known as *diplacusis* (Davis et al. 1950; Ward 1955).

3.14 The Link Between Micromechanics and Macromechanics

In discussing macromechanical vibration, it has so far been assumed that each segment along the length of the cochlear partition was a simple mechanical resonator, as discussed in Section 2.4. The driving point impedance offered by each incremental segment along the length of the cochlea in this case would be that of a simple resonator as depicted in Figure 4.2, and the simple model can be complicated by the addition of a force generation element. It is clear from the discussion of micromechanical motion, however, that the OC may represent a combination of at least two major mechanical resonators, and that the impedance of such resonators can be transformed by the micromechanics of the OC. In other words, the acoustic impedance presented by the cochlear partition to the cochlear fluids is not a simple resonator at all. The effect of such complicated micromechanical motion of the OC on the macromechanical motion of the cochlear partition requires quite complicated mathematical modeling of fluid hydrodynamics and partition mechanics (see de Boer, Chapter 5). These calculations are complicated even further because the OHCs seem to produce forces of their own that act as additional "sound" sources within the cochlea. As described in more detail in Holley (Chapter 7), it is most likely that the OHCs change shape in synchrony with the electrical signals within them, most likely applying forces to the organ of Corti along their long or axial direction. This is depicted in Figure 4.5B,C,F by the inclusion of a force transducer within the micromechanical models. The magnitude of these forces and their direction of application are discussed in detail in de Boer (Chapter 5) and Holley (Chapter 7). The indirect evidence from mechanical, hair cell, and neural recordings that OHCs enhance cochlear vibration is described below.

4. The Active Process and Mechanical Nonlinearlity

4.1 Evidence for an Active Influence in Macromechanics

Although it is beyond the scope of this chapter to summarize all of the evidence supporting the active assistance of OHCs, there are a number of observations that together provide a strong case for an active contribution to vibration:

1. OHCs are known to move when stimulated electrically, either in vitro or in vivo. It seems inescapable that their receptor currents and/or receptor potentials would continually stimulate them, producing movements that would modify the passive vibration in some way. Indeed, it has been demonstrated that isolated OHCs change their length upon displacement of their ciliary bundle (Evans and Dallos 1993).

2. Sounds are produced spontaneously within the cochlea and are emitted into the external ear canal. Known as spontaneous otoacoustic emissions, they can be monitored with sensitive low-noise microphones, occur in a majority of people, and are reduced by manipulations of the cochlea known to reduce cochlear vibration. A passive cochlea would be incapable of producing such sounds.

3. Normal cochlear vibration is much more sensitive and more localized along the cochlear length than predicted by most passive mathematical models using physiologically reasonable parameters. When the passive models are pushed to produce sufficiently sensitive vibration, often using physiologically unreasonable parameters, they either predict large reflections and standing waves along the cochlear length, or a traveling wave that is *too* localized (that is, tuning of the FTC is too sharp). In other words, most passive models seem to miss one or more of the crucial components of mechanics in the real cochlea, which is relatively stable and tolerant of minor variations in its physical and electrical properties (as one would expect from a system that has evolved over millions of years).

4. Many disruptions that affect cochlear metabolism reduce mechanical sensitivity, specifically of that part of the vibration profile that most passive models fail to predict (the sensitive vibration near the CF). This suggests that the active component of cochlear mechanics is vulnerable to disruption, whereas the passive aspects of vibration are relatively tolerant of minor variations in mechanical, electrical, or biochemical factors.

Although these findings can be readily explained by the presence of a labile active process, no passive model of cochlear vibration has yet come close to integrating all these findings.

Figure 4.16 summarizes the "active" view of cochlear transduction outlined so far, wherein the transduction processes at a particular cochlear location are divided into those associated with OHC active force generation and the enhancement of vibration (*motor processes*), and those associated with the IHCs and the generation of action potentials in the primary afferent neurons (*sensory processes*). The loss of neural sensitivity following disruption of one or more events in the transduction chain depicted in Figure 4.16 would result in a hearing loss. Depending on which of these events was disrupted, the hearing loss could have quite different symptoms. Just as it is convenient to divide the stages of auditory transduction into motor and sensory categories, the broad range of possible hearing losses can be divided into a number of subdivisions according to which stages

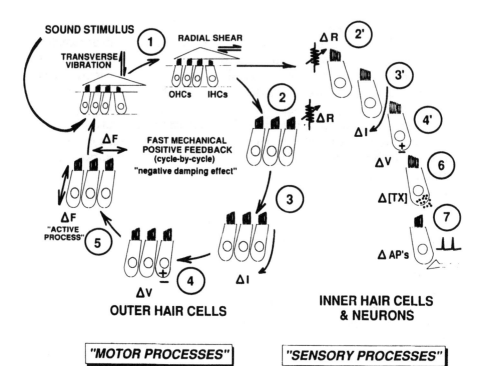

FIGURE 4.16 Transduction processes within the cochlea can conveniently be divided into "motor processes" that involve OHCs and enhance vibration of the cochlear partition, and "sensory processes" that involve IHCs and detect vibration by producing firing of primary afferent fibers. Many transduction processes are shared by the motor and sensory pathways: (1) Macromechanical and micromechanical vibration produce a relative shear stimulus to the hair bundles of the IHCs and OHCs. (2, 2') Deflection of hair bundles modulates electrical conductance (ΔR) at the apex of hair cells by opening and closing MET channels. (3, 3') Receptor currents (ΔI) flow due to this conductance change and the voltage across the apical membranes of the hair cells (endocochlear potential minus hair cell membrane potential). (4, 4') Receptor currents produce changes in cell membrane potentials [receptor potentials (ΔV)]. (5) In OHCs, the changes in membrane potential seem to control active forces (ΔF). (6) In IHCs, the membrane potentials control the release of neurotransmitter (Δ[TX]) and ultimately neural firing (ΔAPs) (7).

of cochlear transduction are disrupted: a motor loss, a sensory loss, a mixed cochlear loss (with motor and sensory components), and finally a mixed cochlear loss with morphological changes.

4.2 How Sensitive Is Vibration to Disruption of the Active Process?

How would vibration of the cochlear partition change if the active assistance of all OHCs could be reduced by some percentage? It is possible to estimate this relationship between the loss of active assistance and the reduction in mechanical sensitivity empirically using a few simple measurements and a modicum of assumptions. First, Figure 4.14C indicates that under some circumstances, a drop in neural sensitivity is due almost solely to a drop in mechanical sensitivity. In these cases, the gross compound action potential (CAP) evoked from the auditory nerve by tone-burst stimuli can be used to monitor mechanical sensitivity. Second, the cochlear microphonic (CM) potential produced by a low-frequency tone in the fluid surrounding the OHCs near the round window is dominated by the receptor currents through them (Dallos 1973a; Russell and Sellick 1983; Patuzzi, Yates, and Johnstone 1989a). As a result, this low-frequency microphonic potential can be used as a simple assay of the ability of the OHCs to produce receptor current, before and after various cochlear manipulations. Finally, moderate acoustic trauma appears to be one manipulation that produces graded changes in the active vibration of the cochlear partition by disrupting OHCs only: after moderate acoustic trauma, morphological changes are minimal (see Liberman and Mulroy 1982; Schmiedt 1984; Liberman and Dodds 1987) and the IHCs and neurons seem unaffected. This seems so because the *tip* of the neural FTC, near the CF, is elevated without elevating the low-frequency tail, as shown in Figure 4.14A. If the IHCs or neurons were themselves disrupted, threshold elevation would be expected for all stimulus frequencies (i.e., across the full FTC, as in Fig. 4.14A for prolonged acoustic trauma). This lack of change below the CF also indicates that the passive mechanical properties of the partition, such as stiffness, are affected little. This interpretation of the changes in the neural FTC is supported by direct observation of BM vibration before and after acoustic trauma (Fig. 4.14E,F): changes occur near the CF of the mechanical FTC, but not on the low-frequency tail.

These findings are very useful because they indicate that the change in the low-frequency CM after moderate acoustic trauma is due *not* to changes in vibration but to disruption of the ability of the OHCs to produce receptor current (Cody and Russell 1985), presumably due to inactivation of MET channels (Patuzzi, Yates, and Johnstone 1989b). Figure 4.17A shows the relationship between the reduction in the ability of the OHCs to produce receptor current after acoustic trauma (as measured by the 200-Hz CM) and the associated reduction in mechanical sensitivity (measured using the CAP response). Clearly the vibration is extremely sensitive to changes in OHC receptor current, and presumably to changes in the active process in general. In fact, a 1% drop in receptor current in a normal animal appears to produce a 1-dB change in mechanical sensitivity. A 50% drop produces a 40-dB or 100-fold reduction in mechanical sensitivity. This extreme sensitivity to changes in OHC receptor current would be expected from a system that enhanced its own vibration using positive feedback (negative damping) driven by the OHC receptor current (see. Fig. 4.4D).

Although this is the simplest interpretation of the data, it is possible that

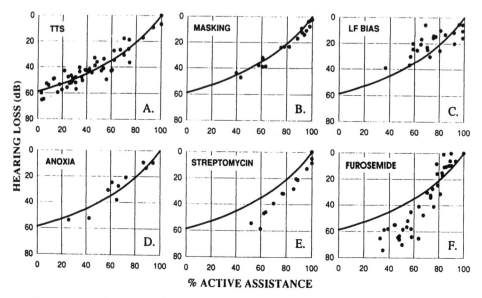

FIGURE 4.17 Many disruptions of cochlear function reduce OHC receptor currents (measured with 200-Hz cochlear microphonic) and presumably reduce active forces from OHCs (% active assistance). When this occurs, neural sensitivity is decreased (hearing loss, as measured by decibel elevation of visual detection threshold of the compound action potential from the auditory nerve). Changes in OHC and neural sensitivity are highly correlated for many cochlear disruptions. TTS: Temporary threshold shift due to overstimulation with an intense pure tone (Patuzzi and Da Cruz, unpublished data; see also Patuzzi, Yates, and Johnstone 1989b). Masking: Reduction in receptor currents and elevation of neural threshold due to a nontraumatic pure tone that jams the generation of receptor current (Patuzzi, Yates, and Johnstone 1989c). LF bias: Slow movement of OC by low-frequency tones (Patuzzi, Yates, and Johnstone 1989c). Anoxia: Drop in endocochlear potential and OHC receptor currents due to lack of oxygen (Patuzzi, Yates, and Johnstone 1989c). Streptomycin: Acute blockage of MET channels by streptomycin in the scala media (Patuzzi and Da Cruz, unpublished data). Furosemide: Reduction of the endocochlear potential by systemic injection of the loop diuretic furosemide (Patuzzi and Da Cruz, unpublished data). The *solid curve* in each panel represents the mean of data for TTS (probably only OHC function impaired). In those cases where data fall below this curve for TTS, IHCs and synaptic transmission are also known to be affected, and a "mixed cochlear loss" has occurred.

more complicated changes occur. This seems unlikely, however: the same relationship between the drop in OHC receptor current and loss of neural sensitivity is observed for a range of cochlear manipulations (Patuzzi, Yates, and Johnstone 1989c). Examples of these are shown in Figure 4.17B-D. Admittedly, in some cases the loss of neural sensitivity is greater than would be expected from a simple loss of OHC receptor current (Fig. 4.17C,D), but in these cases it is known that IHC transduction and/or synaptic sensitivity is also affected, simply because the tail of the neural FTC is elevated under these circumstances.

4.3 Which OHCs Along the Cochlea Are Active?

Mathematical models and a variety of experimental observations suggest that not *all* OHCs encountered by the traveling wave are equally effective in adding to its energy. This is clearly suggested by two-tone suppression experiments in which the neural response to a low-intensity probe tone is suppressed by the presentation of a second suppressor tone. In these experiments, the probe tone is most often set to a frequency equal to the CF of a particular neuron, and at a level a few decibels above neural firing threshold. If the suppression tone is then set to a particular frequency and slowly increased in intensity, it is found that the neural firing to the probe tone is suppressed, even before the suppressor reaches a level high enough to cause firing of its own in that neuron. In other words, the suppression boundary lies below the response boundary in intensity (Fig. 4.18E,F; see Schmiedt 1982; Geisler et al. 1990). This is the case at almost all frequencies for which suppression can be demonstrated, as discussed later.

It is clear that the suppressor tone at these levels is not disrupting the detection of vibration by the IHCs, or the neurotransmission from IHCs to auditory neurons. First, and most obviously, the suppressor tone is not at a high enough intensity to produce stimulation of the IHCs or neurons. Second, the suppressor tone at this level does not suppress neural firing caused by a probe tone presented a few decibels above threshold at frequencies much lower than the CF. It is also unlikely that this suppression is mediated by extracellular electrical or chemical signals: the extracellular potentials are far too small (less than 1 mV), and the suppression phenomenon is too fast for the diffusion of a chemical or ionic intermediate. Rather, the presence of the suppressor tone seems to disrupt the active process, changing the active vibration, without altering passive vibration at frequencies much below the CF. This view is supported by recent direct observations of suppression in vibration of the BM (Patuzzi, Sellick, and Johnstone 1984; Robles, Ruggero, and Rich 1991; Cooper and Rhode 1992). Just how a second tone could disrupt or "jam" the active process is discussed later.

Whatever the details of the jamming mechanism, this two-tone suppression can be used to trace out the longitudinal extent of the active OHCs. Our only assumptions need be that only some of the OHCs along the cochlea are assisting the traveling wave produced by the low-intensity probe tone, and that if the vibration produced by the suppressor tone is large enough in the region of active hair cells, then it can disrupt or jam their contribution in some way and reduce vibration due to the probe tone. We can also assume, for simplicity, that equal jamming is produced by equal displacements of the OHCs, regardless of the frequency of the jamming

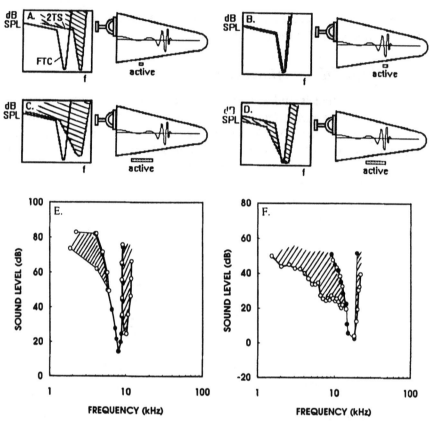

FIGURE 4.18 Expected neural iso-suppression tuning curves (*shaded areas*) are compared with the expected iso-response tuning curves (*black curves*) predicted on the basis of (A) a very localized active process located basal to the vibration peak, (B) a very localized active process located very near the vibration peak, (C) a distributed active process located basal to the vibration peak, and (D) a distributed active process located near the vibration peak. Iso-response (●) and iso-suppression (○) FTCs for (E) action potential generation of a single primary afferent neuron (Arthur, Pfeiffer, and Suga 1971) and (F) depolarization of an IHC (Sellick and Russell 1980).

tone. Figure 4.18A–D present four hypothetical outcomes of such a jamming experiment in which a probe tone of fixed frequency and intensity produces an active traveling wave, and a suppressor tone of variable frequency and intensity is used to jam the active wave associated with the probe tone. In one case (Fig. 4.18A), the active hair cells might exist as a small patch of hair cells located some considerable distance toward the stapes from the peak of the probe-tone traveling wave (its characteristic place). In this case, we would expect equal suppression for equal vibration at this active site, and the suppression boundary should be an iso-displacement tuning curve with its tip at the frequency corresponding to the

CF of the active site: that is, above the frequency of the probe tone (Fig. 4.18A). In the second case (Fig. 4.18B), if the active OHCs were near the peak of the probe-tone traveling wave, then the suppression tuning curve would be very similar to the iso-response FTC itself, assuming that the displacements required to produce a neural response and suppression were similar (Fig. 4.18B). In the third case (Fig. 4.18C), if the active hair cells covered a broad extent of the cochlear length and were situated between the peak of the probe-tone traveling wave and the stapes, we would expect a broad suppression tuning curve centered over a frequency above the probe-tone frequency. In the fourth case (Fig. 4.18D), if the active hair cells were centered around the peak of the traveling wave but extended over a considerable length of the cochlea, we would expect a suppression tuning curve centered over the probe-tone frequency, but much broader than the neural iso-response FTC.

Examples of IHC (Sellick and Russell 1980) and neural suppression contours (Arthur, Pfeiffer, and Suga 1971) are shown in Figure 4.18E,F. Clearly, the suppression boundaries are similar to that expected for a relatively broad distribution of active OHCs located between the peak of the probe-tone traveling wave and the stapes (similar to Fig. 4.18D). This slightly modifies the view of active vibration presented in Figure 4.16: the OHCs and IHCs depicted are probably not located at precisely the same cochlear location, but the OHCs responsible for optimal vibration of a group of IHCs at some particular cochlear region are actually located basally (toward the stapes from the IHCs), "upstream" from the IHCs. By comparing the high-frequency edge of iso-response and iso-suppression FTCs of neurons with high CFs (18 kHz in the guinea pig), the patch of active OHCs begins about 650 μm toward the stapes from the probe tone's characteristic place (calculated from the 2.5 mm/octave place–frequency map and the 0.2 octave difference between the high-frequency slopes of the response and suppression FTCs of Fig. 4.18).

4.4 Active Hair Cells and High- and Low-Frequency Suppression Areas

Figure 4.19 illustrates how the putative distribution of active OHCs might combine with the vibration profiles produced by high- and low-frequency tones to produce two-tone suppression. Figure 4.19A depicts the vibration profile produced by a single tone (the probe tone) and the distribution of the active OHCs producing this vibration. The *horizontal bar* beneath the cochlea indicates the patch of active hair cells. Figure 4.19B depicts a similar distribution of vibration and active hair cells for a tone (the suppressor tone) of higher frequency. If the two tones are presented simultaneously (Fig. 4.19C), the higher-frequency tone vibrates the active region of the lower-frequency probe tone, jamming the active hair cells in

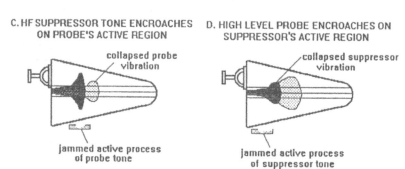

FIGURE 4.19 A panoramic view of high- and low-frequency suppression. (A) A low-level probe tone produces vibration that is large and localized in a normal cochlea, due to the action of the basally located active process (*shaded bar* below cochlear icon). (B) Presentation of a higher-frequency (HF) suppressor tone also produces a large and localized vibration in a more basal region, corresponding to the area of the active process for the probe tone. (C) With both tones, the suppressor "jams" the active process of the probe, and the probe vibration collapses (high-tone suppression). (D) If the probe is increased greatly in intensity, its vibration is largely passive and independent of the active process (Fig. 4.9). If its level is high enough, it could jam the active process of the suppressor, which is then itself suppressed. This reversal of circumstances (the probe becomes the suppressor and the suppressor becomes the probe) is known as low-tone suppression.

some way (Section 4.3) and causing the probe tone vibration near the peak of the probe tone's traveling wave to collapse. This would be described as "high-frequency" or "high-side" suppression. If the intensity of the probe tone were increased sufficiently (Fig. 4.19D), the passive vibration it produced could "jam" the active hair cells of the suppressor tone and reduce the vibration at the peak of the high-frequency suppressor tone's traveling wave. This would be a case of mutual suppression, or if we were to focus only on the suppression of the high-frequency tone's vibration, we would call it "low-frequency" or "low-side" suppression. The point here is that the complicated frequency and level dependence of two-tone suppression (for example, Robertson and Johnstone 1981) is most easily understood in terms of the spatially distributed nature of the active OHCs.

One problem with this view of "low-frequency" suppression often cited is that suppression can be observed even when the low-frequency suppressor is not sufficiently intense to produce a neural or IHC response when presented alone. How can a vibration produce a suppression when it does not produce a response? The answer may be simple: the response normally observed is either the DC or tonic depolarization of the IHCs, or the increase in neural firing rate it produces. It is easily possible that the BM displacements required to "jam" the OHC active process are smaller than those required to produce a depolarization of the IHCs and neural firing. Measurements of low-frequency suppression contours (Schmiedt 1982) suggest that jamming or suppression can occur for vibrations about $\frac{1}{10}$ of the vibration (20 dB below) required for significant depolarization of IHCs and the initiation of afferent firing. Cheatham and Dallos (1995) produced a detailed explanation of this phenomenon.

4.5 Which Nonlinear Processes Account for Mechanical Nonlinearity?

As already mentioned, cochlear vibration can be highly distorted, not faithfully following the original sound stimulus. There are four main manifestations of this mechanical nonlinearity that can be produced by single- or two-tone stimulation. For stimulation with a single pure tone, the vibration amplitude may not grow in proportion to the sound level, and *nonlinear compression* can be observed (Figs. 4.10A, 4.14E,F). Distortion of a single pure-tone stimulus can also produce complicated nonsinusoidal vibration containing harmonics of the stimulus frequency. In particular, if the distortion is asymmetric with larger displacement in one direction than in the other, a net shift of cochlear structures can result while an oscillatory stimulus is applied (Brundin et al. 1991, 1992). This is known as *rectification,* or colloquially as a "DC offset" or "DC shift" (Le Page 1987). When two pure tones are presented to a nonlinear system, superposition does not apply: the resultant vibration is not necessarily the sum of the vibrations produced by each tone presented alone, and the tones are said to interact in a process known as *intermodulation*. For example, an intense low-frequency tone can cyclically modulate the vibration produced by a less intense high-frequency tone (Patuzzi, Sellick, and Johnstone 1984). If both tones are at relatively high frequencies, the cyclic nature of such suppression of one tone by the other may not be apparent, and an overall *suppression* of one tone's response may be observed (Robles, Ruggero, and Rich 1986, 1991; Cooper and Rhode 1992). As just described, this interaction or *two-tone suppression* may be mutual, with each tone suppressing the vibration produced by the other, or it may be one-sided, with one tone (the *suppressor tone*) reducing the vibration produced by the *probe tone,* but the probe tone being powerless to suppress the suppressor tone vibration. Finally, the other side of intermodulation distortion is *distortion tone generation,* where the presentation of two pure tones produces a complicated vibration waveform that contains frequency components that were not present in the original stimulus (Robles, Ruggero, and Rich 1991).

These additional frequency components can be considered to be additional tones, since they are, to some extent, indistinguishable from extrinsic tones that could have been presented in the original stimulus. As described earlier, they are known variously as *distortion tones, combination tones, difference tones,* or *intermodulation products.*

There are many possible causes for this nonlinearity in cochlear mechanics: (1) the passive components of the OC may be mechanically nonlinear so that the partition presents a nonlinear acoustic load to the sound stimulus and/or active process; (2) the passive mechanical components may be linear while the active motor is inherently nonlinear, either because the electrical signal driving it is distorted, the active force generator per se is nonlinear, or both; (3) mechanical components of the OC may be linear while the transverse vibration of the OC and the radial shear of the hair bundles are nonlinearly related by virtue of their geometry. Mechanistically, there are clearly five major steps in cochlear transduction that might introduce distortion into mechanics: the pressure to displacement transformation including the nonlinear trigonometry associated with the transformation from macromechanical vibration to hair bundle displacement (stage 1 of Fig. 4.16); the displacement to current transformation (stages 2 and 3); the current to voltage transformation (stage 4); the voltage to OHC length transformation (stage 5); and, finally, OHC length change to effective force or equivalent pressure on the OC and partition (not shown). Each is known to be nonlinear to some extent and could contribute to the observed nonlinear mechanics.

Although the trigonometric transformation from transverse BM displacement to hair bundle displacement is nonlinear for large displacements (Johnstone and Johnstone 1966), it is not so for the small displacements present for physiological sound levels, and can be ignored. As for the remaining sources of nonlinearity, the link between OHC length change and force production is poorly understood and will not be discussed further. On the other hand, a great deal is known of the nonlinearity of the remaining processes. The relative importance of each nonlinear process is illustrated in Figure 4.20, which shows the nonlinear relationships between hair bundle displacement and hair bundle stiffness (Fig. 4.20C), between hair bundle displacement and electrical conductance at the apical membrane of the hair cells (Fig. 4.20A), between hair cell receptor current and cell membrane potential (Fig. 4.20B), and between cell membrane potential and change in OHC length (Fig. 4.20D). These curves are arranged to emphasize the cascaded relationship between them in the whole cochlea and the relative importance of their nonlinearity.[4] The origin and relative importance of

---

[4] In Figure 4.20C an additional stiffness has been added to the measured hair bundle stiffness of isolated hair cells because the stiffness of the BM and of other cochlear structures must be added to that of the hair bundles, after taking due account of the various acoustic impedance transformations produced by the OC. For the sake of simplicity, a stiffness equal to the maximum hair bundle stiffness has been added.

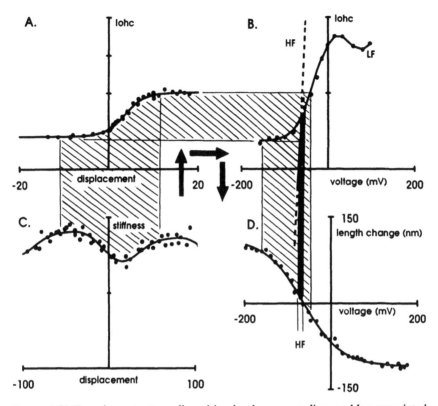

FIGURE 4.20 Four important nonlinearities in the mammalian cochlea associated with OHCs that may produce mechanical nonlinearity. (A) OHC receptor current versus displacement of the hair bundle. (B) The dominant K^+ current through the basolateral wall of OHCs versus low-frequency (LF) OHC membrane potential (redrawn from Housley and Ashmore 1992). At very high frequencies (HF), the curve approaches a vertical line as membrane capacitance dominates the electrical impedance of the membrane (*dashed line*). (C) OC stiffness versus hair bundle angle, estimated from the "gating stiffness" of the hair bundles of isolated hair cells. Data from Howard and Hudspeth (1988) have been normalized in the horizontal and vertical directions to produce an "average" variation of hair bundle stiffness with hair bundle angle, and an additional stiffness has been added to effectively (and somewhat arbitrarily) double the maximum stiffness. This accounts to some extent for the stiffness of other components in the OC. (D) OHC length change versus voltage across the basolateral wall of isolated OHCs (redrawn from Dallos, Evans, and Hallworth 1991). The four nonlinear curves have been arranged to highlight the cascaded nature of the variables (C to A to B to D). The scales have been chosen to allow an approximate comparison between the normal operating range across each nonlinearity (*shaded areas*). Under normal conditions, the saturation of the OHC receptor current (A) and the nonlinear stiffness of the OC (C) are likely to dominate nonlinearity and the production of distortion (see text).

these four nonlinearities in contributing to mechanical nonlinearity in the cochlea are discussed below.

4.6 Mechanoelectrical Transduction and Gating Stiffness

The magnitude of the receptor current through hair cells depends on the number of MET channels that are open at the apex of the hair cells, and on the electrochemical gradient driving the dominant K^+ ions forming the receptor current. The relationship between the deflection of the hair bundles and the receptor current follows a sigmoidal *transfer curve* known as a *Boltzmann function* (Fig. 4.20A; Kros, Chapter 6) which arises from the statistical thermodynamics of the channel gating. As the hair bundle displacement biases the relative potential energies of the open and closed states of the MET channels, they shuffle rapidly into the state with the lowest potential energy (Holton and Hudspeth 1986). The resulting nonlinear Boltzmann transfer curve is relevant to cochlear mechanics because (1) distortion in the MET process would produce electrical and therefore mechanical distortion through the active process, and (2) the shuffling of the transduction channels between the open and closed states requires work to be done and must therefore be "palpable," affecting the mechanical properties of the hair bundles directly, without an electrical intermediate (Howard and Hudspeth 1988; Russell, Kössl, and Richardson 1992; Jaramillo, Markin, and Hudspeth 1993).

Indeed, the hair bundles in individual hair cells with voltage-clamped membrane potentials are relatively compliant in their normal upright position, but become stiffer when deflected from side to side (Fig. 4.20C). This deflection-dependent change in stiffness can be abolished reversibly with the application of the antibiotics gentamicin or neomycin to the solution bathing the hair bundle (Howard and Hudspeth 1988; Russell, Kössl, and Richardson 1992). These antibiotics also block the generation of the receptor current in hair cells, indicating an interaction with the MET channels. As a result, the change in hair bundle stiffness with deflection is attributed to the opening and closing of the MET channels themselves, and is called the hair bundle's *gating stiffness*.

Clearly the hair bundles must offer a nonlinear acoustic load to the pressure fluctuations in the cochlear fluids, even without electrical distortion and any involvement of the active process. Since distortion tones can even be observed in animals post mortem, when the active process is assumed to be highly impaired (if not totally inoperative), we can assume that some of the distortion within the cochlea must be due to a nonlinear passive load presented to the pressures in the cochlear fluids, and therefore to the tympanum via the various acoustic transformations described earlier. Just how much the passive nonlinear properties of the transduction channels contribute to this passive nonlinearity is not clear. Jaramillo, Markin, and Hudspeth (1993) have suggested that the nonlinear gating stiffness of the hair bundles dominates the generation of distortion tones. Although it is possible, this is not the only explanation for distortion in cochlear mechanics.

4.7 Adaptation of Hair Bundle Stiffness and Transduction

In the last few years it has been shown that the MET process at the apex of some hair cells can also adapt to slow or static displacement of their hair bundle (Howard and Hudspeth 1987; Assad, Hacohen, and Corey 1989; Crawford, Evans, and Fettiplace 1989). For example, if the hair bundle of a vestibular hair cell from the frog sacculus is displaced from its rest position, the cell's receptor current may increase initially, but over the course of tens of milliseconds the current decreases to an intermediate level, just above its initial value. Similarly, a decrease in the receptor current produced by a displacement of the hair bundle in the opposite direction produces a transient decrease in the receptor current, which adapts with a similar time course. It has been suggested that this adaptation occurs because the so-called *tip links* that run through the extracellular fluid to connect adjacent stereocilia (Slepecky, Chapter 2; Kros, Chapter 6) and appear to apply the mechanical energy bias to the flickering transduction channels migrate through the cell membranes covering the individual stereocilia of the hair bundle. More surprisingly, one end of the tip links appears to be attached to a molecular myosin motor that runs along the actin cytoskeleton of the stereocilium under calcium control, effectively "reeling in" the tip link, like a fishing line (Kros, Chapter 6). Of course the amount of movement of this myosin motor is limited, but it does seem capable of changing the operating point on the Boltzmann transfer curve of the MET channels sufficiently to explain the operating point migration observed. The location of the MET channels near the myosin adaptation motor presumably allows modulation of the calcium entry into the stereocilium via the MET channels to act in a self-regulatory role, automatically adjusting the operating point on the Boltzmann transfer curve. Although the role of such hair bundle adaptation in the mammalian cochlea is not clear, if adaptation is significant in the mammal it could contribute to mechanical nonlinearity.

4.8 Current-Voltage Relationship of the OHC Basolateral Wall

The relationship between the magnitude of a current injected into a cell and the change in its membrane potential is determined by the electrical properties of the cell membrane. Most simply, the membrane can be represented by the parallel combination of a resistor and a capacitor. A more complicated model of the membrane must include the nonlinear nature of the ionic flux through membrane channels and the voltage-dependent nature of these channels (see Holley, Chapter 7). In general, the relationship between the current injected and the potential change produced can be represented at low frequencies by a nonlinear current-voltage or I-V curve (Figure 4.20B; Kros, Chapter 6). However, whereas the membrane current is dominated by current through the nonlinear membrane channels at low frequencies, above about 1000 Hz the current through the membrane begins to be dominated by the shuffling current that charges and discharges the membrane capacitance on a cycle-by-cycle basis. At these high frequencies, the I-V curve is best approximated by an almost vertical line (Figure 4.20B), and the nonlinearity of the low-frequency I-V curve becomes irrelevant. For example, at 20 kHz the capacitive membrane current is about 20 times the channel current, and closing the membrane channels entirely or doubling their numbers would only produce a change in AC receptor potential of 1 in 21, or about 5%. Such a small change in AC receptor potential would probably produce at most a 5-dB change in mechanical sensitivity (100%-95% on the horizontal axis of Fig. 4.17A).

4.9 Membrane Potential Versus Outer Hair Cell Length

As described in more detail in Holley (Chapter 7), the OHCs change their length as their membrane potential is altered: depolarization causes the cells to become shorter and hyperpolarization causes the cells to lengthen. As shown in Figure 4.20D, the length change is not proportional to the voltage change. When judging the contribution of this nonlinearity to nonlinear mechanics, two points should be considered: it is not yet clear whether the motility observed in OHCs in vitro is definitely the active process responsible for enhancing the cochlear vibration in vivo, and changes in membrane potential of tens of millivolts must occur before significant nonlinearity is encountered (Fig. 4.20D; Holley, Chapter 7). Although such large potential changes are not produced at high frequencies because of the membrane capacitance effect described above, slow membrane potential changes of tens of millivolts could modify the sensitivity of force production either by driving the force generation stage into partial saturation or by causing changes in the length of the OHCs and deflection of the hair bundles, producing partial saturation of the MET process.

5. Input Admittance of the Cochlea and Otoacoustic Emissions

The cochlear input admittance is now important clinically because it contributes to the total acoustic admittance looking into the ear canal and is therefore accessible to objective, noninvasive measurement for the clinical diagnosis of cochlear disorders. It is also intimately linked to the phenomenon of otoacoustic emissions, which play a part in the disorder of tinnitus (ringing in the ear). Finally, the cochlear input admittance is an indirect measure of the mechanical events within the cochlea and is therefore useful experimentally.

5.1 Tympanometry and the Cochlear Input Admittance

A chapter on cochlear mechanics would not be complete without some reference to the acoustic input impedance of the cochlea and the associated topic of otoacoustic emissions (e.g., Whitehead, Longsbury-Martin, and McCoy 1996). The cochlear input admittance, otoacoustic emissions, and the audiological test called *acoustic impedance tympanometry* are inextricably linked, simply because the cochlear input impedance accounts for a small but significant component of the ear canal input impedance. As far as the cochlear input impedance is concerned, the combination of admittances due to the sound source, ear canal, and middle ear can be represented as a single equivalent source admittance (y_s^* of Fig. 4.21). At low frequencies, the perilymph simply pumps backwards and forwards through the helicotrema and displaces the round window, and the cochlear input admittance (y_{coch}) is due almost solely to the admittance of the aperture of the helicotrema (y_h) in series with the round window and stapes admittances (y_{rw}, y_{st}; Fig. 4.21C). At higher frequencies, the admittance of the helicotrema is reduced by the inertia and friction of the fluid within it so that a significant oscillatory pressure difference develops between the scale vestibuli and scala tympani. This difference produces movement of the cochlear partition which, at low frequencies, moves in phase along its length, like a simple spring-loaded piston, albeit with one end (the apex) moving more than the other (the base). In this case, the cochlear input admittance is represented by the parallel combination of the helicotrema and cochlear partition (y_{bm} of Fig. 4.21D) in series with the round window admittance. As the stimulus frequency is increased further, the fluid movement through the helicotrema becomes insignificant, and the cochlear input admittance simplifies to that of the cochlear partition alone (y_{bm} of Fig. 4.21E), in series with the round window and stapes admittances. At even higher frequencies, the admittance of the partition become highly complicated (Fig. 4.21F). As the inertia of individual segments of the cochlear partition becomes significant, and the cochlear partition ceases to move in phase along its entire length. That is, the travel time of the traveling wave becomes significant, and a significant phase difference develops between vibration at one end of the cochlea and at the other. As a result of this complexity, the clinical problem of determining the normality or otherwise of cochlear vibration from the cochlear input admittance becomes *almost* impossible: although a particular distribution of driving point

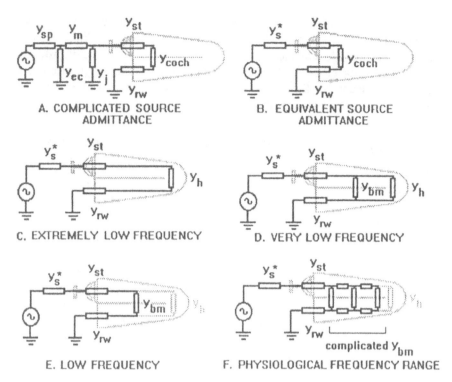

FIGURE 4.21 Components of the cochlear input admittance. (A) The complicated admittances representing the sound source (y_{sp}), ear canal (y_{ec}), tympanum (y_m), and middle ear (y_j) can be replaced by an equivalent source admittance y_s^* (transformation from A to B). (B) The input admittance of the cochlea includes a stapes component (y_{st}), a cochlear component (y_{coch}), and a round window component (y_{rw}). (C) At very low frequencies, the cochlear component is dominated by the high admittance of the helicotrema (y_h). (D) At slightly higher frequencies, both the partition admittance and the helicotrema admittance need to be considered. (E) At even higher frequencies, the helicotrema admittance can be ignored, and (F) at even higher frequencies in the physiological range, the admittance of the cochlear partition becomes highly complicated due to the influence of mass, damping, active hair cells, and the phase delays associated with the hydrodynamics of the cochlear fluids.

admittance along the cochlear length leads *unambiguously* to a particular cochlear input admittance, a particular input admittance does *not* tell us unambiguously about the distribution of driving point admittance (and therefore vibration amplitude) along the length of the cochlea.

One puzzling aspect of cochlear mechanics is that despite this complexity, the input impedance looking into the cochlea is most often described as "resistive," meaning that the ratio of sound pressure to volume velocity does not change markedly with frequency (Lynch, Nedzelnitsky, and Peake 1982). This is quite amazing, and may say more about the evolution of hearing than the simplicity of cochlear vibration. It is very likely that each small partition segment along the length of the cochlea contributes most to the total input admittance of its own CF, but when all segments of the cochlea act together, the total admittance looking into the cochlea appears remarkably dull. Perhaps this result should not be so startling. The same could be said of the normal pure-tone audiogram or threshold contour, which appears simple until it is investigated in great detail. When this is done (Elliot 1958; Long 1984), it is found to contain much greater variation and therefore much more information than would be imagined from the normal clinical audiogram, which consists of (at most) eleven threshold determinations at as many frequencies. In fact, it was precisely the close investigation of the small irregularities in the pure-tone audiogram (Elliot 1958) that lead Kemp to discover otoacoustic emissions in 1977. Indeed, many of the otoacoustic emissions that are observed, including "distortion tones," the "Kemp echo," and "single-tone emissions," are likely to be irregularities in the input impedance of the cochlea, caused by a less than seamless fusion of the contributions from each cochlear region. Although the summation of the contributions from each cochlear region leads to an approximately resistive or absorptive cochlear input, there are small deviations from this ideal that are often described as otoacoustic emissions. This relationship between idiosyncracies in the input admittance and otoacoustic emissions is discussed below.

5.2 Cochlear Input Admittance and Otoacoustic Emissions

At one level, the topic of otoacoustic emissions is quite simple: any sound observed in the ear canal that was not in the original sound stimulus can be termed an otoacoustic emission or OAE (Zurek 1985; Probst, Lonsbury-Martin, and Martin 1991). Otoacoustic emissions fall into two broad categories: spontaneous and evoked emissions. Spontaneous otoacoustic emissions (SOAEs) are sounds present in the ear canal when no external stimulus is applied. Evoked otoacoustic emissions (EOAEs) fall into a number of subcategories, based on the stimulus used to evoke them. They can be evoked either by a single pure tone, in which case the emission emerges at the stimulus frequency (SFOAEs), or by a pair of tones that interact nonlinearly to produce the extra tones known as distortion products (DPOAEs). Emissions can also be evoked by transient stimuli (TEOAEs) such as clicks (CEOAEs) or tone bursts.

Although some of these sounds may be due to middle ear mechanisms (middle ear muscle activity, blood vessels, or fluid in the middle ear), others clearly arise from the cochlea itself. This is most clear because many spontaneous emissions can be modified by tones with frequencies in a very narrow frequency band near the emission frequency. Since the relatively broad frequency-dependence of middle ear vibration cannot explain this highly tuned suppression, but the very highly tuned vibration within the cochlea can, it is most likely that the emissions are generated within the cochlea. Most OAEs are also affected by manipulations that are known to reduce cochlear sensitivity but do not affect middle ear function.

5.2.1 Transient Evoked Otoacoustic Emissions: Reflections from an Imperfect Cochlea

The first OAE observed was the echo evoked by a click stimulus (Kemp 1978). Its existence was not, of itself, an indication of anything particularly new in cochlear physiology. It should have been expected (and was by Kemp!) from a cochlea with a nonideal grading in its mechanical properties along its length. Its presence certainly did not imply an active process. What was most significant was that it grew nonlinearly with sound level. At low sound levels, it also appeared too large to be explained simply by a passive reflection of the sound delivered to the cochlea (Kemp and Chum 1980). Neither of these aspects of the echo was consistent with the view of cochlear vibration that was popular at the time.

The mechanism by which the TEOAEs are produced is not totally clear. It is still a matter of debate whether a perfect cochlea would produce any echoes or reflections at all. It seems more likely that they are due to imperfections in the construction of each individual cochlea, such as abrupt discontinuities in the mechanical properties along the cochlear partition (just as a ripple on the surface of a pond may be partially reflected when it encounters a pond lily). These local imperfections might be caused by the absence or addition of a few hair cells, or by some other localized discontinuity in the partition's mechanical properties. It certainly seems the case that the spectrum of the echo from a particular ear is a unique and stable feature of that ear over many years (Norton and Widen 1991). This stability is at least consistent with the idea that the echoes are due to stable morphological irregularities, rather than to transient electrical or biochemical ones.

5.2.2 Spontaneous Otoacoustic Emissions: Too Much of a Good Thing?

In 1980 Wilson described an even more startling emission phenomenon than the Kemp echo: in some ears, a transient stimulus could produce an echo that refused to die away with time. This was not at all consistent with a simple passive vibration within the cochlea, and was reminiscent of the experiments of both Helmholtz and Gold. Such long-lasting ringing could only be caused by a resonator that possessed little or no friction, or was driven continuously by an external signal. That is, the cochlea must possess some sort of active process, even if it was not associated with the normal traveling wave vibration within the cochlea. Wilson also pointed out that some individuals had previously been observed to produce such a ringing response, even without a click stimulus to initiate it. Since that time, it has been found that these SOAEs are most commonly emitted at a single frequency, or at least over a very narrow band of frequencies, and about 40% of ears produce at least one of them. Since most people possess two ears, most humans possess at least one such SOAE. They are known to be audible to many people who possess them (and sometimes to others too); they interact with each other in the same ear; they are suppressed by externally applied sound in a frequency-dependent manner; they are modified slightly by postural and temperature changes; and they are abolished by a range of cochlear manipulations that are also known to affect vibration of the cochlear partition. They are probably produced by an oscillation of some regions of the cochlear partition due to overactivity of the active process (Section 2.4). Interestingly, the amplitudes of the SOAEs appear to be self-limiting. When the cochlea oscillates, vibration grows to a stable amplitude and remains at that level. Aspects of this activity of SOAEs have been modeled successfully by a series of nonlinear oscillators that interact with each other in a mathematically chaotic fashion (Keefe et al. 1990). This may explain why some tinnitus sufferers experience great difficulty in describing their tinnitus.

5.2.3 Stimulus Frequency Otoacoustic Emissions: Level-Dependent Input Admittance

When a pure tone is presented to the ear from a small speaker, the amplitude of the sound pressure level measured in the ear canal does not grow proportionally with the electrical stimulus applied to the speaker, even though the speaker itself does not produce distortion. This nonlinear behavior is easily explained if the sound source is not an ideal pressure source, but has a finite acoustic output impedance associated with it. In this case, the output impedance of the sound source and the input impedance of the ear canal form a "pressure divider," much as two resistors in series form a "voltage divider" in electrical circuits (Fig. 4.21). The nonlinear growth of ear canal sound pressure with stimulus level is easily explained by a nonlinear input admittance to the ear canal, due to the nonlinear input admittance of the cochlea. Because the sound level in the ear canal does *not* scale proportionately with the stimulus level, the sound level predicted at low stimulus levels on the basis of linear extrapolation from high stimulus levels does not agree with that actually measured. The difference between the expected and measured acoustic waveforms is often attributed to the presence of a "stimulus frequency OAE," when it is just as easily explained by a nonlinear cochlear input admittance and an error of extrapolation. The measurement of SFOAEs can be viewed as an analysis of the first harmonic of the nonlinear input admittance. At a mathematical level, the difference is semantic. At a physiological level, however, it is important to realize that the presence of an SFOAE does *not* imply in any sense an active process within the cochlea. It merely indicates a nonlinear cochlear input admittance. It has been shown that these SFOAEs are modified by a variety of cochlear manipulations, including electrical stimulation of the efferent neurons to the cochlea (see Guinan, Chapter 8). These changes are most easily explained as changes in the nonlinear cochlear input impedance due to the action of the efferent fibers. More recently, similar changes have been produced by sound stimulation of the opposite ear, and have been used to assay the integrity of the efferent neural pathways, in a manner similar to the audiological test of *acoustic impedance reflexometry*. Although the normal role of the efferent neurons in modifying cochlear vibration is not clear, and the changes observed are very small, this technique may prove useful in the clinical diagnosis of lesions of the auditory brain stem.

5.2.4 Distortion Product Otoacoustic Emissions: Nonlinear Input Admittance

Just as the SFOAEs represent the nonlinear growth of the first harmonic of the cochlear input admittance, the DPOAEs can be viewed as two-tone interactions due to the nonlinearity of the cochlear input admittance. As described earlier, the frequencies of these distortion tones are mathematically related to the frequencies of the pure tones used to produce them (f_1 and f_2), and they can be heard by the subject. In the mammalian cochlea, the distortion products at frequencies $2f_1 - f_2$ and $f_2 - f_1$ are most prominent and are called the *cubic distortion tone* and *simple* or *quadratic distortion tone*, respectively. Whatever the generation mechanism at the level of the OC (which depends on the dominant nonlinearity in OC mechanics as discussed in Sections 4.6 to 4.9), it is likely that $f_2 - f_1$ and $2f_1 - f_2$ are larger than other distortion components, because other tones produced at higher frequencies at a particular cochlear region cannot propagate away from the generation site. For example, although high-frequency distortion tones such as $2f_1 + f_2$ or $2f_2 + f_1$ may also be produced at a cochlear region partway between the characteristic places for the frequencies f_1 and f_2, their frequencies may be above the cutoff frequency of the partition at the site of generation.

6. Summary

It is now believed by most auditory researchers that the vibration of the cochlear partition in the mammal can explain most of the response properties of the hair cells and primary afferent neurons of the organ of Corti, including their sensitivity, nonlinearity, frequency discrimination, and lability. The debate in the last few years has centered on two main issues: (1) the degree to which macromechanical vibration without extra micromechanical filtering can account for these response properties and (2) how much passive vibration of the cochlear partition without the motile activity of the OHCs can account for the observed vibration. The first question has become less divisive as it has become more clear that the micromechanical properties of the OC significantly affect the impedance the cochlear partition presents to the cochlear fluids: it is no longer possible to consider the OC as a flea on the back of the BM elephant, because micromechanical and macromechanical vibration are inextricably linked. As for the role of active OHCs, most researchers accept that the motility of OHCs must influence the vibration of the cochlear partition in some way, and the arguments are more commonly over how much they can do so, whether the force generation processes known to exist are adequate to assist vibration at frequencies as high as 100 kHz in some specialized mammals (bats, for example), and why only some OHCs assist vibration as the traveling wave passes. As for cochlear nonlinearity, it is becoming increasingly difficult to find hair cell and neural responses that cannot be explained by a combination of the known properties of partition vibration, IHC current generation, and neurotransmission at the primary afferent synapse. The same holds for cochlear trauma and hearing loss. Most hearing losses seem explicable in terms of known changes in cochlear electrophysiology or OHC motility, and those cochlear hearing losses that remain unexplained have yet to be investigated in detail.

Despite this understanding of cochlear hearing loss, it is unlikely that many causes of cochlear dysfunction can be reversed simply: the cochlear is too small, too deeply buried within the temporal bone, or too vulnerable to intervention, or the pharmaceuticals available are not sufficiently specific. Nevertheless, the understanding of cochlear transduction developed over the years suggests that *some* cochlear lesions may be reversible to *some* extent. In any case, the development of our understanding of cochlear vibration and its remote monitoring using the otoacoustic emissions allows us to understand much of sensorineural hearing loss and will allow the simple objective detection of cochlear hearing loss, as long as middle ear function is normal. Despite these recent advances, there is still a great deal to understand about the details of cochlear micromechanical vibration in the high- and low-frequency regions of the cochlea, about the changes in vibration with specific types of cochlear lesions, and about the complex homeostatic mechanisms (biochemical and neural) that must exist to maintain the cochlea's astounding mechanical properties within a narrow range over a lifetime.

Acknowledgments I would like to thank the many people who read early drafts of this chapter and offered their ideas on how the content could be presented in a logical and digestible form. As usual, the responsibility for oversights or errors lies with the author alone. I would also like to dedicate the chapter to two men who have inspired me over the years: Prof. BM Johnstone and the late Prof. E Zwicker. During their careers, both men had brilliant insight, and Prof. Zwicker in particular had a wicked sense of humor.

References

Allen JB (1980) Cochlear micromechanics – a physical model of transduction. J Acoust Soc Am 68:1660–1670.
Allen JB, Neely ST (1992) Micromechanical models of the cochlea. Physics Today 45:40–47.
Arthur RM, Pfeiffer RR, Suga N (1971) Properties of 'two-tone inhibition' in primary auditory neurons. J Physiol 212:593–609.
Ash EA, Paige EGS (eds) Wave Phenomena: Rayleigh-Wave Theory and Applications. Berlin: Springer, 1985.
Ashmore JF (1990) Forward and reverse transduction in the mammalian cochlea. Neurosci Res Suppl 12:S39–S50.
Assad JA, Hacohen N, Corey DP (1989) Voltage dependence of adaptation and active bundle movement in bullfrog saccular hair cells. Proc Natl Acad Sci USA 86:2918–2922.
Brownell WE (1983) Observations on a motile response in isolated outer hair cells. In: Webster WR, Aitkin LM (eds) Mechanisms of Hearing. Clayton, Australia: Monash University Press, pp. 5–10.
Brownell WE, Bader CR, Bertrand D, Ribaupierre Y (1985) Evoked mechanical responses of isolated cochlear outer hair cells. Science 227:194–196.
Brundin L, Flock Å, Khanna SM, Ulfendahl M (1991) Frequency-specific position shift in the guinea pig organ of Corti. Neurosci Lett 128:77–80.
Brundin L, Flock Å, Khanna SM, Ulfendahl M (1992) The tuned displacement response of the hearing organ is generated by the outer hair cells. Neuroscience 49:607–616.
Cheatham MA, Dallos P (1995) Origins of rate versus synchrony suppression: an evaluation based on two-tone interactions observed in mammalian IHCs. In: Manley GA, Klump GM, Köppl C, Fastl H, Oeckinghaus H (eds) Advances in Hearing Research. Singapore: World Scientific, pp. 117–126.
Cody AR, Johnstone BM (1980) Single auditory neuron response during acute acoustic trauma. Hear Res 3:3–16.
Cody AR, Johnstone BM (1981) Acoustic trauma: single neuron basis for the "half-octave shift." J Acoust Soc Am 70:707–711.
Cody AR, Mountain DC (1989) Low-frequency responses of inner hair cells: evidence for a mechanical origin of peak splitting. Hear Res 41:89–99.
Cody AR, Russell IJ (1985) Outer hair cells in the mammalian cochlea and noise-induced hearing loss. Nature 315:662–665.
Cooper NP, Rhode WS (1992) Basilar mechanics in the hook region of cat and guinea pig cochlea: sharp tuning and nonlinearity in the absence of baseline position shifts. Hear Res 63:163–190.
Crawford AC, Evans MG, Fettiplace R (1989) Activation and adaptation of transducer currents in turtle hair cells. J Physiol 419:405–434.
Dallos P (1970) Low frequency auditory characteristics: species dependence. J Acoust Soc Am 48:489–499.
Dallos P (1973a) The Auditory Periphery. New York: Academic Press.
Dallos P (1973b) Cochlear potentials and cochlear mechanics. In: Møller AR (ed) Basic Mechanisms in Hearing. New York: Academic Press, pp. 335–372.
Dallos P (1985) Response characteristics of mammalian cochlear hair cells. J Neurosci 5:1591–1608.

Dallos P (1986) Neurobiology of cochlear inner and outer hair cells: intracellular recordings. Hear Res 22:185–198.
Dallos P (1988) Cochlear neurobiology: some key experiments and concepts of the past two decades. In: Edelman GM, Gall WE, Coan WM (eds) Auditory Function: Neurobiological Bases of Hearing. John Wiley and Sons, pp. 153–188.
Dallos P, Billone MC, Durrant JD, Wang C-Y, Raynor S (1972) Cochlear inner and outer hair cells: functional differences. Science 177:356–358.
Dallos P, Cheatham MA (1989) Nonlinearities in the cochlear receptor potentials and their origins. J Acoust Soc Am 86:1790–1796.
Dallos P, Evans BN, Hallworth R (1991) Nature of the motor element in electrokinetic shape changes of cochlear outer hair cells. Nature 350:155–157.
Dallos P, Santos-Sacchi J (1983) AC receptor potentials from hair cells in the low-frequency region of the guinea pig cochlea. In: Webster W, Aitkin LM (eds) Mechanisms in Hearing. Clayton, Australia: Monash University Press, pp. 11–16.
Dancer A, Franke R (1980) Intracochlear sound pressure measurements in guinea pigs. Hear Res 2:191–206.
Davis H, Morgan CT, Hawkins JE, Galambos R, Smith FW (1950) Temporary deafness following exposure to loud tones and noise. Acta Oto-Laryngol Suppl. 88.
de Boer E, Viergever MA (1984) Wave propagation and dispersion in the cochlea. Hear Res 13:101–112.
Duifhuis H, Horst JW, van Dirk P, van Netten SM (eds) Biophysics of Hair Cell Sensory Systems. Proceedings of the International Symposium, Patterswolde, The Netherlands, World Scientific, 1993.
Elliot E (1958) A ripple effect in the audiogram. Nature 181:1076.
Evans BN, Dallos P (1993) Stereocilia induced somatic motility of cochlear outer hair cells. Proc Natl Acad Sci USA 90:8347–8351.
Evans EF (1972) The frequency response and other properties of single fibers in the guinea pig cochlear nerve. J Physiol Lond 226:263–287.
Evans EF, Klinke R (1982) The effects of intracochlear and systemic furosemide on the properties of cochlear nerve fibers in the cat. J Physiol Lond 331:409–428.
Evans EF, Wilson JP (1975) Cochlear tuning properties: concurrent basilar membrane and single nerve fiber measurements. Science 190:1218–1221.
Fay RR (1988) Hearing in Vertebrates: A Psychophysics Databook. Winnetka, IL: Hill-Fay Associates.
Flock Å (1965) Transducing mechanisms in lateral line canal organ receptors. Cold Spring Harbor Symp Quant Biol 30:133–145.
Freeman DM, Weiss TF (1990a) Hydrodynamic forces on hair bundles at low frequencies. Hear Res 48:17–30.
Freeman DM, Weiss TF (1990b) Hydrodynamic forces on hair bundles at high frequencies. Hear Res 48:31–36.
Geisler CD (1986) A model of the effect of outer hair cell motility on cochlear vibrations. Hear Res 24:125–131.
Geisler CD, Yates GK, Patuzzi RB, Johnstone BM (1990) Saturation of outer hair cell receptor current causes two-tone suppression. Hear Res 44:241–256.
Gold T (1948) Hearing II: the physical basis of the action of the cochlea. Proc R Soc B 135:492–498.
Goldstein JL, Kiang NYS (1967) Neural correlates of the aural combination tone. Proc IEEE 56:981–992.

Goodman DA, Smith RL, Chamberlain SC (1982) Intracellular and extracellular responses in the organ of Corti of the gerbil. Hear Res 7:161–179.

Guinan JJ, Gifford ML (1988) Effects of electrical stimulation of efferent olivocochlear neurons on cat auditory-nerve fibers. I. Rate level functions. Hear Res 33:97–114.

Guinan JJ, Peake WT (1967) Middle ear characteristics of anaesthetized cats. J Acoust Soc Am 41:1237–1261.

Gummer AW, Johnstone BM (1983) State of stress within the basilar membrane: a re-evaluation of the membrane misnomer. Hear Res 12:353–366.

Gummer AW, Johnstone BM, Armstrong NJ (1981) Direct measurements of basilar membrane stiffness in the guinea pig cochlea. J Acoust Soc Am 70:1298–1309.

Helmholtz H (1863) On the Sensations of Tone. New York: Dover Publications.

Holton T, Hudspeth AJ (1986) The transduction channel of hair cells from the bull-frog characterized by noise analysis. J Physiol 375:195–227.

Housley GD, Ashmore JF (1992) Ionic currents of outer hair cells isolated from the guinea-pig cochlea. J Physiol 448:73–98.

Howard J, Hudspeth AJ (1987) Mechanical relaxation of the hair bundle mediates adaptation in mechanoelectrical transduction by bullfrog's saccular hair cell. Proc Natl Acad Sci USA 84:3064–3068.

Howard J, Hudspeth AJ (1988) Compliance of the hair bundle associated with gating of mechanoelectrical transduction channels in the bullfrog's saccular hair cell. Neuron 1:189–199.

Hudspeth AJ, Corey DP (1977) Sensitivity, polarity, and conductance change in the response of vertebrate hair cells to controlled mechanical stimuli. Proc Natl Acad Sci USA 74:2407–2411.

Jaramillo F, Markin VS, Hudspeth AJ (1993) Auditory illusions and the single hair cell. Nature 364:527–529.

Johnstone BM, Boyle AJF (1967) Basilar membrane vibration examined with the Mossbauer technique. Science 158:389–390.

Johnstone BM, Sellick PM (1972) The peripheral auditory apparatus. Q Rev Biophys 5:1–57.

Johnstone BM, Patuzzi R, Yates GK (1986) Basilar membrane measurements and the traveling wave. Hear Res 22:147–153.

Johnstone BM, Gleich B, Mavadat N, McAlpine D, Kapadia S (1990) Some properties of the cubic distortion tone emission in the guinea pig. In: Hoke M (ed) Advances in Audiology 7, pp. 57–62.

Johnstone JR, Johnstone BM (1966) Origin of the summating potential. J Acoust Soc Am 40:1405–1413.

Keefe DH, Burn EM, Ling R, Laden B (1990) Chaotic dynamics of otoacoustic emissions. In: Mechanics and Biophysics of Hearing. Madison, Wisconsin, June 25–29, pp. 164–171.

Kemp DT (1978) Stimulated acoustic emissions from within the human auditory system. J Acoust Soc Am 64:1386–1391.

Kemp DT, Chum R (1980) Properties of the generator of stimulated acoustic emissions. Hear Res 2:213–232.

Khanna SM, Leonard DGB (1982) Basilar membrane tuning in the cat cochlea. Science 215:305–306.

Kiang NYS, Liberman MC, Levine RA (1965) Auditory-nerve activity in cats exposed to ototoxic drugs and high-intensity sounds. Ann Rhinol Laryngol 85–86:752–768.

Kim DO (1980) Cochlear mechanics: implications of electrophysiological and acoustical observations. Hear Res 2:297–317.

Kolston PJ (1988) Sharp mechanical tuning in a cochlear model without negative damping. J Acoust Soc Am 83:1481–1487.

Kössl M, Vater M (1995) Cochlear structure and function in bats. In: Popper AN, Fay RR (eds) Hearing by Bats. New York: Springer-Verlag, pp. 191–234.

Le Page EL (1987) Frequency-dependent self-induced bias of the basilar membrane and its potential for controlling sensitivity and tuning in the mammalian cochlea. J Acoust Soc Am 82:139–154.

Le Page EL, Johnstone BM (1980) Nonlinear mechanical behavior of the basilar membrane in the basal turn of the guinea pig cochlea. Hear Res 2:183–189.

Lewis ER, Leverenz EL, Bialek WS (1985) The Vertebrate Inner Ear. Boca Raton, FL: CRC Press.

Liberman MC, Dodds LW (1987) Acute ultrastructural changes in acoustic trauma: serial-section reconstruction of stereocilia and cuticular plates. Hear Res 26:45–64.

Liberman MC, Mulroy MJ (1982) Acute and chronic effects of acoustic trauma: cochlear pathology and auditory nerve pathophysiology. In: Hamernik R, Henderson D, Salvi R (eds) New Perspectives on Noise-Induced Hearing Loss. New York: Raven Press, pp. 105–135.

Lim DJ (1980) Cochlear anatomy related to cochlear micromechanics. A review. J Acoust Soc Am 67:1686–1695.

Long GR (1984) The microstructure of quiet and masked thresholds. Hear Res 15:73–88.

Lonsbury-Martin BL, Martin GK, Probst R, Coats AC (1987) Acoustic distortion products in rabbit ear canal. I Basic features and physiological vulnerability. Hear Res 28:173–190.

Lynch TJ, Nedzelnisky V, Peake WT (1982) Input impedance of the cochlea in cat. J Acoust Soc Am 72:108–130.

Main IG (1993) Vibrations and waves in Physics (3rd Ed.). Cambridge University Press.

Mammano F, Nobili R (1993) Biophysics of the cochlea: linear approximation. J Acoust Soc Am 93:3320–3332.

Martin GK, Ohlms LA, Franklin DJ, Harris FP, Lonsbury-Martin BL (1990) Distortion product emissions in humans. III Influence of sensorineural hearing loss. Ann Otol Rhinol Laryngol 99(Suppl 147):30–42.

McFadden D (1986) The curious half-octave shift: evidence for a basalward migration of the traveling-wave envelope with increasing intensity. In: Salvi RJ, Henderson D, Hamernik RP, Colletti V (eds) Basic and Applied Aspects of Noise-Induced Hearing Loss. pp. 295–312.

Moore BCJ (1986) Frequency Selectivity in Hearing. London: Academic Press.

Mountain DC (1980) Changes in the endolymphatic potential and crossed olivocochlear bundle stimulation alter cochlear mechanics. Science 210:71–72.

Nedzelnisky V (1980) Sound pressures in the basal turn of the cat cochlea. J Acoust Soc Am 68:1676–1689.

Neely ST, Kim DO (1983) An active cochlear model showing sharp tuning and high sensitivity. Hear Res 9:123–130.

Neely ST, Kim DO (1986) A model for active elements in cochlear biomechanics. J Acoust Soc Am 79:1472–1480.

Norton SJ, Widen JE (1991) Evoked acoustic emissions in infants and young children. International Symposium on Otoacoustic Emissions. Kansas City, MO (Abstract).

Nuttall AL, Brown MC, Masta RI, Lawrence M (1981) Inner hair cell responses to the velocity of basilar membrane motion in the guinea pig. Brain Res 211:323-336.

Olson ES, Mountain DC (1991) In vivo measurement of basilar membrane stiffness. J Acoust Soc Am 89:1262-1275.

Olson ES, Mountain DC (1993) Probing the cochlear partition's micromechanical properties with measurements of radial and longitudinal stiffness variations. In: Duifhuis H, Horst JW, van Dijk, van Netten SM (eds) Biophysics of Hair Cell Sensory Systems. World Scientific, pp. 280-287.

Patuzzi RB (1986) Mechanical correlates of noise trauma in the mammalian cochlea. In: Salvi RJ, Henderson D, Hamernik RP, Colletti V (eds) Basic and Applied Aspects of Noise-Induced Hearing Loss. pp. 122-136.

Patuzzi RB, Robertson D (1988) Tuning in the mammalian cochlea. Phys Rev 68:1009-1082.

Patuzzi RB, Sellick PM (1983) A comparison between basilar membrane and inner hair cell receptor potential input-output functions in the guinea pig cochlea. J Acoust Soc Am 74:1734-1741.

Patuzzi RB, Yates GK (1987) The low frequency response of inner hair cells in the guinea pig cochlea: implications for fluid coupling and resonance of the stereocilia. Hear Res 30:83-98.

Patuzzi RB, Sellick PM, Johnstone BM (1983) The modulation of the sensitivity of the mammalian cochlea by low frequency tones: its origin and possible mechanism. In: Webster WR, Aitkin LM (eds) Mechanisms in Hearing. Clayton, Australia: Monash University Press.

Patuzzi R, Johnstone BM, Sellick PM (1984) The alteration of the vibration of the basilar membrane produced by loud sound. Hear Res 13:99-100.

Patuzzi RB, Sellick PM, Johnstone BM (1984) The modulation of the sensitivity of the mammalian cochlea by low frequency tones. III Basilar membrane motion. Hear Res 13:19-28.

Patuzzi RB, Yates GK, Johnstone BM (1989a) The origin of the low-frequency microphonic in the first cochlear turn of guinea-pig. Hear Res 39:177-188.

Patuzzi RB, Yates GK, Johnstone BM (1989b) Changes in the cochlear microphonic and neural sensitivity produced by acoustic trauma. Hear Res 39:189-202.

Patuzzi RB, Yates GK, Johnstone BM (1989c) Outer hair cell receptor current and sensorineural hearing loss. Hear Res 42:47-72.

Probst R, Lonsbury-Martin BL, Martin GK (1991) A review of otoacoustic emissions. J Acoust Soc Am 89:2027-2067.

Rayleigh JW (1896) Theory of Sound. New York: Dover, reprint 1945.

Rhode WS (1971) Observations of the vibration of the basilar membrane using the Mossbauer technique. J Acoust Soc Am 49:1218-1231.

Rhode WS (1973) An investigation of post-mortem cochlear mechanics using the Mossbauer effect. In: Møller AR (ed) Basic Mechanisms in Hearing. New York: Academic Press, pp. 49-63.

Rhode WS (1980) Cochlear partition vibration—recent views. J Acoust Soc Am 67:1696-1703.

Rhode WS, Geisler CD (1967) Model of the displacement between opposing points on the tectorial membrane and reticular lamina. J Acoust Soc Am 42:185-190.

Robertson D (1974) Cochlear neurons: frequency selectivity altered by removal of perilymph. Science 186:623-628.

Robertson D, Johnstone BM (1979) Aberrant tonotopic organization in the inner ear damaged by kanamycin. J Acoust Soc Am 66:466-469.

Robertson D, Johnstone BM (1981) Primary auditory neurons: nonlinear responses altered without changes in sharp tuning. J Acoust Soc Am 69:1096-1098.

Robles L, Ruggero MA, Rich NC (1986) Basilar membrane mechanics at the base of the chincilla cochlea. I. Input-output functions, tuning curves, and phase responses. J Acoust Soc Am 80:1364-1374.

Robles L, Ruggero MA, Rich NC (1991) Two-tone distortion in the basilar membrane of the cochlea. Nature 349:413-414.

Ruggero MA (1980) Systematic errors in indirect estimates of basilar membrane travel times. J Acoust Soc Am 67:707-710.

Ruggero MA (1994) Cochlear delays and traveling waves: comments on 'Experimental look at cochlear mechanics.' Audiology 33:131-142.

Ruggero MA, Rich NA (1983) Chinchilla auditory nerve responses to low frequency tones. J Acoust Soc Am 73:2096-2108.

Ruggero MA, Rich NC (1991a) Application of a commercially-manufactured doppler-shift laser velocimeter to the measurement of basilar-membrane vibration. Hear Res 51:215-230.

Ruggero MA, Rich NC (1991b) Furosemide alters organ of Corti mechanics: evidence for feedback of outer hair cells upon the basilar membrane. J Neurosci 11:1057-1067.

Ruggero MA, Robles L, Rich NC (1986a) Cochlear microphonics and the initiation of spikes in the auditory nerve: correlation of single-unit data with neural and receptor potentials recorded from the round window. J Acoust Soc Am 79:1491-1498.

Ruggero MA, Robles L, Rich NC (1986b) Basilar membrane mechanics at the base of the chinchilla cochlea. II Responses to low-frequency tones and relationship to microphonics and spike initiation in the VIII Nerve. J Acoust Soc Am 80:1375-1383.

Ruggero MA, Rich NC, Recio A (1992) Basilar membrane responses to clicks. In: Cazals T, Demany L, Horner K (eds) Auditory Physiology and Perception. Oxford: Pergamon Press, pp. 85-92.

Russell IJ, Sellick PM (1978) Intracellular studies of hair cells in the mammalian cochlea. J Physiol 284:261-290.

Russell IJ, Sellick PM (1983) Low frequency characteristics of intracellularly recorded receptor potentials in mammalian hair cells. J Physiol Lond 338:179-206.

Russell IJ, Kössl M, Richardson GP (1992) Nonlinear mechanical responses of mouse cochlear hair bundles. Proc R Soc Lond B 250:217-227.

Sachs MB, Kiang NYS (1968) Two-tone inhibition in auditory-nerve fibers. J Acoust Soc Am 43:1120-1128.

Salvi RJ, Henderson D, Hamernik RP (1979) Single auditory nerve fiber and action potential latencies in normal and noise-tested chinchillas. Hear Res 1:237-251.

Schmiedt RA (1982) Boundaries of two-tone rate suppression of cochlear-nerve activity. Hear Res 7:335-351.

Schmiedt RA (1984) Acoustic injury and the physiology of hearing. J Acoust Soc Am 76:1293-1317.

Sellick PM, Patuzzi RB, Johnstone BM (1982a) Modulation of responses of spiral ganglion cells in the guinea pig cochlea by low frequency sound. Hear Res 7:199-221.

Sellick PM, Patuzzi RB, Johnstone BM (1982b) Measurement of basilar membrane motion in the guinea pig using the Mossbauer technique. J Acoust Soc Am 72:131-141.

Sellick PM, Patuzzi RB, Johnstone BM (1983) Comparison between the tuning properties of inner hair cells and basilar membrane motion. Hear Res 10:93-100.

Sellick PM, Russell IJ (1978) Intracellular studies of cochlear hair cells: filling the gap between basilar membrane mechanics and neural excitation. In: Naunton RF, Fernandez C (eds) Evoked Electrical Activity in the Auditory Nervous System. Academic Press New York, 113-139.

Sellick PM, Russell IJ (1980) The responses of inner hair cells to basilar membrane velocity during low-frequency auditory stimulation in the guinea-pig. Hear Res 2:439-445.

Sellick PM, Yates GK, Patuzzi R (1983) The influence of Mossbauer source size and position on phase and amplitude measurements of the guinea pig basilar membrane. Hear Res 10:101-108.

Smoorenburg GF (1972) Combination tones and their origin. J Acoust Soc Am 52:615-631.

Spoendlin H (1969) Innervation pattern of the organ of Corti of the cat. Acta Otolaryngol 67:239-254.

Steele CR, Zais JG (1985) Effect of coiling in a cochlear model. J Acoust Soc Am 77:1849-1852.

Tasaki I (1954) Nerve impulses in individual auditory nerve fibers of guinea pig. J Neurophysiol 17:97-122.

Tonndorf J (ed) (1981) Physiological Acoustics. Benchmark Papers in Acoustics, Vol. 15. Hutchinson Ross Publishing Company.

Ulfendahl M, Khanna SM (1993) Mechanical tuning characteristics of the hearing organ measured at the sensory cells in the gerbil temporal bone preparation. Pflügers Arch 424:95-104.

Viergever MA (1978) Basilar membrane motion in a spiral shaped cochlea. J Acoust Soc Am 64:1048-1053.

Voldrich L (1978) Mechanical properties of the basilar membrane. Acta Otolaryngol 86:331-335.

von Békésy G (1960) Experiments in Hearing. New York: McGraw-Hill.

Ward WD (1955) Tonal monaural diplacusis. J Acoust Soc Am 27:365-372.

Weiss TF (1982) Bidirectional transduction in vertebrate hair cells: a mechanism for coupling mechanical and electrical vibrations. Hear Res 7:353-360.

Wever EG (1947) Theory of Hearing. New York: John Wiley & Sons.

Wever EG (1978) The Reptile Ear. Princeton, NJ: Princeton University Press.

Wever EG, Lawrence M (1954) Physiological Acoustics. Princeton, NJ: Princeton University Press.

Whitehead ML, Longsbury-Martin B, McCoy MJ (1996) Otoacoustic emissions: animal models and clinical observations. In: Van De Water T, Popper AN, Fay RR (eds) Clinical Aspects of Hearing. New York: Springer-Verlag (in press).

Wilson JP (1980) Evidence for a cochlear origin for acoustic re-emissions, threshold fine structure and tonal tinnitus. Hear Res 2:233-252.

Wilson JP, Johnstone JR (1975) Basilar membrane and middle ear vibration in guinea pig measured by capacitive probe. J Acoust Soc Am 57:705-723.

Winter IM, Robertson D, Yates GK (1990) Diversity of characteristic frequency rate-intensity functions in guinea pig auditory nerve fibers. Hear Res 45:191-202.

Yates GK (1976) Basilar Membrane Tuning Functions. PhD dissertation, University of Western Australia.

Yates GK (1990) Basilar membrane nonlinearity and its influence on auditory nerve rate-intensity functions. Hear Res 50:145-162.

Yates GK, Winter IM, Robertson D (1990) Basilar membrane nonlinearity determines auditory nerve rate-intensity functions and cochlear dynamic range. Hear Res 45:203-220.

Zurek PM (1985) Acoustic emissions from the ear: a summary of results from humans and animals. J Acoust Soc Am 78:340-344.

Zwicker E (1979) A model describing nonlinearities in hearing by active processes with saturation at 40 dB. Biol Cybern 35-34:243-250.

Zwicker E (1986) A hardware cochlear nonlinear preprocessing model with active feedback. J Acoust Soc Am 80:146-153.

Zwislocki JJ (1979) Tectorial membrane: a possible sharpening effect on the frequency analysis in the cochlea. Acta Otolaryngol 87-3-4, 267-279.

Zwislocki JJ, Ceferatti LK (1989) Tectorial membrane II: Stiffness measurements in vivo. Hear Res 42:211-227.

Zwislocki JJ, Kletsky EJ (1979) Tectorial membrane: a possible effect on frequency analysis in the cochlea. Science 204:639-641.

Zwislocki JJ, Chamberlain SC, Slepecky NB (1988) Tectorial membrane I: Static mechanical properties in vivo. Hear Res 33:207-222.

5

Mechanics of the Cochlea: Modeling Efforts

EGBERT DE BOER

> Mercutio
> *True, I talk of dreams;*
> *Which are the children of an idle brain,*
> *Begot of nothing but vain fantasy,*
> *Which is as thin of substance as the air,*
> *And more inconstant than the wind, who wooes*
> *Even now the frozen bosom of the north,*
> *And, being anger'd, puffs away from thence,*
> *Turning his face to the dew-dropping south.*
> Shakespeare, *Romeo and Juliet*, I, 5.

1. Introduction and Outline

1.1 Goals of Modeling

Compared to reality, models of the cochlea appear ridiculously simple. To many it may seem strange that a very crude model would tell us something essential about how the real organ works, or, what is more, would predict the outcome of experiments yet to be done. This chapter has been written with the desire to prove that modeling is a useful and enjoyable exercise. Unfortunately for many, this cannot be done without the use of mathematics. However, even for those who can follow and understand mathematical derivations, the field of cochlear modeling may seem confusing because it is too wide. The present chapter is intended to be a useful guide in both these respects.

Why are models constructed? There are two main reasons: to represent within one framework the results from a large variety of experiments, and to explain the functioning of the system. The goal of the present chapter is to illustrate these two aspects on the basis of a series of models of the cochlea as they have been developed during recent decades. It is the hope of the author that the reader will become convinced why all this had to be done and also will get insight into the way in which the parts of the system contribute to the overall response.

Wherever possible, I try to convey the physiological meaning and significance behind all that is done. Furthermore, I intend to guide the non-mathematically trained reader as far as possible into the mathematical treatment. Mainly for this reason, unnecessary details (for instance, numerical solution methods) have been omitted entirely. Although the functioning of the cochlea in "lower" animals is highly interesting by itself, only the *mammalian* cochlea is treated.

Models of the cochlea have been formulated and constructed in many forms. They range from mechanical structures built with metal and plastic parts submerged in Plexiglas fluid containers, via electrical networks consisting of inductors, resistances, capacitors, diodes, and amplifiers, to abstract structures that are put into mathematical form and solved by computer. The latest form of cochlear model has appeared in the form of chips (the "silicon cochlea"; see Lyon and Mead 1988) that are designed to simulate the cochlea in electrical form. Such a device would be needed in a more practical study of all processes involved in hearing and understanding. Our goal in the present chapter will be more modest and will concentrate upon unraveling the functioning of that tiny link: the cochlea.

1.2 Outline

In the course of this chapter, many subjects will be treated. Relevant information has to be extracted from experimental evidence. The central question is: *How can we be sure that we are extracting the "true" information or drawing the "right" conclusions?* More specifically: Which are the anatomical-physiological properties that are the most important for the functioning of the cochlea? This topic is treated in Section 2. After the description of the "classical" models of the cochlea (Sections 3–5) the question arises whether elements of the organ of Corti are able to amplify the waves in the cochlea. Arguments in favor of the "cochlear amplifier" are the subject of Section 6. The *outer hair cells* (OHCs) are, according to our present knowledge, crucial in this respect. Models in which these cells, directly or indirectly, play an important part are described in Section 7. The OHCs are also thought to be the main contributors to cochlear nonlinearity, and this forms the subject of Section 8. The topic of activity returns in this section, after which the present state of affairs is summarized in Section 9.

Writing a review chapter means leaving out much material that is interesting by itself but does not fit within the general structure or the desired size of the chapter. Yet such material has contributed to the development of the field and to the insight of the author. Writing a review chapter also means making enemies, because authors may feel that their work has been neglected in the treatment and they may begin to feel a grudge. The only thing the present author can say is that the work reported here appears the most relevant with respect to the main line of development as he sees it. In other review papers (de Boer 1980a, 1984, 1991), the balance may well be different (the mathematical treatment is certainly more detailed), and readers who want to fill in details are referred to these papers.

2. Harvesting the Data

2.1 The Input Line

This section gives a highly condensed survey of experimental findings on the mechanics and physiology of the cochlea, findings that have served to sharpen present thinking about cochlear mechanics. First the mechanics: particularly relevant are direct measurements of *movements of the basilar membrane (BM)* in mammals. We might mention fundamental papers by Rhode (1971, 1978); Sellick, Patuzzi, and Johnstone (1982); Sellick, Yates, and Patuzzi (1983); Robles, Ruggero, and Rich (1986a,b); Nuttall, Dolan, and Avinash (1990, 1991); and Cooper and Rhode (1992). Over time, nonlinear phenomena have turned out to be more and more important (Patuzzi, Sellick, and Johnstone 1984b; Robles, Ruggero, and Rich 1986a; Ruggero, Robles, and Rich 1992; Ruggero et al. 1992; Nuttall and Dolan 1993). More detailed reviews of these findings are given by Patuzzi, Chapter 4, and de Boer (1991). Four basic features emerge from BM responses to stimuli comprising one or two sinusoidal components:

1. At the lowest levels of stimulation, the BM is mechanically very frequency-selective and sensitive; its iso-response function resembles in several respects the neural frequency-threshold curve.
2. The mechanical response in the response-peak region for the higher intensities is compressed by approximately 30 dB with respect to that for low intensities.
3. Nonlinear effects arise and are also frequency-specific (e.g., two-tone suppression and generation of distortion products such as combination tones).
4. Deterioration of the preparation causes the sharp tip of the iso-response curve to disappear, the sensitivity to diminish, and the degree of nonlinearity to decrease.

Otoacoustic emissions from the cochlea reflect the basic physiology of the cochlea and are important to us for several reasons. Emissions were discovered and analyzed by Kemp (1978) and have been used as a tool for investigation of the cochlea by many researchers. In the field of cochlear modeling, relatively little work on emissions has been reported, perhaps because the physics and physiology of the nonemitting cochlea are still not fully understood. Cochlear emissions are clear indicators of nonlinearity (Kemp 1978, 1979; Kemp and Brown 1983), and this forms an additional complication. Useful references and elaborations on nonlinear interaction effects of emissions can be found in van Dijk (1990) and van Dijk and Wit (1990).

There seems to be a good deal of consensus on how important the OHCs are for proper functioning of the cochlea. Of late, very specific mechanical response properties have been found: slow motility (Zenner, Zimmermann, and Schmitt 1985), fast motility (Brownell et al. 1985; Ashmore 1987; Gitter and Zenner 1988; Reuter et al. 1992), modulation and nonlinearity of the mechanical stiffness of the stereocilia (Howard and Hudspeth 1988), and frequency selectivity (Brundin, Flock, and Canlon 1989). Physiological aspects of these cells are being actively studied at present. Descriptions are found in Kros, Chapter 6, and Holley, Chapter 7. Patuzzi, Yates, and Johnstone (1989b) describe how several properties of the pathological cochlea can be ascribed to malfunctioning of OHCs.

In the field of micromechanical properties of the cochlea, the work done by Miller (1985b), Gummer, Johnstone, and Armstrong (1981), and Olson and Mountain (1991) on the stiffness of the BM, that of Flock and Strelioff (1984) on the stiffness of hair cell stereocilia, and that of Crawford and Fettiplace (1985), Ashmore (1987), Holley and Ashmore (1988), Iwasa and Chadwick (1992), and Zenner et al. (1992) on the mechanical properties of hair cells should be mentioned.

Particularly puzzling are the large mechanical DC shifts of the position of the BM that accompany the presentation of tones (Brundin et al. 1991). One should be reminded that DC shifts in the position of the BM were reported by LePage (1987), and furthermore, that van Netten and Khanna (1993) confirmed that similar phenomena occur in the fish lateral line organ (a much simpler hair cell fluid-column system). Cooper and Rhode (1992) maintain that these DC shifts are not important for sharp tuning of the cochlea.

2.2 Sketch of an Interpretation

Interpretation cannot always be separated from speculation. Newly gathered data do not always reinforce the concepts we already have, and this forms part of the fun of doing theoretical work. Von Békésy (1960) directly observed the *cochlear traveling wave,* and this concept has been confirmed in all later mechanical results. However, the mechanical sharpness of tuning appeared to be larger and larger over the years. This sharpness of tuning has been particularly puzzling to modelers, as we will see. One possible solution invokes the assumption that certain elements inside the cochlea are capable of *selectively amplifying* the cochlear wave. The OHCs are then identified as the elements involved.

The nonlinear effects associated with cochlear functioning complicate an already bewildering picture. However, we may assume that only one nonlinear effect is causing them, namely, *nonlinear transduction* in OHCs. On this basis it becomes possible and fruitful to study relations among the various nonlinear effects.

3. Peeling off the Fruit's Skin

3.1 General Simplifications Needed for Modeling

The form of the cochlea and the structure of the cell aggregates within it are very complicated (Slepecky, Chapter 2). To study its basic mode of functioning, we should "rebuild" the structure into a very simple one that simulates essential elements of the actual cochlea yet admits of a quantitative analysis. Several simplifications that have been invoked for making the *real cochlea* into a *tractable model* are described below (for more details, see Viergever 1986).

The cochlea is a fluid-filled organ with immobile walls situated in hard bone. The influence of the fluid ducts (ductus reuniens and endocochlear duct) (see Slepecky, Chapter 2) is neglected. The spiral shape of the fluid channels is simplified to a straight one, and the coordinate x will be used to designate the distance along the length of the model. The mechanical influence of Reissner's membrane is neglected, and therefore cochlear models usually contain two fluid channels instead of three. What we will call the *upper channel* consists of the scala vestibuli and the scala media; the *lower channel* is the scala tympani. Usually the cross-sectional areas and shapes of the two channels are assumed to be equal.

In many models, the *organ of Corti* is considered as a single entity that by its movements displaces the cochlear fluids. In the field of cochlear modeling, it is usually called the *cochlear partition (CP)*, and it is, though incorrectly, often referred to as the *basilar membrane (BM)* because, on the basis of von Békésy's (1960) findings, it is the BM that dominates the mechanics of the organ of Corti. In the case in which all parts of the CP move in unison in any one section of the cochlea, we will speak of a *macromechanical model* (Viergever 1986). Only in more recent models are independent movements of parts of the organ of Corti taken into account; we then speak of *micromechanical models*.

It is logical and convenient to divide the pressure p of the fluid into two components, a symmetrical one, called p_c, which is the same in both channels, and an antisymmetrical one, called p_d, which has opposite signs in the two channels:

$$p = p_d + p_c. \qquad (3.1.1)$$

We have omitted here the coordinates (e.g., x, y, and z) on which the physical variables depend. The component p_c refers to a compressional wave, a wave that invades all of the fluid and all of the tissues. Now, it is assumed that the fluid in the channels—and the fluid contained in the cochlear tissues—is *incompressible,* and this implies that the compressional wave travels with infinite velocity. Hence, the value p_c of the pressure associated with it is the same everywhere; this component thus is totally uninteresting, and from here on it will not be considered. We will be concerned solely with the component p_d, which forms the actual *cochlear wave,* and we repeat that it will be *antisymmetrical* in the two channels. That is

$$p_d(x,y,-z,t) = -p_d(x,y,z,t), \qquad (3.1.2)$$

for a Cartesian coordinate system with the coordinates x, y, and z (where z is perpendicular to the BM). The index d will be omitted from here on.

We now come to a series of simplifications that are of a more specific nature in the realm of cochlear modeling. First, the cochlear fluid is considered to have negligible viscosity; all acoustic power is assumed to be dissipated in the cochlear partition. Displacements in the cochlea are very small compared with the dimensions of the smallest cells, and velocities are much smaller than the propagation velocity of the cochlear waves. Thus, it is well justified to consider the fluid as *linear.* This simplifies the hydrodynamics considerably, since there are only two fundamental laws to be used for the fluid: the conservation of volume (because of the incompressibility) and the linear kinetics of the movement of small fluid elements (force equals mass times acceleration).

A most important point, which still has not been clarified satisfactorily, concerns the mutual mechanical coupling of elements of the organ of Corti in the longitudinal direction (i.e., the x direction). In most of the models described in the literature, longitudinal rigidity is completely neglected. It is invariably found that introduction of minute amounts of longitudinal mechanical coupling in the organ of Corti degrades the performance of the model to an unacceptable level (e.g., Allen and Sondhi 1979). One argument in favor of neglecting longitudinal stiffness is that forces associated with transverse bending of the BM will dominate forces associated with longitudinal bending, because the width of the BM is much smaller than the length. In this chapter, we will refer to a model in which longitudinal coupling is neglected and the mechanical properties of the cochlear partition are described by local parameter functions as a "classical" model.

3.2 The Cochlear Partition

As stated earlier, the CP is often considered as an entity and is functionally identified with the BM. One step further is the explicit assumption that the BM is *linear.* In a linear system, movements with different frequencies can simply be added without influencing each other. In this and the next few sections, we will be concerned only with linear cochlear models, and it is useful to dwell on the immediate consequences of this concept for a while. In linear mechanics, all variables are assumed to vary with time t as the function $\exp(i\omega t)$, where ω is the radian frequency, implying that there is an excitation with a sinusoidal waveform with the frequency $\omega/2\pi$. From here on, any variable or parameter will be represented by a *complex number,* the magnitude of which describes the amplitude and the argument the phase. This number should be considered as being multiplied by $\exp(i\omega t)$, and the actual physical variable corresponding to it (for instance, a pressure or a velocity varying with time t) is the real part of that product. More complicated types of movement will simply entail the addition of more than one of such frequency components. In the complex number representation, the associated numbers are added.

A linear mechanical structure can be characterized by a *mechanical impedance,* defined as the quotient of force and velocity; since both are complex numbers, the impedance is a complex number too. The impedance will always be a function of the radian frequency ω. When we really neglect longitudinal coupling in the organ of Corti, it is as if that organ consisted of a large set of narrow strips or bars (all lying in the radial or y direction) that could move independently of one another. Consider one such bar. When the center point of the bar is displaced from its equilibrium position (in the z direction, i.e., from the lower to the upper channel or vice versa), the bar will produce a restoring force due to its stiffness. For fast movements, resistive and inertial forces come into play too. The force divided by the vertical velocity yields the mechanical impedance of the bar; this impedance will be a function solely of x, the location of the bar, and the radian frequency ω. In the next step, the width of the bar is made to approach zero. It is then logical to consider the force density, that is, the force per unit length in the x direction. Finally, we will use the average force, i.e., the force averaged over the length of the bar (which is equal to the width of the BM). Thus, we have a force averaged over the width of the BM and over a short distance in the x direction; in other words, a pressure (force per unit area). We will therefore characterize the BM by the *BM impedance,* defined as the complex quotient of the *transmembrane pressure difference* and the average *velocity* of the BM. This impedance will also be a function solely of location x and radian frequency ω. In the field of acoustics, this type of impedance—pressure divided by point velocity—is known as "specific acoustical impedance," but in what follows we will be speaking simply of "BM impedance."

Constituent parts of the BM impedance $Z_{BM}(x,\omega)$ are:

1. The *stiffness.* The BM has a stiffness that starts with a large value at $x = 0$ (the stapes region) and decreases very fast with increasing x. In most models, this property is expressed by an exponential function $\exp(-\alpha x)$ (Zwislocki 1948). In the impedance, the stiffness component appears as

a negative imaginary term that is inversely proportional to the radian frequency ω.
2. The *mass* (actually, the mass of the organ of Corti per unit area of the BM). In many models, the mass is taken as larger than the actual mass of the organ of Corti, thus accounting for inertial effects of the fluid that moves with the partition. For computational simplicity, the mass is often assumed to be constant and independent of x. This component of $Z_{BM}(x,\omega)$ appears as a positive imaginary term that is proportional to ω.
3. The *resistance* (i.e., the real part of the impedance). When the cochlear fluid is assumed to be inviscid, the resistance component of the BM impedance is the only device by which the cochlear wave can dissipate energy. To achieve a "standard" situation where the relative bandwidth (in the frequency domain) is constant over the length, the resistance component is made to vary with x as the function $\exp(-\frac{1}{2}\alpha x)$.

To achieve uniformity in our description of cochlear modeling, the following expression will be used for $Z_{BM}(x,\omega)$:

$$Z_{BM}(x,\omega) = i\omega M_0 + \delta(M_0 S_0)^{1/2}\exp(-\tfrac{1}{2}\alpha x) + (S_0/i\omega)\exp(-\alpha x). \quad (3.2.1)$$

The parameter M_0 denotes the mass (in fact, the mass per unit of BM area), S_0 is the stiffness at $x = 0$, and δ is the damping constant. It is useful to realize that $Z_{BM}(x,\omega)$, being a function of ω, can have different values for components of cochlear waves with different frequencies. Similarly, the distribution of its values over x can be widely different for different frequencies. This is a point to keep in mind when considering complex signals in the cochlea (i.e., signals containing more than one frequency component).

Every transverse strip of the BM, when isolated, shows resonance at a specific frequency, namely, at the frequency where positive (mass) and negative (stiffness) imaginary components add up to zero. The resonance radial frequency ω_r for location x is readily found from

$$\omega_r = (S_0/M_0)^{1/2} \exp(-\tfrac{1}{2}\alpha x). \quad (3.2.2)$$

The actual resonance frequency is equal to $\omega_r/2\pi$. We will see later that the model response tends to have a peak near the location where the resonance frequency is equal to the frequency of the stimulus. The correlation between frequency and location is known as the *cochlear map*. In our case, this map is regular: a constant distance corresponds to a given frequency ratio. The BM impedance expressed by eq. (3.2.1) will be used in later subsections as the basic parameter function for cochlear models. In this case, the relative bandwidth of the BM resonance is equal to $(1/\delta)$, and the width of the cochlear response peak will be the same everywhere. As such, the computed response represents an "average" or a "stylized" situation; the actual cochlea is less selective for low than for high frequencies.

4. Long Waves

4.1 Fundamentals

A three-dimensional cochlear model obtained after the simplifications described in Section 3 is shown in Figure 5.1. The longitudinal coordinate x goes to the right, and the model is driven on the left side. The cross section of the model is rectangular, and this remains constant in size and shape over the length of the model. In order to show more details, the size of the cross section is shown as enlarged with respect to the length. The CP divides the model into two channels. It has a flexible part, the BM, which occupies only a fraction of the width, and a part assumed to remain stationary, the spiral lamina and the spiral ligament (*stippled* in the figure). Over its own width,

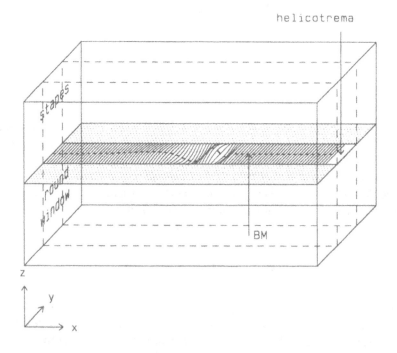

FIGURE 5.1. A simple stylized three-dimensional model of the cochlea with coordinate system x,y,z. The origin for x is the stapes, that for y is the center of the basilar membrane (BM) (*curved dashed line*), and that for z lies in the plane of the BM. The BM is suggested to consist of transverse strips; the stationary part of the CP is stippled.

in the *y* direction, the BM vibrates nonuniformly. This is suggested by the *curved thin lines* running diagonally (which represent the radial strips mentioned in the preceding section). The wave propagating along the BM (in the *x* direction) is suggested by the *thick dashed waving* line. The "hole" in the BM at the right indicates the helicotrema.

Fluid will be able to move in all directions, so the "true" treatment would be a three-dimensional one. Two particular simplifications of the hydrodynamics lead to two-dimensional and one-dimensional (longwave) models. Movements of the fluid in the plane perpendicular to the figure (the *y* direction in the model and the radial direction in the real cochlea) result mainly from the fact that the BM is oscillating only over a part of the total width of the partition. If the kinetic effect of the fluid movements in the *y* direction is neglected, one effectively assumes that the BM is moving uniformly *over the entire width of the model*. Then the fluid can only flow in two dimensions, *x* and *z*, and we have arrived at a *two-dimensional* model. This type of model has been quite popular in the history of cochlear modeling. One thing should be kept in mind, though. For a given velocity at the center of the BM, much more fluid is displaced in a two-dimensional than in a three-dimensional model. We will see later how to take this into account.

As a further simplification, we can neglect the dependence of all fluid variables on the vertical coordinate *z*. However, this leads to a contradiction: over the entire height of the model there would be no fluid movement in the *z* direction, yet at the level of the BM, at $z = 0$, the fluid must move up and down with the BM. To avoid the paradox, we will assume that the fluid velocity is distributed more gradually over the height. The result of the procedure will be a model in which the main variation of the fluid variables occurs in one dimension (the *x* dimension), often called a *one-dimensional model*. For reasons that will become clear later, this model should be known as the longwave model.

We start the derivation with the three-dimensional case. All variables are complex functions depending on *x*, *y*, and *z* as well as on ω. However, for reasons of clarity, we omit these variables wherever possible. The fluid velocity is a vector **v**, with components v_x, v_y, and v_z. Since the fluid is assumed to be incompressible, its divergence is zero. In Cartesian coordinates, this is expressed by

$$\text{div } \mathbf{v} \equiv (\partial v_x/\partial x) + (\partial v_y/\partial y) + (\partial v_z/\partial z) = 0 \qquad (4.1.1)$$

Newton's equation (force equals mass times acceleration) applied to a small fluid element is expressed in the linear domain (where we are only concerned with sinusoidally varying variables) in the following way:

$$\text{grad } p = -i\omega\rho\mathbf{v}, \qquad (4.1.2)$$

where p is pressure, ρ is fluid density, ω is radian frequency, and grad p is the vector with components $(\partial p/\partial x)$, $(\partial p/\partial y)$, and $(\partial p/\partial z)$. Combination with eq. (4.1.1) immediately leads to the Laplace equation for the pressure p:

$$\Delta p \equiv (\partial^2 p/\partial x^2) + (\partial^2 p/\partial y^2) + (\partial^2 p/\partial z^2) = 0. \qquad (4.1.3)$$

To solve for the fluid dynamics, we have to solve the Laplace equation with the proper boundary conditions. Actually, there exist a limited number of coordinate systems in which this equation is *separable*, i.e., can be split up into separate equations for each of the coordinates. The Cartesian and cylindrical coordinate systems are among these, and we have chosen the simpler of the two. Such a choice greatly simplifies the mathematical analysis, and this, finally, is *the major reason why models of the cochlea are so simple*.

If we neglect the dependence on y, as we do in the two-dimensional model, the Laplace equation becomes simpler:

$$\Delta p \equiv (\partial^2 p/\partial x^2) + (\partial^2 p/\partial z^2) = 0. \qquad (4.1.4)$$

4.2 Long-Wave Approximation

Let us start with a two-dimensional model by neglecting the influence of the y dimension. To achieve a further simplification, we assume that the vertical component $v_z(x,z)$ of the fluid velocity decreases linearly from its maximum value at $z = 0$ to zero at the walls. For the upper channel, this leads to

$$v_z(x,z) = v_{BM}(x)\,(H-z)/H \text{ (valid for } z \geq 0\text{)}, \qquad (4.2.1)$$

where $v_{BM}(x)$ is the BM velocity (assumed constant over the width of the BM) and H is the height of the channel (also assumed constant). In view of eq. (4.1.4), this implies that over the stated z domain, $(\partial^2 p/\partial z^2)$ is constant. Substitution in the two-dimensional Laplace eq. (4.1.4) produces

$$(\partial^2 p/\partial x^2) = (-i\omega\rho/H)\,v_{BM} \text{ (valid at } z = 0\text{)}, \qquad (4.2.2)$$

where we have omitted the variable x. With this relation, we can derive the BM velocity from the pressure. So far we have been describing only properties of the fluid. The next thing to do is to provide the link to the impedance of the BM (which is, we remember, trans-BM pressure divided by velocity). The boundary condition imposed by the BM is written as

$$v_{BM} = (-2p)/Z_{BM} \text{ (at } z = 0\text{)}, \qquad (4.2.3)$$

where Z_{BM}, actually $Z_{BM}(x,\omega)$, is the BM impedance we have come to know in Section 3. The factor of 2 results from the facts that the pressure in the lower channel is $-p$ and the pressure difference across the BM is 2 times p. The minus sign is due to the convention that upward movements (from lower to upper channel) are counted as positive. Combination of eqs. (4.2.2) and (4.2.3) gives

$$(d^2p/dx^2) - 2i\omega\rho/(HZ_{BM})\, p = 0 \text{ (at } z = 0), \qquad (4.2.4)$$

the well-known *wave equation* for what is known as the one-dimensional model. All parameters and variables in this equation are functions of only one spatial coordinate x. An important issue concerns the height H. A careful derivation starting from a three-dimensional model shows that H actually is the ratio of the channel's cross-sectional area and the width of the BM. Instead of H, we should use the symbol H_{eff}, meaning the "effective height." Actually, it should also be taken into account that over its width the BM moves as a cosine function (cf. Diependaal and Viergever 1983), but we will omit such refinements here. We will usually consider H or H_{eff} as constant, but as Shera and Zweig (1991a) point out, there is every reason to take H as increasing when nearing the stapes region.

We must now make a small digression, necessary to provide a link to derivations in the next section. We retrace our steps and consider eq. (4.2.2). Let us assume a wave of the type

$$p(x) = p_0 \exp(-ikx) \qquad (4.2.5)$$

(where p_0 is a constant and k is real) is propagating in the fluid. The parameter k is the wave number; since the argument changes by 2π for every wavelength, the wave number equals 2π divided by the wavelength. When $p(x)$ varies with x as eq. (4.2.5), the velocity $v_{BM}(x)$ will vary in the same way; let it have the (complex) amplitude v_{BM0}. In view of eq. (4.2.2), the complex quotient of p_0 and v_{BM0} will be given by

$$p_0/v_{BM0} = i\omega\rho/(Hk^2) \qquad (4.2.6)$$

This quantity represents the *impedance of the fluid channel* as seen from the side of the BM, and applies to the case where a wave with wave number k is propagating in the fluid. When k is real, the fluid impedance is masslike. Apart from the constant ($i\omega\rho$), the quotient (p_0/v_{BM0}) has been given a name: $h_{eq}(\lambda)$ by Steele and Taber (1979a,b) and $Q(k)$ by the present author (de Boer 1980b). For the one-dimensional model, it is given by

$$Q(k) = 1/(Hk^2), \qquad (4.2.7)$$

where, for more generality, H may be replaced by H_{eff}. In a cochlear model, k is not always restricted to being real; it may be complex, because the actual wave is dictated by the properties of the BM as well as the fluid. We will come to this shortly.

It remains to be understood why the wave equation (4.2.4) refers to "long waves." To see this in a simple way, assume that (HZ_{BM}) is constant and negative imaginary. The wave equation then admits a solution of the form of eq. (4.2.5), and k must be a root of the equation

$$k^2 = -2i\omega\rho/(HZ_{\text{BM}}). \qquad (4.2.8)$$

The two roots (both real, one positive, one negative) refer to waves in the positive and negative directions. When v_z is to be a linear function of z, as eq. (4.2.1) prescribes, p must be a quadratic function of z. Then, the first term of eq. (4.1.4) also varies with z, and this contradicts the assumption that $(\partial^2 p/\partial z^2)$ should be constant over z. Therefore, a limit has to be put to the variations of $(\partial^2 p/\partial x^2)$ over the height. If we limit those variations, quite arbitrarily, to a factor of two, judicious use of the Laplace equation leads to the following condition:

$$kH < 1. \qquad (4.2.9)$$

In words, the wavelength should be greater than $(2\pi H)$. The model is valid for waves with wavelengths greater than this limit; the one-dimensional model is, in fact, a *longwave model*. The reader should know that the criterion (4.2.9) is quite arbitrary; slightly different criteria result from different types of derivation.

4.3 History: The Liouville-Green (LG) [or Wentzel-Kramers-Brillouin (WKB)] Method

The history of the longwave model is a long and varied one and will not be detailed here. Zwislocki (1948) was the first to formulate a *realistic model of the cochlea*: he applied the required number of simplifications and omissions to the structure, formulated a central (longwave) equation, solved it (analytically!), and showed that the model predictions matched the experimental findings of that period, i.e., those of von Békésy (later collected in his 1960 book). Later developments came from Fletcher (1951) and Dallos (1973). As we will see further on (Sections 6–8), the longwave model still stands on a pedestal: many elaborations on activity, nonlinearity, and cochlear emissions are done on it.

One method of solution that has attracted great interest in the past is the *LG* or *WKB approximation*. It is an asymptotic solution method that is valid whenever reflection of waves can safely be neglected. Because the method is an asymptotic one, its accuracy cannot be stated beforehand, yet it has turned out to be quite useful in cochlear mechanics. Early applications of the method have been given by Schroeder (1973) and Zweig, Lipes, and Pierce (1976). For explanation and derivation of the LG method for longwave models, the reader can consult de Boer (1980a, Section 6.5). An important asset of the LG method is its *interpretation*, and that is sufficient reason to expound on it here.

In the LG (WKB) approximation, the cochlear wave is described by an expression with two factors, an exponential factor not unlike that in eq. (4.2.5) and a correction factor. The expression for the pressure can be written as

$$p(x) = a_p(x) \exp[-i\phi(x)], \tag{4.3.1}$$

where $a_p(x)$ is the correction factor, and $\phi(x)$ (known as the "phase integral") expresses the accumulated phase:

$$\phi(x) = \int_0^x k(x')dx' \tag{4.3.2}$$

To each location, a *local wave number* $k(x)$ is associated, a complex variable of which the real part describes variations of phase and the imaginary part variations of amplitude of the second factor of eq. (4.3.1). These play an important part in the variations of $p(x)$, as we will see. The speed with which the power is propagated in the wave is equal to $\omega/k(x)$. The correction factor $a_p(x)$ is a complex function contributing to both amplitude and phase of the pressure $p(x)$. This correction factor actually ensures the proper balance between propagation and dissipation of power. The reader should note the difference between k, a real constant in the preceding discussion, and $k(x)$, a complex variable depending on x. Actually, $k(x)$ also depends on ω, as we will see soon.

The most important thing about this solution is that the local wave number $k(x)$ is a root of the following equation:

$$k^2(x) = -2i\omega\rho/[H(x)Z_{\text{BM}}(x)]. \tag{4.3.3}$$

This equation, known as the *eikonal equation*, is clearly the x-dependent counterpart of eq. (4.2.8), the reason that we have explicitly reintroduced the dependence of the parameters on x here. Because $Z_{\text{BM}}(x)$ depends on ω, $k(x)$ should be written as $k(x,\omega)$, but this will be omitted. It is repeated that $k(x)$ does not express all phase and amplitude variations of $p(x)$, since the factor $a_p(x)$ contributes to these variations. In effect, for the longwave model, the factor $a_p(x)$ for the pressure turns out to be of the form

$$a_p(x) = a_{p0}k^{-1/2}(x), \tag{4.3.4}$$

whereas, if we write the LG expression for the BM velocity as

$$v_{BM}(x) = a_v(x) \exp[-i\phi(x)], \qquad (4.3.5)$$

the correction factor $a_v(x)$ for the BM velocity is

$$a_v(x) = a_{v0} k^{3/2}(x). \qquad (4.3.6)$$

It is the latter factor that causes the BM velocity to rise in amplitude as the cochlear wave propagates from the window region on. This is easily seen by assuming the BM impedance to be dominated by a stiffness term varying with x as $\exp(-\alpha x)$. The local wave number $k(x)$ will then vary as $\exp(\frac{1}{2}\alpha x)$ and the amplitude factor $a_v(x)$ of eq. (4.3.6) as $\exp(\frac{3}{4}\alpha x)$.

Finally, let it be noted that *the LG method can only be used when the parameters of the model vary so slowly with x that the relative variation of k(x) over one wavelength is negligible*:

$$|[1/k^2(x)][dk(x)/dx]| \ll 1. \qquad (4.3.7)$$

Loosely speaking, in this case the waves are not reflected.

4.4 Response Patterns: A "Panoramic View"

For the actual solution of the longwave eq. (4.2.4), we still need boundary conditions at the two ends of the model. First, the fluid moving in the upper channel must match the oscillations of the stapes ossicle. Assuming the stapes has an area equal to that of the channel and is oscillating sinusoidally with radial frequency ω and amplitude v_{st}, we can write the condition as

$$dp(0)/dx = -i\omega\rho v_{st}. \qquad (4.4.1)$$

For high enough frequencies, the cochlear wave will be nearly extinguished before it reaches the helicotrema. In that case, the boundary condition at the end of the model ($x = L$) can be formulated as

$$p(L) = 0. \qquad (4.4.2)$$

Other cases will not be considered here.

The wave equation, eq. (4.2.4), can easily be solved numerically. The results are presented in Figure 5.2. The variations of the BM impedance $Z_{BM}(x)$ with x are as prescribed by eq. (3.2.1), and the parameter values are given in the legend. The stiffness parameter is chosen in a range consistent with experimental findings, and the damping is chosen small enough to yield a clear response peak. We shall come to speak about the mass constant

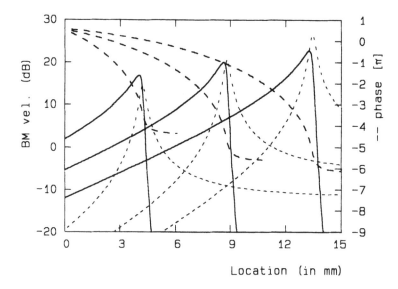

FIGURE 5.2 Cochlear response patterns ("panoramic view") for a longwave model. *Abscissa*: location x. Figure shows the BM velocity $v_{BM}(x)$ relative to the stapes velocity v_{st}. *Solid lines*, magnitude; *dashed lines*, phase. The *finely dashed lines* indicate the magnitude of the BM admittance, $1/Z_{BM}(x)$. Frequencies: 3 (*right-hand curves*), 6 and 12 kHz. Parameters: $S_0 = 1.0 \, 10^9$, $M_0 = 0.05$, $\delta = 0.05$, $H = 0.1$, $\alpha = 3$ (cgs units).

in the next section. The figure shows a cochlear pattern, i.e., the BM velocity $v_{BM}(x)$ as a function of x, for constant frequency. This type of picture has been termed a "panoramic view" by Patuzzi (Chapter 4). If we had computed the velocity at a fixed place as a function of frequency, producing a *frequency-response curve*, the function would have been similar in shape to the cochlear pattern (Patuzzi, Chapter 4).

Let us follow the wave as it propagates from the window region and use the LG approximation as a guide. Near the windows, $Z_{BM}(x)$ is dominated by its stiffness term (negative imaginary), and the local wave number $k(x)$ is nearly real, so that a propagating wave arises. Gradually the stiffness becomes smaller and $k(x)$ increases, and in the figure we observe the slope of the phase lag to increase. The amplitude of the BM velocity increases due to the mechanism described in connection with eq. (4.3.6). In its primary course, the cochlear wave hardly loses any power by dissipation.

Near the location where the BM resonates, we observe a peak in the velocity curve. As x becomes larger, $k(x)$ acquires an ever-increasing (negative) imaginary part, and this eventually causes the wave to be attenuated rapidly (see eq. 4.3.2 and the second factor in eq. 4.3.1). Still further on in the cochlea, the impedance $Z_{BM}(x)$ is dominated by its mass term. As a result, $k(x)$ is nearly purely (negative) imaginary and the magnitude of the wave decreases even more rapidly. For different frequencies, the location of the peak is different so that the cochlear model *maps frequency to place*. In the stylized case treated here, the cochlear frequency-to-place map is regular (see eq. 3.2.2).

In Figure 5.2, the *finely dashed lines* depict the variations of the magnitude of $[1/Z_{BM}(x)]$. Near their peak, the admittance and velocity curves are similar in curvature, but the velocity function is "tilted" so that it drops off rapidly to the right of the peak. When we combine eqs. (4.3.3) and (4.3.6) for the LG expression, we see that the factor $a_v(x)$ is inversely proportional to the ¾th power of $Z_{BM}(x)$. Clearly, the variations in the velocity response $v_{BM}(x)$ are mainly due to the factor $a_v(x)$. It is the exponential factor of the LG expression that accounts for the "tilt" of the response pattern in the longwave case.

4.5 Reflections

We finish this section about the longwave model with a few brief remarks about reflections. In Section 4.2, it was observed that the height H should not be taken as constant, especially in the region of the basal turn. This has to do with reflections. In a model with constant H equipped with a stiffness function varying as $\exp(-\alpha x)$, low-frequency waves will have very long wavelengths in the basal turn and will be partially reflected there (in fact, because eq. 4.3.7 is violated). This is not compatible with experimental data on the input impedance of the cochlea. A possible remedy is to let H increase near $x = 0$ (Shera and Zweig 1991a; see also Puria and Allen 1991). On its further course, *the cochlear wave has negligible reflection*. This holds also for *retrograde* waves (Shera and Zweig 1991b). Models may be equipped with local inhomogeneities to allow for otoacoustic emissions to arise.

For extremely low frequencies, the cochlear wave may arrive at the helicotrema with an appreciable amplitude and will be reflected there (cf. Dallos 1973). Puria and Allen (1991) proved that reflections from the helicotrema are unimportant for the cochlea's input impedance, provided the cochlear fluid is endowed with the proper viscosity.

5. Dimensionality

5.1 A Two-Dimensional Model

In the preceding section, the longwave model was derived as a simplified solution of the two-dimensional case. Actually, near the response peak the computed phase varies so rapidly that the wavelength is *not* large compared to $(2\pi H)$ so that *the condition for long waves is not met*. Therefore, it is necessary to solve the model equation for two- and three-dimensional models in order to find out to what extent the long-wave approximation is accurate enough. This is a task that turns out to be quite formidable; a comprehensive step-by-step treatment is available (de Boer 1984, Sections 4 and 5), but the reader will be guided here along the simplest possible path, in which many of the mathematical details must be omitted.

It should be realized at the outset that the Laplace equation describes a *linear* fluid. Therefore, we can always start by assuming that, for any fixed value of ω, there will be a *single* wave (called a partial solution) in the fluid that travels in the x direction according to the function $\exp(-ikx)$, with k, the wave number, real. A positive value of k refers to a wave going to the right (in the direction of increasing x), and a negative value of k describes a wave going to the left. Such a single wave with a constant wave number cannot be the universal solution to our cochlear problem. Therefore, we will treat the actual case by *integrating the partial solutions over* k, from minus to plus infinity (all for the same ω).

Let us apply this idea to the two-dimensional model. We have a pressure function $p(x,z)$ of x and z that (because the Cartesian coordinate system is separable) can be written as the product of the wave function $\exp(-ikx)$ (where k is an arbitrary real number) and another function $p^{(2)}(z)$ which is *solely* a function of the coordinate z. Substitution in the Laplace equation (eq. 4.1.4) gives for $p^{(2)}(z)$

$$d^2 p^{(2)}(z)/dz^2 = k^2 p^{(2)}(z). \quad (5.1.1)$$

This equation has, for a given k, two independent solutions, one of the form $\exp(+kz)$, the other of the form $\exp(-kz)$. A linear combination of these two must be sought that obeys the boundary condition at the upper (or lower) wall where the vertical velocity component must be zero. One finds that $p^{(2)}(z)$ must be of the form

$$p^{(2)}(z) = \cosh k(H-z)/\cosh kH \text{ (valid for } z \geq 0\text{)}. \quad (5.1.2)$$

The denominator is included solely to make the function equal to unity at the BM ($z = 0$). The function $p^{(2)}(z)$ describes how the pressure p varies with z. For the lower cochlear channel, the same function applies, but the sign of $p^{(2)}(z)$ is inverted and z is replaced by $-z$.

We can now write the pressure $p(x,z)$ as an integral over k, using $P(k)$ as the strength of the component with wave number k:

$$p(x,z) = \int_{-\infty}^{+\infty} P(k) \left[\cosh k(H-z)/\cosh kH\right] \exp(-ikx)\, dk. \quad (5.1.3)$$

To make $p(x,z)$ obey the BM condition, we use eqs. (4.1.2) and (4.2.3), and obtain the fundamental equation for the two-dimensional model:

$$-2i\omega\rho p(x) = Z_{BM}(x) \int_{-\infty}^{+\infty} P(k)\, k \tanh kH \exp(-ikx)\, dk. \quad (5.1.4)$$

This equation has been termed the *Siebert equation* in honor of W.M. Siebert, who revived interest in the issue of short versus long waves in the early 1970s (Siebert 1974). It is convenient to isolate the factor (k tanh kH) and to define the function

$$Q(k) = 1/(k \tanh kH) \qquad (5.1.5)$$

as the descriptor of the two-dimensional model. In terms of $Q(k)$, the Siebert equation reads

$$-2i\omega\rho p(x) = Z_{BM}(x) \int_{-\infty}^{+\infty} [P(k)/Q(k)] \exp(-ikx) \, dk. \qquad (5.1.6)$$

If the quotient $P(k)/[i\omega\rho Q(k)]$ is redefined as $V(k)$, the equation assumes the form

$$-2p(x) = Z_{BM}(x) \int_{-\infty}^{+\infty} V(k) \exp(-ikx) \, dk, \qquad (5.1.7)$$

and it is clear that $V(k)$ is the (inverse) Fourier transform of the BM velocity $v_{BM}(x)$. We will later have occasion to observe that for a three-dimensional model the equations remain the same, and only the function $Q(k)$ changes.

In the most general case, the equation cannot be solved analytically. Numerical solutions were pioneered by Lesser and Berkley (1972) and refined by Allen (1977), Sondhi (1978), Allen and Sondhi (1979), Viergever (1980), and Neely (1981, 1985). The finite-element solution method was applied by Viergever (1980) and Miller (1985a).

For the two-dimensional model, there also exists an LG (or WKB) approximation that has been analyzed and applied by Steele and Taber (1979a). In terms of the function $Q(k)$, the equation from which the local wave number $k(x)$ must be determined reads

$$-2i\omega\rho Q[k(x)] = Z_{BM}(x), \qquad (5.1.8)$$

with $Q(k)$ given as in eq. (5.1.5). The meaning of k has changed from a real integration variable to a *complex* quantity describing the local value of the wavelength as a function of x, just as in the longwave LG (WKB) case (Section 4.3). Continuing what has been stated there, this equation is the *eikonal equation* for the two-dimensional case. Relation (5.1.8) is universal because it has the same form for longwave, two-dimensional and three-dimensional models. Only the function $Q(k)$ varies.

The LG solution also involves a correction factor $a_v(x)$. The general expression for $a_v(x)$ is (de Boer and Viergever 1984)

$$a_v(x) = a[dQ(k)/dk]^{-1/2}. \qquad (5.1.9)$$

Again, the same form applies to all types of model. The general validity of the LG approximation method for the two-dimensional model was checked by Steele and Taber (1979a) and by Viergever and Diependaal (1986). The important fact that in the two-dimensional case, eq. (5.1.8) has not two but infinitely many roots is treated in some depth by de Boer and Viergever (1982).

5.2 Long and Short Waves

It is not difficult to see the connection of the two-dimensional with the longwave case, but there are some intricacies. For very small values of k ("long waves"), the factor (k tanh kH) can be approximated by (k^2H). In this case, the integral in the right-hand member of eq. (5.1.7) can be reduced to a form with the second derivative of $p(x)$. In this way, the longwave equation, eq. (4.2.4), arises naturally. In a more refined approximation, the function $Q(k)$ has the following asymptotic behavior:

$$Q(k) \rightarrow [1/k^2H] + \tfrac{1}{3}H \qquad (5.2.1)$$

When this form is substituted in eq. (5.1.7), the result is a modified form of the longwave equation in terms of the pressure (see eq. B3 in Neely 1985). When, however, the BM velocity is computed, it will be exactly the same as the one resulting from eq. (4.2.4), provided $Z_{BM}(x)$ is replaced by $Z_{BM}'(x)$, which includes an extra mass term:

$$Z_{BM}'(x) = Z_{BM}(x) + \tfrac{2}{3}i\omega\rho H. \qquad (5.2.2)$$

Note that this property is *only* valid for the BM velocity. This is the reason that work on the longwave model is always done with a BM impedance function that includes a fairly large mass term, one that is larger than can be accounted for by the mass of the fluid contained in the organ of Corti (cf. Steele and Taber 1979a).

The other extreme, the case where (kH) is so large that the tanh function saturates, is the *shortwave* case. Short waves are also known as "deep-water waves." In general, this case does not lead to an easily solvable equation; only in one special case can a very simple solution be obtained. Assume that *there are only waves going to the right* (k is restricted to being positive). Replacing (k tanh kH) by k in eq. (5.1.5) leads to the following shortwave equation for the pressure:

$$dp(x)/dx + 2\omega\rho p(x)/Z_{BM}(x) = 0. \qquad (5.2.3)$$

A similar equation with a minus sign is valid for *left-going* waves. For any function $Z_{BM}(x)$, eq. (5.2.3) admits of a simple and direct solution. We will not detail this any further. From eq. (5.1.5), the shortwave $Q(k)$ function in the general case (no restriction on k) is easily found to be

$$Q(k) = 1/|k|. \tag{5.2.4}$$

Note, however, that this is not an analytic function of k.

Let us now look at the contrast between long and short waves. We recall Figure 5.2, which shows representative longwave responses. Figure 5.3 shows the corresponding shortwave solution in the same "panoramic view." For the sake of pure comparison, the *same* BM impedance function is used. The vertical level of the magnitude curve is arbitrary in this case. Because the channel is assumed to be infinitely high, there is no shortwave counterpart to the longwave boundary condition at the stapes. The magnitude of the BM admittance $[1/Z_{BM}(x)]$ is indicated by the *finely dashed line*. Three conspicuous differences from the longwave case emerge:

1. The response peak for the BM velocity is less pronounced in the shortwave than in the longwave case.
2. The response peak is located "far" to the left of the region of the peak of the admittance function (i.e., the resonance region).
3. In the region of the response peak, there are much more rapid phase variations.

All three properties express the principle of *boundary-layer absorption* as formulated by Lighthill (1981): in deep-water waves, nearly all of the wave

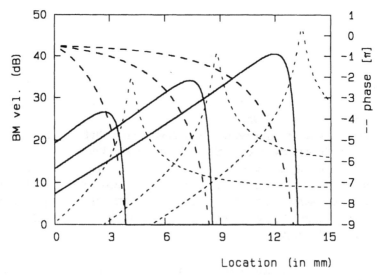

FIGURE 5.3 Cochlear response patterns ("panoramic view") for the shortwave model. Layout as in Figure 5.2. Frequencies: 3, 6, and 12 kHz. Reference for response magnitude is arbitrary. Note the large difference in position and shape between the $v_{BM}(x)$ response (*solid lines*) and the BM admittance functions (*finely dashed lines*).

energy is spent before the wave arrives at the resonance location. In terms of the LG solution, the interpretation of the shortwave case is different: it is the exponential factor that dominates the velocity response, whereas in the longwave case it is the correction factor $a_v(x)$. As a matter of fact, in the shortwave case the local wave number $k(x)$ varies twice as fast with x as in the longwave case (this can be derived from eqs. 5.2.3 and 4.3.3). It must be admitted, finally, that Figure 5.3 presents an exaggerated case. We will encounter a better form of the shortwave approximation when we have covered the three-dimensional model.

Finally, one physical property inherent in the shortwave solution should be stressed. When kH is larger than unity, the wave amplitude at the outer walls of the model ($z = H$) will be much smaller than that near the BM. This is easily seen from eq. (5.1.2). We conclude that *in short waves the bulk of the wave movement occurs in the immediate vicinity of the BM.*

5.3 The Three-Dimensional Model

The derivation for a three-dimensional model is more complicated than that for a two-dimensional model. It leads to a model equation of the same form as eq. (5.1.6) but with a different $Q(k)$ function. The form of $Q(k)$ has been worked out in Cartesian coordinates (Steele and Taber 1979b; de Boer 1981) as well as in cylindrical coordinates, which correspond better to the anatomical shape (de Boer 1980b). Especially in the three-dimensional case, numerical accuracy of the solution is difficult to predict or to control. This explains the desire to use the LG (WKB) method throughout (Steele and Taber 1979b, 1981; Taber and Steele 1981; Novoselova 1987, 1989, 1993).

Now we turn to general properties of three-dimensional models and the connection to other models. First, the longwave approximation is considered. It turns out that the longwave limit of $Q(k)$ for the three-dimensional case (apart from the mass correction) is

$$Q(k) = 1/(k^2 H_{\text{eff}}), \tag{5.3.1}$$

where H_{eff} is the *effective height* of the channel. In the case where ϵ is the fraction of the width W of the channel that is occupied by the BM, and the BM is moving uniformly over its own width, H_{eff} is equal to the cross-sectional area of the channel divided by the width of the BM (i.e., H/ϵ). We may omit the (small) correction for nonuniform motion of the BM over its width (cf. Diependaal and Viergever 1983).

In Figure 5.4, a typical three-dimensional response pattern, computed for $\epsilon = 0.1$, is shown (one frequency only). Most of the parameters are the same as before, but the mass constant is smaller. The figure shows the BM velocity, computed with a simplified solution method developed by de Boer and van Bienema (1982). In this method the $Q(k)$ function is approximated by a rational function of k. The *thickest lines* show the response of the

280 Egbert de Boer

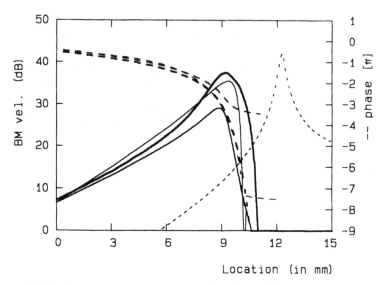

FIGURE 5.4 Cochlear response pattern ("panoramic view") for the three-dimensional model. Layout similar to Figure 5.2. Frequency: 8 kHz. Extra parameter: $\epsilon = 0.1$; the BM occupies 0.1 of the channel width. *Solid lines*, magnitude; *dashed lines*, phase. *Finely dashed line*, magnitude of the BM admittance. *Thick lines*, BM velocity response of three-dimensional model. Mass parameter: $M_0 = 0.010$; damping parameter: $\delta = 0.05$. Real height H of the channel: 0.1 cm. *Thin lines*, response of (matched) longwave model, with parameters: $H_{\text{eff}} = H/\epsilon = 1.0$, $M_0 = 0.025$, $\delta = 0.15$. *Very thin lines*: response of a matched shortwave model; $M_0 = 0.010$, $\delta = 0.05$, $\epsilon_1 = 0.202$.

three-dimensional model, *solid* and *coarsely dashed lines* depicting amplitude and phase, respectively. The *thinner lines* show the response of a longwave model in which the height is taken as (H/ϵ). To get the peak at the right place, some mass is added, and to get a wider peak, the damping is increased (see the legend). It is clear that the longwave approximation is deficient in several respects: its peak is too low and its phase slope too small.

With the same computation method, it is possible to find the response of a two-dimensional model modified and matched to the three-dimensional case. It turns out that the response peak of the three-dimensional model can be excellently represented by that of the matched two-dimensional model. This result substantiates the claim of researchers that a suitably modified two-dimensional model would provide all that is necessary for cochlear modeling. In a similar way, a shortwave model can be considered. It has been modified to have the following $Q(k)$ function:

$$Q(k) = \epsilon_1/|k|, \qquad (5.3.2)$$

where ϵ_1 has been chosen to provide the best fit (see legend for Fig. 5.4). In view of the eikonal equation (eq. 5.1.8 applied to a shortwave model), this modification is equivalent to overall scaling of the impedance. The response of this model is shown by the *thinnest lines* in Figure 5.4. Remarkably enough, in the peak region this response coincides well with that of the three-dimensional model, especially with respect to the phase. This justifies our preoccupation with short waves, although we now realize that the shortwave model must be duly modified.

We conclude this subsection with a short description of the sequence of physical events for waves propagating in a three-dimensional model:

1. Near the stapes, the cochlear wave starts as a long wave; the oscillations of the fluid occupy the full cross section (height and width) of the cochlear channel(s).
2. On nearing the location of the response peak, the wave changes its character. Its phase variations become more rapid, but the main aspect is that the fluid oscillations are going to occupy only the parts of the channel(s) near the BM. In this region, the wave shows shortwave character.
3. After having passed the peak, the wave amplitude drops off at a rate that exceeds that exhibited by the longwave model.

5.4 A Dilemma

One property is common to the BM velocity responses computed for longwave, shortwave, two-dimensional, and three-dimensional models: in no case does the response have a peak of more than 10 to 15 dB above the rising trend according to the function $\exp(\tfrac{3}{4}\alpha x)$. The three-dimensional peak is slightly higher and it has been shown that in more complicated three-dimensional models it can be higher still (Novoselova 1987, 1989, 1993). This does not agree at all with the trend of recent mechanical data on BM vibrations. We might manipulate the parameters of the model, but *either the amplitude of the peak remains too low or the phase variations in the peak region are too fast*. The reason is, no doubt, that adjacent sections of the partition interact via the cochlear fluid. A piano or a carillon when put under water will not sound as beautiful—despite the sonorities suggested by Claude Debussy's prelude *La Cathédrale Engloutie* (The Submerged Cathedral).

In the cochlear models described thus far, the value of δ is positive, the models are *passive*, and their elements can only *dissipate* energy. In 1980 Kim and associates (Kim et al. 1980) made the daring proposal that the cochlear partition could be *locally active*, which means that over a part of the x domain the real part of $Z_{BM}(x)$ could be negative. Over that region, *the BM delivers energy instead of dissipating it*. That extra energy will be dissipated elsewhere in the model (where the real part of $Z_{BM}(x)$ is positive) so that no spontaneous oscillations will arise. Following Davis's (1983) apt description, we may speak of the *cochlear amplifier*. The idea met with much skepticism, but it gradually gained ground, although up to the present day there is considerable debate over it. In the next section, the arguments in favor of this idea are reviewed, and in Section 7 recent locally active models are described.

It goes without saying that a structure containing energy-producing active elements is likely to become unstable, namely, when the energy produced locally is larger than that dissipated elsewhere. This has been and remains the major concern of modelers of the locally active cochlea. Theoretically, it is not too difficult to verify whether a (linear) model is stable, but physiologically, it remains a mystery how nature should keep energy production and dissipation in balance.

6. Locally Active Models I: The Inverse Problem

6.1 Activity

Before starting on the topic of this section, it is necessary to restate what is meant by "active." In biology, a physiological subsystem is called active when it is involved in a noticeable manner in the functioning of the system. In bioacoustics, this should be the case when it produces forces (sound pressures) that are *comparable in size* to other, nonbiological forces (pressures) in the system. In the field of cochlear modeling, we will call a subsystem active whenever the oscillating force (pressure) produced by it (1) causes the *dissipation* of acoustic power to *decrease,* or (2) causes the local acoustic power to *increase.*

In the first case, we will speak of *undamping;* in the second case, we will call the overall system *locally active.* If in the latter case the model is to remain stable, it must be passive in most of its regions so as to dissipate the extra acoustic power generated. In both cases, the active subsystem derives its power from an external source of energy (see Dallos, Chapter 1; Kros, Chapter 6; Holley, Chapter 7).

An active subsystem will need a kind of sensor to assess the state of the system, and it produces an oscillating force. Since that force influences the system, some sort of feedback is involved, and in the case under consideration that feedback is *positive* because it tends to enhance the response. We will confine ourselves to stable systems here. For further simplicity, we will continue to assume that the system is linear.

For the most general linear stable system (not necessarily a cochlear model), it is true that, when it is tested by way of external observation, there is *no* way of finding out whether or not it contains active subsystems. The main question now is: *Is this also true for a specialized structure like a model of the cochlea?*

We will try to derive the BM impedance function $Z_{BM}(x)$ from the properties of the response $v_{BM}(x)$ or $v_{BM}(f)$. A procedure of this kind is known as an inverse solution. An important point is that this solution forms an *ill-posed problem*. That is, small variations in the desired response (v_{BM}) can bring about very large variations in the recovered parameters (e.g., Z_{BM}), and how are we to know whether the obtained set of parameters is the "correct" one?

6.2 The "Standard Response": Shortwave Case

One way to avoid the subtleties of an ill-posed problem is to use a mathematically defined, continuous response function $v_{BM}(x)$ and to search for the impedance function $Z_{BM}(x)$ that will produce it in a given model structure. We will first use the following type of idealized response function (de Boer 1983):

$$v_{BM}(x) = v_0/[1 + i(x - x_0)/X_0]^\nu. \quad (6.2.1)$$

The parameters v_0, x_0, X_0, and ν are constants; the exponent ν is assumed to be equal to 4 or larger; x_0 designates the location of the response peak; and X_0 is chosen such that the width of the peak occupies a few millimeters of cochlear length. This function gives, in the x domain, a crude representation of the response of the real cochlea in the peak region (in amplitude as well as phase).

Let us restrict ourselves to the *shortwave* case because in the peak region the response is of the shortwave type. Assuming that we can safely neglect reflected waves, we use the shortwave equation (5.2.3) and integrate it to find the pressure $p(x)$, and next compute the BM impedance $Z_{BM}(x)$. The result is

$$Z_{BM}(x) = 2\omega\rho[(x - x_0) - iX_0]/(\nu - 1). \quad (6.2.2)$$

According to this result:

1. The imaginary part of $Z_{BM}(x)$ is negative and constant.
2. *The real part of $Z_{BM}(x)$ should be negative to the left of the location of the response peak and positive to the right of it.*

Clearly, this is in essence the type of impedance modification that was introduced by Kim et al. (1980) in their endeavor to explain the peculiar rise in response in Rhode's (1978) findings. The present author wants to stress, at this point, that the type of activity required to produce the response of eq. (6.2.1) is local activity; the desired response can *not* be produced by undamping.

A more realistic cochlear pattern is obtained when the peak of the response is prescribed as a similar function in the frequency domain:

$$F(\omega) = F_0/[1 + i(\omega - \omega_0)/\Omega]^{\nu}. \tag{6.2.3}$$

Here, ω_0 refers to the resonance frequency corresponding to location x_0, and Ω designates the bandwidth. The use of this type of function to describe frequency selectivity has been discussed by de Boer and Kruidenier (1984). We will be very brief in our description of the application of this expression. First, the function of ω must be transformed to the x domain. The main step of the transformation is the logarithmic transformation from frequency to place (as expressed by eq. 3.2.2). Just one thing is different: normally, low frequencies scale to the right, but in our transformation they should project to the left. In fact, x and ω should vary in such a way that the impedance stays constant. One more step is involved. A cochlear response pattern has, in the longwave approximation, an initial rise according to the function $\exp(\tfrac{3}{4}\alpha x)$ (see Section 4.3). The response must be scaled to show the same course as a function of x. After this mapping-and-scaling procedure, the response function has to be integrated as before to find the pressure $p(x)$. This time the integration is performed numerically. The conclusions are the same as before:

1. Around the response peak, the imaginary part of $Z_{BM}(x)$ is negative and approximately constant.
2. To the left of the peak, the real part of $Z_{BM}(x)$ is negative, and to the right of the peak, it is positive.

6.3. *Processing Actual Response Data*

Encouraged by these results, we go on with actual BM data from the literature. Because of the ill-posed nature of the problem, we cannot use raw data but must select a well-smoothed form of them. R.J. Diependaal performed *regularization* on several data sets from the literature, using what is known as the method of generalized cross-validation (Diependaal, Viergever, and de Boer 1986). This method of smoothing ensures that the resulting function is close to the data points, but also that it has a limited second derivative. For data collected in terms of the sound pressure at the eardrum, some kind of correction for the middle-ear transfer function has been incorporated (see Diependaal, Viergever, and de Boer 1986). As before, in deriving the BM impedance, we will use the shortwave approximation.

We present two results for representative cases from the recent literature (Sellick, Yates, and Patuzzi 1983; Robles, Ruggero, and Rich 1986a,b). Figure 5.5 shows the result for a guinea pig and Figure 5.6 that for a chinchilla. Parts (A) and (B) of the figures show the real and imaginary parts of $Z_{BM}(x)$, respectively. Impedance values are plotted on a nonlinear scale: small values are plotted linearly around zero and large values are compressed logarithmically. The *thick vertical bars* on the right indicate the

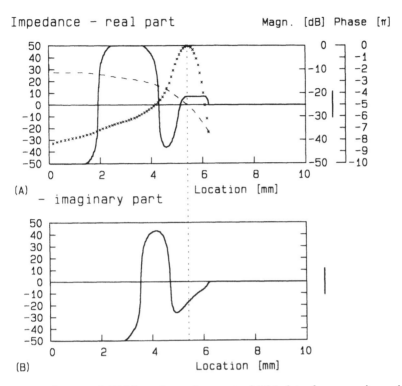

FIGURE 5.5 Computed BM impedance from actual BM data from a guinea pig; shortwave case. Data source: Sellick, Yates, and Patuzzi (1983), animal 90, interpolated and regularized by Rob Diependaal. (A) Real part, (B) imaginary part of $Z_{BM}(x)$. Impedance values are plotted nonlinearly (see text), and the ordinate scale is arbitrary. *Crosses* in (A) mark the magnitude, and the *dashed line* shows the phase of the BM velocity response (transformed to the x domain).

boundary between "small" and "large." The shape of the actual BM response function is indicated in panel (A).

The reader can observe that at the location of the response peak, the real part of $Z_{BM}(x)$ goes from negative to positive, whereas close to the peak the imaginary part remains negative and (approximately) constant. This fully agrees with what has been derived from the artificial response functions used earlier. It should be noted that further away to the left from the peak, the impedance shows rather "wild" excursions. First, the imaginary part changes sign before the real part becomes positive again. It is good to remember this feature. Still further to the left, more irregular variations occur; these can partly be attributed to the poor match to a shortwave model in this (actually, the longwave) region of x and partly to the intrinsic properties of an ill-posed problem.

Zweig (1991) derived a stylized cochlear response function from data obtained by Rhode (1978, animal 73-104). He extrapolated the data to

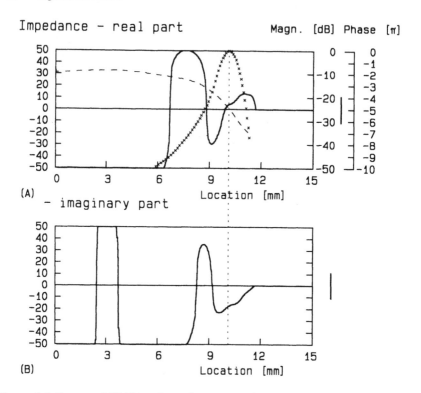

FIGURE 5.6 Computed BM impedance from actual data from a chinchilla; SW case. Layout as in Figure 5.5. Data source: Robles, Ruggero, and Rich (1986a,b), animal 328, interpolated and regularized by Rob Diependaal.

values that would have been obtained at very low levels of stimulation and applied a longwave form of the inverse solution. In general terms, the result was similar: the real part of $Z_{BM}(x)$ must be large and negative in the region to the left of the response peak, just where the cochlear response pattern is steepest. Because his extrapolated response is much sharper than the ones actually measured (at low intensities), it may be questioned whether the extrapolation method used by Zweig is justified.

To illustrate the case of a possibly "passive" cochlea, we present Figure 5.7, based on older data (Johnstone and Yates 1974) that probably came from a considerably damaged cochlea. We applied the inverse method corresponding to the longwave case by integrating eq. (4.2.4) twice. In this case, the real part of the BM impedance is seen to remain positive over the entire region of the peak.

The question of whether there is power amplification inside the cochlea has quite recently been taken up again. From general considerations regarding the flow and concentration of power in the cochlear channels, Brass and Kemp (1993) inferred that a gain of 15 to 40 dB should occur in the cochlea. From the work described in this section, we may well conclude

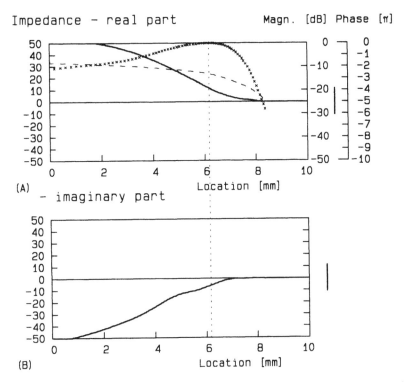

FIGURE 5.7 Computed BM impedance from actual data; longwave case. Layout as in Figure 5.5. Data source: Johnstone and Yates (1974), Figure 5.3, guinea pig, interpolated and regularized by Rob Diependaal.

that "modern" mechanical measurements indicate that a cochlear model explaining the data obtained should be *locally active* near the response peak.

7. Locally Active Models II. Micromechanics

7.1 A Second Degree of Freedom

In principle, all cochlear models that have been described thus far are "classical" *macromechanical* models, i.e., they belong to the class of models in which two fluid columns interact with a CP of which all parts move in unison with the BM and that has no longitudinal mechanical coupling (see Section 3 and Viergever 1986). In the present section, *micromechanical* models will be discussed in which the elements of the organ of Corti can move with different velocities and are given specific parts to play. The description will still be restricted to "classical" models, i.e., models in which there is no longitudinal mechanical coupling in the cochlear partition.

In recent articles, it has been proposed that a *secondary resonance* controls local activity in the organ of Corti. The idea of a secondary resonance, a second degree of freedom (DoF), is not new. In the "older" literature, models in which the *tectorial membrane* (TM) is endowed with an independent DoF have been described. These models were originally developed to explain the difference between tuning as measured in neural and inner hair cell (IHC) responses on the one hand (Evans 1972, 1975; Russell and Sellick 1977, 1978), and that of the mechanical BM response, as it was known in those days (Johnstone and Boyle 1967; Johnstone, Taylor, and Boyle 1970), on the other. As a start, Zwislocki and Kletsky (1979) described a model in which the TM was free to execute (radially directed) movements whereby longitudinal coupling in that membrane remained the principal restraining factor. Allen (1980) went further: he described the TM so as to allow *independent radial movements in each cross section* (this is just what is assumed for the BM, but here the movements are occurring in a different direction). Neely (1981) used the same 2-DoF concept for an extended *locally active* model of the cochlea. Later the resulting model became known as the *Neely-Kim model* (Neely and Kim 1986).

We will first describe the second resonator, the second DoF. Remember that the first DoF is formed, as usual, by the mass and stiffness of the BM. Figure 5.8 shows, in a cross section of the cochlea, a much simplified diagram of elements of the organ of Corti. Figure 5.8A shows how the TM and BM are assumed to *rotate* about two different connections to the bone of the modiolus; the hinges are shown as half-circles. As a consequence, the cilia of the hair cells are bent when the BM is deflected from its equilibrium position. When the vertical distance between the hinges is h and the radial distance from the hinges to a hair cell is b, the radial deflection Δy_c of the cilia of that cell will be related to the deflection Δz_{BM} of the BM as

$$\Delta y_c = (h/b)\Delta z_{BM}. \qquad (7.1.1)$$

The average value of the factor (h/b) is known as the shear gain (cf. Allen 1980) and will be denoted by g. In this conception, the TM is acting, *in each cross section*, like a *stiff bar;* it can only rotate but not bend.

The second DoF arises when the TM is considered as a mass that can freely slide in the *radial* direction of the cochlea (the y direction in our model) and is restrained only by its elastic attachments. In this respect, a central attachment to the modiolus is essential, otherwise the stimulation mechanism indicated in Figure 5.8A cannot work, a point made by Allen (1980). This attachment is given the form of an additional elastic connection between the body of the TM and the modiolar wall. Other attachments are to the reticular lamina (RL) at the outer margin (Slepecky, Chapter 2) and to the organ of Corti via the OHC cilia. The resulting configuration is shown in Figure 5.8B. There are two impedances involved in radial movements of the TM: (1) Z_{TM}, formed by the stiffness k_{TM} and resistance

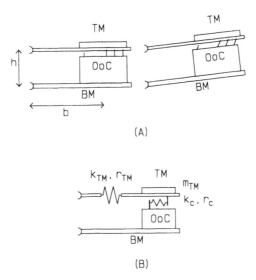

FIGURE 5.8 Illustrating the second degree of freedom (2nd DoF). The organ of Corti (OoC) in cross section, schematically; the modiolus is on the left. (A) Principle of excitation of the stereocilia of the hair cell. *Left*: equilibrium position; *right*: deflected state. The IHC stereocilia (only one is shown) do not touch the tectorial membrane (TM); the IHC cilium is "dragged along" by the fluid. The upper ends of the OHC cilia are embedded in the TM. (B) The TM as a mass suspended on two springs. Movements of the TM occur in the radial direction (from left to right and back). For symbols see text. After Allen (1980).

r_{TM} of the connection of the TM to the wall, and the mass m_{TM} of the TM; (2) Z_c, formed by the stiffness k_c and resistance r_c of the connection via the cilia to the main part of the CP. Note that in the cross section, *the TM is still infinitely stiff against bending: it can rotate about its hinge and oscillate in the radial direction, but can make no other movements.* Furthermore, longitudinal coupling in the TM is neglected, as in the BM.

So much for the second DoF by itself. In the Neely-Kim model, it is assumed that bending of OHC cilia produces a force (or pressure) that *acts directly on the BM*. The pressure p_c is the active pressure, and its complex amplitude is assumed to be proportional to the shear velocity v_c of the cilia (which is equal to the relative radial velocity of the TM with respect to the RL). This is expressed in the following form:

$$P_c = \gamma g Z_{NK} v_c. \qquad (7.1.2)$$

The parameter γ is introduced to let the model go from the passive state ($\gamma = 0$) to the fully active state ($\gamma = 1$). The parameter Z_{NK} is a *transfer impedance* that relates the pressure exerted on the BM to the shear velocity.

In theory, Z_{NK} should not depend on place or frequency, but in reality it is made to contain a resistance and a stiffness component, mainly introduced to obtain the desired phase shift between v_c and P_c. The filtering action brought about by the second DoF is due to the actual relation between v_{BM} and v_c. The factor g is used in eq. (7.1.2) because we go from radial to vertical velocity. The pressure P_c is to be added to the trans-BM pressure. The total effective BM impedance $Z_{BM}^{(eff)}$ then contains contributions from three origins, the passive Z_{BM}, the combination of Z_{TM} and Z_c, and the influence of the active source P_c. Calculation produces the following expression (cf. Neely and Kim 1986):

$$Z_{BM}^{(eff)} = Z_{BM} + g^2[Z_{TM}/(Z_c + Z_{TM})][Z_c - \gamma Z_{NK}]. \quad (7.1.3)$$

For the action of the BM as the source of cochlear amplification, the last term is the most important one. When resonance occurs in the denominator of this term, i.e., when the reactive part of $(Z_c + Z_{TM})$ is zero, the influence of the active term with Z_{NK} is largest. This resonance, then, produces the required distribution of activity over the x axis. It is made to occur a few millimeters basalwards from the location of resonance for the first DoF (or, for a given location, the resonance of the second DoF lies approximately one octave below that of the first DoF). It should, furthermore, be noted that γ does not occur in the denominator of eq. (7.1.3). This demonstrates that there is *no feedback in an isolated cross section of the model;* the feedback in the Neely-Kim model is *global,* that is, it arises because the sections interact via the cochlear fluid.

The BM responses of the Neely-Kim model replicate the actual data very well (Neely and Kim 1986). We will present one typical example of a BM-velocity response pattern computed with the longwave model (Fig. 5.9A) and the course of the corresponding BM impedance (Fig. 5.9B). Both variables are plotted as functions of x for a fixed frequency (in the "panoramic view"). The *thick lines* refer to the maximal value of the activity parameter γ for which the model remains stable ($\gamma = 1.0$), and the *thin lines* represent the "passive" case ($\gamma = 0.0$). To accommodate the large response peak in the velocity curves, the vertical scale has been changed to encompass 100 dB, against 50 dB for previous figures. After an initial small decrease, a substantial rise of the BM velocity is observed, and this illustrates well what is produced by the "cochlear amplifier" (a term borrowed from Davis 1983). As an aside, it should be pointed out that Patuzzi, Yates, and Johnstone (1989b) argued that the BM velocity gain has been overestimated by Neely and Kim by at least 15 dB. Indeed, locally active models developed later show smaller gains.

Figure 5.9B shows the effective BM impedance (for the same two values of γ). The *solid* and *dashed lines* depict real and imaginary parts, respectively. Both are plotted so that small values around zero are represented faithfully, but large values are compressed logarithmically (in the same way

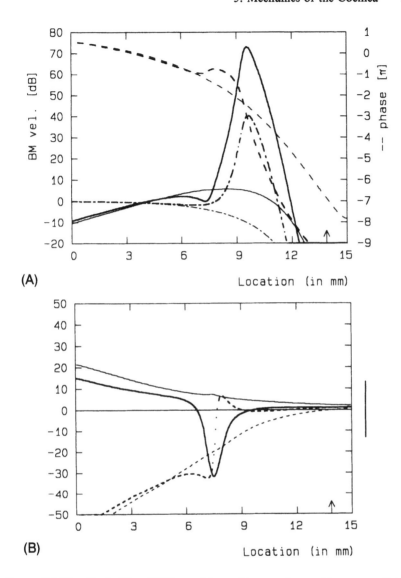

FIGURE 5.9 The Neely-Kim (1986) model, cochlear patterns. Frequency: 6 kHz. *Arrow*: location of resonance in the passive case. (A) BM velocity. *Solid lines*, magnitude; *dashed lines*, phase. *Dash-dot lines*, power flux function, see text. *Thick lines*, γ = 1; *thin lines*, γ = 0. (B) BM impedance. *Thick lines*, γ = 1.0; *thin lines*, γ = 0. *Solid lines*, real part; *dashed lines*, imaginary part. Impedance values are shown as in the figures of Section 6: values within the limits of the *thick vertical bar* are shown linearly, larger values are compressed.

as in the impedance plots of Section 6). Consider the case for γ equal to 1.0 (*thick solid line*). The real part has a pronounced *negative excursion* in the region to the left of the response peak that reaches far into the domain of negative values. This clearly shows the effect of the second resonance. The behavior of the impedance is actually more complicated. In the region where the real part shows its activity peak, the imaginary part also shows pronounced variations. In the light of general physical properties of electrical networks, this behavior is certainly not surprising, since real and imaginary parts of an electrical impedance are tightly connected (cf. Koshigoe and Tubis 1982), and the same is true for mechanical and acoustic impedances. As a matter of fact, we encountered similar linked variations in real and imaginary parts in the preceding section (and we did well to remember them). For γ equal to zero (*thin lines*), the course of the impedance function is more gradual; in the real part, a small trace is seen of the influence of the micromechanics (the combination of Z_{TM} and Z_c).

The *dashed-dotted lines* in Figure 5.9A show the course of the *power flux* carried by the fluid. In the longwave case, power flux and power density (i.e., the power flux per unit of cross-sectional area) are proportional. The per channel power density $P_x(x)$ is defined by

$$P_x(x) \equiv \tfrac{1}{2}\mathrm{Re}[p(x)v_x^*(x)], \qquad (7.1.4)$$

where Re means "real part of," $v_x(x)$ is the velocity of the fluid in the x direction, which can be found from eq. (4.1.2), and the asterisk denotes the complex conjugate. The variable $P_x(x)$ is normalized to start with 0 dB at $x = 0$. In the active case, the power flux first decreases somewhat—as it should do in a passive model—but then it starts to rise gradually. Remarkably, the power gain is much less than the actual rise of the BM velocity, approximately 43 versus 70 dB. We will come back to the cause of this effect later.

7.2 Outer Hair Cells

After the discovery of fast OHC motility, in the form of *cyclic lengthening and shortening of the cells* (Brownell et al. 1985), the source of cochlear activity was assumed to be found in this mechanism. It should be noted right away that some researchers hold a different view: they think that OHCs influence cochlear mechanics by way of mechanical effects exerted by their stereocilia. It would lead us too far afield to discuss this point more deeply; we will only review some models that have appeared in the literature. Although Neely and Kim (1986) considered the OHCs as the source of activity, their model (as described in the preceding section) is abstract, in the sense that it does not specify in what manner the OHC force acts directly on the BM. The details on this issue were filled in by Geisler (1986, 1991) who realized that because the hair cells are suspended between the RL and (via the Deiters cells) the BM, the motile force should always be represented as a *pair of forces* inside the organ of Corti. That is, on a cycle-to-cycle basis, the OHCs will alternately pull the RL and BM together

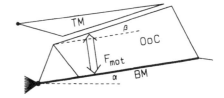

FIGURE 5.10 Micromechanics: a pair of OHC forces inside the organ of Corti (After Geisler, 1986).

and push them apart. Figure 5.10 illustrates this basic principle. The BM is deflected and the tunnel of Corti rotated with respect to its central attachment (the spiral lamina) over an angle α. The OHCs produce a pair of motile forces F_{mot}, shown in the form of a *double arrow*. In the situation shown, the force pair would actually be contractile for low frequencies (Ashmore 1987), but for clarity of illustration, it is shown as expanding and causing a further dilation of the RL with respect to the BM over an angle β. The TM follows the total displacement of the RL – in this case, the TM need not have a mode of movement of its own – and it is therefore logical to assume that the radial drive to the hair bundles of the OHCs is approximately proportional to the total RL deflection ($\alpha + \beta$). The lower force of the pair acts directly on the BM (the Deiters cells form an incompressible mass) and participates in the macromechanics. The upper force acts against the TM. A reaction force will arise when the RL moves and slides along the underside of the TM, causing the stereocilia to "feel" a radially directed counterforce. This force is not directed at the modiolar hinge point of the BM and thus has a nonzero torque with respect to that point. The torque is compensated by a force on the BM, and in this way the radial force is converted into a force in the z direction. Actually, the mechanism is somewhat more complicated (see Geisler 1991 or 1993 for more details). We may note, finally, that in this model there is *feedback within each cross section of the model*, the reason being that the action of F_{mot} on the relative motion of RL and BM has a direct repercussion on the drive to the OHCs and, therefore, on F_{mot} itself. Apparently, in this type of model there is local as well as global feedback.

Several recent models have been constructed on this basis. We will describe only their principal features so as to give the reader an idea of the amount of variation that is possible. Furthermore, we will simplify the argument for these models somewhat. In Geisler's (1991) model, there is no second resonator. All frequency dependence of the active force is concentrated in the OHCs. The strength F_{mot} of the active source is assumed to depend on the (vertical) velocity v_{RL} of the RL, the velocity associated with the RL deflection ($\alpha + \beta$). This dependence is written as follows:

$$F_{mot} = \gamma Z_{HC} v_{RL}, \qquad (7.2.1)$$

where γ has the same meaning as before, and the parameter Z_{HC} is the *transfer impedance* of the OHCs. Influenced by considerations put forward by Zweig (1990, 1991), Geisler included an intrinsic delay in OHC transduction which is approximately inversely proportional to frequency (for a refinement leading to a fully realizable model, see Geisler 1993). The size of the computed BM response peak is more modest than in the Neely-Kim model, but it is comparable with the trend of modern mechanical data.

A combination of the two themes discussed—the second DoF and the concept of a force pair—can be discerned in a later model of Neely (1993). In this model, independent radial motion of the TM (the second DoF) is possible. The concept of OHCs tending to move RL and BM with respect to one another is put into an extreme form: it is proposed that the effect of OHC contraction and expansion is to change the separation between the RL and the BM, *irrespective of the impedance encountered*. The high-frequency cutoff effects in OHCs are represented by a second-order transfer function. Neely worked out his model in great detail, and the reader is referred to the original article for further subtleties and results.

Another variation is found in the model described by Mammano and Nobili (1993). These authors start from a slightly different conception of the statics of the organ of Corti and reason that the required amount of phase shift for positive feedback would be mainly due to the inertia of the TM combined with the low-pass filtering action of the OHCs. They too achieve quite realistic model response functions.

It is necessary to point out one obstacle to proper operation of OHC motility, as it has been described in the foregoing. Take a look again at Figure 5.10 depicting the micromechanics of the organ of Corti and consider the counterforce that the OHCs experience at their tops. This is, in principle, directed vertically, in the z direction. In the models described, the actual counterforce (directed radially) was always converted to the z direction via the torque exerted on the OHC stereocilia (leading to the shear gain g). It is then inevitable that the impedance Z_c (see Fig. 5.8B) will play a crucial part. It should not be too small, because then the OHC force will create large vertical movements of the RL and little movement of the BM. Yet the stiffness component of Z_c is unlikely to be comparable to the BM stiffness. In the author's new model (de Boer 1993), this problem is removed: the counterforce is assumed to be produced *directly in the z direction,* and this occurs by the TM. At first sight, the TM would be able to rotate freely about its central attachment to the modiolus. At closer sight, this is not so because of the fluid surrounding the TM. Any rotation of the TM (remember, it is stiff against bending) causes a displacement of the fluid in the *internal spiral sulcus* (for the anatomy, see Slepecky, Chapter 2), and the displaced fluid has to be compensated by fluid from elsewhere. Because of the narrow aperture of the sulcus, a high impedance is associated with lengthwise fluid flow in the sulcus, and this effect is found to restrain the upper ends of the OHCs sufficiently. In this light, the Neely-Kim model (Section 7.1) is not as abstract as it seems: hair cells can really exert forces that almost exclusively operate on the BM.

We conclude this section with a brief and perhaps incomplete account of alternative micromechanical models. The general problem areas have recently been reviewed by Allen and Neely (1992), so here we can be more specific. In Kolston's (1988) first model the fact that in the radial direction, the BM consists of two parts, the pars tecta and pars pectinata (Slepecky, Chapter 2) was utilized. The pars pectinata was assumed to partake in the macromechanics of the cochlea, and a frequency-specific transformation was hypothesized to act between the two parts. The pars tecta (which includes the floor of the tunnel of Corti) was then assumed to be the principal factor in effecting neural stimulation. Novoselova (1987, 1989, 1993) treated a related radial inhomogeneity, namely, that due to the irregular distribution of mass in the radial direction. In a sense, these approaches are related, in that both authors sought to implement some sort of "mechanical second filter" in the cochlear partition of a passive model. Later work (Kolston et al. 1990) defended the idea that BM vibrations can be most efficiently enhanced by assuming that it is the *imaginary part* of the BM impedance that is the most affected by the OHC motile mechanism. An actual model on this principle showed quite realistic response curves. We should realize, now, that we have encountered the same principle—but in a disguised form—in the Neely-Kim model described earlier in this section. In the active region, we found the increase in the BM velocity to be larger than the power gain (see Figure 5.9A). We can now attribute this difference to concentration of acoustic energy accompanying variations of the imaginary part of the BM impedance.

The idea of a "cochlear amplifier" has recently been put into a novel form by Hubbard (1993). In analogy to the operation of an electronic device called a "traveling-wave tube," he conceives of *two* structures in the cochlea that are stretched out over the full length and that both can support traveling waves. The coupling of these structures is given a form in which energy is transferred from one to the other and back in such a way that the power of the main cochlear wave increases as it is propagated. It should be clear that this type of model is not a "classical" model in the sense used thus far: its operation cannot be simply understood from a quantity like the BM impedance. Because the development of this new type of model seems to be still in its infancy, we must refrain from delving further into its principles of operation.

8. Nonlinearity

8.1 Introduction: The Essence of Cochlear Nonlinearity

What is *linearity*? In fact, since Section 3 we have been using linearity without being aware of its true nature at all. Let us just state the definition (without specifically saying so, we always consider a time-invariant system—a periodic stimulus leading to a periodic response). Then, paraphrasing a well-known expression:

linearity is . . . *The response to the sum of two stimulus signals is the sum of the responses to the stimuli separately.*

This property has the consequence that response magnitude is proportional to stimulus magnitude. From the same property it can be proven that the response to a sinusoidal stimulus is sinusoidal too.

In a nonlinear system, this property does not hold. When the response to a sine wave is not a sine wave, but is still periodic, the response will contain *harmonics*, i.e., sinusoidal components with frequencies that are integral multiples of the stimulus frequency. When the stimulus signal consists of more than one sine wave, the response will contain components with the primary frequencies, their harmonics, but also *intermodulation products* (in acoustics known as *combination tones*) with frequencies that are integral combinations of the primary frequencies. In this chapter, intermodulation products will often be referred to as *distortion products* (DPs). For instance, when the stimulus contains components with frequencies f_1 and f_2 (where f_2 is larger than f_1), DPs have frequencies $(n_1 f_1 \pm n_2 f_2)$ where n_1 and n_2 are integers. An important DP is the one with frequency $(2f_1 - f_2)$.

A common property of nonlinear systems is that the response amplitude is not proportional to the stimulus amplitude. This is also true for the cochlea. In fact, in the cochlea, the response amplitude is generally *less* than proportional to the stimulus amplitude; the cochlea is a *compressive system*.

In the theory of linear systems as we have used it throughout this chapter, a central property is that one component does not affect in any way what happens to another component. In a nonlinear system, this is no longer valid. Therefore, we cannot use a complex function like $a \exp(i\omega t)$ (with a constant) to describe a time-varying quantity like a stimulus or response component. Unfortunately, there is no comprehensive theory of nonlinear systems that is as simple as the theory of linear systems and as useful for providing insight. That makes the treatment of cochlear nonlinearity fundamentally complex and seemingly noncoherent.

Let us start by reviewing the nature of nonlinear effects found in the peripheral auditory system when stimuli like simple sinusoidal signals or combinations of two sinusoidal signals with frequencies f_1 and f_2 (where f_2 is larger than f_1) are used. Nonlinear effects in *neural responses* have been known for a long time. Response measures are rates of firing of auditory nerve fibers or Fourier components of temporal response patterns of such fibers. Typical neural nonlinear effects are:

1. With strong stimulation, the response appears as reduced or compressed, at and above the characteristic frequency.
2. A Two-tone suppression: one stimulus component can suppress the response to another one (Sachs and Kiang 1968; Sachs 1969; Abbas

and Sachs 1976; Sachs and Abbas 1976; Schmiedt 1982; Fahey and Allen 1985; Prijs 1989).
B A special case of two-tone suppression is flattening of the tuning curve when a low-frequency noise band or a low-frequency tone is presented (Kiang and Moxon 1974; Fahey and Allen 1985).
C Another manifestation of two-tone suppression is modulation by a strong tone of very low frequency (e.g., 40 Hz) (see Patuzzi, Sellick, and Johnstone 1984a).
3. DPs are produced by the cochlea, for instance, components with frequencies $(2f_1 - f_2)$, $(2f_2 - f_1)$, and $(f_2 - f_1)$. The first of these DPs can be detected by a primary nerve fiber when neither of the primary tones is excitatory when presented alone (Goldstein and Kiang 1968; Kim, Molnar, and Matthews 1980). The same is true for the third DP (the "difference tone").

We have excluded here nonlinear effects such as saturation, adaptation, refractoriness, and masking that can safely be attributed to the synapse and not to the mechanics. Equivalent nonlinearities have been observed in electrical responses of IHCs (e.g., Sellick and Russell 1979; Cheatham and Dallos 1989, 1990a,b; Nuttall and Dolan 1993).

In *psychophysics,* nonlinear effects are also known to exist:

1. Psychophysical tuning curves differ when different detection criteria are used, because of asymmetry and dependence of masking patterns on level.
2. To detect two-tone suppression, a special measuring technique is required (the classical paper is Houtgast 1972).
3. The DP with frequency $(2f_1 - f_2)$ is easily observable, and it is found to have "unusual," i.e., compressive, amplitude behavior (Zwicker 1955; Goldstein 1967; Smoorenburg 1972a,b). Moreover, this DP has a reduced amplitude when there is a hearing loss for one of the primary frequencies (Smoorenburg 1972b, 1980).

Later, *evoked otoacoustic emissions* were recognized and analyzed, and were found to be nonlinear in their amplitude behavior (Kemp 1978, 1979) and to include DPs for multicomponent stimuli (e.g., Kim, Molnar, and Matthews 1980). This evidence can be interpreted as resulting from the fact that the BM is nonlinear in its vibration.

In *BM motion,* nonlinearity has always been observed since the epoch-making experiments of Rhode (1971, 1978). The most noteworthy features are the following (the correspondence with the properties mentioned above is all too evident): compression at and around the peak of the response, level dependence of the shape of impulse responses, two-tone suppression, modulation by low-frequency tones, and generation of DPs (see Section 2 for references). Many data from the literature imply that nonlinearity is physiologically vulnerable, just as the neural (or the BM) response

(Robertson 1976; Kim, Siegel, and Molnar 1979; Dallos et al. 1980; Kim, Molnar, and Matthews 1980; Schmiedt, Zwislocki, and Hamernik 1980).

A final note: two-tone suppression and the typical amplitude behavior of cochlear DPs can only occur in a nonlinear system when it is *compressive*. Also the phase behavior of the psychophysically observed DP with frequency ($2f_1 - f_2$) is typical for a compressive nonlinearity (Goldstein 1967; Smoorenburg 1972a,b). A comprehensive phenomenological model tying together all known nonlinear effects has been developed by Goldstein (1990). The first steps toward integration with properties of a wave-propagating structure have been taken (Goldstein 1993).

8.2 Nonlinear Damping in a Mechanical Model of the Cochlea

Before 1970, the mechanical response of the cochlea was found to be linear and much less frequency-selective than the neural response. Hence, a *sharpening mechanism* or *second filter* was assumed to exist between the BM and the excitation of the (inner) hair cells (Evans 1975), and all nonlinearity was assumed to reside in the same mechanism. In cases of cochlear pathology, the sharpening mechanism was generally assumed to be affected, which explained why nonlinear effects diminished with increasing pathology.

In the period after 1970, the mechanical response of the cochlea turned out to be sharper and sharper over the years, but also more and more nonlinear. Two explanations are possible: (H1) basic cochlear mechanics is nonlinear and the sharpening mechanism is linear, or (H2) the sharpening mechanism is nonlinear but it loads the BM. For reasons that are not completely clear now, hypothesis H1 was the first to be adopted for quantitative evaluation. As early as 1972, Hubbard and Geisler started to develop nonlinear versions of longwave cochlear models. Because the frequency response flattens when loud stimuli are used (Rhode 1971), it was assumed that *BM resistance increases with increasing velocity*. A curious fact is that physical or physiological reasons for this assumption have never been given, nor has the relation to cochlear pathology been clarified. In any event, a nonlinear cochlear model with this property will behave as a compressive system, as it should.

The first really well-developed model in which this assumption was used has been described by Hall (1974). Hall assumed the BM resistance—now written as $R_{BM}(x,t)$—to be a nonlinear function of the BM velocity $v_{BM}(x,t)$, as follows:

$$R_{BM}(x,t) = R_0[1 + \alpha v_{BM}(x,t) + \beta v_{BM}^2(x,t)]. \qquad (8.2.1)$$

The nonlinear function is assumed to be instantaneous, which is why the time t figures as an additional independent variable. Hall developed programs for computing the model response in the *time domain* and used the Fourier transform to extract the components of the response. If α equals 0 and β differs from 0, the resistance is an even function of $v_{BM}(x,t)$, and only odd-order DPs will be generated. An odd-order term ($\beta = 0$, $\alpha \neq 0$) in the resistance function produces even-order DPs.

We will now briefly outline a few typical results that have retained their value. We take the first case ($\alpha = 0$ but $\beta > 0$). Let the stimulus consist of two components with frequencies f_1 and f_2 ($f_2 > f_1$). In the model, a DP with the frequency ($2f_1 - f_2$) was found to arise, present as a wave originating in the region where both the f_1 and the f_2 tones have substantial amplitudes and traveling toward the site where it shows its own response peak. In this way Hall could explain the experimental results of Goldstein and Kiang (1968). This ($2f_1 - f_2$) DP wave corresponds to the DP component commonly observed in psychophysical experiments. The model response also included a *reverse* (retrograde) wave for this DP; we would now recognize this as the origin of a DP emission (similar retrograde waves have invariably been found in more modern nonlinear models). Another type of DP, the DP with frequency ($2f_2 - f_1$), arose in the model too. This DP, however, does not generate a wave to its "own" location, because it arises in a region tuned to lower frequencies than its own. Neither does it produce a retrograde wave.

In his next papers, Hall (1977a,b) treated two-tone suppression. He confirmed that one strong tone, by causing the resistance to increase, could cause another tone to be damped in its propagation. To harmonize his model results with experimental data (in particular, responses of auditory nerve fibers), Hall had to introduce a (linear) sharpening mechanism after (nonlinear) cochlear filtering. This makes his work on two-tone suppression less interesting for us now.

Further work with this type of model was done by Matthews (1983). He recognized the need for a good representation of the middle ear and the acoustic driver because the destiny of retrograde DP waves is determined by reflection occurring at the level of the stapes.

More recently, the mechanical selectivity of the cochlea has been found to be higher and higher. This led to the thought that there would be a much closer and more mutual coupling between BM movements and neural excitation (hypothesis H2) so that mechanical and neural responses would show similar frequency selectivity and similar degrees of nonlinearity (arguments in favor of this hypothesis were presented by Goldstein as early as 1967, and later by Kim, Molnar, and Matthews 1980, and Weiss 1982). Linear models with such bidirectional active elements have been described in the preceding section; nonlinear models of this class will be treated in the next two sections.

8.3 Nonlinear Undamping Versus Nonlinear Local Activity

From here on we will consider models of the cochlea in which an *active* subsystem affects the BM impedance: undamped and locally active models. The active component of the BM impedance is produced by a specific physiological mechanism that, in itself, *may become overloaded by strong signals*. Then, for strong stimuli, the gain produced by the active process is likely to diminish, with the result that the frequency response becomes flatter, two-tone suppression may occur, and DPs are generated. In very general terms, we can say that nonlinear effects are caused by *deactivation* (diminution of activity). Note that now *the nonlinear element itself is compressive,* which is physiologically much more likely than the expansive behavior of a resistance as considered by Hall and others. For undamped as well as locally active systems (recall Section 6 for the distinction between these two cases), we will find that relatively small degrees of nonlinearity of the active elements have a large effect upon the degree of nonlinearity of the overall system. This property is a direct consequence of the fact that the response of a system with positive feedback is also quite sensitive to the degree of feedback (cf. Patuzzi, Yates, and Johnstone 1989b).

If the input to the active element (for instance, the OHCs) is simply proportional to BM velocity, activity only produces *undamping*. In the nonlinear case, it is not damping which increases, but undamping which decreases, with strong stimulation. The first endeavor in this field was by Mountain, Hubbard, and McMullen (1983), but the simplest form of the hardware model described by Zwicker (1986a,b,c) also belongs to this category.

In Section 6, we argued that it is not undamping but *local activity* that is required in a cochlear model (and we elaborated on such models in Section 7). This is the principle we are going to follow now. But we also must mention a fundamental difference between these two types of activity, one that specifically concerns nonlinearity. To make a model locally active, some sort of frequency-place-selective filter must be included in the path from the BM to the active element (see Section 7). That filter will have a frequency-dependent phase response. When now the amount of activity varies as a result of overloading the active element, the *reactive* as well as the *resistive* component of the BM impedance will be affected. This is an effect which, in the modeling of cochlear nonlinearity, has not yet been analyzed in great depth. We must caution ourselves that we are speaking of impedances, where this is formally not allowed in a nonlinear system. Yet, as we shall see soon, the use of the impedance concept is legitimate in a weakly nonlinear system such as the cochlea.

8.4 Nonlinear Local Activity: the Quasi-Linear Method

Models of the cochlea containing nonlinear elements have usually been solved "in the time domain." This means that the fundamental equations are written in terms of place and time coordinates and that waveforms of the physical variables are computed. For a concise description of the time-domain solution method, the reader is referred to de Boer (1991). A robust computation scheme for time-domain computations has been developed by Diependaal et al. (1987). Most of the work in the time domain is done on a longwave model. It should be clear that the time-domain solution is more complicated than the frequency-domain solution, especially in locally active systems because of the positive feedback and power gain involved; furthermore, numerical stability is difficult to control.

Time-domain solution methods require very much computer time. Although with modern fast computers this aspect is gradually becoming less important, the time-domain solution provides only limited insight. This is so because concepts we are familiar with, such as *amplitude, frequency, phase angle, impedance, power flux, wavelength, wave number, wave velocity,* etc., cannot be used any more. Every segment of the waveform computed has to be analyzed meticulously before any useful information can be extracted. An alternative solution method has been developed (Kanis and de Boer 1993a) which actually embodies the concepts mentioned above. A concise description follows.

In the model, OHCs are considered to produce an active pressure across the BM and are also considered to be the *sole sites of nonlinearity.* Let the input to an OHC be the radial stereociliary velocity $v_{\text{cilia}}(x,t)$ and the output the pressure $p_{\text{OHC}}(x,t)$; the latter is assumed to be acting directly on the BM. It is assumed that between these two signals there exists a nonlinear memoryless relation. A linear but frequency- and place-dependent transformation is assumed to be involved from the BM velocity $v_{\text{BM}}(x,t)$ to the velocity $v_{\text{cilia}}(x,t)$. This transformation ensures the proper distribution of activity over frequency and place.

For a weak sinusoidal stimulus to the model, the signals $v_{\text{cilia}}(x,t)$ and $p_{\text{OHC}}(x,t)$ will be (nearly) sinusoidal, and the relation between BM velocity $v_{\text{BM}}(x,t)$ and pressure $p_{\text{OHC}}(x,t)$ can be expressed in the form of a *transfer impedance* (pressure divided by velocity), a complex function of x and radian frequency ω. Let this be called $Z_{\text{OHC}}^{(\text{lin})}(x,\omega)$. When the stimulus to the model is made stronger (but remains sinusoidal), the hair cell will produce a distorted (in fact, a compressed and rectified) response, $p_{\text{OHC}}(x,t)$. Because the BM velocity is closely and linearly linked to the input $v_{\text{cilia}}(x,t)$, it is reasonable to suppose that the BM response $v_{\text{BM}}(x,t)$ will remain nearly sinusoidal. This implies that for the BM velocity, only *the first-order Fourier term* of the OHC pressure waveform $p_{\text{OHC}}(x,t)$ is important. Likewise, for the input $v_{\text{cilia}}(x,t)$ to the hair cell, *the first-order*

Fourier term of the BM velocity $v_{BM}(x,t)$ is the main contributor. The essence of the quasi-linear method is that it is assumed that higher-order terms do not interact with first-order terms. The ratio of the first-order term of $p_{OHC}(x,t)$ and the first-order term of $v_{BM}(x,t)$ again yields a transfer impedance, $Z_{OHC}^{(NL)}(x,\omega)$. This impedance is to be added to the (passive) BM impedance, but note that it depends on the level of vibration in the model. This determines that the ultimate solution can only be obtained in a series of *iterations*. Since in every iteration step the model is considered as linear, the method is called the *quasi-linear method*.

This method has a number of advantages, and its application can greatly contribute to our insight. For instance, *deactivation* (i.e., reduction of activity), which occurs when strong stimuli start to saturate the hair cells, leads to *self-suppression* of the response. And as we will see soon, antero- and retrograde waves can easily be distinguished. Thus, the chain of events and effects occurring in a nonlinear model can be decomposed into its elements.

8.5 Distribution of Distortion Products (Combination Tones)

We now discuss two figures showing the distribution of components over the length of a locally active model in the case where two sinusoidal signals form the input signal (adopted from Kanis and de Boer 1993b). Figure 5.11 shows the case where the primary frequencies f_1 and f_2 are wide apart, and Figure 5.12 the case where they are close together. In fact, the primary frequencies are selected so that the frequency of the principal DP, the one with frequency $2f_1 - f_2$, is the same in the two figures. The responses shown are *BM velocities* for the two primary tones as well as for this DP. The influence of other components is neglected. The computations are done with the quasi-linear method on a variation of the Neely-Kim model.

We start the description with Figure 5.11. Figure 5.11A shows the magnitude of BM velocity for each of the components. Primary components are shown by *thick lines,* the DP by a *thin line*. To offset its small size, the DP is depicted at a 40-dB-higher level. For this figure, the ratio f_2/f_1 is equal to 1.4. The two *vertical arrows* indicate the locations of maximal (velocity) response for the two primary components. There are three regions where we observe one component to undergo a rapid increase of its amplitude; there is an "active region" for each of the three frequencies. For the case illustrated here, the active regions are well separated. Figure 5.11B shows the phase response of the DP. Quite arbitrarily, the phase is depicted as starting with zero at the stapes location. The effective "source" of the DP component (this is the point where the OHC-generated pressure is maximal) is indicated with a *cross*. To the left of this point, the DP phase lag is seen to increase when the distance from the locus of origin increases, just as in

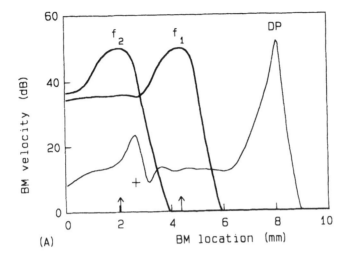

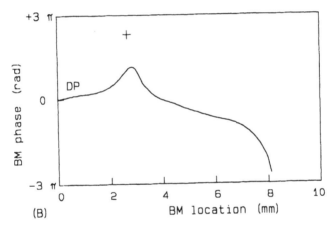

FIGURE 5.11 Response of a nonlinear cochlear model to a two-component stimulus. (A) BM velocity patterns (re 10^{-3} mm/s) produced by presenting two sinusoidal signals at 60 dB SPL. For this figure, f_1 and f_2 are 10 and 14 kHz, respectively. *Vertical arrows* indicate response maxima for these components. *Thick lines* show primary components (magnitude only). The *thin line* shows the $2f_1 - f_2$ DP plotted at a 40-dB higher level. The *cross* marks the source of this component (see text). (B) Phase response of the DP component. At $x = 0$, the phase is normalized to zero. Model used: see Kanis and de Boer (1993a); a most simplified model of the middle ear is included.

a wave traveling to the left. The left-going wave is seen to reach the stapes smoothly. As a matter of fact, at the location of the stapes, a near-perfect match has (artificially) been created between the driving source and the cochlear model so that the retrograde DP wave is not reflected at that point. To the right of the cross, there is a wave that travels to the right, and this

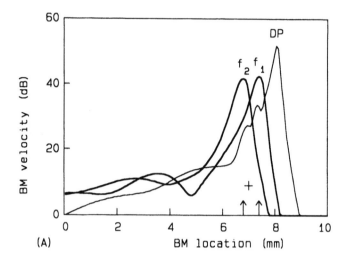

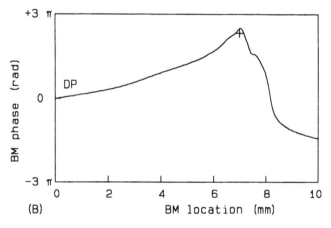

FIGURE 5.12 As in Figure 5.11, but the frequencies f_1 and f_2 are 6.6 and 7.2 kHz, and the primary component levels are 36 dB SPL. Note that the DP frequency is the same (6 kHz) as in Figure 5.11 and the DP component reaches the same amplitude in its response peak.

DP wave reaches its maximal amplitude at location $x = 8.05$ (mm). Note that it passes through a region where it is amplified.

Figure 5.12, laid out in the same manner, illustrates the case where the ratio f_2/f_1 is equal to 1.09. This is closer to 1, so that the three active regions are partially overlapping. We see that in Figure 5.12 the level of the primary components is lower than in the case of Figure 5.11. This has been done to make the DP level at its "own" destination site the same in the two figures. As a result of the high level in Figure 5.11, a good deal of self-suppression is seen: the response peaks for the primary components are much broader than the peak for the DP. This is much less the case in Figure 5.12.

Quite recently, an ingenious experiment was performed aimed at measuring the extent of "cochlear gain" (Allen and Fahey 1992). These authors simultaneously recorded neural and acoustic responses from the cochlea of a cat (see also Fahey and Allen 1985). Briefly, their experiment went as follows. They made contact with one fiber of the auditory nerve, measured its frequency-threshold curve, and noted the characteristic frequency f_{CF} of the fiber. Then they applied two tones with frequencies f_1 and f_2 ($f_2 > f_1$), at equal levels, and ensured that f_1 and f_2 always had such a relation that the DP frequency, f_{DP} (which equals $2f_1 - f_2$), remained equal to f_{CF}. In this case, the two primary tones generate a DP that travels as a cochlear wave in two directions. Apically, it reaches the location of the neuron under test (the DP destination site) where it gives rise to a neural response. On its basal tour, it reaches the stapes and, via the middle ear, the external ear canal, where it can be detected by a tiny microphone as an otoacoustic emission.

In Allen and Fahey's experiment, the ratio f_2/f_1 was varied and the intensities of the primary components were adjusted so that the neural DP response (at the DP destination site) remained constant (just above the neural threshold). Then they monitored the level of the DP emission in the ear canal. The authors based their reasoning upon the comparison of the two cases just mentioned: f_2/f_1 large, and f_2/f_1 close to unity. In the former case, the site where the DP is generated is to the left of the active region for the DP frequency, which means that the apically going DP wave is amplified before it reaches the DP destination site, and the basally directed wave is not amplified. In the second case, the generation site of the DP would be more or less "inside" the active region for the DP frequency. The apically going DP wave will be amplified much less than before, and the DP source must be proportionally stronger to yield the same neural response at the DP destination site. The basally going wave will now undergo some nonzero amplification, with the result that the DP sound pressure in the external ear canal is expected to be considerably larger than in the first case.

The result of the experiment was negative: there is less than 10 dB of difference between the emission pressures in the two cases (f_2/f_1 equaling 1.8 and 1.12, respectively), and the authors concluded that there is no cochlear amplifier. Let us now turn to the theoretical results of Figures 5.11 and 5.12, which closely relate to the type of experiment done by Allen and Fahey. The results have been obtained for a cochlear model with a gain of over 40 dB. Yet, in the two cases shown, the DP waves *at the stapes* are not more than 9 dB different, and the same will be true for the emission pressures measured in the external ear canal.

Why is the difference in emission levels so small? The main reason is that even for f_2/f_1 as small as 1.09, the effective source of the DP (see the *cross*) is still in the left half of the DP's active region. Hence the right-going wave undergoes nearly the full cochlear amplification, and there is little difference from the case where the components are more widely separated. It appears that the Allen-Fahey experiment is (unfortunately) not suited to determination of the cochlear gain (Kanis and de Boer 1993b).

9. General Conclusions

9.1 Solved and Unsolved Problems

In the preceding sections, an attempt was made to present a coherent picture of the subject of modeling the mechanics of the cochlea. The general tone was optimistic, because in the process, more and more mechanical and physiological properties of the cochlea were found to be explainable by models. It is now time to reflect on the question whether or not this optimism was justified. Did we really select the most important anatomical and physiological properties of cochlear functioning as the basis of our model? And did we choose the most appropriate way of modeling? Or are there still insurmountable obstacles? We will do well to review a few problem areas. First, take the question of *traveling waves* in the cochlea. Events taking place in the cochlea all have a delay that typically depends on frequency: for low frequencies the delay is longer than for high frequencies. An exception seems to be formed by the extracochlear electrical potential known as the cochlear microphonic (CM) potential, which appears to have an exceptionally small latency (latency is the time between the start of a stimulus and the start of a physiological response). This potential is measured by a single electrode on the wall of the cochlea, near or on the round window. It can be regarded as being generated by the OHCs (Dallos and Cheatham 1976), and thus it could be expected to have a latency corresponding to the time it takes for the traveling wave to reach the location of the response peak. However, in the peak region, the cochlear response has, as we have seen in Sections 5–7, a rapidly varying phase angle, and because the recording electrode is far from the hair cells, the contributions from cells in the peak region will cancel. This is the reason why the CM potential appears to be predominantly generated by hair cells in the basal turn (cf. Patuzzi, Yates, and Johnstone 1989a). Indeed, when the CM potential is measured not globally but locally, that is, with a pair of "differential electrodes" very close to the organ of Corti, its latency corresponds to the delay of the traveling wave to the location of the electrodes (Dallos and Cheatham 1971). It should be noted that the effective delay of the BM response is much longer than the traveling-wave delay, because the vibrations of the BM take many cycles to build up to their maximum amplitude.

The next point at issue concerns the *geometry* of the model: we selected our models to be extremely simple so as to be able to solve the model equations. Excursions into the third dimension did not produce major deviations in the response (Section 5); hence we may expect that more complicated geometries may only produce marginal deviations. The same is true when we allow more gradual variations in geometric parameters over the length of the cochlea.

Our models were further simplified in the sense that they contained a cochlear partition (CP) that divides the cochlear channel into two parts and that can be described (in the linear case) by a point impedance (Section 3). This implies that *longitudinal coupling* in the CP has been neglected, one of the requirements for a model to be called "classical." The next stage in cochlear modeling should go deeper into this question. Would this assumption really be justified? Such a quest has become possible because present-day computer and programming techniques will allow us to find solutions for much more complex mechanical structures than those considered up till now.

The models that were constructed earlier (before 1980) showed a response that falls short of the "truth" dictated by more modern mechanical measurement results, and this shortcoming proved to be a major point of concern. One way out is to assume that the cochlea is *locally active,* i.e., that it contains sections where the cochlear partition is producing more acoustic power than is absorbed. Various forms of this type of model have been treated in Section 7. In all of these, the region where the real part of the effective BM impedance is negative has a length of a few millimeters. One problem involved here is that Cody (1992) has argued (on the basis of physiological experiments with acoustic overstimulation) that an "active region," if it exists, could not be much longer than 0.5 mm.

One central problem concerning locally active models is stability. By the choice of parameters, the models treated in Section 7 have been made zero-point stable. But this is only theory. When the real cochlea *is* locally active in the same sense, it is a problem how it can remain stable in view of anatomical and physiological variations. We really do not know the answer. Nor do we know whether isolated spontaneous otoacoustic emissions (SOAEs) are indicators of lack of stability at certain locations or are products of mechanisms we have not yet taken into account. The former view is more likely. Anyway, for a model to explain spontaneous emissions with a finite amplitude, it has to be nonlinear (compressive, of course).

Then, in view of activity and nonlinearity, the question poses itself: Haven't we left out something essential? Don't we have to take "chaos" into account? Don't we have to take quantum mechanical events into account (Bialek and Wit 1984)? Remember, we are still short of good insight into the erratic behavior of isolated SOAEs when external tones are presented (cf. Jones et al. 1986). The best possible answer at present seems to be that as far as the propagation of cochlear waves is concerned, we can do without these complications; for SOAEs, as well as for explaining more details of OHC functioning and so forth, we certainly need more complicated models.

More questions: In the realm of locally active models, is it really true that OHCs produce the active forces? Are length changes or forces produced by stereocilia involved? We cannot tell at the moment. Are there no alternatives? Yes, there is at least one, described in de Boer (1989). Assume that a state of steady tension (or stress) is maintained somewhere in the CP, and that in each period of the vibration, and over a limited phase region, a tiny part of that tension is released and given to the BM. The temporal modulation of the tension could provide acoustic energy to the BM. The mechanism of energy transfer would be analogous to the action of the bow on a violin string or the air pressure in an organ pipe, where a uniform motion is converted into an oscillatory one. If this would be the case, it seems likely that chaotic processes play a direct part in the functioning of active elements.

Returning once more to the point of stability, let us consider the situation for DC displacements. When the BM moves upward, the stereocilia of the OHCs are deflected outward, and the cells depolarize and contract. When the upper end of the hair cells is more restrained than the lower end, the original displacement of the BM is enhanced. This is the view taken in constructing all or most present-day cochlear models. In effect, measures have been taken to neutralize the potential DC instability in these models. However, we are confronted with experimental evidence that seems to indicate a different course of events: Khanna, Flock, and Ulfendahl (1989) and Mammano and Ashmore (1993) found, using in vitro preparations, that the RL moved with a larger amplitude than the BM. In view of experimental uncertainties and the drastically altered mechanical conditions in these experiments, it is not at all clear whether these findings represent the "truth" or not.

The major part of our treatment of *nonlinearity* has been tightly interwoven with the activity problem. In particular, saturation of the active system would be the cause of nonlinearity. The same can be said of the rectification that is assumed to occur in the excitation of hair cells. Saturation gives rise to odd-order nonlinear effects such as two-tone suppression and generation of cubic distortion products. Rectification not only produces DC components in hair-cell potentials but also gives rise to even-order distortion products (of which the difference tone is the best-known example). Remember that there is hardly any alternative for nonlinear effects in the cochlea (cf. Viergever 1980). Then what if we have been wrong in assuming that OHCs produce local activity by generating a pressure? We return once more to Allen (1990), one of the authors opposed to the idea of cochlear activity, and note that he proposes all nonlinear effects (and all effects of cochlear pathology, too) to be due to variable stiffness of the BM that would be under control of the OHCs (see also Allen and Fahey, 1992). Unfortunately, this idea has not yet been worked out in any mathematical detail, so at present there is no ground for discussion or quantitative evaluation. Still, for the benefit of further scientific discussion, we should keep our minds and ears open.

Associated with this topic is the question of how "important" are nonlinear effects, both for the functioning of the cochlea and for the analysis and encoding of sounds by the brain. As has been remarked in connection with the quasi-linear solution method in Section 8, nonlinear effects can generally be treated as *perturbations*. There are, in the case of stimulation by two tones, many places where the primary tones dominate and where the relative distortion is small (maximally, say, 10%), but only few locations where the distortion products dominate. Still, it has been claimed (e.g., Sachs and Young 1980) that two-tone suppression is important for proper neural encoding of speech sounds at high levels. It remains to be seen whether there is a solid basis for this assumption in cochlear mechanics. In the same vein, isn't there a comparable case with stimulation by noise signals, where it has been found experimentally that nonlinear effects in neural responses are difficult to detect because they are so small (Breed, Kanis, and de Boer 1992)?

9.2 The "Classical" Model

To go somewhat deeper into the roots of cochlear modeling, our work has invariably been based on what has been termed the "classical" model of the cochlea. Recall the definition in Section 3.1: *longitudinal coupling* in the organ of Corti has been neglected and the partition was always described by a *local* parameter function (in the linear case, by an impedance). In many respects, work in the field of cochlear modeling has been quite successful. Yet there remain subtle discrepancies between theory and experiment. For instance, experimental impulse responses of the cochlea show an oscillatory waveform in which a typical upward frequency shift with time can be detected (data were kindly put at the author's disposal by Mario Ruggero and Fred Nuttall). This feature—present at low levels of stimulus energy— is not simulated by any recent locally active model of the cochlea. If it turns out to be impossible to explain this frequency shift, what then? Has the end of the period of the "classical" model come into sight? Or is this too rash a conclusion? In the interest of generality, it is well to note that nonclassical models are already beginning to appear. Hubbard's model (1993) is a first example; that model contains two wave-propagating structures, and one of these may well include longitudinal coupling. Furthermore, Steele et al. (1993) have considered a model in which the output of OHC motility is not applied at the same longitudinal position as that at which the input to the OHCs occurs. By this type of "derivative" drive, their model achieved remarkable properties, quite different from those of "classical" models. Finally, remember Allen's (1990) proposal that OHCs may tend to stiffen the BM.

We have to conclude that modeling of the cochlea has not nearly come to an end. Several important questions and many challenges remain. Maybe some authors who have not been put in the foreground in this chapter have already found answers to several of them. The present author can only repeat and extend the apologies he started out with in Section 1. All the same, the author hopes he has not overlooked or left out absolutely relevant material and, more importantly, that he has retained all his friends.

Acknowledgments Writing this chapter has been made possible by the help and support of many friends. The author has had many fruitful discussions and has also accumulated an extensive file of communications, by mail and e-mail, which testify to this help and the ensuing interactions. He is most grateful to Luc-Johan Kanis, who carefully read the manuscript and provided many helpful comments, to Peter Dallos for his patient guidance and critical reading, and to all who contributed their views and criticisms, discussed the various items, advanced arguments against those of the author, pointed out omissions or inaccessible publications, etc. To be mentioned are (in alphabetical order): Jont Allen, Jonathan Ashmore, Arjan Breed, Peter Dallos, Diek Duifhuis, Dick Fay, Dan Geisler, Matthew Holley, Al Hubbard, Brian Johnstone, Paul Kolston, Dick Lyon, Charles Molnar, Dave Mountain, Steve Neely, Stella Novoselova, Fred Nuttall, Rob Patuzzi, Vera Prijs, Mario Ruggero, Ruurd Schoonhoven, Chris Shera, Arnold Tubis, and Graeme Yates. For the better part of the text, Karin van Merkesteyn functioned as a vital link between the author and the computer disk.

> Ferdinand
> *This music crept by me upon the waters,*
> *Allaying both their fury and my passion*
> *With its sweet air: thence I have follow'd it,*
> *Or it hath drawn me rather. But 'tis gone.*
> *No, it begins again.*
> Shakespeare, *The Tempest*, V, 1.

References

Abbas PJ, Sachs MB (1976) Two-tone suppression in auditory-nerve fibers: extension of a stimulus-response relationship. J Acoust Soc Am 59:112–122.

Allen JB (1977) Two-dimensional cochlear fluid model: new results. J Acoust Soc Am 61:110–119.

Allen JB (1980) Cochlear micromechanics—a physical model of transduction. J Acoust Soc Am 68:1660–1670.

Allen JB (1990) Modeling the noise damaged cochlea. In: Dallos P, Geisler CD, Matthews JW, Ruggero MA, Steele CR (eds) The Mechanics and Biophysics of Hearing. Berlin: Springer-Verlag, pp. 324–331.

Allen JB, Fahey PF (1992) Using acoustic distortion products to measure the cochlear amplifier gain on the basilar membrane. J Acoust Soc Am 92:178–188.

Allen JB, Neely ST (1992) Micromechanical models of the cochlea. Physics Today 45:40–47.

Allen JB, Sondhi MM (1979) Cochlear macromechanics—time domain solutions. J Acoust Soc Am 66:123–132.

Ashmore JF (1987) A fast motile response in guinea-pig outer hair cells: the cellular basis of the cochlear amplifier. J Physiol (Lond) 388:323–347.

Bialek W, Wit HP (1984) Quantum limits to oscillator stability: theory and experiments on acoustic emissions from the human ear. Phys Lett 104A:173–178.

Brass D, Kemp DT (1993) Analysis of Mössbauer mechanical measurements indicate that the cochlea is mechanically active. J Acoust Soc Am 93:1502–1515.

Breed AJ, Kanis LJ, de Boer E (1992) Cochlear nonlinearity for complex stimuli. In: Cazals Y, Demany L, Horner K (eds) Auditory Physiology and Perception. Oxford: Pergamon Press, pp. 189–195.

Brownell WE, Bader CR, Bertrand D, de Ribaupierre Y (1985) Evoked mechanical responses of isolated cochlear outer hair cells. Science 227:194–196.

Brundin L, Flock A, Canlon B (1989) Tuned motile responses of isolated cochlear hair cells. Acta Otolaryngol (Stockholm) Suppl 467:229–234.

Brundin L, Flock A, Khanna SM, Ulfendahl M (1991) Frequency-specific position shift in the guinea pig organ of Corti. Neurosci Lett 128:77–80.

Cheatham MA, Dallos P (1989) Two-tone suppression in inner hair cell responses. Hear Res 40:187–196.

Cheatham MA, Dallos P (1990a) Two-tone interactions in inner hair cell potentials: AC versus DC effects. Hear Res 43:135–139.

Cheatham MA, Dallos P (1990b) Comparison of low- and high-side two-tone suppression in inner hair cell and organ of Corti responses. Hear Res 50:193–210.

Cody AR (1992) Acoustic lesions in the mammalian cochlea: implications for the spatial distribution of the 'active process.' Hear Res 62:166–172.

Cooper NP, Rhode WS (1992) Basilar mechanics in the hook region of cat and guinea pig cochlea: sharp tuning and nonlinearity in the absence of baseline position shifts. Hear Res 63:163–190.

Crawford AC, Fettiplace R (1985) The mechanical properties of ciliary bundles of turtle cochlear hair cells. J Physiol (Lond) 364:359–379.

Dallos P (1973) The Auditory Periphery. Biophysics and Physiology. New York: Academic Press.

Dallos P, Cheatham MA (1971) Travel time in the cochlea and its determination from cochlear-microphonic data. J Acoust Soc Am 49:1140–1143.

Dallos P, Cheatham MA (1976) Production of cochlear potentials by inner and outer hair cells. J Acoust Soc Am 60:510–512.

Dallos P, Harris PM, Relkin E, Cheatham MA (1980) Two-tone suppression and intermodulation distortion in the cochlea: effect of outer hair cell lesions. In: van den Brink G, Bilsen FA (eds) Psychophysical, Physiological and Behavioural Studies in Hearing. Delft: Delft University Press, pp. 242–252.

Davis H (1983) An active process in cochlear mechanics. Hear Res 9:79–90.

de Boer E (1980a) Auditory physics. Physical principles in hearing theory. I. Phys Rep 62:87–174.

de Boer E (1980b) A cylindrical cochlea model—the bridge between two and three dimensions. Hear Res 3:109–131.

de Boer E (1981) Short waves in three-dimensional cochlea models: solution for a 'block' model. Hear Res 4:53–77.

de Boer E (1983) No sharpening? A challenge for cochlear mechanics. J Acoust Soc Am 73:567–573.

de Boer E (1984) Auditory physics. Physical principles in hearing theory. II. Phys Rep 105:141–226.

de Boer E (1989) Outer hair cell motility and wave amplification in the inner ear. In: Pravica P, Draculic G (eds) Proc. 13th International Congress on Acoustics. Sabac, Yugoslavia: Dragen Press, pp. 499–502.

de Boer E (1991) Auditory physics. Physical principles in hearing. III. Phys Rep 203:127–229.

de Boer E (1993) The sulcus connection. On a mode of participation of outer hair cells in cochlear mechanics. J Acoust Soc Am 93:2845–2859.

de Boer E, van Bienema E (1982) Solving cochlear mechanics problems with higher-order differential equations. J Acoust Soc Am 72:1427–1434.

de Boer E, Kruidenier C (1990) On ringing limits of the auditory periphery. Biol Cybern 63:433–442.

de Boer E, Viergever MA (1982) Validity of the Liouville-Green (or WKB) method for cochlear mechanics. Hear Res 8:131–155.

de Boer E, Viergever MA (1984) Wave propagation and dispersion in the cochlea. Hear Res 13:101–112.

Diependaal RJ, Viergever MA (1983) Point-impedance characterization of the basilar membrane in a three-dimensional cochlea model. Hear Res 11:33–40.

Diependaal RJ, Viergever MA, de Boer E (1986) Are active elements necessary in the basilar membrane impedance? J Acoust Soc Am 80:124–132.

Diependaal RJ, Duifhuis H, Hoogstraten HW, Viergever MA (1987) Numerical methods for solving one-dimensional cochlear models in the time domain. J Acoust Soc Am 82:1655–1666.

Evans EF (1972) The frequency response and other properties of single fibres in the guinea-pig cochlear nerve. J Physiol (Lond) 226:263–287.

Evans EF (1975) The sharpening of cochlear frequency selectivity in the normal and abnormal cochlea. Audiology 14:419–442.

Fahey PF, Allen JB (1985) Nonlinear phenomena as observed in the ear canal and at the auditory nerve. J Acoust Soc Am 77:599–612.

Fletcher H (1951) On the dynamics of the cochlea. J Acoust Soc Am 23:637–645.

Flock A, Strelioff D (1984) Studies on hair cells in isolated coils from the guinea pig cochlea. Hear Res 15:11–18.

Geisler CD (1986) A model of the effect of outer hair cell motility on cochlear vibrations. Hear Res 24:125–131.

Geisler CD (1991) A cochlear model using feedback from motile outer hair cells. Hear Res 54:105–117.

Geisler CD (1993) A realizable model of the effect of outer hair cell motility on cochlear vibrations. Hear Res 68:253–262.

Gitter AH, Zenner H-P (1988) Auditory transduction steps in single inner and outer hair cells. In: Duifhuis H, Horst JW, Wit HP (eds) Basic Issues in Hearing. London: Academic Press, pp. 32–39.

Goldstein JL (1967) Auditory nonlinearity. J Acoust Soc Am 41:676–689.

Goldstein JL (1990) Modeling rapid waveform compression on the basilar membrane as multiple-bandpass-nonlinear filtering. Hear Res 49:39–60.

Goldstein JL (1993) Exploring new principles of cochlear operation: bandpass filtering by the organ of Corti and additive amplification on the basilar membrane. In: Duifhuis H, Horst JW, van Dijk P, van Netten SM (eds) Biophysics of Hair-Cell Sensory Systems. Singapore: World Scientific, pp. 315–322.

Goldstein JL, Kiang NY-S (1968) Neural correlates of the aural combination tone $2f_1 - f_2$. Proc IEEE 56:981–992.

Gummer AW, Johnstone BM, Armstrong NJ (1981) Direct measurements of basilar membrane stiffness in the guinea pig cochlea. J Acoust Soc Am 70:1298–1309.

Hall JL (1974) Two-tone distortion products in a nonlinear model of the basilar membrane. J Acoust Soc Am 56:1818–1828.

Hall JL (1977a) Two-tone suppression in a nonlinear model of the basilar membrane. J Acoust Soc Am 61:802–810.

Hall JL (1977b) Spatial differentiation as an auditory "second filter": assessment on a nonlinear model of the basilar membrane. J Acoust Soc Am 61:520-524.

Holley MC, Ashmore JF (1988) On the mechanism of a high-frequency force generator in outer hair cells isolated from the guinea pig cochlea. Proc R Soc Lond B 323:413-429.

Houtgast T (1972) Psychophysical evidence for lateral inhibition in hearing. J Acoust Soc Am 51:1885-1894.

Howard J, Hudspeth AJ (1988) Compliance of the hair bundle associated with gating of mechanoelectrical transduction channels in the bullfrog's saccular hair cell. Neuron 1:189-199.

Hubbard AE (1993) A traveling wave-amplifier model of the cochlea. Science 259:68-71.

Hubbard AE, Geisler CD (1972) A hybrid-computer model of the cochlear partition. J Acoust Soc Am 51:1895-1903.

Iwasa KH, Chadwick RS (1992) Elasticity and active force generation of cochlear outer hair cells. J Acoust Soc Am 92:3169-3173.

Jones K, Tubis A, Long GR, Burns EM, Strickland EA (1986) Interactions among multiple spontaneous otoacoustic emissions. In: Allen JB, Hall JL, Hubbard A, Neely ST, Tubis A (eds) Periperhal Auditory Mechanisms. Berlin: Springer-Verlag, pp. 266-273.

Johnstone BM, Boyle AJF (1967) Basilar membrane vibration examined with the Mössbauer technique. Science 158:389-390.

Johnstone BM, Yates GK (1974) Basilar membrane tuning curves in the guinea pig. J Acoust Soc Am 55:584-587.

Johnstone BM, Taylor KJ, Boyle AJ (1970) Mechanics of the guinea pig cochlea. J Acoust Soc Am 47:504-509.

Kanis LJ, de Boer E (1993a) Self-suppression in a locally active nonlinear model of the cochlea: a quasi-linear approach. J Acoust Soc Am 94:3199-3206.

Kanis LJ, de Boer E (1993b) The emperor's new clothes: DP emissions in a locally-active nonlinear model of the cochlea. In: Duifhuis H, Horst JW, van Dijk P, van Netten SM (eds) Biophysics of Hair-Cell Sensory Systems. Singapore: World Scientific, pp. 304-311.

Kemp DT (1978) Stimulated acoustic emissions from within the human auditory system. J Acoust Soc Am 64:1386-1391.

Kemp DP (1979) Evidence of mechanical nonlinearity and frequency selective wave amplification in the cochlea. Arch Otorhinolaryngol 224:37-45.

Kemp DT, Brown AM (1983) An integrated view of cochlear mechanical nonlinearities observable from the ear canal. In: de Boer E, Viergever MA (eds) Mechanics of Hearing. Delft: Delft University Press, pp. 75-82.

Khanna SM, Flock A, Ulfendahl M (1989) Comparison of the tuning of outer hair cells and the basilar membrane in the isolated cochlea. Acta Otolaryngol (Stockholm) Suppl 467:151-156.

Kiang NY-S, Moxon EC (1974) Tails of tuning curves of auditory nerve fibers. J Acoust Soc Am 55:620-630.

Kim DO, Siegel JH, Molnar CE (1979) Cochlear mechanics: physiologically-vulnerable nonlinear behavior in two-tone responses. Scand Audiol Suppl 9:63-81.

Kim DO, Molnar CE, Matthews JW (1980) Cochlear mechanics: nonlinear behavior in two-tone responses as reflected in cochlear-nerve-fiber responses and in ear-canal sound pressure. J Acoust Soc Am 67:1704-1721.

Kim DO, Neely ST, Molnar CE, Matthews JW (1980) An active cochlear model with negative damping in the partition: comparison with Rhode's ante- and post-mortem observations. In: van den Brink G, Bilsen FA (eds) Psychophysical, Physiological and Behavioural Studies in Hearing. Delft: Delft University Press, pp. 7-14.

Kolston PJ (1988) Sharp mechanical tuning in a cochlear model without negative damping. J Acoust Soc Am 83:1481-1487.

Kolston PJ, de Boer E, Viergever MA, Smoorenburg GF (1990) What type of force does the cochlear amplifier produce? J Acoust Soc Am 88:1794-1801.

Koshigoe S, Tubis A (1982) Implications of causality, time-translation invariance, linearity, and minimum phase behavior for basilar membrane response functions. J Acoust Soc Am 71:1194-1200.

LePage EL (1987) Frequency-dependent self-induced bias of the basilar membrane and its potential for controlling sensitivity and tuning in the mammalian cochlea. J Acoust Soc Am 82:139-154.

Lesser MB, Berkley DA (1972) Fluid mechanics of the cochlea. J Fluid Mech 51:497-512.

Lighthill MJ (1981) Energy flow in the cochlea. J Fluid Mech 106:149-213.

Lyon RF, Mead C (1988) An analog electronic cochlea. IEEE Trans ASSP 36:1119-1134.

Mammano F, Ashmore JF (1993) Reverse transduction measured in the isolated cochlea by laser Michelson interferometry. Nature (Lond) 365:838-841.

Mammano F, Nobili R (1993) Biophysics of the cochlea: linear approximation. J Acoust Soc Am 93:3320-3332.

Matthews JW (1983) Modeling reverse middle ear transmission of acoustic distortion signals. In: de Boer E, Viergever MA (eds) Mechanics of Hearing. Delft: Delft University Press, pp. 11-18.

Miller CE (1985a) VLFEM analysis of a two-dimensional cochlear model. J Appl Mech 52:1-9.

Miller CE (1985b) Structural implications of basilar membrane compliance measurements. J Acoust Soc Am 77:1465-1474.

Mountain DC, Hubbard AE, McMullen TA (1983) Electromechanical processes in the cochlea. In: de Boer E, Viergever MA (eds) Mechanics of Hearing. Delft: Delft University Press, pp. 119-126.

Neely ST (1981) Finite difference solution of a two-dimensional mathematical model of the cochlea. J Acoust Soc Am 69:1386-1393.

Neely ST (1985) Mathematical modeling of cochlear mechanics. J Acoust Soc Am 78:345-352.

Neely ST (1993) A model of cochlear mechanics with outer hair cell motility. J Acoust Soc Am 94:137-146.

Neely ST, Kim DO (1986) A model for active elements in cochlear biomechanics. J Acoust Soc Am 79:1472-1480.

Novoselova SM (1987) A three-chamber model of the cochlea. J Soviet Math 38:1655-1663.

Novoselova SM (1989) A possibility of sharp tuning in a linear transversally inhomogeneous cochlear model. Hear Res 41:125-136.

Novoselova SM (1993) An alternative mechanism of sharp cochlear tuning. In: Duifhuis H, Horst JW, van Dijk P, van Netten SM (eds) Biophysics of Hair-Cell Sensory Systems. Singapore: World Scientific, pp. 338-344.

Nuttall AL, Dolan DF (1993) Two-tone suppression of inner hair cell and basilar membrane responses in the guinea pig. J Acoust Soc Am 93:390-400.

Nuttall AL, Dolan DF, Avinash G (1990) Measurements of basilar membrane tuning and distortion with laser doppler velocimetry. In: Dallos P, Geisler CD, Matthews JW, Ruggero MA, Steele CR (eds) The Mechanics and Biophysics of Hearing. Berlin: Springer-Verlag, pp. 288-295.

Nuttall AL, Dolan DF, Avinash G (1991) Laser doppler velocimetry of basilar membrane vibration. Hear Res 51:203-214.

Olson ES, Mountain DC (1991) In vivo measurement of basilar membrane stiffness. J Acoust Soc Am 89:1262-1275.

Patuzzi RB, Sellick PM, Johnstone BM (1984a) The modulation of the sensitivity of the mammalian cochlea by low frequency tones. I: Primary afferent activity. Hear Res 13:1-8.

Patuzzi RB, Sellick PM, Johnstone BM (1984b) The modulation of the sensitivity of the mammalian cochlea by low frequency tones. III Basilar membrane motion. Hear Res 13:19-27.

Patuzzi RB, Yates GK, Johnstone BM (1989a) The origin of the low-frequency microphonic in the first cochlear turn of guinea pig. Hear Res 39:177-188.

Patuzzi RB, Yates GK, Johnstone BM (1989b) Outer hair cell receptor current and sensorineural hearing loss. Hear Res 42:47-72.

Prijs VF (1989) Lower boundaries of two-tone suppression regions in the guinea pig. Hear Res 24:73-82.

Puria S, Allen JB (1991) A parametric study of cochlear input impedance. J Acoust Soc Am 89:287-309.

Reuter G, Gitter AH, Thurm U, Zenner H-P (1992) High frequency radial movements of the reticular lamina induced by outer hair cell motility. Hear Res 60:236-246.

Rhode WS (1971) Observations of the vibration of the basilar membrane in squirrel monkeys using the Mössbauer technique. J Acoust Soc Am 49:1218-1231.

Rhode WS (1978) Some observations on cochlear mechanics. J Acoust Soc Am 64:158-176.

Robertson D (1976) Correspondence between sharp tuning and two-tone inhibition in primary auditory neurones. Nature (Lond) 259:477-478.

Robles L, Ruggero MA, Rich N (1986a) Mössbauer measurements of the mechanical response to single-tone and two-tone stimuli at the base of the chinchilla cochlea. In: Allen JB, Hall JL, Hubbard A, Neely ST, Tubin A (eds) Peripheral Auditory Mechanisms. Berlin: Springer-Verlag, pp. 121-127.

Robles L, Ruggero MA, Rich NC (1986b) Basilar membrane mechanics at the base of the chinchilla cochlea. I. Input-output functions, tuning curves, and response phases. J Acoust Soc Am 80:1364-1374.

Ruggero MA, Robles L, Rich NC (1992) Two-tone suppression in the basilar membrane of the cochlea: mechanical basis of auditory-nerve rate suppression. J Neurophysiol 68:1087-1099.

Ruggero MA, Robles L, Rich NC, Recio A (1992) Basilar membrane responses to two-tone and broadband stimuli. Phil Trans R Soc Lond B 336:307-315.

Russell IJ, Sellick PM (1977) Tuning properties of cochlear hair cells. Nature (Lond) 267:858-860.

Russell IJ, Sellick PM (1978) Intracellular studies in the mammalian cochlea. J Physiol (Lond) 284:261-290.

Sachs MB (1969) Stimulus-response relation for auditory-nerve fibers: two-tone stimuli. J Acoust Soc Am 45:1025-1036.

Sachs MB, Abbas PJ (1976) Phenomenological model for two-tone suppression. J Acoust Soc Am 60:1157-1163.

Sachs MB, Kiang NY-S (1968) Two-tone inhibition in auditory-nerve fibers. J Acoust Soc Am 43:1120–1128.

Sachs MB, Young ED (1980) Effects of nonlinearities on speech encoding in the auditory nerve. J Acoust Soc Am 68:858–875.

Schmiedt RA (1982) Boundaries of two-tone rate suppression of cochlear-nerve activity. Hear Res 7:335–351.

Schmiedt RA, Zwislocki JJ, Hamernik RP (1980) Effects of hair cell lesions on responses of cochlear nerve fibers. I. Lesions, tuning curves, two-tone inhibition, and responses to trapezoidal-wave patterns. J Neurophysiol 43:1367–1389.

Schroeder MR (1973) An integrable model for the basilar membrane. J Acoust Soc Am 53:429–434.

Sellick PM, Russell IJ (1979) Two-tone suppression in cochlear hair cells. Hear Res 1:227–236.

Sellick PM, Patuzzi RB, Johnstone BM (1982) Measurement of basilar membrane motion in the guinea pig using the Mössbauer technique. J Acoust Soc Am 72:131–141.

Sellick PM, Yates GK, Patuzzi R (1983) The influence of Mössbauer source size and position on phase and amplitude measurements of the guinea pig basilar membrane. Hear Res 10:101–108.

Shera CA, Zweig G (1991a). A symmetry suppresses the cochlear catastrophe. J Acoust Soc Am 89:1276–1289.

Shera CA, Zweig G (1991b) Reflection of retrograde waves within the cochlea and at the stapes. J Acoust Soc Am 89:1290–1305.

Siebert WM (1974) Ranke revisited—a simple short-wave cochlear model. J Acoust Soc Am 56:594–600.

Smoorenburg GF (1972a) Audibility region of combination tones. J Acoust Soc Am 52:603–614.

Smoorenburg GF (1972b) Combination tones and their origin. J Acoust Soc Am 52:615–632.

Smoorenburg GF (1980) Effects of temporary threshold shift on combination-tone generation and on two-tone suppression. Hear Res 2:347–355.

Sondhi MM (1978) Method for computing motion in a two-dimensional cochlear model. J Acoust Soc Am 63:1468–1477.

Steele CR, Taber LA (1979a) Comparison of WKB and finite difference calculations for a two-dimensional cochlear model. J Acoust Soc Am 65:1001–1006.

Steele CR, Taber LA (1979b) Comparison of WKB calculations and experimental results for a three-dimensional cochlear model. J Acoust Soc Am 65:1007–1018.

Steele CR, Taber LA (1981) Three-dimensional model calculations for guinea pig cochlea. J Acoust Soc Am 69:1107–1111.

Steele CR, Baker G, Tolomeo J, Zetes D (1993) Electro-mechanical models of the outer hair cell. In: Duifhuis H, Horst JW, van Dijk P, van Netten SM (eds) Biophysics of Hair-Cell Sensory Systems. Singapore: World Scientific, pp. 207–214.

Taber LA, Steele CR (1981) Cochlear model including three-dimensional fluid and four modes of partition flexibility. J Acoust Soc Am 70:426–436.

van Dijk P (1990) Characteristics and Mechanisms of Spontaneous Otoacoustic Emissions. Thesis, University of Groningen, the Netherlands.

van Dijk P, Wit HP (1990) Synchronization of spontaneous otoacoustic emissions to a $2f_1 - f_2$ distortion product. J Acoust Soc Am 88:850–856.

van Netten SM, Khanna SM (1993) Mechanical demodulation of hydrodynamic stimuli performed by the lateral line organ. Prog Brain Res 97:45-51.
Viergever MA (1980) Mechanics of the Inner Ear—A Mathematical Approach. Thesis, Technical University of Delft, Delft University Press.
Viergever MA (1986) Cochlear macromechanics—a review. In: Allen JB, Hall JL, Hubbard A, Neely ST, Tubis A (eds) Peripheral Auditory Mechanisms. Berlin: Springer-Verlag, pp. 63-72.
Viergever MA, Diependaal RJ (1986) Quantitative validation of cochlear models using the Liouville-Green approximation. Hear Res 21:1-15.
von Békésy G (1960) Experiments in Hearing. New York: McGraw-Hill.
Weiss TF (1982) Bidirectional transduction in vertebrate hair cells: a mechanism for coupling mechanical and electrical vibrations. Hear Res 7:353-360.
Zenner H-P, Zimmermann U, Schmitt U (1985) Reversible contraction of isolated cochlear hair cells. Hear Res 18:127-133.
Zenner HP, Gitter AH, Rudert M, Ernst A (1992) Stiffness, compliance, elasticity and force generation of outer hair cells. Acta Otolaryngol (Stockholm) 112:248-253.
Zweig G (1990) The impedance of the organ of Corti. In: Dallos P, Geisler CD, Matthews JW, Ruggero MA, Steele CR (eds) The Mechanics and Biophysics of Hearing. Berlin: Springer-Verlag, pp. 362-369.
Zweig G (1991) Finding the impedance of the organ of Corti. J Acoust Soc Am 89:1229-1254.
Zweig G, Lipes R, Pierce JR (1976) The cochlear compromise. J Acoust Soc Am 59:975-982.
Zwicker E (1955) Der ungewöhnliche Amplitudengang der nichtlinearen Verzerrungen des Ohres. Acustica 5:67-74.
Zwicker E (1986a) A hardware cochlear nonlinear preprocessing model with active feedback. J Acoust Soc Am 80:146-153.
Zwicker E (1986b) "Otoacoustic" emissions in a nonlinear cochlear hardware model with feedback. J Acoust Soc Am 80:154-162.
Zwicker E (1986c) Suppression and $(2f_1-f_2)$-difference tones in a nonlinear cochlear preprocessing model with active feedback. J Acoust Soc Am 80:163-176.
Zwislocki J (1948) Theorie der Schneckenmechanik: Qualitative und quantitative Analyse. Acta Otolaryngol Suppl 72.
Zwislocki JJ, Kletsky EJ (1979) Tectorial membrane: a possible effect on frequency analysis in the cochlea. Science 204:639-641.

6
Physiology of Mammalian Cochlear Hair Cells

CORNÉ J. KROS

1. Introduction

The scope of this chapter is to review what is currently known about the workings of the two types of auditory hair cells in the mammalian cochlea, the inner hair cells (IHCs) and outer hair cells (OHCs). Mammalian hair cells have for a long time been impenetrable to the electrophysiologist's microelectrodes, necessitating the extrapolation of results from more easily accessible and less vulnerable hair cells of nonmammalian vertebrate preparations to explain mammalian cochlear physiology. To assess the validity of this approach, we need to consider first whether there are common elements shared by all vertebrate hair cells.

1.1 The Minimal Hair Cell

Hair cells are sensory receptors used by all vertebrates to transduce mechanical stimuli into electrical responses. The organ in which the hair cells are situated determines to which type of mechanical stimulus they respond best: for example, hair cells in the lateral line organ of fishes and amphibians signal motion of fluid around the animal, vestibular hair cells are most sensitive to linear acceleration or rotational velocity, and auditory hair cells detect sound. Moreover, taking the hearing organs as an example, there is an enormous variation in structure among different classes of vertebrates, and even among different species. This variation is matched by large differences in sensitivity and frequency range, and both intensity and frequency discrimination (Stebbins 1983; Fay 1992). As judged by behavioral audiograms, mammals as a group are on average more sensitive than other classes of vertebrates, although some individual nonmammalian species approach the performance of the most sensitive mammals. The most salient distinguishing feature of mammalian auditory function is a considerably increased limit of high-frequency hearing (11-150 kHz, depending on the species) compared with birds (6–12 kHz) and other nonmammalian vertebrates (0.2–4 kHz). The architecture of the mammalian inner ear is quite unlike that of other vertebrates, with a coiled cochlea containing a longer basilar membrane, and two distinct types of hair cells with different afferent and efferent innervation patterns (Slepecky, Chapter 2). The mammalian basilar membrane acts as a sharply tuned frequency analyzer (Patuzzi, Chapter 4). The cochlea anatomy of birds and Crocidilia is more or less intermediate between that of other nonmammalian vertebrates and mammals, with tall hair cells comparable to IHCs, and short hair cells comparable to OHCs, also in their innervation. There is, however, a gradual transition from tall to short hair cells across the width of the cochlea, rather than two very distinct cell types. Basilar membrane tuning in birds is also apparently considerably less sharp than in mammals (Gummer, Smolders, and Klinke 1987).

Despite these differences in the anatomical environment in which the hair cells are situated and the functional requirements made on them, the hair cells themselves appear morphologically quite similar. This raises the question how many of the functional differences that exist among the organs containing hair cells can be attributed to the accessory structures, and how many to physiological differences in the hair cells themselves. This is an important question, because much information about hair cell physiology is derived from experiments on the sacculus, a vestibular organ responding to linear acceleration, of the bullfrog, *Rana catesbeiana,* and the basilar papilla, a primitive hearing organ, of the red-eared turtle, *Pseudemys scripta*. Likely differences between cochlear hair cells in mammals and those of other vertebrates are associated with the extended high-frequency limit of the mammalian cochlea.

The common elements expected for all hair cells are illustrated in Figure 6.1. First, displacements of the stereociliary bundle need to be transduced into a change in the electrical current flowing through mechanosensitive ion channels. By flowing across the electrical resistance (R) of the basolateral cell membrane, this transducer current (I_T) brings about a receptor potential. The size and shape of the receptor potential are determined by the properties of the transducer current itself, by the low-pass filter of the cell's membrane capacitance (C) and input resistance, and by any time- and voltage-dependent or ligand-gated ion channels that may be present in the basolateral cell membrane, provided they are activated over the relevant range of potentials. As a minimum, a K^+ current (I_K) is required to help set the resting potential and input conductance (g, the inverse of resistance) of the cell. Such a role could, in principle, also be played by a Cl^- conductance, but it will become clear that this would not be appropriate for a hair cell (Section 3.2.). Finally, the cell needs to exert an effect on its environment. The most common effect is excitation of the afferent nerve fibers by release of neurotransmitter. As in nerve fibers, a Ca^{2+} current

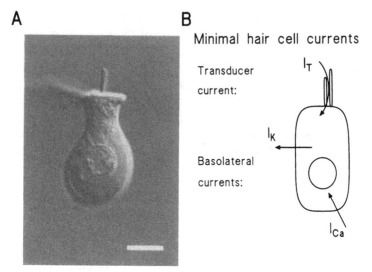

FIGURE 6.1. (A) Photomicrograph of an IHC isolated from the guinea pig cochlea. Tallest row of stereocilia to the *left*. Scale bar 10 μm. (From Kros and Crawford 1990, Fig. 1.) (B) Schematic representation of a hair cell, indicating the three membrane conductances that a hair cell is expected to possess as a minimum requirement for it to function.

(I_{Ca}) is expected to induce transmitter release. OHCs are unusual in that they appear to affect their environment mainly by exerting a mechanical force (discussed by Holley, Chapter 7).

In addition, most hair cells, with the exception of IHCs and vestibular type I hair cells, receive, aside from a mechanical input, a neural input from the brain through efferent synapses. The physiological effects of modulation of OHC responses by activity of the efferent nerve fibers are discussed by Guinan, Chapter 8, and the molecular mechanisms of synaptic transmission are dealt with by Sewell, Chapter 9.

1.2 Comparison of Different Approaches Used to Study Hair Cell Physiology

1.2.1 Gross Cochlear Potentials

The first evidence for electrical activity associated with the functioning of the mammalian cochlea came from extracellular recordings of potentials evoked by sound, with large electrodes in the vicinity of the cochlea, by Wever and Bray (1930), who thought the signals originated from the auditory nerve. These potential variations followed the waveform of the sound, so that, if presented to a loudspeaker, the sounds were reproduced with (for the time) remarkably high fidelity. Adrian (1931) correctly pointed to the cochlea as the main source of these potentials. More and more refined versions of this approach were developed in the following decades, with microelectrodes positioned in the different scalae of the cochlea. These more refined experiments made it likely that certain components of these "gross" potentials (the frequency-following "cochlear microphonic" and the "summating potential," which is a baseline shift in the records) originated in the reticular lamina, probably from the hair cells. Unfortunately, the relative technical ease of doing these experiments is amply offset by the difficulty of interpreting the results in terms of inferred hair cell physiology. For example, these potentials are likely to originate from the summed activity of many hair cells, thus smudging information about the tuning of the individual cells, and the shape of the potentials is determined by the complicated three-dimensional electrical impedance network consisting of the hair cells and supporting cells in the organ of Corti and the fluid spaces and membranes of the cochlea. Modeling these gross potential responses involves a large number of arbitrary variables. For those interested in the historical importance of extracellular recordings of gross potentials in the cochlea, there is an excellent account by Dallos (1973). Experiments involving recordings of gross potentials are still sometimes used, for example, to assay the effects of various drugs on hair cell function. Apart from the above-mentioned difficulties, interpretation of such experiments also suffers from unknown effects of these drugs on the nonsensory cells, which may change the distribution of impedances in the cochlea.

1.2.2 Intracellular Recordings with Microelectrodes from Nonmammalian Vertebrate Hair Cells

The failure of early attempts to record receptor potentials from cochlear hair cells, and the difficulties in interpreting the gross cochlear potentials, prompted investigators to take a lateral approach and investigate suitable, more robust, hair cell systems in nonmammalian vertebrates. The earliest successful attempts were in vivo recordings from lateral line hair cells in the mudpuppy, *Necturus maculosus,* yielding small receptor potentials of less than 1 mV from unknown resting potentials (Harris, Frishkopf, and Flock 1970). The first account of in vivo recordings of receptor potentials (up to 3 mV) in response to sound in an auditory organ, the basilar papilla of the alligator lizard, *Gerrhonotus multicarinatus,* was given by Weiss, Mulroy, and Altmann (1974). Again, no resting potentials were reported.

Significant advances were made with the use of isolated, in vitro, preparations, presumably in part because a much improved mechanical stability caused less damage to the cells by the microelectrodes. Recordings from vestibular hair cells in the isolated frog sacculus (Hudspeth and Corey 1977) and auditory hair cells in the isolated half-head of the turtle (Fettiplace and Crawford 1978) showed large receptor potentials of up to 40 mV, from healthy resting potentials around -50 mV. Both of these articles also gave the first hints that time- and voltage-sensitive conductances may be involved in shaping the receptor potentials. Further work over the years on these two preparations has provided a large part of our current knowledge about hair cells.

1.2.3 Intracellular Recordings with Microelectrodes from Mammalian Hair Cells

The first successful intracellular recordings of receptor potentials of IHCs in vivo were reported by Russell and Sellick (1977). Sharply tuned receptor potentials of up to 17 mV were recorded from resting potentials of around -40 mV. Since then a large amount of data has been collected in similar preparations, detailing properties of the receptor potentials of both IHCs and OHCs (Dallos, Santos-Sacchi, and Flock 1982).

Microelectrode recordings from acutely isolated mammalian cochlear coils (Flock and Strelioff 1984) or isolated hair cells (Brownell 1984; Brownell et al. 1985) have been much less successful than those from other vertebrates. IHCs had reasonable testing potentials, as in vivo, of around -40 mV, but OHC resting potentials were generally smaller than -15 mV, and the cells were very leaky (i.e., they had a very low input resistance), indicative of severe damage upon isolation. Receptor potentials recorded with microelectrodes in vitro have been reported for neither IHCs nor OHCs. The advantage of potential stability in vitro is apparently offset by the fragility of mammalian hair cells, especially OHCs, which may cause them to survive isolation less well (for example, as a result of hypoxia or mechanical trauma), or may make them more prone to injury by the microelectrode than the sturdier cells from nonmammalian vertebrates.

A radically different approach to the physiology of mammalian hair cells in vitro came with the use of organ cultures (Russell, Richardson, and Cody 1986; Russell and Richardson 1987). Cochleae were taken form newborn mice and the isolated organs of Corti were kept alive for up to 5 days on collagen-coated coverslips, to which they adhered tightly. The cells had good resting potentials of about -40 mV for IHCs and -60 mV for OHCs and responded with receptor potentials of up to about 10 mV when their hair bundles were mechanically stimulated by a glass rod. This preparation presents a number of advantages as well as disadvantages. At birth, the mouse cochlea is anatomically and functionally immature, and hearing only commences around 11 days later (Romand 1983). The tectorial membrane has not yet reached or contacted the neonatal OHCs, making their hair bundles readily accessible for mechanical stimulation. This may also be one of the reasons that their transduction mechanism is less damaged during isolation than that of mature OHCs, in which the tectorial membrane is ripped off the hair bundles. A conceivable disadvantage is that properties of the transduction mechanism and of ion channels in the basolateral membrane may change during maturation.

1.2.4 Voltage-Clamp Recordings of Nonmammalian Vertebrate Hair Cells

The approaches for studying hair cells described in Sections 1.2.2 and 1.2.3 all rely on using a single high-resistance microelectrode, with which it is possible to record membrane potentials and inject currents into cells, but not to control the voltage of the cell and record membrane currents. Experimental control of the membrane potential by voltage clamping [using two microelectrodes or, more recently, patch clamping (Sakmann and Neher 1995)] is necessary to investigate the ionic conductances underlying the receptor potential. When the voltage in a cell is changing freely, the currents flowing are always a mixture of ionic and capacitative currents. Time- and voltage-dependent ion channels, for example Na^+ channels, change their probability of being open as a function of voltage, thereby themselves changing the potential across the cell membrane, and indirectly affecting their own opening probability, as well as that of other ion channels, such as those permeable to K^+. It is therefore not easy to relate voltage responses to the behavior of ion channels in the cell membrane. The voltage clamp allows the experimenter to control the voltage of the cell (usually by changing it in a stepwise fashion) to separate capacitative currents from ionic currents, to control and estimate the opening probability of ion channels, and (with some luck) to study the properties of different classes of ion channels in isolation (e.g., Hille 1992). To sum it up: voltage responses show *what* a cell does electrically, currents recorded under voltage clamp help unravel *how* the cell does it. With the patch-clamp technique, it is possible to study either the summed, stochastic behavior of all the channels of a particular class that are present in a cell (whole-cell recording), or just a single channel molecule (single-channel recording).

Whereas single-microelectrode recordings can be made by advancing the pipette into the cochlea in vivo without visual control, this does not work if two microelectrodes have to be advanced into the same cell. The patch-clamp technique requires a clean tip of the patch-pipette and an exposed cell membrane, again requiring good visual control. Voltage-clamp experiments on hair cells have therefore been performed only in vitro. Again, because cells from poikilotherms are metabolically less demanding and generally more robust than their homoiotherm counterparts, hair cells from cold-blooded vertebrates have been a natural choice for studying these fundamental aspects of hair cell physiology. Two-electrode voltage-clamp recordings were first reported by Corey and Hudspeth (1979) in hair cells from the bullfrog's sacculus, providing the first direct recordings of transducer currents and time- and voltage-dependent basolateral K^+ currents. This paper also describes a technique that works particularly well for the sacculus, in which the whole epithelium is voltage-clamped with large extraellular electrodes by positioning it between two chambers and stimulating all the hair bundles en masse by moving the otolithic membrane. This method avoids most of the problems of gross-potential recording, but because the hair cells themselves are not properly voltage-clamped, these experiments are not as informative as single-cell voltage clamp (Corey and Hudspeth 1983a,b). An advantage is that the normal environment of the cells, with endolymph bathing the apex and perilymph bathing the basolateral membrane, can be easily simulated. Massed transducer currents can thus be studied under more physiological conditions than is generally possible with isolated cells, for which it is difficult to mimic the endolymph-perilymph separation normally preserved by the tight junctions between the apical poles of the cells in the reticular lamina. Exposing the basolateral membrane to the high K^+ concentration of endolymph would depolarize and kill the cells in the bath that are not voltage-clamped, and the low Ca^{2+} (30 μM) and Mg^{2+} (10 μM) concentrations of endolymph (Bosher and Warren 1978) would also limit the cells' survival in vivo. Although high-Na^+, high-divalent-cation media are probably appropriate for studying the normal functioning of the basolateral conductances (see Section 3), they are not a priori suitable for studying transducer currents.

Patch-clamp recordings of whole-cell currents in frog saccular hair cells were initiated by Lewis and Hudspeth (1983). This technique allows a better quality of voltage clamp than the two-electrode clamp for small cells such as hair cells, and also gives more versatility because the intracellular as well as the extracellular solution can be determined by the experimenter. This has become the method of choice for voltage clamping hair cells, and a lot of information about properties of the transducer and basolateral conductances has emerged from work on nonmammalian vertebrates, but some of these findings might not be directly relevant to mammalian hearing.

1.2.5 Voltage-Clamp Recordings of Mammalian Hair Cells

Since the mid 1980s, a veritable explosion of work on isolated mammalian hair cells has occurred. To some extent, this was triggered by the surprising finding that isolated OHCs lengthen and shorten rapidly in response to intra- and extracellular currents (Brownell 1984; Brownell et al. 1985). This observation, which has not been matched in either IHCs or any nonmammalian vertebrate hair cells, provided the first direct evidence that OHCs might act as motors and be the mystery ingredient that might somehow provide the energy for the extremely high sensitivity and sharp tuning of the healthy cochlea, manifested in the similarly sharp tuning curves in recordings from auditory nerve fibers, IHCs, OHCs, and the basilar membrane.

The patch-clamp technique allowed longer and more stable recordings from the fragile mammalian cells that was possible with microelectrodes. Direct information about basolateral membrane currents in OHCs (Ashmore and Meech 1986) and IHCs (Kros and Crawford 1988) became available for the first time. OHCs, which are so susceptible to damage on isolation, could to some extent be "resuscitated" by an appropriate intracellular solution in the patch-pipette, making the resting potential more negative. It remains unclear to what extent damage to the cells may have irreversibly altered their ionic conductances. Large transducer currents in OHCs and IHCs were first observed by patch clamping cells in cochlear cultures of neonatal mice (Kros, Rüsch, and Richardson 1992).

In this chapter, the emphasis will be on a critical evaluation of the electrophysiological aspects of these recent recordings of mammalian hair cells in vitro. It will be explored to what extent the ionic currents found in vitro can explain the properties of the receptor potentials recorded in vivo. This will be seen to be complicated by interactions between the hair cells in vivo, and by the OHCs being part of an intracochlear feedback loop. Comparisons with currents found in a variety of other vertebrates will serve to diminish between mechanisms that appear to be similar in different types of hair cell, and those that have evolved specifically to partake in producing a high sensitivity and frequency selectivity over the extended frequency range of the mammalian auditory system.

2. The Transduction Mechanism in Mammalian Hair Cells

The study of transduction mechanisms includes transducer currents that flow when the hair bundle is mechanically stimulated, as well as the mechanics of the bundle. In vivo, the hair bundles are either stimulated by *displacement* of the tectorial membrane relative to the reticular lamina (OHCs), or by *force* due to motion of the fluid around the hair bundle (IHCs).

Similarly, in vitro, the experimenter can choose to stimulate the bundle with displacement, usually by coupling a rigid glass fiber to the hair bundle and moving the glass fiber over a known distance by a piezoelectric bimorph (e.g., Hudspeth and Corey 1977), or with force. An approximation of constant force can be delivered by a flexible glass fiber attached to the bundle, provided the fiber is much less stiff than the hair bundle. If the stiffness of the fiber is known and the movements of the end of the fiber attached to the bundle can be observed, the bundle's stiffness can be calculated (e.g., Crawford and Fettiplace 1985). A disadvantage of the flexible fiber method is loading of the bundle by the fiber's stiffness, viscous damping, and mass, which may interfere with the bundle's mechanical properties. This can be avoided by using a fluid jet to stimulate the bundle, which can generate force steps (Kros, Rüsch, and Richardson 1992). The disadvantages of the fluid jet include the fact that the magnitude of the force exerted on the hair bundle cannot be measured, and that measuring the movements of the hair bundle itself is much more difficult than monitoring the motion of a glass fiber coupled to the bundle. A refinement of the flexible fiber method, the displacement clamp, appears at present the best way to measure force exerted by the hair bundle, but again, the bundle mechanics are unavoidably loaded by the fiber (Jaramillo and Hudspeth 1993). Information about the transduction mechanism can also be obtained by analysis of the brownian motion of freestanding hair bundles and the cells' voltage noise, and their correlation (Denk and Webb 1992).

The different methods mentioned above all have their good and bad points, and the main consideration is to try to choose the most appropriate method for the question to be answered experimentally. Displacement by a rigid fiber is suitable as a controlled way to induce transducer currents but gives no information about the mechanical properties of the hair bundle. The other methods, when combined with electrical recording, are all more or less suitable to correlate transducer currents or receptor potentials to bundle mechanics.

2.1 In Vitro Recordings of Transducer Currents

The transducer currents that underlie the receptor potential have been elusive in isolated mammalian hair cells, perhaps due to mechanical disruption of the mechanosensitive hair bundles during the isolation procedure. Large transducer currents have recently been recorded from cochlear cultures of neonatal mice (Kros, Rüsch, and Richardson 1992). Some recordings of transducer currents from adult OHCs are now available for comparison (Ashmore, Kolston, and Mammano 1993).

2.2.1 Transfer Function

Figure 6.2A shows transducer currents recorded from a neonatal mouse OHC. Sinusoidal stimuli of increasing amplitude resulted in increasingly large transducer currents, which eventually saturated. The cell was stimulated with force from an oscillating fluid jet. Force that moved the bundle in the direction of the kinocilium (still present in these neonatal cells) and the tallest stereocilia increased the transducer current; force in the opposite direction closed to 2% to 15% of the transducer channels that were open at rest. The transfer function of Figure 6.2B shows the relation between the transducer conductance and the driver voltage to the fluid jet for the cell of Figure 6.2A. The force exerted by the fluid jet was proportional to the driver voltage, so the force sensitivity of the transducer conductance (the current divided by the driving force) should have the same shape as the transfer function of Figure 6.2B. This transfer function is not rotation symmetrical around the point at which 50% of the maximum transducer

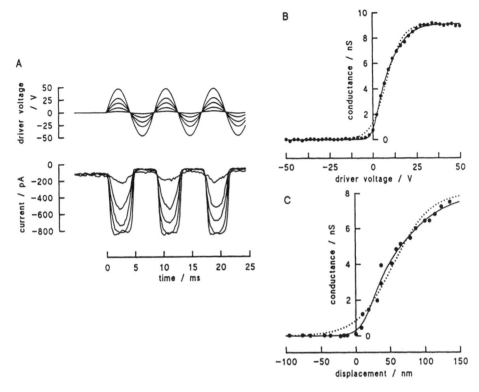

FIGURE 6.2 (A) Five records (unaveraged) of transducer currents of an OHC in a culture of a neonatal mouse cochlea, in response to different stimulus intensities. A piezoelectric disk was used to drive a fluid jet that was directed at the hair bundle sinusoidally. The force exerted on the hair bundle was inferred to be proportional to the driver voltage to the disk (*above*). A positive driver voltage corresponded to a fluid flow that moved the bundle in the direction of the kinocilium, and opened transducer channels, causing inward transducer currents (*below*). Fluid movement in the other direction closed the transducer channels that were open at rest. The cell membrane potential was clamped to -84 mV. (B) transfer function of the transducer conductance (current divided by the driving voltage) of this cell, derived from the maxima and minima of the transducer currents elicited at 22 different stimulus intensities. (Modified from Kros, Rüsch, and Richardson 1992, Fig. 3A,B.) (C) Transducer conductance as a function of bundle displacement for another OHC, with a 4-μm-tall bundle. This cell was stimulated with force steps, and currents were measured before onset of adaptation. Holding potential -84 mV. (Modified by permission from Kros, Lennan, and Richardson 1995, Fig. 5.) In B and C, the fitted continuous lines are second-order Boltzmann functions. For comparison, first-order Boltzmann functions are also fitted (*dotted lines*). Transducer current reversal potential $+3$ mV.

conductance is activated: the current varies more steeply with force near the resting position of the bundle than where the current approaches saturation. Functions of similar shape (which are best fitted with a second-order Boltzmann function: *solid lines* in Fig. 6.2B,C) were also observed in the relation between transducer current and bundle displacement (rather than force) for hair cells of nonmammalian vertebrates (Crawford, Evans, and Fettiplace 1989) and, indeed, for neonatal OHCs (Kros et al. 1993a; Kros, Lennan, and Richardson 1995). An example is shown in Figure 6.2C. In vivo this nonlinearity in the transducer current as a function of force or displacement, which is probably similar in IHCs and OHCs, would contribute to the ear's having a high sensitivity and good intensity discrimination at low sound intensities combined with a large dynamic range. A symmetrical, first-order Boltzmann process would result in less sensitivity around the resting position of the bundle and a smaller dynamic range (*dotted lines* in Fig. 6.2B,C).

So-called compressive nonlinearities similar to that produced by saturation of the transducer currents can be measured almost anywhere in the cochlea (e.g., basilar membrane vibration, itself determined in part by OHC transduction) and indeed in IHC receptor potentials (e.g., Patuzzi and Sellick 1983). The fit to the data of Figure 6.2B,C by second-order Boltzmann functions indicates that the transducer channel has at least two closed states and one open state. The steep activation of the current near the resting position of the bundle arises because the transition between the closed states is about four times more force- or displacement-sensitive than the opening transition (Kros, Rüsch, and Richardson 1992; Kros, Lennan, and Richardson 1995). In an adult OHC, the current-displacement curve was described by a first-order Boltzmann function, suggesting the presence of just one closed state (Ashmore, Kolston, and Mammano 1993), but the currents were very small (less than 50 pA). More and better records from adult OHCs are needed to establish whether this represents a real difference.

2.1.2 Total Transducer Conductance and Sensitivity to Bundle Displacement

The total available transducer conductance that can be activated is an important quantity. In combination with knowledge about the size of the basolateral conductances, it allows for an estimate of the maximal size of the receptor potential in vivo. In neonatal IHCs and OHCs, the maximal transducer conductances were 4.2 nS and 9.2 nS, respectively (Kros, Rüsch, and Richardson 1992). One would predict that in vivo, with endolymph surrounding the bundle, the conductances would be about three times larger, up to 13 nS in IHCs and 28 nS in OHCs (see Section 2.1.4.1). In adult OHCs, the maximum conductance reported so far is 2 nS (Ashmore, Kolston, and Mammano 1993). Quoted estimates for the single-channel conductance of the transducer-channel at room temperature vary from 17 pS (Holton and Hudspeth 1986) to about 100 pS (Crawford, Evans, and Fettiplace 1991). The former value is derived from the variance of the noise caused by the random opening and closing of all transducer channels in a hair bundle. This fluctuation analysis tends to underestimate the single-channel conductance (Hille 1992). The latter value comes from direct measurements, under whole-cell voltage clamp in perilymph-like solutions, of the currents through a single transducer channel that was left over after all other transducer channels had been disabled by perfusion with a solution containing a very low free Ca^{2+} concentration of $1\mu M$. A unit conductance of 50 or 100 pS had also been deduced from steplike variations in the transducer currents of isolated chick hair cells, which generally seem to possess no more than about 10 functioning transducer channels (Ohmori 1984, 1985). Recent measurements of the single-channel conductance in mammalian hair cells confirm that it is about 100 pS (Lennan, Géléoc and Kros, in preparation). Neonatal OHCs would then contain up to 90 functioning transducer channels, IHCs up to 40. A lot of evidence suggests that the transducer channels are pulled open by an elastic "gating spring" which may be identical to the tip link, one of which connects each pair of adjacent stereocilia in neighboring rows (reviewed by Pickles and Corey 1992). Neonatal IHCs and OHC appear to have one transducer channel for every such pair of stereocilia (based on counts of stereocilia in scanning electron micrographs; C.J. Kros and G.P. Richardson, unpublished observations), even though the tip links may apparently contain transducer channels at both ends (Denk et al. 1995). The largest transducer conductance reported in an isolated adult OHC suggests it had 20 functioning channels, or about one for every four tip links. The simplest explanation is that most tip links and/or channels were damaged during the isolation procedure.

In neonatal OHCs with bundle heights of 4 μm, 90% of the total available transducer conductance was activated by displacement of the top of the bundle over 150 to 200 nm, or a rotation of just 2° to 3° (Fig. 6.2C; Kros et al., 1993a; Kros, Lennan, and Richardson 1995), probably defining the operating range in vivo. The maximum sensitivity in the same cells in terms of conductance was 75-100 pS nm^{-1}, or 5.5-7 nS per degree. The maximum sensitivity in isolated adult guinea pig OHCs was more than 50 times smaller. Unfortunately the bundle height was not given, but it can be at most 7 μm (Fig. 6.5B). Since the maximum transducer currents were only about five times smaller than in the neonatal cells, this indicates that the coupling of bundle rotation to transducer channel gating was an order of magnitude less effective in isolated adult OHCs. The sensitivity of neonatal OHCs is larger than that found in the most sensitive hair cells of nonmammalian vertebrates; that of adult OHCs is smaller. The reasons for this discrepancy are at present unclear, but it may be due to damage to the bundles of adult OHCs, combined with their inability to maintain their normal intracellular ionic milieu upon isolation (Housley and Ashmore 1992; see Section 3.3.2). For the purpose of modeling hair cell responses, it is therefore perhaps more realistic to use the values derived from experiments on neonatal OHCs until this problem with isolated adult OHCs is effectively tackled.

The sensitivity to small displacements or forces around the resting position of the bundle is only about half the maximum sensitivity: the resting position is not at the steepest part of the transfer function (where about 30% of the transducer channels are open), but close to the point of the transfer function where the rate of change of sensitivity (the second derivative of the transfer function) is maximal (Kros, Rüsch, and Richardson 1992). A similar observation was subsequently also reported for the receptor potentials of neonatal hair cells (Russell, Kössl, and Richardson 1992). What is lost in sensitivity is made up for by the ability to produce asymmetrical receptor potentials with a larger depolarizing than hyperpolarizing phase, also for the smallest stimuli. This is important for the generation of steady depolarizing receptor potentials at frequencies above the cutoff frequency of the low-pass filter formed by the cell membrane (see Sections 3.2.2 and 3.4).

The limits of the frequency response of mammalian transducer currents have not been explored yet for technical reasons, but the transducer channels should be able to open and close up to frequencies between 11 and 150 kHz (Section 1.1), otherwise no receptor potentials would be generated at all in response to high-frequency stimuli. From the temperature dependence of the time constants of transducer currents in turtle auditory hair cells, an f_{3dB} (frequency at which the power of the output signal is halved, see, for example, Horowitz and Hill 1989) of only 2.5 kHz was predicted at 38°C (Fettiplace and Crawford 1989), clearly insufficient for the generation of transducer currents at high frequencies. This suggests a difference in design of the transduction mechanism in mammals compared to other vertebrates, to cope with the extended high-frequency limit of mammalian hearing. It would be interesting to find out whether this redesign only applies to cells in the basal (high-frequency) turns, or also to cells with best frequencies below a few kilohertz.

2.1.3 Adaptation and Modulation of the Transducer Current

A time-dependent reduction in the transducer current following a displacement step or an (approximate) force step to the bundle has been found in various nonmammalian vertebrates (Eatock, Corey, and Hudspeth 1987; Assad, Hacohen, and Corey 1989; Crawford, Evans, and Fettiplace 1989). This reduction is not due to inactivation of the transducer channels, because a larger stimulus applied after the current has adapted can still elicit the maximum transducer current. The mechanism of transducer current adaptation, which may or may not involve a mechanical rearrangement within the hair bundle, is discussed in detail in Section 2.2.2.2.

Adaptation could be abolished by reducing the concentration of free Ca^{2+} in the intracellular solution, by introducing a fast, powerful Ca^{2+} buffer in the intracellular solution, and by depolarization to positive potentials. The last two effects suggest that it is in fact a rise in intracellular Ca^{2+}, probably close to the transducer channel, that promotes adaptation. Depolarization to positive potentials near the Ca^{2+} equilibrium potential would reduce the electrochemical driving force for Ca^{2+} ions flowing in through the transducer channels. A sudden increase in the intracellular free Ca^{2+} concentration by flash photolysis of an intracellular Ca^{2+} "cage" increased the speed of adaptation in voltage-clamped cells, but only if sufficient Ca^{2+} was flowing in through the transducer channels as well (Kimitsuki and Ohmori 1992). Measures that abolished adaptation also increased the fraction of the transducer channels open with the bundle in its resting position (normally about 10%), demonstrating that the channels' resting opening probability is in part determined by adaptation.

Adaptation of the transducer currents disappears when voltage-clamped cells are superfused with solutions containing realistic (i.e., low: see Section 1.2.4) endolymphatic Ca^{2+} and Mg^{2+} concentrations (Crawford, Evans, and Fettiplace 1991). This raises the question whether this form of adaptation could occur in vivo, because insufficient Ca^{2+} might enter through the transducer channels. Evidence for this is sparse: in vivo recordings of microphonic potentials in the frog sacculus suggested adaptation to a constant acceleration, but there were difficulties with the interpretation because it could not be observed whether the mechanical input to the hair cells was really constant (Eatock, Corey, and Hudspeth 1987). Adaptation was, however, convincingly demonstrated in vitro in the transepithelial microphonic currents recorded in the isolated frog sacculus, a preparation which, as discussed before in Section 1.2.4, does allow perilymph- and endolymph-like solutions to be maintained around the basolateral and apical parts of the cells, respectively (Corey and Hudspeth 1983a; Eatock, Corey, and Hudspeth 1987). It is possible that adaptation is normally, in vivo, caused by Ca^{2+} influx not through the transducer channels, but through the Ca^{2+} channels in the basolateral membrane (Section 3.5), which would be opened by the depolarizing receptor potential in cells that are not voltage-clamped (Crawford, Evans, and Fettiplace 1991).

There is some evidence for the presence of transducer current adaptation in neonatal mouse OHCs bathed in artificial perilymph (Kros, Rüsch, and Richardson 1992), with a time constant of about 4 ms for small force steps, slowing down and eventually disappearing for larger steps. Depolarization abolished this time-dependent decline in transducer current (Figure 6.3). Similar results, in response to displacement steps, have been obtained in adult guinea pig OHCs (Ashmore, Kolston, and Mammano 1993). Russell, Richardson, and Kössl (1989) had already demonstrated adaptation of receptor potentials in neonatal mouse OHCs, with a similar time constant to transducer current adaptation. Adaptation of the receptor potential might also be caused by a time-dependent increase in basolateral K^+ currents (see

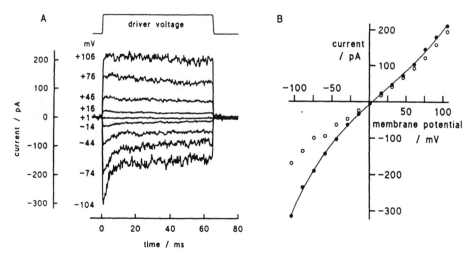

FIGURE 6.3 Adaptation of transducer currents in a neonatal mouse OHC. (A) Unaveraged transducer currents at different holding potentials, indicated next to the current traces. Adaptation is manifested as a time-dependent reduction in size of the transducer currents at hyperpolarized potentials, and virtually disappears at potentials positive to -14 mV. The time course of the driver voltage to the piezoelectric disk is shown above. (B) Current-voltage curves of this cell, for early (●) and adapted (○) transducer currents. Note, for the unadapted currents, the inward rectification at negative potentials and the outward rectification at positive potentials. The continuous line is based on a simple, single-energy-barrier model, to quantify the degree of rectification. (Modified from Kros, Rüsch, and Richardson 1992, Fig. 4A,B.)

Section 3.1.3), but its persistence at membrane potentials around -70 mV, where these currents are not activated, suggested that the adaptation was indeed due to the transducer mechanism.

Although transducer current adaptation is conceptually useful for vestibular hair cells, whose bundles may be subjected to prolonged static displacements around which sensitivity to transient stimuli must be maintained, its function in the mammalian cochlea is less obvious. Adaptation acts to some extent like a high-pass filter, with an f_{3dB} of about 40 Hz for a (room temperature) time constant of 4 ms (Kros, Rüsch, and Richardson 1992). There is not much information to date about the temperature sensitivity of the adaptation process, but from one turtle hair cell studied at 3°C (Crawford, Evans, and Fettiplace 1989) one can calculate, from applying the Arrhenius equation (Hille 1992, p. 272), a $Q_{10(°C)}$ (the ratio of the time constants at temperatures 10°C apart) close to 3 around room temperature, quite a steep temperature dependence that is typical for the kinetics of channel gating. If one assumes a similar temperature dependence for the mammalian transducer channel, the fastest time constant of adaptation would become about 1 ms at 37°C, equivalent to an f_{3dB} of 160 Hz. This frequency sets a limit above which transducer current adaptation would be hard to detect in hair cells in vivo. Thus, for all but the lowest frequencies mammals can hear, adaptation would appear to play no role. This adaptation mechanism may nevertheless be necessary to regulate the amount of transducer current activated at rest and maintain the cell's sensitivity within relatively narrow limits. In this respect, it is interesting to note that in neonatal mouse cochlear cultures, OHCs with very immature, cone-shaped bundles had transducer currents whose properties were indistinguishable from those of OHCs with more mature-looking bundles. This suggests that during the developmental rearrangement of the bundles' anatomy, which may well involve changes in the forces exerted on the channels within the bundle, the adaptation mechanism may keep the functional properties of the transducer channels relatively constant.

A possible modulation (at present of unknown physiological significance) of transduction may occur by endolymphatic ATP. Apart from activating a large, nonselective cation conductance near the apex of the OHC (see Section 3.7), 50 μM ATP also appears to directly reduce the total available transducer conductance by about half, without changing the shape of the current–displacement relation (Kolston and Ashmore 1993; Ashmore, Kolston, and Mammano 1993). A caveat is that if the series resistance R_s in the patch-pipette were underestimated in these experiments, the same result would have been produced by the voltage drop across R_s due to the much larger nonselective current that is also activated by ATP.

2.1.4 Ionic Selectivity and Blockers of the Transducer Channel

2.1.4.1 Ionic Selectivity

A great deal of what we know about the ion permeation pathway and the gating process of ion channels has been learned from experiments with blocking agents (Hille 1992). Although in vivo most of the transducer current will be carried by K^+ ions, experiments on nonmammalian vertebrates have shown the channel to be a nonselective cation channel that has very similar conductances for the alkali cations, and has a larger permeability (deduced from reversal potentials) for divalent than for monovalent metal ions (Ohmori 1985; Howard, Roberts, and Hudspeth 1988). This, as well as the permeability sequence for the alkali metal ions (Eisenman sequence XI; see Hille 1992), suggests that the selectivity filter in the channel has a high negative charge density (corresponding to a strong field), and that anions will be very impermeant (Ohmori 1985). Among other nonselective cation channels, the cyclic nucleotide-gated channels of rod and cone photoreceptors have a qualitatively similar strong field strength filter (Furman and Tanaka 1990; Menini 1990; Haynes 1993), and that of olfactory receptors is slightly less so (Frings, Lynch, and Lindemann 1992). The endplate channel at the frog neuromuscular junction, by contrast, has a weak field strength filter (Eisenman sequence I) and is less permeable for divalents (Adams, Dwyer, and Hille 1980). Because the transducer channel can be permeated by the tetraethylammonium ion (TEA), the narrowest part of the permeation pathway has a relatively large diameter of at least 0.7 nm (Howard, Roberts, and Hudspeth 1988), comparable to that of the motor endplate channel (Dwyer, Adams, and Hille 1980) but larger than that of photoreceptors (Menini 1990; Haynes 1993).

Because Ca^{2+} and, to a lesser extent, Mg^{2+} ions have such a high affinity for the transducer channel's selectivity filter, they are more permeant but have a lower conductance than monovalent cations (Howard, Roberts, and Hudspeth 1988): they tend to "linger around" in the pore and act as permeant blockers (Crawford, Evans, and Fettiplace 1991), as they do in the cyclic nucleotide-gated channels of rod and cone photoreceptors (Colamartino, Menini, and Torre 1991; Zimmerman and Baylor 1992; Haynes 1993) and (for Ca^{2+} but not Mg^{2+}, which was impermeant) olfactory receptor neurons (Zufall and Firestein 1993). In both photoreceptors and olfactory receptors, divalent block of the cyclic nucleotide-gated channels serves to increase the receptors' signal-to-noise ratio. These receptor cells have about half a million stimulus-activated channels, with an extremely small single-channel conductance of about 0.1 pS due to the nature of the divalent cation block (Yau and Baylor 1989; Zufall, Firestein, and Shepherd 1991). This enormously reduces the voltage noise due to the random opening and closing of the ion channels (Yau and Baylor 1989; Zufall and Firestein 1993) and allows reliable detection of single photons by rod photoreceptors. In hair cells, this strategy would not work. First, in cells such as IHCs with freestanding hair bundles, the voltage noise is to a large extent caused by transduced brownian motion of the hair bundle (Denk and Webb 1992); therefore, reducing the single-channel conductance would not improve the signal-to-noise ratio (noise due to brownian motion in OHCs, however, is likely to be much reduced because their bundles are partially embedded in the tectorial membrane). Second, the hair cell has only about 100 transducer channels, and there are good reasons for that (see Section 2.2.2.1 below). Thus, the low Ca^{2+} and Mg^{2+} concentrations in the endolymph (Bosher and Warren 1978) serve to maximize the transducer conductance, making it about three times larger than when it is recorded in perilymph-like solutions (Crawford, Evans, and Fettiplace 1991), probably by pushing the single-channel conductance to 300 pS, close to the theoretical maximum (Hille 1992, p. 333). The function of the high K^+ concentration in the endolymph seems mainly to be to avoid the need for the cell to spend a lot of energy extruding Na^+ in maintaining its normal K^+-rich intracellular milieu (discussed by Wangemann and Schacht, Chapter 3).

Experiments on neonatal mouse OHCs have confirmed that also in the mammal the transducer current is equally readily carried by Na^+, K^+, and Cs^+ ions (Kros, Rüsch, and Richardson 1992). The large organic monovalent cation N-methyl-D-glucamine (NMDG) passes through the channel with a permeability similar to that of TEA in the chick. Thus, the ion permeation pathway seems similar to that of other vertebrate transducer channels. A difference is that in a perilymph-like extracellular solution, the voltage dependence of the current showed inward rectification at hyperpolarized potentials and outward rectification at depolarized potentials (Fig. 6.3B). In other vertebrates, the current–voltage curve of the transducer channels has been described as linear under comparable conditions. The double rectification is certainly consistent with extracellular divalent cations and possibly also intracellular Mg^{2+} acting as permeant blockers, which are "pushed through" the channel at extreme potentials (Yau and Baylor 1989; Zufall and Firestein 1993). This different shape of the current–voltage curve may be due to some difference in the position or the nature of the selectivity filter between mammals and other vertebrates. Isolated adult OHCs pass comparatively little outward transducer current (Ashmore, Kolston, and Mammano 1993): the current–voltage curve shows a quite strong inward rectification (Ashmore 1988). The question may be asked whether this reflects a developmental change in the properties of the channel, or whether it may be due to different experimental conditions. Because adult OHCs are leaky, the small transducer currents are superimposed on much larger leak currents, which cannot be blocked. These leak currents may introduce voltage errors by the voltage drop they cause across the pipette's series resistance. If these errors are not correctly estimated, the voltage in the current–voltage curves may be quite distorted (discussed by Santos-Sacchi 1989). This is almost certainly the cause of the early suggestion (Ashmore 1988), based on a reversal potential of about +30 mV, that the OHC transducer channel was much more permeable to Na^+ than to K^+. More recently (Ashmore, Kolston, and Mammano 1993), the reversal potential (with an unspecified intracellular solution) was near 0 mV, suggesting that these channels, too, are nonselective cation channels after all. The inward rectification thus needs another explanation. One possibility is that the elevated concentration of free intracellular Ca^{2+} in isolated adult OHCs (Housley and Ashmore 1992) may partially block outward transducer currents. At present, it seems therefore safest to assume that the permeation pathway in the transducer channel does not change during development.

2.1.4.2 Block by Aminoglycosides

The actions of aminoglycoside antibiotics on the cochlea have been extensively studied, originally because ototoxicity is a prominent side effect in patients treated with these drugs. One of the many effects that the aminoglycoside antibiotics have on hair cells is that they reversibly block the transducer current, with half-blocking concentrations between 2 and 95 μM for frog saccular hair cells (Kroese, Das, and Hudspeth 1989). A physiological pH, they are large cations (MW 300–800) with multiple positively charged groups, which are likely to be essential for the blocking mechanism. The block is voltage-dependent, in that is relieved at positive potentials (Ohmori 1985; Kroese, Das, and Hudspeth 1989). Binding of one molecule seems sufficient to block a channel, and the drug binding site can be reached only from the extracellular solution bathing the apical surface of the hair cells (Kroese, Das, and Hudspeth 1989). The binding site appears to be situated inside the pore, at about 20% of the channel's electrical distance (from the extracellular face). The drug molecules seem too large (about 1 nm) to pass the selectivity filter, which is therefore situated further inside the pore. In mammalian cochlear hair cells, too, aminoglycosides reversibly block transducer currents (Kros, Rüsch, and Richardson 1992), but the mechanism of the block has yet to be investigated.

2.1.4.3 Mechanism of Block by Pyrazinecarboxamides

More is known about the action of another group of drugs, the thousand or so (because of their commercial pharmacological interest) compounds derived from the diuretic amiloride (collectively called pyrazinecarboxamides; reviewed by Kleyman and Cragoe 1988), on transducer currents in mammalian hair cells, although clinical treatment with amiloride is not known to affect the auditory system. Amiloride and most of its derivatives are weak bases, with most molecules carrying a single positive charge at physiological pH, the rest being neutral. The range of half-blocking concentrations (equal to the dissociation constant K_D) of various amiloride derivatives on transducer currents in OHCs of neonatal mice was 2–53 μM (Rüsch, Kros, and Richardson 1994), comparable to the aminoglycosides, with amiloride itself being the least potent blocker. The K_D for amiloride was very similar to that previously found for chick cochlear and/or vestibular hair cells (Jörgensen and Ohmori 1988). It is not known whether the charged or neutral forms, or both, are able to block the channel. The mechanism of block is completely different from that proposed for the aminoglycosides. Two molecules rather than one are required to block the channel. The drug binding site can only be reached when the channel is open, and the channel cannot close as long as the drug is bound. The block is, like that of the aminoglycosides, relieved at positive potentials, but the concentration-dependent incompleteness of the block at extremely negative potentials strongly suggests that the binding sites are not in the electrical field of the pore (Rüsch, Kros, and Richardson 1994). It is instead likely that the voltage dependence of the block derives from a voltage-dependent accessibility of the drug binding sites on the extracellular face of the channel (no block occurred when the drug was introduced on the intracellular side), revealing the presence of two open states of the transducer channel (Fig. 6.4), in addition to the two closed states described in Section 2.1.1. The existence of two open states of the transducer channel has previously been proposed for turtle hair cells, based on experimental observations involving adaptation, but in this case the transition between the states was Ca^{2+}-dependent (Fettiplace, Crawford, and Evans 1992). It is possible that the same phenomenon underlies both sets of observations.

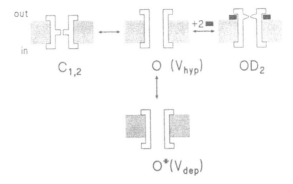

FIGURE 6.4. A possible model for the action of amiloride on the transducer channel. The channel has two closed states (simplified as $C_{1,2}$), which do not bind amiloride, and two open states O and O*. At hyperpolarized potentials, state O, which exposes two amiloride binding sites, prevails. Amiloride binding results in an allosteric block of the transducer channel (OD_2). For the blocked channel to close, the drug molecules have to be released first, so that the normal, conducting open state O is momentarily reached. Depolarization favors the second open state O*, in which the binding sites are not accessible.

2.1.4.4 Pyrazinecarboxamide Block Suggests Genetic Similarities Between Transducer Channels and Stretch-Activated Channels

There are considerable differences in the ways the pyrazinecarboxamides interact with different channels and ion carriers. The channel to which amiloride binds most avidly ($K_D < 1\ \mu M$) is the voltage-independent Na^+ channel on the apical membrane of cells in Na^+-absorbing epithelia. The drug's action on the distal and collecting tubules of the kidney explains its diuretic effect. The positive charge is essential for the block of the epithelial Na^+ channel, and the binding site is probably within the electrical field of the pore (reviewed by Smith and Benos 1991). The way in which the addition of various aliphatic or aromatic side chains at different positions of the amiloride molecule affects the binding affinity is also quite different for the transducer channel and this amiloride-sensitive Na^+ channel (Rüsch, Kros, and Richardson 1994). Therefore, there appears to be no molecular similarity between the amiloride binding sites of the transducer channel and the epithelial Na^+ channel.

Among the various channels and transporters that are blocked by pyrazinecarboxamides, the so-called structure–activity sequence most similar to that of the transducer channel is that of a mechanosensitive cation channel in frog oocytes, although the affinity of amiloride for the latter is an order of magnitude smaller (Lane, McBride, and Hamill 1992; Rüsch, Kros, and Richardson 1994). The blocking mechanisms of the stretch-activated channel and the OHC transducer channel also appear similar, although the question whether the stretch-activated channel has to be open for the drug binding sites to be exposed has not yet been addressed (Lane, McBride, and Hamill 1991). Nevertheless, the study of block by amiloride has shown suggestive molecular similarities between a class of stretch-activated channels present in abundance on frog oocytes (about 1-10 million per cell, estimated from data by Methfessel et al. 1986), and the rare and elusive hair cell transducer channels (up to 100 per cell). Since, therefore, there are probably at least as many mechanosensitive channels on one oocyte as there are in an entire mammalian cochlea, isolation and purification of the channel, as yet not achieved, may be somewhat less daunting using oocytes in first instance, and then searching for similarities in the transducer channel.

A surprising twist is that the recently cloned gene for the epithelial Na^+ channel turns out to be related to a family of genes that may have something to do with mechanosensitivity in the nematode *Caenorhabditis elegans* (Canessa, Horisberger, and Rossier 1993; Lingueglia et al. 1993). Perhaps the epithelial Na^+ channel is after all related to mechanosensitive channels, which may also explain the putative labeling of transducer channels by antibodies raised against the amiloride-sensitive Na^+ channel found in the kidney, on the grounds that their amiloride binding sites might be similar (Hackney et al. 1992). Ironically, as shown above, the binding sites and the mechanism of amiloride binding appear quite different for the two classes of ion channels!

2.2 Mechanics of the Hair Bundle

The stereociliary bundles of the OHCs, by being directly connected to the tectorial membrane, are in a strategic position to alter the mechanics of the coupling between the tectorial membrane and the apical surface of the reticular lamina, and thus the transformation, via fluid coupling, of sound-induced movement in the cochlea to excitation of the IHCs, which are generally assumed to have freestanding bundles. If such a mechanism were employed to achieve the high sensitivity and frequency selectivity of mammalian hearing, it would have to operate, cycle-by-cycle, at the frequencies appropriate for the tonotopic position of the cells along the cochlea. For the guinea pig, this is expected, from psychophysical experiments, to be up to at least 50 kHz (Prosen et al. 1978) for the most basally located hair cells. This is the major unresolved problem in the physiology of the mammalian cochlea. A great deal of work has been done on the rapid motile properties of the cell body of the OHC (Holley, Chapter 7), but nonlinear mechanical behavior of the hair bundle may provide an alternative or perhaps additional candidate mechanism. At present, there is no direct experimental evidence that elucidates the contribution of the OHC bundles to the mechanical impedance of the cochlea, either at the level of the tectorial membrane (Zwislocki and Cefarratti 1989) or of the basilar membrane (Olsen and Mountain 1991). This section describes what is known about the mechanical properties of the hair bundles of OHCs and IHCs, and how they are possibly affected by the transduction process itself. See also Patuzzi, Chapter 4, and Holley, Chapter 7, for different perspectives on the same subject.

2.2.1 Estimates of the Stiffness of the Hair Bundle

Stiffness measurements of mammalian cochlear hair bundles have been made in cochleae acutely isolated from adult guinea pigs (Strelioff and Flock 1984), and in cochlear cultures of neonatal mice (Richardson et al. 1989, Kössl, Richardson, and Russell 1990). Both sets of measurements were made with flexible glass fibers of known stiffness with which the stereocilia were deflected. Both measured steady-state stiffness. In the guinea pig, the translational stiffness was measured in IHCs and OHCs of the second to the fourth turns, with a predicted characteristic frequency (CF; the frequency of the sound stimulus that elicits a responsive at the lowest stimulus intensity) of 3.7-0.3 kHz. The translational stiffness (the force required to deflect the tip of the bundle divided by the displacement) decreased with distance from the stapes, much more pronouncedly in OHCs than in IHCs (Fig. 6.5A). The height of the hair bundles increases with distance from the stapes, more so for OHCs than IHCs (Lim 1980; Fig. 6.5B). Hair bundles of a number of nonmammalian vertebrates behaved to a first approximation as a collection of stiff rods, pivoting around their insertions in the circular plate, and connected higher up by stiff (compared to the pivot) springs that can rotate around their points of attachment (Crawford and Fettiplace 1985; Howard and Ashmore 1986). This would leave the rods free to slide past each other in such a way that an applied force would be equally shared among all stereocilia. Since, for small deflections of a stiff rod pivoting around its base, the translational stiffness is approximately equal to the rotational stiffness (defined as the torque required to rotate it through an angle of one radian) divided by the square of the height of the rod, one may wonder whether the rotational stiffness is simply proportional to the number of stereocilia in a bundle, reflecting a similar structure of the stereociliary pivots. This was not found for OHCs: their rotational stiffness [named *stiffness coefficient* by Strelioff and Flock (1984), and unfortunately described by an erroneous equation, although the calculations seem to be correct] decreased by an order of magnitude with increasing distance from the stapes (Fig. 6.5C). This is much more than would be predicted from the at most twofold difference in the number of stereocilia that might be expected over the range of locations examined (Lim 1986). Moreover, it appears that only a few stereocilia were deflected by the probe, showing that the side-to-side links between stereocilia within a row are either very compliant (and therefore different from the nonmammalian vertebrate hair bundles mentioned above) or easily damaged (Pickles et al. 1989). The deflections

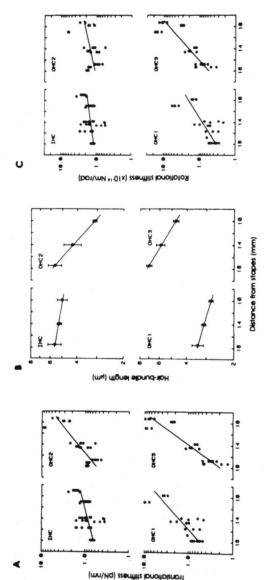

FIGURE 6.5 Translational (A) and rotational (C) bundle stiffness of guinea pig hair cells. Bundles were deflected in the excitatory direction. The hair bundle lengths shown in (B) were used to convert translational into rotational stiffness. For small rotations, the latter approximately equals the translational stiffness multiplied by the square of the bundle length. Lines are best fits. (Modified by permission from Strelioff and Flock 1984, Figs. 3, 4, and 5.)

used by Strelioff and Flock (1984) were indeed very, possibly destructively, large (± 1 μm for bundle lengths between 3 and 7 μm, resulting in rotations of $\pm 8°$-$19°$), well beyond the approximately 3° which opens all transducer channels (see Section 2.1.2). These factors would lead not only to underestimates of the stiffness of the whole bundle, but also to systematic distortions dependent on the length of the bundle: if the side-to-side links within a row are compliant, the larger angular deflections used for shorter bundles would cause more stereocilia to contribute to the measured stiffness, thus overestimating the gradient in rotational stiffness. Estimates of translational and rotational stiffness shown in Figure 6.5A,C apply to the excitatory direction. Stiffness values in the inhibitory direction were about half of those shown in the figure. In most studies of hair bundle stiffness, including those of OHCs from the neonatal mouse cochlea, no such difference has been found.

The stiffness measurements in the neonatal mouse cochlear cultures were less systematic than those in the guinea pig discussed above. The measured translational stiffness [1.04 pN/nm for IHCs, 1.69 pN/nm for OHCs (Richardson et al. 1989)], for small displacements of up to 100 nm, was unfortunately not correlated with hair bundle height, although it was noted that the bundles of OHCs in the apical turn of the cochlea were 1.6 times less stiff than those in the base. Most OHC data were gathered from the apical coil (I.J. Russell, personal communication), where the average bundle height is about 3 μm. This gives an estimate of the OHCs rotational stiffness of about 1.5×10^{-14} Nm/rad. Again, it is not clear how many of the stereocilia were actually contributing to the measured stiffness, although probably more than in the guinea pig experiments, because the axis of the glass fiber was more or less parallel to the cuticular plate rather than perpendicular to it, as in the experiments of Strelioff and Flock (1984). An added level of complexity in neonatal mouse hair cells is that the bundle morphology is still developing, with the basal coil being the most advanced (Furness, Richardson, and Russell 1989).

Given the uncertainties in these stiffness measurements in mammalian cochlear hair bundles, it is perhaps surprising that the estimates of rotational stiffness are quite close to those made by a number of different methods in hair bundles in a variety of other vertebrates (e.g., Crawford and Fettiplace 1985; Howard and Ashmore 1968; Van Netten and Kroese 1987). Many aspects of the mechanics of mammalian hair bundles remain poorly understood. It would, for example, be very useful to collect more information about the contributions of the different kinds of interstereociliary links to the bundle mechanics (Pickles 1993). If the stereocilia within a row are indeed compliantly coupled, then this, combined with the peculiar shape of the mammalian cochlear hair bundles, would make stimulation with glass probes problematic, because displacement or force will not be evenly distributed along the rows, in contrast to what might be expected to occur in vivo.

2.2.2 Gating of the Transducer Channels, Adaptation, and the Membrane Potential May Affect Bundle Mechanics

2.2.2.1 Gating Springs

The above analysis assumes that the stiffness of the hair bundle is mainly determined by the sum of the stiffnesses of the individual stereociliary pivots. There is, however, strong evidence that the transducer channels are situated near the tips of the stereocilia (Denk et al. 1995; Lumpkin and Hudspeth 1995). The rapidity with which the transducer channels are inferred to open following a mechanical step makes it very likely that the channels are directly mechanically gated (Corey and Hudspeth 1983b; Crawford, Evans, and Fettiplace 1989). The Boltzmann relation between transducer current and deflection of the bundle can be explained if the probability of opening of the channel is related to the deflection-dependent tension in an elastic linkage coupled to the channel (Corey and Hudspeth 1983b). These hypothetical linkages, called *gating springs* (Howard and Hudspeth 1987), may well correspond to the tip links that have been found to connect pairs of stereocilia in neighboring rows (Pickles, Comis, and Osborne 1984; Pickles and Corey 1992).

Thus it is a priori possible that the stiffness of these gating springs may contribute a significant fraction of the measured stiffness of the hair bundle. If so, the gating of the transducer channels by these springs may be detectable in the overall mechanics of the hair bundle. Specifically, under the simplifying assumption of a two-state channel, the stiffness contributed by the gating springs and the transducer channels is expected to be at a minimum when an average of 50% of the channels are open (Howard, Roberts, and Hudspeth 1988).

A dip in the hair bundle's stiffness upon moving the hair bundle (with a flexible glass fiber) from its resting position in the excitatory direction has indeed been found in hair cells of the frog sacculus (Howard and Hudspeth 1988). This dip in stiffness disappeared in the presence of gentamicin (Howard and Hudspeth 1988), which appears to lock the transducer channels in their open state (Denk, Keolian, and Webb 1992). Unfortunately, Howard and Hudspeth (1988) measured the receptor potentials, not the transducer currents, but the minimum in the stiffness of the bundles (at 22 nm displacement) was quite close to the displacement where the receptor potential was half-maximal, giving qualitative support for the gating-spring model. Opening half the channels reduced the bundle's stiffness by 42%. Because this stiffness reduction can be only in the gating springs and not in the stereociliary pivots, it can be deduced that the gating springs contribute at least 42% of the bundle's stiffness. Therefore, at a minimum, almost half the force needed to deflect a hair bundle is used to gate the transducer channels, which seems energetically sensible. A disadvantage from the point of view of signaling the impact of sound on the bundle is that mechanical and (secondarily) electrical distortion products will be generated by this mechanical nonlinearity associated with opening the transducer channels (Jaramillo, Markin, and Hudspeth 1993). The force needed to gate the transducer channels in a hair bundle may set a limit to the useful number of channels in a bundle, and may also explain the large single-channel conductance of the transducer channel, compared with, for example, the cyclic nucleotide-gated channel in photoreceptors. It would seem that in the hair cell, a high sensitivity is achieved at the cost of a relatively large noise caused by the probabilistic gating of the channels. The presence of this noise may nevertheless not be a serious problem, at least for freestanding hair bundles (such as that of the IHCs), since in the absence of mechanical stimulation of frog saccular hair bundles, their voltage noise was dominated by transduced brownian motion of the bundle (Denk and Webb 1992; see also Section 2.1.4.1).

Evidence for a displacement-dependent change in stiffness of the hair bundle has also been found in IHCs and OHCs of neonatal mouse cochlear cultures (Russell, Kössl, and Richardson 1992). Hair bundle displacements of 20–30 nm in the excitatory direction caused the stiffness of the bundles to drop by about 14%, a much smaller change than that found in frog saccular hair cells. It is somewhat disconcerting that a similarly sized increase in stiffness was noted for inhibitory displacements of equal magnitude, for which no satisfactory explanation was found. Such inhibitory stiffness increases can also be seen in some of the data of Howard and Hudspeth (1988). In both sets of experiments, the bundle mechanics were not studied under voltage clamp, so it is conceivable that some voltage-dependent motor process independent of the gating compliance (Section 2.2.2.3) contributed to the stiffness changes. These stiffness changes in mammalian hair bundles, in both directions, were reversibly abolished by neomycin and cobalt.

2.2.2.2 Models for Adaptation

Two different models are used at present to explain the adaptation of transducer currents discussed in Section 2.1.3. One of them, based on experiments on frog saccular hair cells, assumes that on one end of the gating spring there is an adaptation "motor" that is continuously trying to climb along a stereocilium (Fig. 6.6). The tension in the gating spring also causes the motor to slip down the stereocilium, and when the bundle is at rest the climbing rate equals the slipping rate, thus keeping the gating springs under a certain resting tension. During adaptation to excitatory displacements, influx of Ca^{2+} through the transducer channels is assumed to increase the slipping rate, thus reducing the tension in the spring and changing the coupling between bundle position and force on the transducer channels (Howard and Hudspeth 1987; Assad and Corey 1992). This "gating-spring model of adaptation" was prompted by the observation that during adaptation the hair bundle's stiffness was reduced with a similar time course (Howard and Hudspeth 1987; Jaramillo and Hudspeth 1993). The

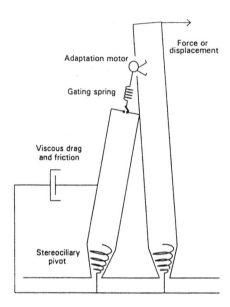

FIGURE 6.6 Gating-spring model. The *arrow* indicates the direction of excitatory stimuli. See text for further details. (Reproduced by permission from Assad and Corey 1992, Fig. 2A.)

extent of this reduction in stiffness led to estimates that the gating springs contributed about 30%–40% of the stiffness of the hair bundle, a figure comparable to that based on the dip in stiffness upon opening half the transducer channels.

The other model, which may be named the "transducer channel model of adaptation," is based on experiments on auditory hair cells from the turtle's basilar papilla. In these cells, it was found that although there were components in the bundle motion with a time course comparable to adaptation, abolition of adaptation by a large depolarization did not alter the motion of the bundle (Crawford, Evans, and Fettiplace 1989). A similar observation was made for mammalian neonatal OHCs (Kros, Lennan, and Richardson 1995). This control experiment has to date not been performed in frog saccular hair cells. It is therefore assumed that the influx of Ca^{2+} through the transducer channels directly affects the relation between the force exerted on the transducer channel and channel gating, by stabilizing the channel in one of its closed states (Crawford, Evans, and Fettiplace 1991; Fettiplace, Crawford, and Evans 1992).

A number of additional observations provide indirect support for the gating-spring model of adaptation: depolarization to near the Ca^{2+} equilibrium potential would reduce the slipping rate of the gating spring, thus increasing the tension on the springs, which would cause the hair bundle to move in the inhibitory direction (the climbing rate is also inferred to be reduced, but less so). Such movements were indeed found, with amplitudes of up to 100 nm at the bundle's tip (Assad and Corey 1992), although the kinetics of the movements were about twice as slow as predicted, and were much slower than the voltage-induced movements discussed below. Lowering extracellular Ca^{2+} or applying streptomycin (which should reduce the entry of Ca^{2+} through the transducer channels) produced, reversibly, the same result and reduced the effect of the membrane potential on the position of the bundle. This makes it unlikely that the movements were directly voltage dependent (see below). Adaptation was also blocked by intracellular ADP, which supports the possibility that myosin is the adaptation motor (Gillespie and Hudspeth 1993).

A finding that cannot be explained by the gating-spring model, and which provides an additional argument in favor of the transducer channel model, is that often the shape of the relation between transducer current and bundle displacement changes considerably with the adaptive state of the cell (Crawford, Evans, and Fettiplace 1989, 1991). The gating-spring model predicts a simple shift of this relation along the displacement axis (Assad and Corey 1992).

OHCs from neonatal mice are the only mammalian hair cells in which the mechanical aspects of adaptation have been studied. Stimulation of the hair bundles with flexible fibers showed that the receptor potentials adapted in some, but not all, cells (Russell, Richardson, and Kössle 1989). In the adapting cells, the stiffness of the hair bundle was reduced to about half of its initial value, with a time course similar to that of the adaptation of the receptor potential. In the cells that failed to show adaptation, the bundle stiffness remained constant during the stimulus. The steady-state stiffness of the bundles of nonadapting cells was about 50% of that of the cells that did show adaptation. These experiments offer some support for the gating-spring model of adaptation. On the other hand, experiments made using the same preparation, but stimulating the hair bundle with constant force from a fluid jet, and recording transducer currents rather than receptor potentials to avoid contributions of the basolateral membrane currents, have so far failed to show changes in the mechanics of the hair bundle when adaptation is abolished by depolarization (Kros et al. 1993a; Kros, Lennan, and Richardson 1995).

At present, therefore, the mechanism of hair bundle adaptation is still uncertain. The fraction of the bundle's stiffness contributed by the gating springs may be different in different species, or there may be a difference between auditory and vestibular hair cells. Different experimental conditions may result in large differences in viscous drag on the bundles. All these factors could contribute to the failure to observe correlates of adaptation in the overall bundle mechanics in some cases, even though an adaptation motor may be causing mechanical rearrangements within the bundle. On the other hand, crucial direct evidence that the mechanical relaxation and adaptation are causally related in frog saccular hair cells is still lacking.

2.2.2.3 Voltage-Dependent Bundle Properties

Voltage-dependent motility and/or stiffness changes of the hair bundle of OHCs could form an alternative mechanism to the electromotility of the cell body, which is now the main candidate for improving sensitivity and frequency selectivity in the cochlea (see also Holley, Chapter 7). Voltage-dependent bundle movements were most striking in hair cells in the isolated basilar papilla of the turtle, where they were correlated with membrane potential oscillations (due to an electrical resonance; see Section 3) at the cells' best frequency (Crawford and Fettiplace 1985). Spontaneous membrane potential oscillations with an amplitude of only about 2.5 mV corresponded to bundle movements with an amplitude of up to 20 nm. Depolarization corresponded to movement in the excitatory direction, toward the kinocilium, signifying positive feedback (opposite to the bundle movements described by Assad and Corey 1992; see above). It was concluded that the hair bundles contained a voltage-dependent force-generating mechanism. In isolated turtle hair cells under voltage clamp, depolarization to large positive potentials sometimes caused rapid, sustained bundle movements of unknown direction with amplitudes of less than 10 nm (Crawford, Evans, and Fettiplace 1989). This suggests that the force-generating mechanism may have been less powerful or less voltage-sensitive in the isolated cells, or that it might saturate for small voltage deviations from the resting potential. Fast, voltage-dependent bundle movements, with small amplitudes of just a few nanometers and sometimes opposite in polarity to those seen in the turtle hair cells, have only occasionally been observed in frog saccular hair cells (Howard and Hudspeth 1987; Denk and Webb 1992). In a feedback loop, it is not surprising that the phase relations may change with frequency and the state of unknown physiological processes that form part of the loop.

In mammalian hair cells, very little is known about voltage-dependent bundle movements or stiffness changes. In OHCs of the neonatal mouse cochlea, depolarization to large positive potentials gave some indication of an increase in bundle stiffness, based on the reduced slope of the relation between inferred force and transducer current. But the measured bundle movements did not support this inference, because little difference in resting position and stiffness was found between hyperpolarized and depolarized potentials (Kros et al. 1993a; Kros, Lennan, and Richardson 1995). A preliminary report in adult OHCs suggests that there may be a stiffness increase on depolarization (Evans and Dallos 1993a). This is clearly a very interesting area for further research.

2.3 Inferences from In Vivo Recordings of Receptor Potentials About the Transduction Mechanism

As noted in Section 1.1, the receptor potential is only in part determined by the underlying transducer current. For example, the membrane's time constant severely attenuates the frequency-following, periodic component ("AC response"; Russell and Sellick 1978) of the receptor potential above a few kilohertz, leaving, in IHCs, a steady depolarization, the "DC response"

(Russell and Sellick, 1978, 1983), which at lower frequencies can be thought of as the (generally depolarized) average of the receptor potential.

One would also expect time- and voltage-dependent basolateral conductances to distort the stimulus–response curve of receptor potential versus sound pressure with respect to that of transducer current versus force or displacement (see Fig. 6.2B,C). Because microelectrode recordings of small cells tend to introduce a leak conductance that shunts out time- and voltage-dependent conductances (see Section 3.1.2), this distortion is probably fortuitously not as large as it would be in an undisturbed cell. This allows some rough estimates to be made about the transducer currents as they occur in vivo.

2.3.1 Estimates of the Transfer Function and Total Transducer Conductance of Inner Hair Cells in Vivo

Microelectrode recordings from IHCs show large, sharply tuned receptor potentials in response to sound. When high-frequency cells from the first turn were stimulated with tones near their CF of 14–32 kHz, large depolarizing DC receptor potentials of up to about 20 mV were recorded (Fig. 6.7A), from fairly small (probably due to injury caused by the microelectrode; see Section 3.1.2) resting potentials of -25 to -45 mV (Russell and Sellick 1978; Cody and Russell 1987). Tones of a few hundred hertz, below the corner frequency of the cell membrane filter, resulted in periodic receptor potentials with larger depolarizing than hyperpolarizing phases (Fig. 6.7B). The relation between maxima and minima of the receptor potential and those of the sound pressure at the eardrum (Fig. 6.8A) looks qualitatively similar to the transfer functions of the transducer currents of isolated OHCs (Fig. 6.2B,C). Peak-to-peak receptor potentials were up to 36 mV (Russell and Sellick 1983).

At frequencies well below the cells' CF, basilar membrane responses are quite linear, so that the mechanical input to the hair bundles should be reasonably proportional to the sound pressure at the tympanic membrane, in contrast to the situation at and above the CF (Patuzzi and Sellick 1983). By observing the voltage responses to current steps injected through the microelectrode, and comparing these responses at the minima and maxima of the receptor potentials, the magnitude of the basolateral and transducer conductances could be estimated (Russell 1983; Russell and Kössl 1991). In some cases, in cells that appeared to have lost all their time- and voltage-dependent K^+ currents, it was possible to reverse the receptor potentials around $+80$ mV, the value of the endocochlear potential (Russell 1983), indicating that in vivo the transducer current is carried by K^+ ions, and that the difference between the endocochlear potential and the membrane potential provides the electrical driving force. The total available transducer conductance was estimated as about 8 nS, with 10% of the channels open at rest (Russell and Kössl 1991). These estimates, as

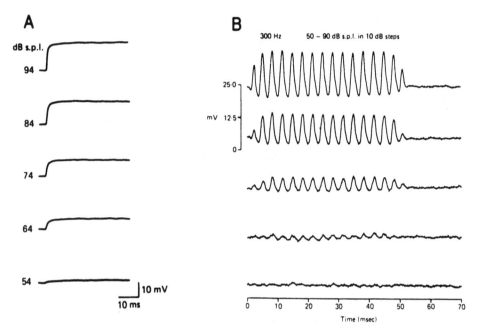

FIGURE 6.7 Receptor potentials of first-turn guinea pig IHCs in vivo. (A) Voltage responses to 16-kHz tones of different intensities, close to the CF of 18 kHz. The responses are depolarizing and consist almost completely of a DC component. (Modified by permission from Cody and Russell 1987, Fig. 4A.) (B) Voltage responses of a different first-turn IHC to 300-Hz tones of different intensities. The AC component is prominent, but the average voltage response (DC component) is depolarizing. (Reproduced by permission from Russell and Sellick 1983, Fig. 1A.) All responses are single traces.

well as the shape of the transfer function (Fig. 4B in Russell and Kössl 1991) agree quite well with the direct measurements made in neonatal hair cells in vitro (Section 2.1.1 and 2.1.2). A surprising deduction from the experiments of Russell and Kössl is that *hyperpolarizing* the IHC to about −80 mV would seem to double the fraction of the transducer channels open at rest. In experiments with isolated cells, by contrast, *depolarization*, by reducing adaptation, tends to increase the channels' resting open probability (see Section 2.1.3). Given the simplifications in the analysis used by Russell and Kössl (1991) to deduce the transducer functions (essentially, a steady-state analysis was used to model dynamic responses), their surprising conclusion needs further substantiation.

Periodic receptor potentials of about 30 mV have also been recorded in IHCs from the other three turns of the guinea pig's cochlea, with CFs centered around 0.3, 1.0, and 3.5 kHz, respectively (Dallos 1985a; Cheatham and Dallos 1993). At all stimulus frequencies, the depolarizing phase of the receptor potential was larger than the hyperpolarizing one (Dallos 1985a), as in the first-turn IHCs. No numerical estimates of the transducer

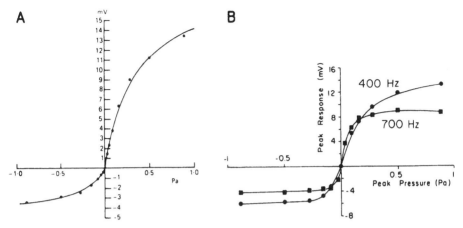

FIGURE 6.8 Transfer functions of the voltage responses of IHCs. (A) Peak depolarizing and hyperpolarizing voltage responses of the first-turn IHC of Fig. 6.7B, in response to 300-Hz tones. The *fitted* line is composed of two rectangular hyperbolae. (Reproduced by permission from Russell and Sellick 1983, Fig. 1B.) (B) Similar transfer functions for a third-turn IHC, obtained from peak depolarizing and hyperpolarizing voltage responses. Note that for tones of 700 Hz, near the CF, the dependence of the voltage on peak pressure is steeper, and the size of the maximum and minimum voltage responses is smaller, than for tones below the CF. Continuous lines drawn through points by eye. (Reproduced by permission from Dallos and Cheatham 1989, Fig. 2.) 1 Pa peak pressure corresponds to 91 dB SPL.

conductance were obtained, but some qualitative comparisons with the first-turn data can be made. Current-injection experiments (Dallos and Cheatham 1990) suggest that in these low-frequency cells, the transducer transfer function of the IHCs is not affected by current injections of comparable magnitudes to those used by Russell and Kössl (1991), but that the changes in the receptor potential can be fully explained by changes in activation of time- and voltage-dependent basolateral K^+ currents (see Section 3.1.2). Figure 6.8B illustrates very well how the relation between sound pressure and receptor potential is affected by other factors than the transfer function of the transducer current. The relation at 400 Hz may largely reflect the transducer's transfer function, with some distortion due to basolateral K^+ currents. At 700 Hz, near the CF of this third-turn cell, the slope around zero pressure is much steeper, and the receptor potential saturates for smaller pressures. The increased slope probably reflects a much increased mechanical input to the cell, due to whatever mechanism increases the sensitivity near the CF of a particular location along the length of the reticular lamina. That both the maximum and the minimum of the receptor potential are reduced at a frequency where the cell is more sensitive to low sound intensities can only be explained by a process occurring after the generation of the receptor current: low-pass filtering due to the cell membrane (Section 3.2.2).

2.3.2 Speculations About the Transfer Function of the Outer Hair Cells' Transducer Conductance In Vivo

Obtaining stable recordings from OHCs in vivo has proven extremely difficult (Dallos 1985a), and relatively few data are available. OHC resting potentials were much more hyperpolarized than those of IHCs [the mean resting potential of the OHCs was −54 mV in the 3rd turn (Dallos 1985a) and −83 mV in the first turn (Cody and Russell 1987)], thus providing a greater driving force for the transducer currents. Their maximal AC receptor potentials (generally up to 15 mV or so) were, however, only 25% to 50% of those of the IHCs (Russell and Sellick 1983; Dallos 1985a; Cody and Russell 1987). This is somewhat surprising, since OHCs probably have more transducer channels per hair bundle than IHCs (see Section 2.1.2), and indeed occasional OHC receptor potentials of over 30 mV have been reported (Dallos 1986). Whether this discrepancy occurs because recordings from OHCs with microelectrodes are of generally poorer quality (Dallos and Cheatham 1992) or because there may be more basolateral K^+ conductance turned on at rest in OHCs cannot be decided at the moment.

It is widely assumed, on numerous lines of indirect evidence, that the process in the OHC that mechanically enhances frequency selectivity and sensitivity in the cochlea is driven in vivo by its receptor potential. It is then expected that this mechanical process, whether it consists of electromotility of the cell body or of the hair bundle, will affect the motion of the hair bundle of the OHC and will thus modify the receptor potential that generated it in the first place, and so on. This feedback action is expected to be most evident at the CF and at low sound intensities.

The DC component of the OHC receptor potential, reflecting to some extent the point on the transducer transfer function around which the hair bundle operates, had a quite different level and frequency dependence than that of the IHC receptor potential (Fig. 6.9A,B). Whereas the DC component in third-turn IHCs was always positive, that of the OHC responses could be negative at low sound intensities at frequencies below the CF, becoming positive for louder sounds. At the CF, the OHC DC component was always positive. Similar observations have been made in second- and fourth-turn OHCs (Dallos and Cheatham 1992). High-frequency OHCs from the first turn appear to have no DC component in their receptor potentials at the CF, at least for sound intensities below 90 dB SPL (Russell, Cody, and Richardson 1986; Cody and Russell 1987). Above approximately 90 dB, a positive DC component developed (and decayed) slowly over tens of milliseconds, in contrast to the DC component of the receptor potential of IHCs or third-turn OHCs, which appeared and decayed as fast as the envelope of a tone burst (Dallos 1985a; Cody and Russell 1987). This slow component was attributed to accumulation of K^+ ions around the basolateral cell membrane. The first-turn OHCs responded to frequencies below 1 kHz with a hyperpolarizing, instantaneous DC component for medium

FIGURE 6.9 (A) Frequency response functions at different sound levels (in dB SPL next to the traces) of the DC receptor potential of a third-turn IHC. *Ordinate*: receptor potential (mV). *Abscissa*: frequency (kHz). Note the receptor potential is depolarizing at all frequencies and levels. (Reproduced by permission from Dallos 1985a, Fig. 10C.) (B) Same, for a third-turn OHC. The polarity of the DC receptor potential changes with level and frequency, although at the CF it is always depolarizing. (Reproduced by permission from Dallos 1985a, Fig. 11C.)

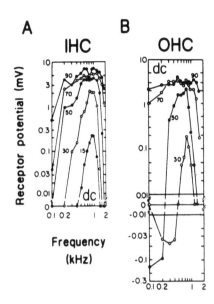

sound intensities (70–110 dB) and a depolarizing, slow DC component for sounds of more than approximately 110 dB (Cody and Russell 1987). Thus, both for first- and third-turn OHCs, the level and frequency dependence of their DC receptor potentials was more complicated than that of IHCs, and cannot be simply predicted from the transfer functions of the transducer currents (see Section 2.1.1 and Fig. 6.2B,C), or the receptor potentials (Russell, Cody, and Richardson 1986; Russell, Richardson, and Cody 1986) of neonatal OHCs.

The most likely possibility would seem to be that the hair bundles of the OHCs are differently biased (away from their "normal" resting position with about 10% of the transducer channels open) at different frequencies and sound intensities, by a force acting between the insertions of their stereocilia in the reticular lamina and the tectorial membrane. A hyperpolarizing DC component would arise if most channels were open in the "forced" resting position around which the bundle moves. The absence of an appreciable DC component could be explained if the bundle operated around the steepest slope of the transducer current's transfer function, where about 30% of the transducer channels are open (Kros, Rüsch, and Richardson 1992; Section 2.1.2), and which does not coincide with the "natural" resting position of the bundle. This would have the effect of maximizing the AC component of the receptor potential, and would be especially important for high-frequency OHCs if their AC receptor potentials themselves drove the mechanism that enhances sensitivity and frequency selectivity in the cochlea. This is because the AC component of their receptor potentials is very severely attenuated by the cell membrane filter (estimated to behave as a first-order low-pass filter with f_{3dB} around 1.2 kHz (Cody and Russell 1987; see also Section 3.4). The most optimistic estimate for the peak-to-peak size near neural threshold of the AC receptor potential in first-turn OHCs close to their CF is about 40 μV (Russell and Kössl 1992), which seems quite insufficient to drive the electromotility of the OHC's cell body (discussed by Holley, Chapter 7) in such a way that it could significantly affect cochlear mechanics: the estimated length change is about 0.03 nm, the radius of a hydrogen atom. This can be compared to an IHC DC receptor potential at neural threshold of 800 μV (Cody and Russell 1987).

The differences in receptor potentials at CF between first- and higher-turn OHCs are quite striking, provoking a lively debate about which (if any) bear most relation to physiology. Dallos (1985b) discussed changes that could occur in the resting and receptor potentials of hair cells during prolonged recording. One such change that occurred regularly in OHCs was an increase in resting potential (i.e., hyperpolarization), together with a decrease and a linearization (i.e., loss of the DC component) of the receptor potential. This was tentatively attributed to abnormal activation of a Ca^{2+}-dependent K^+ current $I_{K(Ca)}$, activated by Ca^{2+} flowing into the cell due to injury caused by the electrode penetration. The finding that the average resting potentials found in first-turn OHCs were about 30 mV more hyperpolarized than in those in the third turn (see above) may perhaps be due to this mechanism (see also Section 3.3.1). Another suggestion is that the presence of the recording electrode could alter the mechanics of (or around) the OHC that is penetrated. For recordings from first-turn OHCs, the microelectrode was pushed up through the basilar membrane from the scala tympani; for the other turns, it was inserted parallel to the reticular lamina from the scala media. Both these techniques might introduce a specific mechanical bias of the OHC (discussed by Dallos and Cheatham 1992). Until recordings from the first turn with the lateral approach become available, or a radically different way of recording voltage responses and/or transducer currents from hair cells in vivo is developed, this issue seems hard to resolve, highlighting perhaps how poorly the role of the OHC receptor potential is understood at present.

3. Basolateral Membrane Currents and Their Physiological Effects

The first indication that time- and voltage-dependent conductances in the basolateral membrane of hair cells might be an important determinant of the properties of their receptor potentials came from experiments on the basilar papilla of a turtle (Crawford and Fettiplace 1981). Using an isolated half-head preparation, they could stimulate the papilla acoustically and record receptor potentials from the hair cells with microelectrodes. It was found that the hair cells were tonotopically organized and sharply tuned, with CFs between 70 and 670 Hz at room temperature (Crawford and Fettiplace 1980). When small current steps were applied through the microelectrode, individual hair cells responded with a voltage oscillation at a frequency close to their CF. Moreover, the sharpness of tuning could be determined in response to acoustic stimuli, and also derived from the decay times of the current-induced voltage oscillations. The two measures correlated well when obtained in the same cell. This showed that in the turtle, frequency selectivity and tuning are governed by an electrical resonance in the hair cells themselves. One of the explanations offered for this electrical resonance was that it might be caused by time- and voltage-dependent ionic conductances, one of them a K^+ conductance (Crawford and Fettiplace 1981).

Upon isolation of the hair cells, the tuned electrical resonance persisted, supporting the idea that frequency selectivity is a property of the turtle's hair cells themselves. When these isolated cells were voltage-clamped, it was found that the interplay between a Ca^{2+}-dependent K^+ current and a sustained Ca^{2+} current could explain the sharply tuned electrical resonance in hair cells from the turtle's basilar papilla (Art and Fettiplace 1987) and the frog's sacculus (Lewis and Hudspeth 1983; Hudspeth and Lewis 1988a,b). Different resonance frequencies were achieved by varying certain aspects of the underlying membrane currents: in high-frequency cells, $I_{K(Ca)}$ had more rapid kinetics and was larger than in low-frequency cells (Art and Fettiplace 1987). The size of I_{Ca} also increased with increasing resonance frequency, keeping the ratio of the maximal available $g_{K(Ca)}$ and g_{Ca} constant across the range of CFs (Art, Fettiplace, and Wu 1993).

To establish whether this electrical tuning mechanism is a general property of vertebrate auditory and vestibular hair cells, membrane currents have been characterized and voltage responses to current injection recorded in hair cells isolated from other classes of vertebrates, including birds and mammals. Sharp electrical tuning has indeed been found in hair cells from the chick's basilar papilla, but only over a restricted frequency range, predicted to be about 400–1000 Hz in vivo (Fuchs, Nagai, and Evans 1988). Cells predicted to have higher CFs have not been tested yet, and very-low-frequency cells fired action potentials in response to current steps applied at their resting potential. The ionic currents underlying the electrically tuned cells were found to be much like those found in turtles and frogs (Fuchs and Evans 1990; Fuchs, Evans, and Murrow 1990). The avian low-frequency cells, by contrast, had little $g_{K(Ca)}$ but instead had a slow delayed rectifier (g_K) and an inward rectifier (g_{IR}) K^+ conductance. A recent review (Fuchs 1992) discusses basolateral membrane conductances found in various vertebrates from a comparative and functional perspective.

In mammalian auditory hair cells, convincing electrical tuning of individual isolated cells has not been demonstrated, and sharp tuning of IHCs is probably brought about in situ by an effect of nearby OHCs on cochlear mechanics. The various time- and voltage-dependent membrane currents that have recently been characterized in mammalian auditory hair cells are therefore likely to have physiological roles different from determining frequency selectivity directly. These currents as well as their possible functions are discussed in this Section. Early work on basolateral membrane currents in mammalian hair cells, as well as inferences made about these currents from voltage recordings with microelectrodes, have been reviewed in more detail than is possible here (Kros 1990).

3.1 Potassium Currents of Inner Hair Cells

When it became possible to isolate and voltage clamp IHCs, they were found to possess very large time- and voltage-dependent K^+ currents (Kros and Crawford 1988, 1989, 1990). The cells were isolated from a restricted region near the apical, low-frequency end of the cochlea, which was expected to be tuned to frequencies below 1 kHz. At present, it is not known whether high-frequency IHCs have different membrane currents. Figure 6.10A shows some of the properties of these currents. Setting the holding potential of the voltage clamp to about -80 mV, more hyperpolarized than the resting potential of around -65 mV, ensured that the K^+ conductances were deactivated. When the potential was stepped to potentials more depolarized than about -60 mV, outward membrane currents developed with a time course that became more rapid for larger voltage steps. The currents in all cells developed with a fast and a slow component, and reached a steady level within 30 ms (for small depolarizing voltage steps) to 5 ms (for large depolarizations). Figure 6.10B shows steady-state current-voltage curves for three IHCs. When the potential was returned to a constant level at the end of the depolarizing voltage steps, the currents elicited from the different depolarized levels relaxed to a common steady-state value, again with distinct fast and slow components (Fig. 6.11A). These current relaxations could be fitted very well with the sum of two exponentials, with time constants an order of magnitude apart (about 0.4 and 7 ms, respectively, for the cell of Fig. 6.11A). These large outward currents were shown to be carried by K^+ ions, because the tail currents that flowed when the membrane potential was returned to different levels after a large depolarization reversed around -76 mV, near the expected K^+ equilibrium potential, and away from the equilibrium potential of other ions in the intra- and extracellular solutions (Kros and Crawford 1990). A further argument that the currents are carried by K^+ ions is that they were abolished when intracellular K^+ ions were replaced by Cs^+ ions (Kros and Crawford 1988), which do not pass through K^+ channels and actually block them (Yellen 1984).

The fast and slow components of the outward currents have been shown to correspond to two distinct K^+ currents, $I_{K,f}$ and $I_{K,s}$, probably flowing through two different classes of ion channels. $I_{K,f}$, the fast K^+ current, could be selectively blocked by TEA in the extracellular solution, whereas $I_{K,s}$, the slow K^+ current, was selectively reduced by 4-aminopyridine

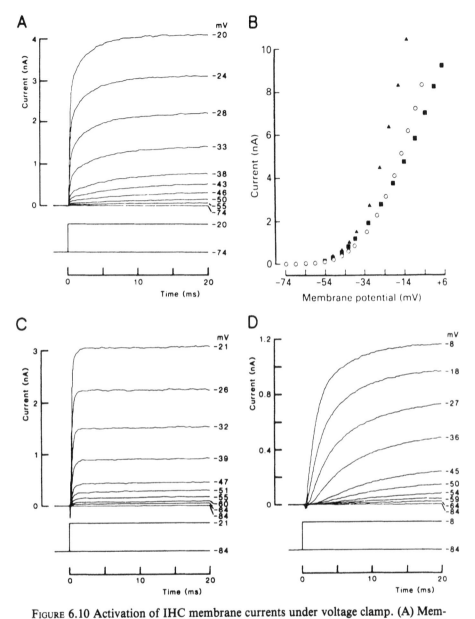

FIGURE 6.10 Activation of IHC membrane currents under voltage clamp. (A) Membrane currents with normal intra- and extracellular solutions. *Lower part*: outline of voltage-step protocol. *Upper part*: onset of currents elicited by steps to absolute membrane potentials shown by each trace. Resting potential −68 mV. Temp. 38°C. (B) Steady-state current–voltage curves for three IHCs studied with normal solutions, measured 40 ms after onset of the voltage steps. *Open circles*: same cell as (A). Temperature 36°–38°C. (C) Membrane currents with 10 mM 4-AP in the intracellular solution. This isolates $I_{K,f}$. Resting potential −67 mV. Temperature 35°C. (D) Membrane currents with 25 mM TEA in the extracellular solution. This isolates $I_{K,s}$. Resting potential −62 mV. Temperature 36°C. All currents are averages from 7 to 27 repetitions of the stimulus and are shown with leakage and capacity currents subtracted. Membrane potentials have been corrected for voltage drop across 3- to 6-mΩ residual series resistance. (A, C, D reproduced by permission from Kros and Crawford 1989, Figs. 1, 2, and 3. B from Kros and Crawford 1990, Fig. 2B.)

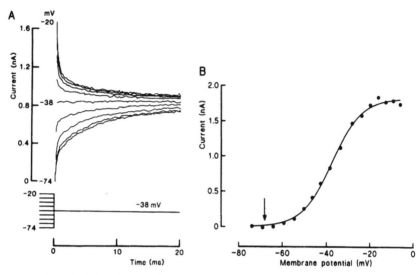

FIGURE 6.11 Deriving the overall activation range of the potassium currents of IHCs. (A) Membrane currents on returning the potential to a constant level after a 40-ms step to a range of potentials in nominal 10-mV increments. Voltage-clamp protocol indicated schematically beneath the current traces. Time zero indicates the instant of stepping the potential to −38 mV; prior to this, the currents had reached a steady state. Twelve repetitions per trace. Same cell and conditions as Fig. 6.10A. (B) Activation curve for this cell, derived from the experiment illustrated in A. *Ordinate*: instantaneous current at the onset of the voltage step to −38 mV. *Abscissa*: membrane potential of the preceding step. The *arrow* indicates the resting potential (−68 mV). The first-order Boltzmann fit shows that the conductances are half-activated at −37 mV, and increase e-fold per 6.8 mV at hyperpolarized potentials. Temperature 36°C. (A from Kros and Crawford 1990, Fig. 4A. B modified from Kros 1989, Fig. 4.7B.)

(4-AP) in the extra- or intracellular solution. Figure 6.10C,D shows pharmacologically isolated fast and slow K^+ currents. Both currents were found to activate with a sigmoidal time course on depolarization, and to deactivate according to a single exponential. Their time courses could be well matched by a model that supposes the underlying ion channels to have two closed states and a single open state (Kros and Crawford 1990). In most cases, neither of the currents displayed any inactivation.

From the onsets of the current relaxations of Figure 6.11A, it is possible to deduce the range of membrane potentials over which the ion channels that underlie the K^+ currents vary their probability of opening: this relation is called the activation curve. This can be done because the current flowing at the instant of a step to a constant potential (and thus a constant driving force for the K^+ ions) is proportional to the number of ion channels opened by the preceding voltage step. The problem that the underlying single-channel conductance may vary with the driving force (precluding direct derivation of the activation curve from the steady-state current–voltage curve) is sidestepped in this manner (Frankenhaeuser 1963). The activation curve in Figure 6.11B shows that almost all the channels are closed at potentials more negative than about −60 mV, and sharply increase their probability of opening over the range of potentials up to about −20 mV. Above this potential, no further channels open, most likely because the channels' probability of being open is approaching 1.

Activation curves were constructed for $K_{K,f}$ and $I_{K,s}$ studied separately under pharmacological isolation, and both were found to be activated over a similar potential range as the total K^+ conductance (Kros and Crawford 1990). This membrane potential range of activation is approximately the same as the range of the receptor potentials measured in vivo. Both $I_{K,f}$ and $I_{K,s}$ should therefore influence the properties of the receptor potential of IHCs (see Section 3.2).

3.1.1 Classification of the Fast and Slow Potassium Currents

The fast K^+ current $I_{K,f}$ has a large number of similarities with the Ca^{2+}-activated K^+ current ($I_{K(Ca)}$) that is responsible for electrical tuning in hair cells from a variety of nonmammalian vertebrates: for example, cells from the bullfrog sacculus (Lewis and Hudspeth 1983; Hudspeth and Lewis 1988a), the turtle basilar papilla (Art and Fettiplace 1987; Art, Fettiplace, and Wu 1993), and the chick basilar papilla (Fuchs and Evans 1990). Similarities include block by TEA and resistance to block by 4-AP, strong outward rectification of the instantaneous current–voltage relationship (i.e., the single-channel conductance depends strongly on the driving force, favoring flow of outward current), sigmoidal activation and exponential deactivation kinetics, absence of inactivation, and a large single-channel conductance [about 120 pS in cell-attached patches on IHCs with a high K^+ concentration in the patch-pipette (Kros 1989; Kros and Crawford 1989)]. Among classes of K^+ channels, such large single-channel conductances of over 100 pS have only been reported for large-conductance K(Ca) channels (Hille 1992, p. 330). However, although $I_{K,f}$ could be reversibly reduced in size by extracellular cadmium, a Ca^{2+} channel blocker, removing extracellular Ca^{2+} did not reduce either of the K^+ currents in guinea pig IHCs (Kros and Crawford 1990), suggesting that the opening probability of the channels may be modulated by Ca^{2+} released within the cell or exclusively by their intrinsic voltage sensitivity (Barrett, Magleby, and Pallotta 1982). Perhaps commensurate with this, there was an e-fold increase in conductance per 6 mV of depolarization at hyperpolarized potentials, a voltage sensitivity of activation only half of the extremely steep activation found in frog and turtle hair cells (information concerning voltage sensitivity is at present not available for avian cochlear hair cells). Recent work in which the voltage of guinea pig IHCs was slowly ramped from −150 to +100 mV demonstrated that the K^+ currents could be reduced in the absence of extracellular Ca^{2+} (Housley et al. 1993). Although this method does not distinguish between effects on $I_{K,f}$ and on $I_{K,s}$, it strengthens the evidence that at least one of the K^+ currents of IHCs is Ca^{2+}-dependent.

In turtle auditory hair cells, the kinetics of $I_{K(Ca)}$ are the most important determinant of the frequency of the electrical resonance, the frequency increasing as the kinetics become more rapid (Art and Fettiplace 1987). Near body temperature, the principal (slowest) time constant describing the activation of $I_{K,f}$ (about 0.3 ms for small depolarizations) was about three to four times faster than that of the fastest (i.e., resonance frequencies above 300 Hz) turtle hair cells when the latter were studied at room temperature (Art and Fettiplace 1987), but when the IHCs were cooled to room temperature, the time constants (about 1.2 ms) became comparable (Kros and Crawford 1990). The kinetics of $I_{K,f}$ in the IHC are thus faster than the fastest $I_{K(Ca)}$ in turtle hair cells only by virtue of the normally higher body temperature of mammals.

The slow K^+ current $I_{K,s}$ has the characteristics of a delayed rectifier (I_K), with a smaller single-channel conductance than $I_{K,f}$ (Kros and Crawford 1989). It has some similarities to the mother of all delayed rectifiers, that of the squid giant axon, first described by Hodgkin and Huxley (1952), in its nearly linear instantaneous current–voltage curve (i.e., the single-channel conductance does not vary with the driving force), its absence of inactivation, its resistance to block by extracellular TEA, and its susceptibility to block by 4-AP, probably acting from the intracellular side of the membrane (summarized by Rudy 1988). $I_{K,s}$ could be reduced in size by dendrotoxin I (or toxin I), which blocks some delayed rectifiers, but was not affected by apamin, which blocks small-conductance K(Ca) channels (Dreyer 1990). Its limiting voltage sensitivity of an e-fold increase in conductance for a depolarization of 9 mV was less steep than that of $I_{K,f}$. A similar, TEA-resistant, delayed-rectifier K^+ conductance has been described in tall hair cells (functionally equivalent to mammalian IHCs) of the chick basilar papilla, where it is the main K^+ conductance in very low-frequency cells situated near the apex of the papilla, which respond with slow, repetitive action potentials to current injection (Fuchs and Evans 1990). Somewhat further removed from the apical end of the papilla, cells with roughly equal proportions of I_K and $I_{K(Ca)}$ were found that exhibited a rather low-quality electrical resonance, whereas tall hair cells located about halfway along the length of the papilla had mostly $I_{K(Ca)}$ and showed a clear electrical resonance. A slow delayed rectifier appeared also to be the main K^+ current in turtle hair cells best responding to very low frequencies (Art, Fettiplace, and Wu 1993). Some vestibular hair cells, including mammalian type II vestibular hair cells, also incorporate a delayed-rectifier K^+ conductance, together with other K^+ conductances (Lang and Correia 1989; Steinacker and Romero 1991; Griguer et al. 1993). In these cells, however, I_K invariably showed a slow, voltage-dependent inactivation that was not usually evident in the auditory cells. Another difference between $I_{K,s}$ in the

IHC and I_K in the mammalian type II vestibular hair cell is that the block by extracellular 4-AP occurs much more rapidly and is much more readily reversible in the vestibular cell (Griguer et al. 1993), suggesting a different 4-AP receptor. The underlying channels are thus probably different subtypes of I_K, probably serving different functions. In conclusion, the complement of K^+ currents in low-frequency IHCs seems to be qualitatively most similar to that of fairly low-frequency tall hair cells in the avian cochlea.

3.1.2 Magnitude of the Fast and Slow Potassium Conductances

The size of the K^+ conductances that can be activated over the membrane potential range of the receptor potential determines how profoundly they affect the shape of the receptor potential. From the activation curve of Figure 6.11B, a maximum conductance, measured at -38 mV, of 48 nS was calculated. By fitting single exponentials to both the activating and the deactivating currents, the contributions of $g_{K,f}$ and $g_{K,s}$ could be estimated separately, as about 60% and 40% of the total conductance, respectively. Up to 86 nS of conductance was found from tail currents at about -30 mV in IHCs, 70% of which was $g_{K,f}$ (deduced from data for Fig. 4B in Kros and Crawford 1990). The magnitudes of these conductances are comparable to maximum conductances measured similarly (from the activation curves) at around -48 mV in turtle auditory hair cells, where the size of $g_{K(Ca)}$ increases systematically with the cells' resonance frequency. From Figure 5B in Art, Fettiplace, and Wu (1993), the largest g_{max} of $g_{K(Ca)}$ is 82 nS, for a cell with a high resonance frequency of about 320 Hz. The very low-frequency turtle hair cells which appear to have mainly I_K, have a maximum conductance of about 15 nS. No comparable activation curves are available for tall hair cells of the chick, but a comparison can be made of the amount of current activated at a particular potential. In IHCs, 8 to 16 nA of outward current flowed near 0 mV, about 80% of which was $I_{K,f}$. In the chick, the currents to similar potentials rarely exceeded 1 nA.

The slope conductances of the steady-state current–voltage curves in IHCs are very large: 250–500 nS near 0 mV, reflecting the steep outward rectification of the instantaneous current–voltage curve of $I_{K,f}$. When the K^+ conductances are not activated, near -80 mV, only 0.5–2.5 nS of linear leak conductance is present. These values can be compared with earlier measurements of slope conductances in current-injection experiments with microelectrodes. Russell (1983) found linear current–voltage curves between -150 and $+40$ mV membrane potentials, and slope conductances (equal to chord conductances in this linear case) of only 10–13 nS. This suggests an increased leak conductance and a complete functional absence of time- and voltage-dependent channels under these recording conditions. An outward rectification seen in some cells was attributed to the recording system. Later, Russell, Cody, and Richardson (1986) reported slope conductances in vivo of 13–40 nS for adult IHCs with quite depolarized resting potentials between −25 and −45 mV, reducing to about half upon hyperpolarization to beyond −60 mV in some cells. No further increase in conductance upon depolarization from the resting potential was noted. This evidence for at least a small fraction of the K^+ conductances found with whole-cell patch recording was in part based on the original data of Russell (1983), presumably by attributing the rectification to the cell membrane on this occasion. Similarly, Russell and Kössl (1991) found slope conductances of 17 to 21 nS negative to −80 mV, increasing to 57 to 65 nS positive to −35 mV.

These discrepancies between whole-cell and microelectrode recordings from IHCs illustrate some problems associated with microelectrode recordings from small cells. In mammalian hair cells the discrepancies between the two techniques seem particularly severe; for less dramatic differences in other cell types, see, for example, Fenwick, Marty, and Neher (1982) and Staley, Otis, and Mody (1992). Upon penetrating a cell with a microelectrode, a large leak conductance is introduced, which would tend to linearize effects of voltage-dependent conductances and may cause a more depolarized resting potential if the leak's reversal potential is close to 0 mV. A large leak conductance and a depolarized resting potential may in turn rapidly cause changes in the ionic composition of the cell, for example, loading by Na^+ and Ca^{2+} ions through the leak itself or through voltage-dependent Na^+ or Ca^{2+} channels. This may reduce the conductance through K^+ channels by reducing the concentration of intracellular K^+. Another possibility is that the gating behavior of channels is altered, for instance, by changes in Ca^{2+}-dependent processes, such as some forms of phosphorylation.

A secondary disadvantage of microelectrode recordings in small cells is that very high-resistance electrodes must be used; otherwise damage would be even larger. For hair cell recordings in vivo, electrode resistances between 80 and 200 MΩ are typical (e.g., Russell and Sellick 1983; Dallos 1985a). Such high-resistance (> 50 MΩ) electrodes may have very nonlinear current--voltage curves themselves for small injected currents of less than 1 nA (Purves 1981), and these are likely to change after penetration of a cell (Dallos 1986). This may partly explain the difficulty in deciding whether IHCs studied in vivo have linear current–voltage curves or show a modest outward rectification. Qualitative support for linear current–voltage curves in vivo was given by Brown and Nuttall (1984), and for a degree of outward rectification by Nuttall (1985), Dallos (1986), and Dallos and Cheatham (1990), all on the basis of changes in the receptor potential with current injection.

3.1.3 Development of Inner Hair Cell Potassium Currents

$I_{K,f}$ was absent in IHCs in cochlear cultures of neonatal mice but was always found in cells isolated from mice after the onset of hearing at about 11 days after birth (Kros, Rüsch, and Richardson 1991). This suggests that $I_{K,f}$ is particularly important for the normal functioning of IHCs. Interestingly, in chick cochlear hair cells, the appearance of $I_{K(Ca)}$ is also correlated with the maturation of auditory function (Fuchs and Sokolowski 1990).

The K$^+$ current in neonatal IHCs ($I_{K,neo}$), activating above about -45 mV, is much slower and smaller than $I_{K,f}$, reaching a size of about 2.5 nA near 0 mV (upon depolarization from a holding potential near -80 mV). Neonatal OHCs seem to possess an identical K$^+$ current, but of a somewhat smaller size (Rüsch et al. 1991). $I_{K,neo}$ is a delayed rectifier, but appears different from $I_{K,s}$ in that it can be blocked by external TEA and exhibits slow inactivation.

The cochlear explant of the neonatal mouse cochlea offers a direct comparison between microelectrode and whole-cell patch-clamp recordings in the same preparation. The resting potential of IHCs was about -62 mV with patch clamp and about -38 mV with microelectrodes. Slope conductances under whole-cell voltage clamp were <1 nS below -80 mV and reached a maximum of about 75 nS around -15 mV (Kros and Rüsch, in preparation). Current injection through the microelectrode showed a conductance varying from 17 nS below -40 mV to a maximum of 31 nS (Russell and Richardson 1987). This reiterates the need for caution in interpreting hair cell responses obtained with microelectrodes.

3.2 Physiological Actions of the Potassium Currents of Inner Hair Cells: Shaping the Receptor Potential

Since the K$^+$ currents in IHCs are most evident in patch-clamp recordings of isolated cells, it is best to use these as a basis for considering what these currents might do to the receptor potential when current is injected through the transducer channels. Recordings of transducer currents or receptor potentials in isolated IHCs are not yet available, but current injection through the patch-pipette gives an approximation to the effect of opening the transducer channels.

The K$^+$ currents will largely determine the negative resting potential of the cells and so contribute (with the positive endocochlear potential of about $+80$ mV) to the driving force for the transducer current. K$^+$ ions flowing into the cell through the transducer channels can also conveniently flow out down their electrochemical gradient through K$^+$ channels. This is why Cl$^-$ channels would be less suitable to set the resting potential in these cells.

Figure 6.12A shows voltage responses to currents steps in an isolated IHC. These can be thought of as corresponding to responses to either steps of transducer conductance, or alternatively auditory stimuli of many kilohertz well above the cutoff frequency of the membrane's low-pass resistance–capacitance (RC) filter, which would cause a net depolarization

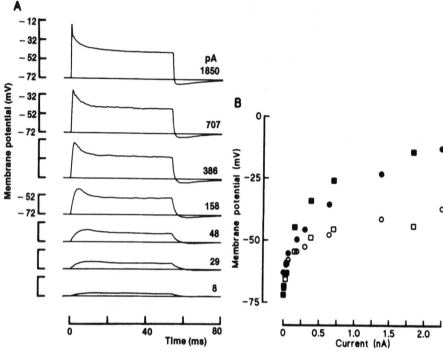

FIGURE 6.12 IHC voltage responses to injections of depolarizing current and nonlinear membrane behavior. (A) Changes in membrane potential in response to current steps starting at time zero. All traces corrected for voltage drop across total series resistance. Records are single traces, except the top one, which is averaged from five repetitions. Resting potential −72 mV, indicated by the *horizontal lines*. Temperature 37°C. (B) Compressive nonlinearity of the voltage responses of the cell of A (*squares*) and another cell (*circles*). *Closed symbols*: peak voltage responses. *Open symbols*: steady-state voltage responses. (Modified from Kros and Crawford 1990, Figs. 13A and 14.)

due to the rectification seen in the transducer transfer function (see Section 2.1.2). Small currents cause a slow depolarization, which reaches a peak within about 10 ms and declines to a steady-state potential within another 10–20 ms. Larger currents reduce the time to peak to under a millisecond, again followed by a slow reduction of the potential to a steady state. For the largest currents, a spike with a small damped oscillation occurred. It has been shown that the reduction of the time constant of the depolarization and the associated spike were caused by $I_{K,f}$ and that the slow decline of the potential following the spike was due to the action of $I_{K,s}$ (Kros and Crawford 1990).

3.2.1 Inner Hair Cells Have Very Little Electrical Resonance

When viewed on an expanded time scale, the spikelike responses of IHCs to large current injections of 0.7–2.3 nA are seen to be strongly damped oscillations, with frequencies between 700 and 1400 Hz. These frequencies are in good agreement with simple predictions based on the properties of $g_{K,f}$ around the voltage level at which the resonance occurs, and are close to the range of best frequencies of less than 1000 Hz which the cells are expected to manifest in vivo (eq. 3 in Kros and Crawford 1990). If IHCs have a maximum transducer conductance of 13 nS (Section 2.1.2), then with a driving force of 100–150 mV (endocochlear potential minus membrane potential), the maximum transducer current would be about 1.3–2 nA, so the damped oscillations are observed for current levels that could be achieved by the transducer.

The quality factor (approximately equal to the number of sinusoidal oscillations elicited by a step or impulse disturbance of a resonator) of the resonance observed in vitro was only about 1.0–1.5 – close to critically damped and much less sharply tuned than turtle hair cells operating at similar frequencies, which had quality factors of up to 5–10 (Q_{3dB}; Crawford and Fettiplace 1980). However, auditory nerve fibers with CFs below about 900 Hz in the guinea pig cochlea are also much less sharply tuned (Evans 1972) than fibers with corresponding CFs in the turtle (Crawford and Fettiplace 1980). Nevertheless, isolated IHCs are clearly less sharply tuned than nerve fibers of similar CF. For example, impulse responses derived for cochlear nerve fibers in the cat showed about four to seven peaks and troughs for fibers with CFs below 1 kHz (Evans 1989), compared with up to three in IHCs. This suggests that the tuning of low-frequency IHCs in vivo results more from the micromechanical properties of the cochlea than from intrinsic properties of the cells themselves.

The electrical tuning in isolated IHCs is not very sharp, for at least two reasons. First, $I_{K,s}$ interferes with the tuning conductance $I_{K,f}$, especially for small depolarizing currents. Tuning for small current steps could be improved by blocking $I_{K,s}$ (Kros 1989). Analogously, avian cochlear hair cells with a mixture of $I_{K(Ca)}$ and I_K showed poor electrical tuning (Fuchs and Evans 1990). Second, I_{Ca} may be reduced in vitro due to isolation damage (see Section 3.5), thus reducing the quality of the resonance (Hudspeth and Lewis 1988b). In this respect it is interesting that Dallos (1984, 1985a, 1986) found some indirect evidence in favor of the possible existence of an electrical resonance in apical IHCs in vivo. He argued that cells with CFs of 800–1000 Hz might possess an electrical resonance with a frequency of about 1200 Hz. Similarly, low-CF auditory nerve fibers often show a secondary minimum about half an octave to an octave above the CF if the tuning curve is determined with sufficiently fine frequency resolution (Liberman and Kiang 1978; Evans 1989). All that these findings really suggest is the conceivable presence of two resonance mechanisms in the apex of the cochlea, one of which might reflect an electrical resonance in the IHC. At present, however, the balance of evidence indicates that electrical resonance in IHCs plays only a minor role, if any, in the frequency selectivity in the apex of the cochlea. For basal, high-frequency IHCs (about whose electrical properties very little is known at present), electrical tuning is very unlikely, because physically unrealizable demands might be placed on their ionic conductances (Roberts, Howard, and Hudspeth, 1988; Wu et al. 1995). It seems instead that frequency selectivity in the mammalian cochlea is achieved in a radically different manner than it is in hearing organs of most other vertebrates. Rather than being private to each hair cell, it is probably a distributed property, due to an interaction of OHCs and cochlear micromechanics, which sharpens up tuning in the (intrinsically rather poorly tuned) IHCs (e.g., Brown and Nuttall 1984; Ruggero and Rich 1991).

3.2.2 Effects of the Potassium Currents on the Membrane Time Constant of the Inner Hair Cells

When the K^+ conductances are not activated, the conductance of the IHC membrane is only 0.5–2.5 nS, some of this probably contributed by the resting transducer conductance. With a membrane capacitance of 10 pF, this results in a low-pass filter with an RC-time constant of 4–20 ms. For low-intensity sounds, this would give a corner frequency (f_{3dB}) of 8–40 Hz, above which AC receptor potentials would be progressively attenuated at a rate of 6 dB/octave (assuming first-order filtering: Horowitz and Hill 1989). Phase locking in auditory nerve fibers (defined as the ability of the fibers to preserve the periodicity of an auditory stimulus in their firing pattern) must depend on periodic neurotransmitter release, following the AC receptor potential, and is already quite good at very low stimulus intensities (e.g., Palmer and Russell 1986). In the guinea pig, phase locking only begins to attenuate at around 600 Hz (Palmer and Russell 1986), as does the AC receptor potential of the IHCs (Russell and Sellick 1983; Palmer and Russell 1986), which suggest phase locking is also limited by the IHC membrane time constant.

The K^+ conductances of the IHCs reduce their membrane time constants in a voltage-dependent manner. Assuming the resting potential is somewhat more depolarized in vivo than in vitro (because the resting transducer current and Ca^{2+} current are likely to be larger in vivo), say -55 mV, the slope conductances of the K^+ currents (including the small linear leak conductance) around this potential are 23–43 nS for the cells of Figure 6.10B. This makes the time constant fall to 170–330 μs, and f_{3dB} to 480–940 Hz (Kros and Crawford 1990), in good agreement with the findings of Palmer and Russell (1986). The K^+ conductances are thus necessary to allow phase locking in the auditory nerve, important for localization of low-frequency sounds, up to reasonably high frequencies, at which intensity differences due to acoustic shadows of the head become detectable. The coding and discrimination of vowel sounds may also require phase locking (Young and Sachs 1979; Palmer 1990).

3.2.3 The Potassium Currents Contribute a Compressive Nonlinearity to the Input–Output Function of the Inner Hair Cells

The voltage-dependent activation of the K^+ currents results in a compression of voltage responses to injections of large depolarizing currents, by increasing the conductance of the cell membrane (Fig. 6.12B). This compression occurs for injected currents well within the range of the expected maximum IHC transducer current of up to 2 nA (Section 3.2.1), thus enhancing the compressive nonlinearity of transducer channel gating (Fig. 6.2B,C; Section 2.1.1). This reinforces the relatively detailed representation of faint sounds in the firing of auditory nerve fibers, combined with a large dynamic range.

Because of the time dependence of the activation, the compression is more pronounced when the potentials have reached a steady state than at the peak of the responses. This would cause receptor potentials in response to long-duration tone bursts to reduce with time, a form of adaptation that should be noticeable in IHC receptor potentials in vivo and in the firing of the auditory nerve fibers. In contrast to transducer current adaptation, which may only occur at frequencies below a few hundred hertz (see Section 2.1.3), this type of adaptation would also affect DC receptor potentials at high frequencies where AC potentials are attenuated. The time constant of this adaptation would mainly be determined by the time course of activation of $I_{K,s}$. For the cell of Figure 6.12A, the time constant varied from 11 ms (48 pA) to 2 ms (1850 pA), in good agreement with the range of principal time constants of activation of $I_{K,s}$ shown by Kros and Crawford (1990). No adaptation of any kind has been described in IHC receptor potentials recorded in vivo, but this is not surprising given the very reduced manifestation of any time- and voltage-dependent conductances in these circumstances (Section 3.1.2). The firing rate of afferent auditory nerve fibers in response to a long tone burst declines with a number of exponential components, two of which occur within the first 100 ms (Westerman and Smith 1984) and are called rapid (time constant <1–10 ms) and short-term (time constant about 60 ms) adaptation. The time constant of rapid adaptation decreased on average with increasing sound intensity, whereas that of short-term adaptation did not. Rapid adaptation in the auditory nerve may thus well be caused by the activation of $I_{K,s}$.

3.3 Potassium Currents of Outer Hair Cells

3.3.1 Inferences from Microelectrode Recordings In Vivo

OHCs studied in vivo have generally very hyperpolarized resting potentials near −70 mV (Dallos, Santos-Sacchi, and Flock 1982; Russell and Sellick 1983; Dallos 1985a; Cody and Russell 1987), in contrast to the IHCs, which tend to be rather depolarized (see Section 3.1.2). Dallos (1985a,b) observed that resting potentials were generally unstable and tended to change after initial penetration of the cells, reaching a steady-state value different from the initial membrane potential. The general trend was for OHCs to become more hyperpolarized during a recording (from -54 to -71 mV), whereas for IHCs the opposite was true (from -32 to -20 mV). Dallos (1985b) correlated changes in the receptor potential with these changes in membrane potential, and the most common correlations were that a reduction in receptor potential size occurred in IHCs when they became further depolarized, and in OHCs when they hyperpolarized. A plausible explanation offered by Dallos (1985b) is that a leak conductance develops with time in both cells, but that of the IHC is nonselective (reversal potential near 0 mV), whereas that of the OHC is mainly K^+-selective, due to an abnormally high injury-induced intracellular Ca^{2+} concentration that causes massive opening of K(Ca) channels. Consistent with this interpretation, he also found that the RC-time constant of the cell membrane low-pass filter became very small when OHCs reached extremely negative potentials. Sunose et al. (1992) presented evidence that leakage of hyperosmotic K^+ salt from the microelectrodes might also contribute to this negative shift in resting potential, by making the K^+ equilibrium potential more negative. Cody and Russell (1987) observed very hyperpolarized resting potentials in basal turn OHCs (mean -83 mV) and did not observe a systematic change in resting potential equivalent to that reported by Dallos, perhaps pointing to a greater fragility or more rapid K^+ loading in the (smaller) OHCs in the basal (first) turn compared to those from more apical regions. Basal-coil OHCs in vivo did indeed have a very large slope conductance at the resting potential: about 100 nS at -70 mV. This could be reduced to 20 nS on hyperpolarization to -75 mV and seemed to increase to about 200 nS on further depolarization to -68 mV (Fig. 13 in Russell, Cody, and Richardson 1986). Interestingly, hyperpolarization to below about -90 mV provided some evidence for an inward rectifier K^+ current (I_{IR}) which is often found in nonmammalian vertebrate hair cells with very negative resting potentials (reviewed by Fuchs 1992). A small I_{IR} can also be demonstrated in neonatal IHCs and OHCs (see Section 3.7).

3.3.2 Voltage-Clamp Recordings In Vitro

Whereas in the case of IHCs, the leak conductance inherent to recording with microelectrodes can be reduced to around 1 nS when using patch-pipettes, the same is apparently not possible for OHCs, suggesting that the latter are more severely damaged than IHCs upon isolation. This suggestion is reinforced by the zero current potential under voltage clamp (equivalent to the resting potential in current clamp), which was about -20 to -30 mV on breaking the patch at the onset of whole-cell recording and slowly reached a more negative (about -45 to -75 mV) steady-state potential (Housley and Ashmore 1992). This suggests that the cells were initially loaded with Na^+, which was gradually replaced by K^+ diffusing in from the patch-pipette (Santos-Sacchi and Dilger 1988). Using a fluorescent Na^+ indicator, Ikeda et al. (1992) found that isolated OHCs contained about 110 mM Na^+. The same group inferred intracellular K^+ concentrations of about 33 mM with microelectrode recordings in vitro (Sunose et al. 1992). In IHCs, by contrast, resting potentials became only about 5 mV more negative during the first few minutes of whole-cell recording (Kros 1989).

Because of these difficulties, and perhaps because more time has been spent investigating the voltage-dependent motility (Holley, Chapter 7), the time- and voltage-dependent currents of OHCs are at present less well characterized and quantified than those of IHCs. This is a pity, because these current are the link between the transducer current induced by stimulating the bundle and the membrane potential changes that set the cells in motion (Evans and Dallos 1993b). Santos-Sacchi and Dilger (1988) found, in cells from the apical coils of the guinea pig cochlea, apart from the above-mentioned leak, evidence for a partially inactivating K^+ current, activating on depolarization above -40 mV, and reaching a size of about 1 nA at large positive potentials. Housley and Ashmore (1992) recorded from OHCs of all coils, and found the conductance dominated by a large leak. Nevertheless, they found some interesting differences between OHCs from basal and apical regions (Fig. 6.13). In Figure 6.13A, currents under voltage clamp, from a holding potential of -50 mV, can be seen. Their time course is rather complicated but is similar for all three cells, which originate from the base, midregion, and apex of the cochlea (from top to bottom). The initial transient could in part be caused by the cells' nonlinear capacitance (Holley, Chapter 7). The current that fails to deactivate at hyperpolarized potentials is a cation-selective leak conductance (I_L) with a reversal potential near 0 mV. The size of this leak increased progressively with time within 3 hours following isolation of the cells and was correlated with pathological increases in intracellular free Ca^{2+} (Ashmore and Ohmori 1990; Housley and Ashmore 1991, 1992). The current that does deactivate below about -90 mV is called $I_{K,n}$ by the authors and is K^+-selective. This unusual current is fully activated at -50 mV and may well be identical to the K^+-selective leak conductance postulated by Dallos (1985b). Finally, a minuscule, noninactivating $I_{K(Ca)}$, reaching a maximum size of about 0.5 nA, activates above -35 mV. This latter current would appear to be similar to that found by Santos-Sacchi and Dilger (1988), although the difference in activation behavior is puzzling. The smaller the cell, the larger are these three currents described by Housley and Ashmore (1992). This can be seen also in the steady-state current–voltage curves of Figure 6.13B. The slope input conductance around -70 mV ranged from about 50 nS for short, high-frequency cells to about 4 nS for tall, low-frequency cells.

It is perhaps surprising that the total size of these three membrane conductances in adult OHCs is actually smaller than the pure K^+ conductance of neonatal OHCs: neonatal cells develop about 3 nA of K^+ current

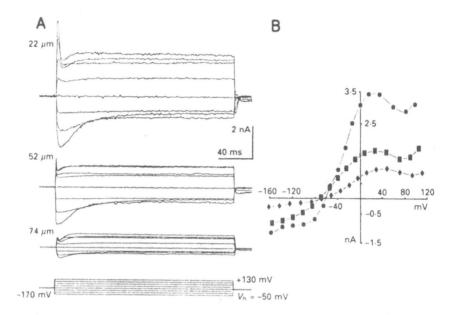

FIGURE 6.13 Membrane currents under voltage clamp for three OHCs, isolated from the basal (length 22 μm, ●), mid (52 μm, ■), and apical (74 μm, ♦) regions of the cochlea. (A) Current records in response to voltage steps, indicated schematically below, from a holding potential of −50 mV. (B) Current-voltage relationships. Leak currents not subtracted. Resting potentials near −60 mV. Room temperature. (Reproduced by permission from Housley and Ashmore 1992, Fig. 3.)

near +20 mV (Rüsch et al. 1991). This is despite the fact that the capacitance (proportional to the cell's surface area) of the neonatal cells is only about 5.5 pF, whereas the smallest cell studied by Housley and Ashmore (1992) had a capacitance of about 12 pF. If this is representative of what happens under physiological conditions, OHCs must actually reduce their ion channel density considerably during development. Patch-clamp recordings from OHCs in the isolated temporal bone of the adult guinea pig have recently provided a promising technical advance (Mammano, Kros, and Ashmore 1995). OHCs studied in situ in the third turn were less leaky and had larger potassium currents than OHCs that had been isolated from the same region of the cochlea.

3.4 Possible Physiological Roles of Potassium Currents in Outer Hair Cells

Given the poor physiological state of isolated OHCs, generally depolarized and loaded with Na$^+$ at the onset of recording, it is difficult to determine the function of the membrane currents that are observed under these conditions. The main effect of the leak conductance I_L and also, because of its very negative activation range, $I_{K,n}$, would be to short out the transducer current, thus reducing the size of the receptor potential, which should drive the cell's motile mechanism. Some speculations will nevertheless be attempted.

The voltage-driven motility of the OHC body is nonlinear: the cell length is a sigmoidal function of the membrane potential, described by a first-order Boltzmann curve with a $V_{1/2}$ near -25 mV (Santos-Sacchi 1992). One consequence of this is that, provided the resting potential is sufficiently far away from $V_{1/2}$, the motile responses to a sinusoidal voltage change manifest a tonic shortening as well as a frequency-following, phasic component (Holley, Chapter 7). This (labile) tonic component may well prove to be functionally important. If it is true that first-turn, high-frequency OHCs produce symmetrical receptor potentials without a DC component at their CF (Section 2.3.2), then the operating point on the *electromechanical* transfer function relating cell length changes to receptor potential alone determines whether a tonic length change occurs. In lower-frequency, third-turn cells, the operating point, at CF, on the *mechanoelectrical* transfer function relating receptor potential to bundle displacement would by itself contribute to a tonic shortening of the OHC. It is therefore important to know the "true" resting potential of OHCs in vivo. Ideally, one would measure the initial potential in a whole-cell patch recording in vivo, to avoid problems due to injury by microelectrodes and isolation of the cells. At present, the potential upon penetration with a microelectrode in vivo is the best estimate available: between about -70 mV, in the best-quality third-turn OHC recordings (Dallos 1985a), and -85 mV, in the first turn (Cody and Russell 1987), thus allowing a tonic component to any OHC length changes that may operate in vivo. If the OHC receptor potentials are really only up to about 15 mV in size (Section 2.3.2), then the small $I_{K(Ca)}$ will have no physiological role if it only activates for potentials more positive than -35 mV (Section 3.3.2).

The current–voltage curves of Figure 6.13B are almost linear between about -80 and $+20$ mV. If this represents "the truth," the transfer functions of the OHC receptor potentials will approximately be a scaled version of those of the transducer currents. The scaling will depend on the input impedance, which varies as a function of stimulus frequency (Figure 6.14). At low frequencies, little current will flow through the membrane capacitance, so the larger resistance of the apical OHC will ensure a larger receptor potential. The greater resistance and capacitance of the apical cell make its RC membrane time constant much slower than that of the basal cell, so that receptor potentials in response to high-frequency transducer currents are more strongly attenuated (Figure 6.14B). This first-order low-pass filter arrangement goes some way toward maximizing the OHC receptor potentials at their CFs, although it is clearly not sufficient to explain the sharp tuning found in vivo.

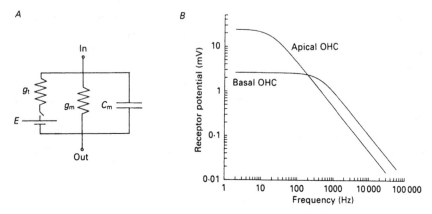

FIGURE 6.14 (A) Simple equivalent circuit for OHCs. (B) Predicted receptor potential magnitudes at different sound frequencies for a basal and an apical OHC when 1 nS of transducer conductance is activated. Parameters used: E (driving force for the transducer conductance g_t) 120 mV; g_m (basolateral membrane conductance) 45 nS (basal), 4 nS (apical); C_m (membrane capacitance) 18 pF (basal), 42 pF (apical). (Reproduced by permission from Housley and Ashmore 1992, Fig. 15.)

3.5 Calcium Currents in Inner and Outer Hair Cells

Hair cells from various nonmammalian vertebrates all possess very similar Ca^{2+} currents (Lewis and Hudspeth 1983; Art and Fettiplace 1987; Hudspeth and Lewis 1988a; Fuchs, Evans, and Murrow 1990; Prigioni et al. 1992; Art, Fettiplace, and Wu 1993). They all activate rapidly at room temperature, reaching their maximum size at a given potential in a few milliseconds, and show little or no inactivation. Sustained Ca^{2+} currents were up to 700 pA in turtle hair cells in the presence of 2.8 mM external Ca^{2+}. In these turtle cells, the size of the Ca^{2+} current, and presumably the number of Ca^{2+} channels, increased with increasing resonance frequency (Art and Fettiplace 1987; Art, Fettiplace, and Wu 1993).

Time- and voltage-dependent Ca^{2+} channels have been found in a large variety of vertebrate cells and have been classified into transient (T-type) channels, which inactivate rapidly and can be activated from potentials positive to about -70 mV, and three groups of channels that inactivate more slowly and can generally only be activated from potentials above about -20 mV. The latter differ among one another in pharmacology and kinetics and are called L-, N-, and P-type channels (Hille 1992). The Ca^{2+} currents in nonmammalian vertebrate hair cells have the pharmacological characteristics of the L-type but have a more hyperpolarized activation range, activating above about -60 mV and reaching an (inward) maximum at around -20 mV, and generally rather more rapid activation kinetics.

In mammalian hair cells, Ca^{2+} currents have been quite hard to demonstrate. In isolated cells, the currents tend to be very small, even when unphysiologically high concentrations of Ca^{2+} are present or when Ba^{2+} ions are used, which carry more current than Ca^{2+} in L-type channels. In

OHCs, in the presence of 50 mM Ba^{2+} in the extracellular solution, Santos-Sacchi and Dilger (1988) found an inward current component activating above about -10 mV, reaching a maximum size of about 150 pA at $+30$ mV. Also in isolated OHCs, Nakagawa et al. (1991) found Ca^{2+} currents, in 5 mM external Ca^{2+}, that activated from about -30 mV and peaked near 0 mV, with a size of less than 100 pA. These inward currents in OHCs showed little or no inactivation.

In IHCs, in solutions designed to minimize the large K^+ currents, and with a physiological Ca^{2+} concentration of 1.3 mM, inward currents developed above -60 mV and peaked near -20 mV, but the currents had a transient and a sustained component (Crawford and Kros 1990). The transient component was up to 100 pA in size, the sustained component up to 25 pA (Kros 1989). It was suggested that the transient component might also be a Ca^{2+} current (like the T-type). Recent work on neonatal hair cells in cochlear cultures suggests that the transient component of the inward current in IHCs is in fact almost certainly a Na^{2+} current (see Section 3.7).

IHCs and OHCs of neonatal mice in organ cultures both possess sustained Ca^{2+} currents like those found in hair cells of nonmammalian vertebrates (Kros et al. 1993b). With 1.3 mM Ca^{2+} in the extracellular solution, they activate above -60 mV and peak at -20 mV, reaching a maximum size within 2 ms (at room temperature) of 75 pA for OHCs and 200 pA for IHCs. Although the possibility remains that Ca^{2+} currents are genuinely reduced in size upon maturation, it seems far more likely that isolated IHCs and OHCs exhibit small Ca^{2+} currents due to damage during the dissociation procedure (Hudspeth and Lewis 1988a).

3.6 Physiological Functions of Calcium Currents

The most obvious function of Ca^{2+} currents in hair cells is to provide the link between the generation of the receptor potential and the release of neurotransmitter onto the afferent nerve fibers. To fulfil this role, the Ca^{2+} currents need to be activated over the range of potentials spanned by the receptor potential in vivo. Ca^{2+} currents in IHCs seem well suited for this role. The activation starting at about -60 mV and the maximum inward currents at about -20 mV correspond rather well with the range of receptor potentials found in recordings made in vivo (see Section 2.3.1) and with predictions from in vitro experiments (Section 3.1.2; Fig. 6.12), and should enable a graded release of neurotransmitter with depolarization. In the absence of deliberate sound stimulation, the type I auditory nerve afferents, which innervate exclusively IHCs, fire irregular action potentials (Kiang et al. 1965), the rate of which is reduced by experimental manipulations that are expected to hyperpolarize the IHC (Sewell 1984). This requires that a standing Ca^{2+} current induce tonic transmitter release at the resting potential of the IHC, which should therefore be about -60 mV, or a little more depolarized. The potential at which the inward Ca^{2+} current reaches it maximal size, about -20 mV, limits the useful size of the DC receptor potential in high-frequency (i.e., above the phase-locking limit; see Section 3.2.2) IHCs in vivo to about 40 mV. Any large receptor potential would actually release less neurotransmitter. For low frequencies, a prominent quadratic distortion component would be produced at the stage of neurotransmitter release for periodic receptor potentials that overshot a membrane potential of -20 mV.

The kinetics of I_{Ca} in IHCs should be sufficiently rapid to support periodic neurotransmitter release and thus phase locking in the auditory nerve (see also Section 3.2.2). The relevant data are at present only available in recordings from neonatal hair cells. At room temperature, the kinetics of I_{Ca} in neonatal hair cells are similar to those of nonmammalian vertebrate hair cells. Activation was sigmoidal, but a single-exponential approximation yielded time constants of less than 800 μs (Kros and Rüsch, in preparation). This would predict an f_{3dB} of more than 200 Hz at room temperature, or 900 Hz at 37°C [conservatively estimating $Q_{10°C}$ for the activation kinetics as 3 (Nobile et al. 1990)], indeed sufficiently fast not to limit phase locking.

Any physiological role of I_{Ca} in OHCs is harder to envisage. In adult OHCs, I_{Ca} seems only to be activated at potentials positive to about -30 mV, whereas the resting potentials in vivo are probably between -70 and -85 mV (Section 3.4). Given the smallish size of OHC receptor potentials, it would appear that I_{Ca} is never activated in vivo. It is in fact at present not known at all whether type II afferents in the auditory nerve, which innervate exclusively OHCs, fire action potentials. If the Ca^{2+} currents in OHCs are to serve some other second messenger function, then either the estimate of the resting potential or the estimate of the activation range of I_{Ca} must be wrong. The 30 mV discrepancy in activation range between neonatal and adult OHCs can be only partially explained by a change in the surface potential associated with channel gating, because of the elevated concentrations of divalent ions used in the experiments of Nakagawa et al. (1991) on adult OHCs (Smith, Ashcroft, and Fewtrell 1993). Superfusing neonatal OHCs with 10 mM rather than 1.3 mM Ca^{2+} shifted activation by only 10 mV in the depolarizing direction (Kros and Rüsch, in preparation).

In neonatal IHCs and OHCs, I_{Ca} surprisingly contributed to the firing of slow, repetitive action potentials (Kros et al. 1993b). This behavior may serve some developmental function and is not observed in adult IHCs (because their Ca^{2+} currents are smaller and their K^+ currents are much larger and faster) or OHCs (because their Ca^{2+} currents are smaller and because of their large leak conductance; Section 3.3.2).

3.7 Other Currents in Inner and Outer Hair Cells and Their Possible Functions

About 50% of neonatal IHCs and OHCs possess a classical transient Na^+ current (up to 700 pA from a holding potential of -84 mV) that is reversibly abolished with 100 nM tetrodotoxin, a selective blocker of Na$^+$ channels (Kros et al. 1993b). In retrospect, because of the identical kinetics, a smaller I_{Na} (up to 100 pA from -105 mV) is probably also present in some adult IHCs, showing up as a rapid, transient inward current (Crawford and Kros 1990). The current persists in a small fraction (about 7%) of mature OHCs and can be larger than 2.5 nA in some cells (Witt et al. 1994).

The first report of an I_{Na} in hair cells was in low-frequency cells of the auditory organ of the mature alligator (*A. mississippiensis*), the basilar papilla (Evans and Fuchs 1987). It has subsequently also been found in cells from the adult goldfish sacculus (an auditory organ in fish; Sugihara and Furukawa 1989) and the vestibular system of both the developing and the mature chick (Sokolowski, Stahl, and Fuchs 1993). I_{Na} is likely to contribute to the firing of action potentials by these cells.

There is no known function for I_{Na} in mammalian auditory hair cells. In contrast to the basilar papilla of the alligator, the presence of I_{Na} is not restricted to hair cells in the low-frequency region of the cochlea. Its absence in half the neonatal cells studied, apparent reduction in size upon maturation in IHCs, and disappearance in most OHCs suggest some developmental role, probably associated with the spiking behavior observed in neonatal cells (Section 3.6).

Another current whose function is at present poorly understood is a large, ATP-activated, nonselective cation current first described in OHCs (Nakagawa et al. 1990) and more recently also in IHCs (Housley et al. 1993). The ion channels carrying this current are located on the apical side of the cells, most likely only on the part of the cell normally in contact with endolymph: the top of the cuticular plate and the hair bundle (Housley, Greenwood, and Ashmore 1992; Housley et al. 1993). The ionic selectivity is remarkably similar to that of the transducer channel (see Section 2.1.4.1), with a larger permeability to Ca^{2+} than to monovalent cations and an Eisenman sequence characteristic of a strong-field-strength selectivity filter (Nakagawa et al. 1990). The ATP-induced current in OHCs is also affected by similar blockers (amiloride, aminoglycoside antibiotics) as the transducer channel, albeit with an order of magnitude lower affinity (Housley, Greenwood, and Ashmore 1992; Lin, Hume, and Nuttall 1993). The currents could be quite large, over 1 nA of inward current at -70 mV. The similarity to the transducer channel ends with the estimated single-channel conductance of only 3.5 pS (from noise analysis). On the basis of this, it was deduced that several thousand ATP-activated channels may be located on the apical surface of an OHC (Lin, Hume, and Nuttall 1993), an order of magnitude more than the number of transducer channels (see Section 2.1.2).

The channels were half-activated by about 12 μM ATP, with a Hill coefficient close to 1, suggesting that binding of a single ATP molecule opens the channel (Nakagawa et al. 1990). If ATP were present in similar concentrations in the endolymph around the hair bundles, for which there is no evidence, a possible physiological role for this current would be humoral modulation of the function of OHCs (by depolarization and shunting of the transducer current), as an alternative to the efferent neurotransmitter acetylcholine (ACh), whose receptor, probably the α9 subunit of the nicotinic ACh receptor gene family (Elgoyhen et al. 1994), may be another nonselective cation channel, situated near the basal pole of the OHC. Ca^{2+} influx through this ACh receptor could cause nearby K(Ca) channels to open, thus inducing a hyperpolarizing, outward current. The best evidence for this model of the action of ACh comes from experiments on short hair cells in the chick (Fuchs and Murrow 1992), but partial support has been gathered in OHCs too (Housley and Ashmore 1991; Kakehata et al. 1993). In IHCs, the effect of ATP seems more complicated and rather like that of ACh on OHCs, in that Ca^{2+} influx through the ATP-activated channels may secondarily activate a Ca^{2+}-activated K^+ conductance, possibly $I_{K,f}$ (Housley et al. 1993). This might not work with endolymphatic Ca^{2+} concentrations, though.

Finally, neonatal IHCs and OHCs possess a small inward rectifier K^+ current with a maximum chord conductance of 2 to 3 nS at normal perilymph-like K^+ concentrations (Géléoc, Lennan, and Kros, in preparation). It is not yet known whether this current remains upon maturation. If it does, its function may be to reduce the membrane time constant during the hyperpolarizing phase of the receptor potential, especially in IHCs where $I_{K,f}$ and $I_{K,s}$ deactivate at potentials negative to -60 mV (Section 3.1).

4. Summary of Membrane Currents in Mammalian Cochlear Hair Cells, Discrepancies, Problems, and Suggestions for Future Research

In this chapter, the membrane currents that have thus far been characterized in mammalian hair cells in vitro have been discussed. They are summarized in Figure 6.15. Immature, neonatal hair cells are very much like the minimal hair cell of Figure 6.1, with the exception of an unexpected Na^+ current and a small I_{IR}. IHCs and OHCs are qualitatively similar at this stage. Upon maturation, IHCs and OHCs become very different, both acquiring diverse K^+ channels not present before the onset of hearing. In IHCs, the total available time- and voltage-dependent K^+ conductance increases dramatically with maturation, mainly due to the acquisition of $I_{K,f}$; in OHCs, the K^+ conductance appears to decrease. Transducer currents are larger in cultured neonatal than in isolated mature mammalian hair cells, most likely because of damage caused by the harsh procedures used to isolate the mature cells. The same probably applies to I_{Ca}, necessary to mediate transmitter release. Ligand-gated channels modulated by micromolar concentrations of ATP and ACh have been studied in mature hair

FIGURE 6.15 Neonatal and adult IHCs and OHCs, with the different membrane currents that have been described in each. The thickness of the arrows is varied to give some indication of the size of the currents, *thin arrows* indicating small currents. The *interrupted arrow* indicates that a particular expected current has not been reported in the literature. The *dotted line* is to suggest the possible secondary activation of $I_{K,f}$ by I_{ATP} in IHCs (Section 3.7). See text for details.

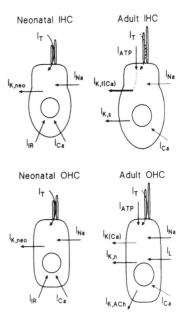

cells but not yet in neonatal cells. This may be a useful thing to do. Uncertainties about the maturity of the mechanisms in neonatal cells may be offset by their better physiological condition.

A full understanding of the physiology of cochlear hair cells requires study of their individual properties, as well as of how they function together in the cochlea. Unfortunately, both approaches involve specific problems that make it impossible to put together a coherent story at present. The microelectrodes used for in vivo recording damage the cells, obliterating evidence of the basolateral membrane currents that are so important in shaping the receptor potentials. On the other hand, isolating the cells to allow the more gentle whole-cell patch recordings is also likely to damage the cells and irreversibly change their properties. Moreover, experiments on isolated mammalian cells are often conducted at room temperature, which may in itself be deleterious and dramatically alter the characteristics of ionic conductances, as has been found in other systems (Walsh, Begenisich, and Kass 1989; Koumi et al. 1994). It is thus naive to expect to find the mechanisms of sharp tuning to be intact in isolated cells. Neonatal hair cells are more resistant and survive for several days in vitro. Long-term survival of hair cells is not possible at present with cells or tissues from the adult mammalian cochlea. The problem with neonatal cells is that it is not known which of their properties will change during development. Cells from nonmammalian vertebrates are also more resistant, but here the question of which of their properties are relevant for mammalian hearing will ultimately have to be answered by repeating experiments in mature mammalian hair cells that are in good shape.

We are thus presently trying to construct a model of the cochlea using data obtained with microelectrodes in vivo, patch-clamp recording in vitro, from neonatal cultures, and using cells from nonmammalian vertebrates: four sets of differently deficient building blocks. No wonder a Babel-like confusion of ears abounds. The future will require tackling the problems inherent to in vitro and in vivo work, and combining the positive aspects of both approaches. Patch-clamping or voltage-sensitive dyes may become applicable in vivo. On the other hand, improved tissue culture techniques may allow the full functional development and maturation of fetal cochleae in vitro, with advantages of stability and accessibility. Our understanding of the biophysics of the membrane currents of hair cells will be deepened by molecular biology. At present, the main difficulty for rapid progress with the latter approach is the remarkably small scale of the design of the cochlea: a mere 30,000 hair cells have to serve the sense of hearing throughout a human lifetime (Ulehlova, Voldrich, and Janisch 1987). The challenge to match this ingenuity of nature should ensure plenty of work in hearing research for the coming years.

Acknowledgments I thank Robert Fettiplace, Gary Housley, Guy Richardson, and Ian Russell for reading an early version of this chapter and for many useful comments and suggestions. The author is a Royal Society University Research Fellow.

References

Adams DJ, Dwyer TM, Hille B (1980) The permeability of endplate channels to monovalent and divalent metal cations. J Gen Physiol 75:493–510.

Adrian ED (1931) The microphonic action of the cochlea: an interpretation of Wever and Bray's experiments. J Physiol 71:xxviii–xxix.

Art JJ, Fettiplace R (1987) Variation of membrane properties in hair cells isolated from the turtle cochlea. J Physiol 385:207–242.

Art JJ, Fettiplace R, Wu Y-C (1993) The effects of low calcium on the voltage-dependent conductances involved in tuning of turtle hair cells. J Physiol 470:109–126.

Ashmore JF (1988) What is the stimulus for outer hair cell motility? In: Duifhuis H, Horst JW, Wit HP (eds) Basic Issues in Hearing. London: Academic Press, pp. 42–47.

Ashmore JF, Meech RW (1986) Ionic basis of membrane potential in outer hair cells of guinea pig cochlea. Nature 322:361–372.

Ashmore JF, Ohmori H (1990) Control of intracellular calcium by ATP in isolated outer hair cells of the guinea-pig cochlea. J Physiol 428:109–131.

Ashmore JF, Kolston PJ, Mammano F (1993) Dissecting components of the outer hair cell feedback loop. In: Duifhuis H, Horst JW, Dijk P van, Netten SM van (eds) Biophysics of Hair Cell Sensory Systems. Singapore: World Scientific, pp. 151–158.

Assad JA, Corey DP (1992) An active motor model for adaptation by vertebrate hair cells. J Neurosci 12:3291–3309.

Assad JA, Hacohen N, Corey DP (1989) Voltage dependence of adaptation and active bundle movement in bullfrog saccular hair cells. Proc Natl Acad Sci USA 86:2918-2922.

Barrett JN, Magleby KL, Pallotta BS (1982) Properties of single calcium-activated potassium channels in cultured rat muscle. J Physiol 331:211-230.

Bosher SK, Warren RL (1978) Very low calcium content of cochlear endolymph, an extracellular fluid. Nature 273:377-378.

Brown MC, Nuttall AL (1984) Efferent control of cochlear inner hair cell responses in the guinea-pig. J Physiol 354:625-646.

Brownell WE (1984) Microscopic observation of cochlear hair cell motility. Scan Elect Microsc 3:1401-1406.

Brownell WE, Bader CR, Bertrand D, Ribaupierre Y de (1985) Evoked mechanical responses of isolated cochlear outer hair cells. Science 227:194-196.

Canessa CM, Horisberger J-D, Rossier BC (1993) Epithelial sodium channel related to proteins involved in neurodegeneration. Nature 361:467-470.

Cheatham MA, Dallos P (1993) Longitudinal comparisons of IHC ac and dc receptor potentials recorded from the guinea pig cochlea. Hear Res 68:107-114.

Cody AR, Russell IJ (1987) The responses of hair cells in the basal turn of the guinea-pig cochlea to tones. J Physiol 383:551-569.

Colamartino G, Menini A, Torre V (1991) Blockage and permeation of divalent cations through the cyclic GMP-activated channel from tiger salamander retinal rods. J Physiol 440:189-206.

Corey DP, Hudspeth AJ (1979) Ionic basis of the receptor potential in a vertebrate hair cell. Nature 281:675-677.

Corey DP, Hudspeth AJ (1983a) Analysis of the microphonic potential of the bullfrog's sacculus. J Neurosci 3:942-961.

Corey DP, Hudspeth AJ (1983b) Kinetics of the receptor currents in bullfrog saccular hair cells. J Neurosci 3:962-976.

Crawford AC, Fettiplace R (1980) The frequency selectivity of auditory nerve fibres and hair cells in the cochlea of the turtle. J Physiol 306:79-125.

Crawford AC, Fettiplace R (1981) An electrical tuning mechanism in turtle cochlear hair cells. J Physiol 312:377-412.

Crawford AC, Fettiplace R (1985) The mechanical properties of ciliary bundles of turtle cochlear hair cells. J Physiol 364:359-379.

Crawford AC, Kros CJ (1990) A fast calcium current with a rapidly inactivating component in isolated inner hair cells of the guinea-pig. J Physiol 420:90P.

Crawford AC, Evans MG, Fettiplace R (1989) Activation and adaptation of transducer currents in turtle hair cells. J Physiol 419:405-434.

Crawford AC, Evans MG, Fettiplace R (1991) The actions of calcium on the mechano-electrical transducer current of turtle hair cells. J Physiol 434:369-398.

Dallos P (1973) The auditory periphery. New York: Academic Press.

Dallos P (1984) Some electrical circuit properties of the organ of Corti. II. Analysis including reactive elements. Hear Res 14:281-291.

Dallos P (1985a) Response characteristics of mammalian cochlear hair cells. J Neurosci 5:1591-1608.

Dallos P (1985b) Membrane potential and response changes in mammalian cochlear hair cells during intracellular recording. J Neurosci 5:1609-1615.

Dallos P (1986) Neurobiology of cochlear inner and outer hair cells: intracellular recordings. Hear Res 22:185-198.

Dallos P, Cheatham MA (1989) Nonlinearities in cochlear receptor potentials and their origins. J Acoust Soc Am 86:1790-1796.

Dallos P, Cheatham MA (1990) Effects of electrical polarization on inner hair cell receptor potentials. J Acoust Soc Am 87:1636–1647.

Dallos P, Cheatham MA (1992) Cochlear hair cell function reflected in intracellular recordings in vivo. In: Corey DP, Roper SD (eds) Sensory Transduction. New York: Rockefeller University Press, pp. 371–393.

Dallos P, Santos-Sacchi J, Flock Å (1982) Intracellular recordings from cochlear outer hair cells. Science 218:582–584.

Denk W, Webb WW (1992) Forward and reverse transduction at the limit of sensitivity studied by correlating electrical and mechanical fluctuations in frog saccular hair cells. Hear res 60:89–102.

Denk W, Keolian RM, Webb WW (1992) Mechanical response of frog saccular hair bundles to the aminoglycoside block of mechanoelectrical transduction. J Neurophysiol 68:927–932.

Denk W, Holt JR, Shepherd GMG, Corey DP (1995) Calcium imaging of single stereocilia in hair cells: localization of transduction channels at both ends of tip links. Neuron 15:1311–1321.

Dreyer F (1990) Peptide toxins and potassium channels. Rev Physiol Biochem Pharmacol 115:93–136.

Dwyer TM, Adams DJ, Hille B (1980) The permeability of the endplate channel to organic cations in frog muscle. J Gen Physiol 75:469–492.

Eatock RA, Corey DP, Hudspeth AJ (1987) Adaptation of mechanoelectrical transduction in hair cells of the bullfrog's sacculus. J Neurosci 7:2821–2836.

Elgoyhen AB, Johnson DS, Boulter J, Vetter DE, Heinemann S (1994) $\alpha 9$: an acetylcholine receptor with novel pharmacological properties expressed in rat cochlear hair cells. Cell 79:705–715.

Evans BN, Dallos P (1993a) Mechanomotility and ciliary stiffness change in cochlear outer hair cells. Abst Assoc Res Otolaryngol 16:116.

Evans BN, Dallos P (1993b) Stereocilia displacement induced somatic motility of cochlear outer hair cells. Proc Natl Acad Sci USA 90:8347–8351.

Evans EF (1972) The frequency response and other properties of single fibres in the guinea-pig cochlear nerve. J Physiol 226:263–287.

Evans EF (1989) Cochlear filtering: a view seen through the temporal discharge patterns of single cochlear nerve fibers. In: Wilson JP, Kemp DT (eds) Cochlear Mechanisms. Structure, Function and Models. New York: Plenum Press, pp. 241–248.

Evans MG, Fuchs PA (1987) Tetrodotoxin-sensitive voltage-dependent sodium currents in hair cells from the alligator cochlea. Biophys J 52:649–652.

Fay RR (1992) Structure and function in sound discrimination among vertebrates. In: Webster DB, Fay RR, Popper AN (eds) Comparative Studies of Hearing in Vertebrates. New York: Springer-Verlag, pp. 229–263.

Fenwick EM, Marty A, Neher E (1982) A patch-clamp study of bovine chromaffin cells and their sensitivity to acetylcholine. J Physiol 331:577–597.

Fettiplace R, Crawford AC (1978) The coding of sound pressure and frequency in cochlear hair cells of the terrapin. Proc R Soc Lond B 203:209–218.

Fettiplace R, Crawford AC (1989) Mechano-electrical transduction in turtle hair cells. In: Wilson JP, Kemp DT (eds) Cochlear Mechanisms. Structure, Function and Models. New York: Plenum Press, pp. 99–105.

Fettiplace R, Crawford AC, Evans MG (1992) The hair cell's mechanoelectrical transducer channel. Ann NY Acad Sci 656:1–11.

Flock Å, Strelioff D (1984) Studies on hair cells in isolated coils from the guinea pig cochlea. Hear Res 15:11-18.
Frankenhaeuser B (1963) A quantitative description of potassium currents in myelinated nerve fibres of *Xenopus laevis*. J Physiol 169:424-430.
Frings S, Lynch JW, Lindemann B (1992) Properties of cyclic nucleotide-gated channels mediating olfactory transduction. Activation, selectivity, and blockage. J Gen Physiol 100:45-67.
Fuchs PA (1992) Ionic currents in cochlear hair cells. Prog Neurobiol 39:493-505.
Fuchs PA, Evans MG (1990) Potassium currents in hair cells isolated from the cochlea of the chick. J Physiol 429:529-551.
Fuchs PA, Murrow BW (1992) Cholinergic inhibition of short (outer) hair cells of the chick's cochlea. J Neurosci 12:800-809.
Fuchs PA, Sokolowski BHA (1990) The acquisition during development of Ca-activated potassium currents by cochlear hair cells of the chick. Proc R Soc Lond B 241:122-126.
Fuchs PA, Nagai T, Evans MG (1988) Electrical tuning in hair cells isolated from the chick cochlea. J Neurosci 8:2460-2467.
Fuchs PA, Evans MG, Murrow BW (1990) Calcium currents in hair cells isolated from the cochlea of the chick. J Physiol 429:553-568.
Furman RE, Tanaka JC (1990) Monovalent selectivity of the cyclic guanosine monophosphate-activated ion channel. J Gen Physiol 96:57-82.
Furness Dn, Richardson GP, Russell IJ (1989) Stereociliary bundle morphology in organotypic cultures of the mouse cochlea. Hear Res 38:95-110.
Gillespie PG, Hudspeth AJ (1993) Adenine nucleoside diphosphates block adaptation of mechanoelectrical transduction in hair cells. Proc Natl Acad Sci USA 90:2710-2714.
Griguer C, Kros CJ, Sans A, Lehouelleur J (1993) Potassium currents in type II vestibular hair cells isolated from the guinea pig's crista ampullaris. Pflügers Arch 425:344-352.
Gummer AW, Smolders JWT, Klinke R (1987) Basilar membrane motion in the pigeon measured with the Mössbauer technique. Hear Res 29:63-92.
Hackney CM, Furness DN, Benos DJ, Woodley JF, Barratt J (1992) Putative immunolocalization of the mechanoelectrical transduction channels in mammalian cochlear hair cells. Proc R Soc Lond B 248:215-221.
Harris GG, Frishkopf LS, Flock Å (1970) Receptor potentials from hair cells of the lateral line. Science 167:76-79.
Haynes LW (1993) Mono- and divalent cation selectivity of catfish cone outer segment cGMP-gated channels. Biophys J 64:A133.
Hille B (1992) Ionic Channels of Excitable Membranes. 2nd ed. Sunderland, MA: Sinauer.
Hodgkin AL, Huxley AF (1952) Currents carried by sodium and potassium ions through the membrane of the giant axon of *Loligo*. J Physiol 116:449-472.
Holton T, Hudspeth AJ (1986) The transduction channel of hair cells from the bull-frog characterized by noise analysis. J Physiol 375:195-227.
Horowitz P, Hill W (1989) The Art of Electronics. 2nd ed. Cambridge: Cambridge University Press.
Housley GD, Ashmore JF (1991) Direct measurement of the action of acetylcholine on isolated outer hair cells of the guinea-pig cochlea. Proc R Soc Lond B 244:161-167.

Housley GD, Ashmore JF (1992) Ionic currents of outer hair cells isolated from the guinea-pig cochlea. J Physiol 448:73-98.

Housley GD, Greenwood D, Ashmore JF (1992) Localization of cholinergic and purinergic receptors on outer hair cells isolated from the guinea-pig cochlea. Proc R Soc Lond B 249:265-273.

Housley GD, Greenwood D, Mockett BG, Muñoz DJB, Thorne PR (1993) Differential actions of ATP-activated conductances in outer and inner hair cells isolated from the guinea-pig organ of Corti: a humoral purinergic influence on cochlear function. In: Duifhuis H, Horst JW, Dijk P van, Netten SM van (eds) Biophysics of Hair Cell Sensory Systems. Singapore: World Scientific, pp. 116-123.

Howard J, Ashmore JF (1986) Stiffness of sensory hair bundles in the sacculus of the frog. Hear Res 23:93-104.

Howard J, Hudspeth AJ (1987) Mechanical relaxation of the hair bundle mediates adaptation in mechanoelectrical transduction by the bullfrog's saccular hair cell. Proc Natl Acad Sci USA 84:3064-3068.

Howard J, Hudspeth AJ (1988) Compliance of the hair bundle associated with gating of mechanoelectrical transduction channels in the bullfrog's saccular hair cell. Neuron 1:189-199.

Howard J, Roberts WM, Hudspeth AJ (1988) Mechanoelectrical transduction by hair cells. Annu Rev Biophys Biophys Chem 17:99-124.

Hudspeth AJ, Corey DP (1977) Sensitivity, polarity, and conductance change in the repsonse of vertebrate hair cells to controlled mechanical stimuli. Proc Natl Acad Sci USA 74:2407-2411.

Hudspeth AJ, Lewis RS (1988a) Kinetic analysis of voltage- and ion-dependent conductances in saccular hair cells of the bull-frog *Rana catesbeiana*. J Physiol 400:237-274.

Hudspeth AJ, Lewis RS (1988b) A model for electrical resonance and frequency tuning in saccular hair cells of the bull-frog *Rana catesbeiana*. J Physiol 400:275-297.

Ikeda K, Saito Y, Nishiyama A, Takasaka T (1992) Na^+-Ca^{2+} exchange in the isolated cochlear outer hair cells of the guinea-pig studied by fluorescence image microscopy. Pflügers Arch 420:493-499.

Jaramillo F, Hudspeth AJ (1993) Displacement-clamp measurement of the forces exerted by gating springs in the hair bundle. Proc Natl Acad Sci USA 90:1330-1334.

Jaramillo F, Markin VS, Hudspeth AJ (1993) Auditory illusions and the single hair cell. Nature 364:527-529.

Jörgensen F, Ohmori H (1988) Amiloride blocks the mechano-electrical transduction channel of hair cells of the chick. J Physiol 403:577-588.

Kakehata S, Nakagawa T, Takasaka T, Akaike N (1993) Cellular mechanism of acetylcholine-induced response in dissociated outer hair cells of guinea-pig cochlea. J Physiol 463:227-244.

Kiang NY-S, Watanaba T, Thomas EC, Clark LF (1965) Discharge Patterns of Single Fibres in the Cat's Auditory Nerve. Cambridge, MA: MIT Press.

Kimitsuki T, Ohmori H (1992) The effect of caged calcium release on the adaptation of the transduction current in chick hair cells. J Physiol 458:27-40.

Kleyman TR, Cragoe EJ (1988) Amiloride and its analogs as tools in the study of ion transport. J Memb Biol 105:1-21.

Kolston PJ, Ashmore JF (1993) Action of ATP at the mechano-electric transducer site in outer hair cells isolated from the guinea-pig cochlea. J Physiol 459:428P.

Kössl M, Richardson GP, Russell IJ (1990) Stereocilia bundle stiffness: effects of neomycin sulphate, A23187 and concanavalin A. Hear Res 44:217–230.

Koumi SI, Sato R, Horkawa T, Aramaki T, Okumura H (1994) Characterization of the calcium-sensitive voltage-gated delayed rectifier potassium channel in isolated guinea-pig hepatocytes. J Gen Physiol 104:147–171.

Kroese ABA, Das A, Hudspeth AJ (1989) Blockages of the transduction channels of hair cells in the bullfrog's sacculus by aminoglycoside antibiotics. Hear Res 37:203–218.

Kros CJ (1989) Membrane properties of inner hair cells isolated from the guinea-pig cochlea. PhD Thesis, University of Cambridge.

Kros CJ (1990) Electrical properties of the basolateral membrane of hair cells – a review. In: Borsellino A, Cervetto L, Torre V (eds) Sensory Transduction. New York: Plenum Press, pp. 51–63.

Kros CJ, Crawford AC (1988) Non-linear electrical properties of guinea-pig inner hair cells: a patch-clamp study. In: Duifhuis H, Horst JW, Wit HP (eds) Basic Issues in Hearing. London: Academic Press, pp. 27–31.

Kros CJ, Crawford AC (1989) Components of the membrane current in guinea-pig inner hair cells. In: Wilson JP, Kemp DT (eds) Cochlear Mechanisms. Structure, Function and Models. New York: Plenum Press, pp. 189–195.

Kros CJ, Crawford AC (1990) Potassium currents in inner hair cells isolated from the guinea-pig cochlea. J Physiol 421:263–291.

Kros CJ, Rüsch A, Russell IJ (1991) Differences between potassium currents of neonatal and mature mouse inner hair cells in vitro. J Physiol 434:51P.

Kros CJ, Rüsch A, Richardson GP (1992) Mechano-electrical transducer currents in hair cells of the cultured mouse cochlea. Proc R Soc Lond B 249:185–193.

Kros CJ, Rüsch A, Lennan GWT, Richardson GP (1993a) Voltage dependence of transducer currents in outer hair cells of neonatal mice. In: Duifhuis H, Horst JW, Dijk P van, Netten SM van (eds) Biophysics of Hair Cell Sensory Systems. Singapore: World Scientific, pp. 141–150.

Kros CJ, Rüsch A, Richardson GP, Russell IJ (1993b) Sodium and calcium currents in cultured cochlear hair cells of neonatal mice. J Physiol 473:231P.

Kros CJ, Lennan GWT, Richardson GP (1995) Transducer currents and bundle movements in outer hair cells of neonatal mice. In: Flock Å, Ottoson D, Ulfendahl M (eds) Acting Hearing. Oxford: Elsevier Science, pp. 113–125.

Lane JW, McBride DW Jr, Hamill OP (1991) Amiloride block of the mechanosensitive cation channel in *Xenopus* oocytes. J Physiol 441:347–366.

Lane JW, McBride DW Jr, Hamill OP (1992) Structure-activity relations of amiloride and its analogues in blocking the mechanosensitive channel in *Xenopus* oocytes. Br J Pharmacol 106:283–286.

Lang DG, Correia MJ (1989) Studies of solitary semicicular canal hair cells in the adult pigeon. II. Voltage-dependent ionic conductances. J Neurophysiol 62:935–945.

Lewis RS, Hudspeth AJ (1983) Voltage- and ion-dependent conductances in solitary vertebrate hair cells. Nature 304:538–541.

Liberman MC, Kiang NY-S (1978) Acoustic trauma in cats. Acta Otolaryngol Suppl 358:1–63.

Lim DJ (1980) Cochlear anatomy related to cochlear micromechanics. A review. J Acoust Soc Am 67:1686-1695.

Lim DJ (1986) Functional structure of the organ of Corti: a review. Hear Res 22:117-146.

Lin X, Hume RI, Nuttall AL (1993) Voltage-dependent block by neomycin of the ATP-induced whole-cell current of guinea-pig outer hair cells. J Neurophysiol 70:1593-1605.

Lingueglia E, Voilley N, Waldmann R, Lazdunski M, Barbry P (1993) Expression cloning of an epithelial amiloride-sensitive Na^+ channel: a new channel type with homologies to *Caenorhabditis elegans* degenerins. FEBS Lett 318:95-99.

Lumpkin EA, Hudspeth AJ (1995) Detection of Ca^{2+} entry through mechanosensitive channels localizes the site of mechanoelectrical transduction in hair cells. Proc Natl Acad Sci USA 92:10297-10301.

Mammano F, Kros CJ, Ashmore JF (1995) Patch clamped responses from outer hair cells in the intact adult organ of Corti. Pflügers Arch 430:745-750.

Menini A (1990) Currents carried by monovalent cations through cyclic GMP-activated channels in excised patches from salamander rods. J Physiol 424:167-185.

Methfessel C, Witzemann V, Takahashi T, Mishina M, Numa S, Sakmann B (1986) Patch clamp measurements on *Xenopus laevis* oocytes: currents through endogenous channels and implanted acetylcholine receptor and sodium channels. Pflügers Arch 407:577-588.

Nakagawa T, Akaike N, Kimitsuki T, Komune S, Arima T (1990) ATP-induced current in isolated outer hair cells of guinea-pig cochlea. J Neurophysiol 63:1068-1074.

Nakagawa T, Kakehata S, Akaike N, Komune S, Takasaka T, Uemura T (1991) Calcium channel in isolated outer hair cells of guinea pig cochlea. Neurosci Lett 125:81-84.

Netten SM van, Kroese ABA (1987) Laser interferometric measurements on the dynamic behaviour of the cupula in the fish lateral line. Hear Res 29:55-62.

Nobile M, Carbone E, Lux HD, Zucker H (1990) Temperature sensitivity of Ca currents in chick sensory neurones. Pflügers Arch 415:658-663.

Nuttall AL (1985) Influence of direct current on dc receptor potentials from cochlear inner hair cells in the guinea pig. J Acoust Soc Am 77:165-175.

Ohmori H (1984) Mechanoelectrical transducer has discrete conductances in the chick vestibular hair cell. Proc Natl Acad Sci USA 81:1888-1891.

Ohmori H (1985) Mechano-electrical transduction currents in isolated vestibular hair cells of the chick. J Physiol 359:189-217.

Olsen ES, Mountain DC (1991) In vivo measurement of basilar membrane stiffness. J Acoust Soc Am 89:1262-1275.

Palmer AR (1990) The representation of the spectra and fundamental frequencies of steady-state single- and double-vowel sounds in the temporal discharge patterns of guinea-pig cochlear-nerve fibres. J Acoust Soc Am 88:1412-1426.

Palmer AR, Russell IJ (1986) Phase-locking in the cochlear nerve of the guinea-pig and its relation to the receptor potential of inner hair cells. Hear Res 24:1-15.

Patuzzi R, Sellick PM (1983) A comparison between basilar membrane and inner hair cell receptor potential input-output functions in the guinea pig cochlea. J Acoust Soc Am 74:1734-1741.

Pickles JO (1993) A model for the mechanics of the stereociliar bundle on acousticolateral hair cells. Hear Res 68:159-172.

Pickles JO, Corey DP (1992) Mechanoelectrical transduction by hair cells. Trends Neurosci 15:254–259.
Pickles JO, Comis SD, Osborne MP (1984) Cross-links between stereocilia in the guinea-pig organ of Corti and their possible relation to sensory transduction. Hear Res 15:103–112.
Pickles JO, Osborne MP, Comis SD, Köppl C, Gleich O, Brix J, Manley GA (1989) Tip-link organization in relation to the structure and orientation of stereovillar bundles. In: Wilson JP, Kemp DT (eds) Cochlear Mechanisms. Structure, Function and Models. New York: Plenum Press, pp. 37–44.
Prigioni I, Masetto S, Russo G, Taglietti V (1992) Calcium currents in solitary hair cells isolated from frog crista ampullaris. J Vest Res 2:31–39.
Prosen CA, Petersen MR, Moody DB, Stebbins WC (1978) Auditory thresholds and kanamycin-induced hearing loss in the guinea pig assessed by a positive reinforcement procedure. J Acoust Soc Am 63:559–566.
Purves RD (1981) Microelectrode Methods for Intracellular Recording and Ionophoresis. London: Academic Press.
Richardson GP, Russell IJ, Wasserkort R, Hans M (1989) Aminoglycoside antibiotics and lectins cause irreversible increases in the stiffness of cochlear hair-cell stereocilia. In: Wilson JP, Kemp DT (eds) Cochlear Mechanisms. Structure, Function and Models. New York: Plenum Press, pp. 57–65.
Roberts WM, Howard J, Hudspeth AJ (1988) Hair cells: transduction, tuning, and transmission in the inner ear. Annu Rev Cell Biol 4:63–92.
Romand R (1983) Development of the cochlea. In: Romand R (ed) Development of Auditory and Vestibular Systems. New York: Academic Press, pp. 47–88.
Rudy B (1988) Diversity and ubiquity of K channels. Neuroscience 25:729–749.
Ruggero MA, Rich NC (1991) Furosemide alters organ of Corti mechanics: evidence for feedback of outer hair cells upon the basilar membrane. J Neurosci 11:1057–1067.
Rüsch A, Kros CJ, Richardson GP, Russell IJ (1991) Potassium and calcium currents in outer hair cells in organotypic cultures of the neonatal mouse cochlea. J Physiol 434:52P.
Rüsch A, Kros CJ, Richardson GP (1994) Block by amiloride and its derivatives of mechano-electrical transduction in outer hair cells of mouse cochlear cultures. J Physiol 474:75–86.
Russell IJ (1983) Origin of the receptor potential in inner hair cells of the mammalian cochlea—evidence for Davis' theory. Nature 301:334–336.
Russell IJ, Kössl M (1991) The voltage responses of hair cells in the basal turn of the guinea-pig cochlea. J Physiol 435:493–511.
Russell IJ, Kössl M (1992) Voltage responses to tones of outer hair cells in the basal turn of the guinea-pig cochlea: significance for electromotility and desensitization. Proc R Soc Lond B 247:97–105.
Russell IJ, Richardson GP (1987) The morphology and physiology of hair cells in organotypic cultures of the mouse cochlea. Hear Res 31:9–24.
Russell IJ, Sellick PM (1977) Tuning properties of the cochlear hair cells. Nature 267:858–860.
Russell IJ, Sellick PM (1978) Intracellular studies of hair cells in the mammalian cochlea. J Physiol 284:261–290.
Russell IJ, Sellick PM (1983) Low-frequency characteristics of intracellularly recorded receptor potentials in guinea-pig cochlear hair cells. J Physiol 338:179–206.

Russell IJ, Cody AR, Richardson GP (1986) The responses of inner and outer hair cells in the basal turn of the guinea-pig cochlea and in the mouse cochlea grown in vitro. Hear Res 22:199-216.

Russell IJ, Richardson GP, Cody AR (1986) Mechanosensitivity of mammalian auditory hair cells in vitro. Nature 321:517-519.

Russell IJ, Richardson GP, Kössl M (1989) The responses of cochlear hair cells to tonic displacements of the sensory hair bundle. Hear Res 43:55-70.

Russell IJ, Kössl M, Richardson GP (1992) Nonlinear mechanical responses of mouse cochlear hair bundles. Proc R Soc Lond B 250:217-227.

Sakmann B, Neher E (1995) Single-Channel Recording. 2nd ed. New York: Plenum Press.

Santos-Sacchi J (1989) Asymmetry in voltage-dependent movements of isolated outer hair cells from the organ of Corti. J Neurosci 9:2954-2962.

Santos-Sacchi J (1992) On the frequency limit and phase of outer hair cell motility: effects of the membrane filter. J Neurosci 12:1906-1916.

Santos-Sacchi J, Dilger JP (1988) Whole cell currents and mechanical responses of isolated outer hair cells. Hear Res 35:143-150.

Sewell WF (1984) The relation between the endocochlear potential and spontaneous activity in auditory nerve fibers of the cat. J Physiol 347:685-696.

Smith PA, Ashcroft FM, Fewtrell CMS (1993) Permeation and gating properties of the L-type calcium channel in mouse pancreatic β cells. J Gen Physiol 101:767-797.

Smith PR, Benos DJ (1991) Epithelia Na^+ channels. Annu Rev Physiol 53:509-530.

Sokolowski BHA, Stahl LM, Fuchs PA (1993) Morphological and physiological development of vestibular hair cells in the organ-cultured otocyst of the chick. Dev Biol 155:134-146.

Staley KJ, Otis TS, Mody I (1992) Membrane properties of dentate gyrus granule cells: comparison of sharp microelectrode and whole-cell recordings. J Neurophysiol 67:1346-1358.

Stebbins WC (1983) The Acoustic Sense of Animals. Cambridge, MA: Harvard University Press.

Steinacker A, Romero A (1991) Characterization of voltage-gated and calcium-activated potassium currents in toadfish saccular hair cells. Brain Res 556:22-32.

Strelioff D, Flock Å (1984) Stiffness of sensory-cell hair bundles in the isolated guinea-pig cochlea. Hear Res 15:19-28.

Sugihara I, Furukawa T (1989) Morphological and functional aspects of two different types of hair cells in the goldfish sacculus. J Neurophysiol 62:1330-1343.

Sunose H, Ikeda K, Saito Y, Nishiyama A, Takasaka T (1992) Membrane potential measurements in isolated outer hair cells of the guinea-pig cochlea using conventional microelectrodes. Hear Res 62:237-244.

Ulehlova L, Voldrich L, Janisch R (1987) Correlative study of sensory cell density and cochlear length in humans. Hear Res 28:149-152.

Walsh KB, Begenisich TB, Kass RS (1989) β-Adrenergic modulation of cardiac ion channels. J Gen Physiol 93:841-854.

Weiss TF, Mulroy MJ, Altmann DW (1974) Intracellular responses to acoustic clicks in the inner ear of the alligator lizard. J Acoust Soc Am 55:606-619.

Westerman LA, Smith RL (1984) Rapid and short-term adaptation in auditory nerve responses. Hear Res 15:249-260.

Wever EG, Bray CW (1930) Action currents in the auditory nerve in response to acoustical stimulation. Proc Natl Acad Sci USA 16:344–350.

Witt CM, Hu H-Y, Brownell WE, Bertrand D (1994) Physiologically silent sodium channels in mammalian outer hair cells. J Neurophysiol 72:1037–1040.

Wu Y-C, Art JJ, Goodman MB, Fettiplace R (1995) A kinetic description of the calcium-activated potassium channel and its application to electrical tuning of hair cells. Prog Biophys Molec Biol 63:131–158.

Yau K-W, Baylor DA (1989) Cyclic GMP-activated conductance of retinal photoreceptor cells. Annu Rev Neurosci 12:289–327.

Yellen G (1984) Ionic permeation and blockade in Ca^{2+}-activated K^+ channels of bovine chromaffin cells. J Gen Physiol 84:157–186.

Young ED, Sachs MB (1979) Representation of steady-state vowels in the temporal aspects of discharge patterns of populations of auditory-nerve fibers. J Acoust Soc Am 66:1381–1403.

Zimmerman AL, Baylor DA (1992) Cation interactions with the cyclic GMP-activated channel of retinal rods from the tiger salamander. J Physiol 449:759–783.

Zufall F, Firestein S (1993) Divalent cations block the cyclic nucleotide-gated channel of olfactory receptor neurons. J Neurophysiol 69:1758–1768.

Zufall F, Firestein S, Shepherd GM (1991) Analysis of single nucleotide gated channels in olfactory receptor cells. J Neurosci 11:3573–3580.

Zwislocki JJ, Cefaratti LK (1989) Tectorial membrane II: Stiffness measurements in vivo. Hear Res 42:211–228.

7

Outer Hair Cell Motility

MATTHEW C. HOLLEY

1. Introduction

Outer hair cells (OHCs) are extremely versatile mechanical components of the cochlea. Sitting above the basilar membrane, they appear able to perceive its vibration through their mechanosensitive hair bundles and to feed back mechanical forces that enhance both its sensitivity and its frequency selectivity. They also receive an efferent nerve supply that may be used to modulate this response, and they may receive additional information from mechanoreceptive elements in their basolateral membrane. During the last 10 years they have been the subject of many experiments designed to discover just exactly how they generate mechanical forces and how these forces contribute to the micromechanics of the cochlea. If all the proposed mechanisms actually occur in vivo, then OHCs can boast a remarkable behavioral repertoire.

The cell body has been the subject of the most extensive studies, and it appears to express two types of motile response. The first is a high-frequency length change of up to 5% that can be driven at acoustic frequencies and that is not directly dependent on ATP or any other chemical intermediate (Fig. 7.1). Direct experimental observations suggest that this mechanism is both fast enough and strong enough to sharpen the sound-induced mechanical displacements of the basilar membrane. The second response is a much larger but slower cell length change that is directly dependent upon ATP and that probably involves several other chemical intermediates. There are many different ways of stimulating this response and there could be more than one mechanism involved. Its function may be to alter the sensitivity of the ear by modulating the mechanical coupling of the high-frequency mechanism. In addition to these responses, it has been suggested that the cuticular plate can generate mechanical forces that rock it back and forth with respect to the rest of the reticular lamina, thus moving the hair bundle that is attached to it. Finally, evidence from experiments with other animals, particularly the amphibia, suggest motile mechanisms within the hair bundle that may apply equally to all hair cells. In one case, hair bundles may be able to generate their own

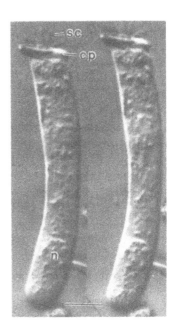

FIGURE 7.1. Apical OHC from the guinea pig dissociated with the enzyme trypsin. The membrane potential is held at +50 mV (*left*) and −150 mV (*right*) via the patch electrode next to the nucleus. The stereocilia (sc), disrupted by the enzyme, project from the surface of the cuticular plate (cp), and the nucleus (n) is located at the base. The basolateral membrane refers to the entire cell membrane except for that surrounding the stereocilia and cuticular plate. Scale bar = 10 μm. (From Holley and Ashmore 1988a.)

intrinsic oscillations. In another, experiments on processes of adaptation during mechanoelectrical transduction suggest that the mechanosensory tip links can actively crawl up the sides of adjacent stereocilia.

Can OHCs realistically express all these motile mechanisms in vivo? For any proposed motility, several issues must be considered. The first concerns the experimental stimulus. Is it within the normal physiological range and is it likely to occur within the normal ear? These questions are particularly important with respect to experiments on single, dissociated OHCs. Other questions relate to the mechanisms of motility. How are the stimulus signals transduced and by what mechanism might the resultant forces be generated? Last, we should ask how the resultant forces or movements might be expressed in vivo and what their function might be. In this chapter these questions will be addressed to all forms of proposed cell motility in OHCs. The functional consequences in terms of cochlear physiology and mechanics are discussed more fully by Patuzzi, Chapter 4, and de Boer, Chapter 5. The largest part of the chapter deals with the mechanism of high-frequency cell length changes, first because it has received most experimental attention by several research groups, and second because it is used here as a frame within which to introduce most of the basic information about cell structure and physiology.

2. High-Frequency Cell Length Changes

The most impressive property of OHCs is their ability to generate length changes at very high frequencies. The phenomenon was first observed in

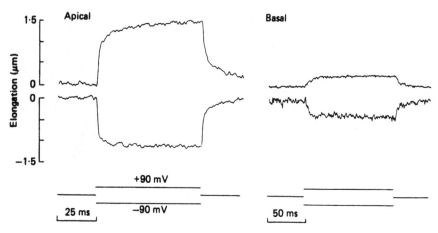

FIGURE 7.2 Length changes recorded with a photodiode detector from apical and basal OHCs during changes of membrane potential driven by a patch electrode. The cells were stimulated by voltage steps of 90 mV from their resting potentials of −44 and −38 mV, respectively. Each record was averaged from between 40 and 128 responses. (Reproduced with permission from Ashmore 1987.)

cells stimulated by either intracellular or extracellular electrodes (Brownell et al. 1985; Kachar et al. 1986), but the first accurate measurements came with the use of patch electrodes for electrical stimulation and a photodiode to record cell length changes (Ashmore 1987; Santos-Sacchi and Dilger 1988). Guinea pig OHCs shorten and lengthen by up to 5% when the plasma membrane is depolarized and hyperpolarized, respectively (Fig. 7.2). This means a displacement of 1 μm for a cell of 20 μm at the high-frequency end of the cochlea and a displacement of 5 μm for a cell of 100 μm at the low-frequency end. The latency is less than 100 μs, and length changes can be driven reversibly to frequencies in excess of 10 kHz. As the stimulus frequency increases above 100 Hz, the magnitude of the length change decreases. The limits of these physical parameters are, however, more likely to reflect the limitations of the patch electrode circuit than those of the cell (Santos-Sacchi 1992). It now seems likely that OHCs can generate significant length changes at much higher frequencies, certainly into the upper range of the human ear at 20 kHz and possibly as high as 40 kHz (Xue, Mountain, and Hubbard 1993). The displacement of the basilar membrane at threshold is only a fraction of a nanometer, and cell length changes should be effective at the same order of magnitude.

2.1 Voltage Dependence

Cell length changes are almost certainly driven directly by voltage changes across the plasma membrane rather than by the passage of ionic current

(Ashmore 1987; Santos-Sacchi and Dilger 1988; Dallos, Evans, and Hallworth 1991). Most of the ion channels can be blocked by applying ions such as cadmium, barium, or tetraethylammonium (TEA) in the extracellular medium, or by putting ions such as cesium into the cell through a patch electrode. This treatment does not inhibit motility. More conclusive evidence comes from cells bathed in 50 mM barium and patched with electrodes containing 140 mM cesium chloride. If these cells are depolarized progressively from a holding potential of -60 mV to values of up to $+90$ mV, they shorten progressively with membrane potential, despite the fact that a large net current reversal occurs between -10 and $+40$ mV (Santos-Sacchi and Dilger 1988; Fig. 7.3).

The sensitivity of the voltage dependence is maximally about 20–25 nm/mV when the cell is stimulated via a patch electrode (Ashmore 1987; Santos-Sacchi 1989). The cell is almost fully elongated at -70 mV, the approximate value of the resting potential in vivo, but shortening is more linear in the depolarizing direction to about $+20$ mV (Santos-Sacchi 1989). Thus a sinusoidal voltage stimulus about the resting membrane potential leads to an asymmetric length change in which the dominant force would be equivalent either to an upward pull on the basilar membrane or a downward

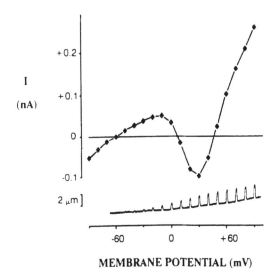

FIGURE 7.3 This graph shows that voltage-dependent length changes are independent of ionic current. Most of the outward ionic currents were blocked by replacing intracellular potassium ions with cesium ions. The magnitude of what would normally be the inward calcium current was increased by replacing extracellular calcium with barium. Under these conditions there was a net reversal of the current as the cell was depolarized to levels between -10 and $+40$ mV. The *lower trace* shows that the length change increased steadily with depolarization to more positive potentials, despite reversal of the current. (Reproduced with permission from Santos-Sacchi and Dilger 1988.)

pull on the reticular lamina. Studies in vitro also show that above stimulus frequencies of 10 Hz there is a measurable, cumulative shortening of the cell. It is not known if this DC shift occurs in vivo, particularly near threshold where the displacements may be only a few nanometers, and it is difficult to predict the consequences for cochlear mechanics.

The relationship between cell length and membrane potential has also been studied in cells mounted in a 10-μm-diameter polished pipette tip called a microchamber (Dallos, Evans, and Hallworth 1991; Evans, Hallworth, and Dallos 1991). The pipette tip can be sealed around the circumference of the cell at any point along its length, and cell length changes can be measured with a photodiode (Fig. 7.4A). The results from this preparation are similar to those obtained with patch electrodes, but they also suggest that the sensitivity of the mechanical response increases as the cell is depolarized. For example, in a recording from one particular cell, the sensitivity changed from about 5 nm/mV at -50 mV to a maximum of

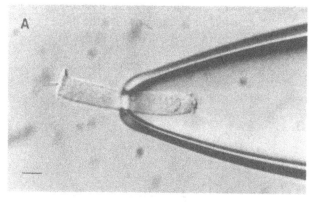

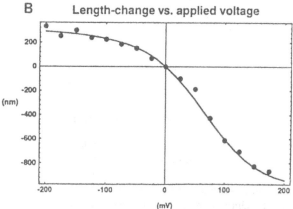

FIGURE 7.4 (A) Guinea pig OHC held in a polished pipette tip. Scale bar = 10 μm. (B) Recording of cell length as a function of applied voltage for the half of the cell protruding from the pipette. (From Dallos and Evans, unpublished.)

about 25 nm/mV at +17 mV, beyond which the response was abruptly saturated (Fig. 7.4B). This means that for *any* sinusoidal oscillation of membrane potential about any given resting potential above −70 mV, the cell should shorten fractionally more than it elongates. It is argued that this could introduce significant functional nonlinearities into the cochlea (Evans, Hallworth, and Dallos 1991). If the asymmetric dependence upon voltage is genuine, then it should reflect a basic property of the cellular mechanism. Although the main disadvantage of the microchamber is that the cell membrane potential cannot be accurately measured or controlled, similar asymmetries are also recorded under voltage clamp (Santos-Sacchi 1989). The microchamber is an elegant preparation that can also solve one major problem of experiments in vitro, that is, to replicate conditions in vivo where the apical surface of the cell is bathed in potassium-rich endolymph and the basolateral surface in sodium-rich perilymph.

Measurements of membrane capacitance from patch electrode recordings suggest that changes in membrane potential are restricted to the plasma membrane and that they are not transmitted directly to the intracellular membrane systems, particularly the large reservoir of membrane in the lateral cisternae. The membrane capacitance can be estimated by measuring the time taken to charge the membrane during a given change in membrane potential. Assuming that the membrane has a specific capacitance of about 1 $\mu F/cm^2$, the recorded capacitance accounts for a membrane that has about the same surface area as that of the cell (Ashmore 1987). This estimate does not take into account the nonlinear relationship between capacitance and membrane potential (Section 2.5.3), but considering the very large area of membrane in the lateral cisternae, we can conclude that the electrical stimulus is likely to be confined to the plasma membrane.

These physiological experiments suggest that the sensors for electromechanical transduction must lie in the plasma membrane. To detect changes in membrane potential efficiently, the sensors are most likely to be embedded within the lipid bilayer. But what is the nature of the motor?

2.2 Calcium and ATP

Many cellular mechanisms of force generation are based upon actin filaments or microtubules in conjunction with a wide range of force-generating molecules or "motor proteins." For example, myosin works against actin to power muscle contraction, dyneins work against microtubules to drive cilia and flagella, and kinesins also work against microtubules as part of a complex system of intracellular vesicle transport. All of these interactions depend upon chemical energy such as ATP and inorganic ions such as calcium (Alberts et al. 1989a).

To find out whether similar mechanisms might drive high-frequency length changes in OHCs, many experiments have been designed to stimulate length changes while simultaneously destroying the actin filaments or microtubules (Holley and Ashmore 1988a). Drugs such as cytochalasins B and D break actin filaments and cause dramatic changes in the stiffness of the cell but do not inhibit cell length changes. Similar results are obtained with colchicine and nocodazole, which act against microtubules. But many stable polymers of actin and tubulin are insensitive to these drugs, and the real test is the dependence upon calcium and ATP.

Calcium can be excluded from both intracellular and extracellular fluids by using calcium-free salines in the experimental chamber and the patch electrode. The remaining calcium can be further reduced to levels of about 1 nM by sequestration with ethylenediaminetetraacetic acid (EDTA) or 1,2-bis(2-aminophenoxy)ethane-N,N,N',N'-tetraacetic acid (BAPTA), but there is no evidence that removal of calcium inhibits cell length changes (Ashmore 1987; Santos-Sacchi 1989).

Excluding ATP is a little more difficult (Kachar et al. 1986; Holley and Ashmore 1988a). Numerous agents that block either aerobic or anaerobic production of ATP, such as iodoacetamide, sodium azide, and the proton ionophores carbonyl cyanide p-trifluoromethoxyphenylhydrazone (FCCP) and carbonyl cyanide m-chlorophenylhydrazone (CCCP), fail to block force generation even after hours of electrical stimulation. Nevertheless, these agents may not eliminate all sources of ATP. A more convincing experiment is to apply a nonhydrolyzable analog of ATP that has a much higher affinity for ATP binding sites than the native form. This does not inhibit the mechanism either. If ATP were involved, one might expect some dependence upon temperature, but the response has a value for Q_{10} of only 1.3 (Ashmore and Holley 1988). After numerous experiments by several research groups, we can reasonably conclude that the mechanism is not directly dependent upon ATP. In the cochlea, it is, of course, ultimately dependent upon the ATP used by the cells of the stria vascularis to maintain the electrochemical gradient between the endolymph and the perilymph (see Slepecky, Chapter 2). In experiments with single cells, the energy is supplied directly via the stimulus.

These experiments tell us that the high-frequency mechanism is quite different from those in other motile systems. The key difference is the independence of ATP. In fact, the speed of the response in itself suggests a direct biophysical mechanism with no intermediate biochemical steps. The latency of less than 0.1 ms is significantly faster than the activation time for muscle fibers, which, including the intermediate calcium step and the hydrolysis of ATP, is about 1 ms.

2.3 Location of the Motor Elements

If the high-frequency mechanism were dependent upon actin filaments or microtubules, it would be relatively easy to study the intracellular organization of these polymers in search of the motor apparatus. But this is not the case, and no obvious intracellular structures suggest themselves. Nevertheless, the speed of the motor response implies that the motor elements must be tightly coupled with the sensors and that they must lie close to the membrane potential field, that is, in the cell cortex either within or very close to the plasma membrane.

Early experiments designed to test this idea depended upon the ability to change intracellular pressure and volume via a patch electrode (Holley and Ashmore 1988a). When fluid is sucked out to reduce intracellular pressure without altering cell shape, the cell fails to elongate or shorten. Instead it generates irregular deformations of the lateral cortex. If a central muscle-like fibril drove the length changes, this response would not occur. Alternatively, the cells can be inflated until they eventually become spherical. Despite shortening by more than 30%, they are still able to generate length changes along the longitudinal axis as long as the intracellular pressure remains above a critical level (Fig. 7.5). Under such circumstances, cytoplasmic structures would be deformed to the extent that they would be unable to generate visible length changes, like a muscle trying to contract when it is bent. Furthermore, the large-scale fluid injection appears to cause substantial cytoplasmic damage with little effect on the mechanism.

Three different experimental approaches have been used to plot the distribution of force-generating elements. The first was based on the patch electrode, which functions both for electrical stimulation and for tethering at a point (Holley and Ashmore 1988a). If the electrode is attached to the base of the cell, the greatest recorded cellular displacement is at the cell apex. But the fractional displacements of structures along the cell are directly proportional to their distance from the base. If the electrode is attached to any point along the cell, the relative displacements of the base

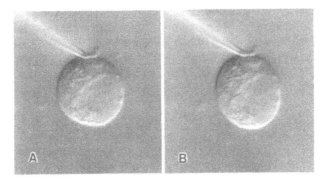

FIGURE 7.5 Guinea pig OHC held at +50 and −150 mV following a fluid inflation via a patch electrode attached beneath the cuticular plate (cp). The depolarized cell (A) is shorter and fatter than the hyperpolarized cell (B). (From Holley and Ashmore 1988a.)

and apex are again directly proportional to their distance from the electrode. These results are consistent with the idea that forces are generated along the full length of the lateral cortex.

The second technique employed the microchamber preparation (Dallos, Evans, and Hallworth 1991). When current is passed through the cell, one end shortens as its membrane depolarizes and the other end simultaneously elongates as its membrane hyperpolarizes. If the cell body is held in the middle, these displacements give the false impression that the cell is being sucked in and out of the pipette tip. By moving the position of the seal along the cell, the motor elements can be dissected electromechanically, thus showing that they can be activated independently along the longitudinal axis. The motor units are thus dissected as a series of cortical rings. Their distribution is accurately defined by this method as extending from the lower edge of the cuticular plate to the level of the nucleus at the base of the cell. The independent activation of the two halves of the cell provides further evidence that the mechanism lies in the cell cortex, since it is hard to imagine a contractile cytoplasmic structure that could produce the same effect. More importantly, the experiment shows that forces do not have to be transmitted via the cytoplasmic fluid to effect a change in cell length. Nevertheless, a positive cytoplasmic pressure is important to maintain cortical tension, otherwise forces could not be transmitted effectively along the cortex (Holley and Ashmore 1990).

The third technique depends upon stimulating small patches of membrane attached to the cell but isolated from it inside a patch electrode (Kalinec et al. 1992; Fig. 7.6). In most of the other experiments, the patch electrode is sucked onto the plasma membrane, and the membrane patch beneath it is then broken so that the pipette contents are continuous with the cytoplasm. In this case the unbroken membrane patch is observed at high magnification using video-enhanced light microscopy. When the membrane is hyperpolarized and depolarized, it is deflected back and forth inside the pipette. The nature of the movement is complex, because the membrane patch is stuck circumferentially to the inner surface of the pipette and the largest movements are recorded in the center. By sampling patches of membrane from along the cell, it is possible to map the distribution of the motors. The results agree with those obtained by the second method. However, they add one more piece of information, in that the motor elements can operate independently around the circumference of the cell as well as along the longitudinal axis.

The behavior of membrane patches adds further information on the nature of the mechanism. When cells elongate they become thinner, and when they shorten they become fatter, consistent with shape changes at constant volume. This means that cell surface area should increase during elongation and decrease during shortening. Volume measurements are not sufficiently accurate to establish this, but the behavior of membrane patches is more convincing. The patch profiles increase their curvature in

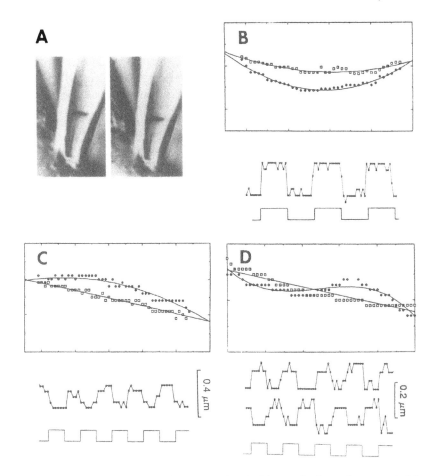

FIGURE 7.6 Movements of membrane patches inside patch electrodes recorded by video-enhanced light microscopy during depolarization and hyperpolarization of the membrane potential to +80 and −80 mV respectively. The outline of the patch (A) was recorded by plotting the pixel density from a digitized image (B–D). The curvature of the membrane profile always increased during hyperpolarization. Beneath each box are traces of the magnitude of the membrane displacement with respect to the square-wave voltage stimulus. (From Kalinec et al. 1992.)

either direction when they are hyperpolarized, that is, they can become concave or convex. But following depolarization they tend to flatten as if being pulled tight like a drum skin. The interpretation is that hyperpolarization increases membrane surface area and depolarization decreases it, a response that is entirely consistent with the behavior of the cell.

2.4 Structure of the Lateral Cortex

Unfortunately, we do not know which cortical structures are present in the patched membrane. The lateral cortex of an outer hair cell is composed of three distinct layers, namely the lateral cisternae, the cortical cytoskeletal lattice, and the plasma membrane. The cytoskeletal lattice is likely to remain attached to the plasma membrane within the patch electrode, and some cisternal membranes may also be present. Attention will now be directed to these structures as possible sites of force generation and transmission.

2.4.1 The Lateral Cisternae

The lateral cisternae are a specialized and substantial fraction of the endoplasmic membrane within the OHC (Fig. 7.7). They form multiple, highly ordered layers that line the lateral cytoplasmic surface of the plasma membrane (Gulley and Reese 1977; Saito 1983) and that broadly resemble smooth endoplasmic reticulum (Kimura 1975; Forge 1991). They often appear to be formed from flattened vesicles or tubes, especially in animals such as the rat and the role rat (*Spalax sp.*), where there is only one layer (von Lubitz 1981; Raphael and Wroblewski 1986). In the guinea pig, as many as 12 layers have been recorded, and they appear to form sparsely fenestrated sheets that have a surface area of many square micrometers

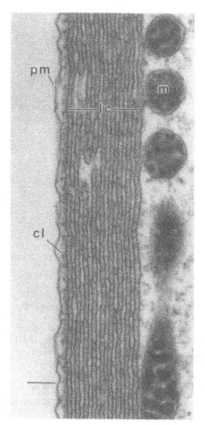

FIGURE 7.7 Longitudinal thin section through the lateral cortex of a guinea pig OHC. The most striking structures beneath the plasma membrane (pm) are the lateral cisternal membranes (lc). Immediately beneath the plasma membrane is the cortical lattice (cl), and inside the cisternae there are numerous mitochondria (m). Scale bar = 200 nm.

(Evans 1990; Forge et al. 1993). The number of layers and the morphology of these membranes vary considerably along a single cell, between different cells, and between different mammals (Furness and Hackney 1990). In considering these factors, it is worth remembering that the endoplasmic reticulum in most cells is dynamic and that its constant restructuring can be observed in real time using video microscopy. For example, in cultured kidney epithelial cells, the endoplasmic reticulum, labeled with the fluorescent dye 3,3'-dihexyloxacarbocyanine ($DiOC_6$), can extend tubular branches at over 1 μm/s (Lee and Chen 1988). Although the same dye has not revealed such dynamic events in OHCs, it does imply that the cisternae are dynamically linked with the rest of the endoplasmic membrane that extends beneath the cuticular plate and down the axis of the cell to the nucleus (Forge et al. 1993).

The endoplasmic reticulum of most cells effectively encloses a large intracellular space that is isolated from the cytoplasm and that can account for as much as 10% of the cell volume (Alberts et al. 1989b). As an integral part of the cellular machinery for protein and lipid synthesis and deployment, the lateral cisternae must at least be involved in the structural and biochemical maintenance of the cell. But they have also been compared directly to the sarcoplasmic reticulum of muscle cells, thus suggesting that they store calcium (Flock, Flock, and Ulfendahl 1986; Brundin et al. 1992). This idea receives strong support from experiments showing that antibodies raised against the calcium-ATPase from rabbit cardiac muscle also label the lateral cisternae (Schulte 1993).

It does not seem likely that the lateral cisternae play an important role in the mechanism of high-frequency cell length changes beyond that of structural maintenance. One theory for the mechanism, based upon electro-osmosis, required the cisternae simply to define the physical space beneath the plasma membrane (Kachar et al. 1986). The idea was that the charge on the inner surface of the plasma membrane would induce fluid flow along the cell of sufficient force to determine cell length. The fluid viscosity is likely to offer considerable resistance to such high-frequency fluid flow in this 25- to 30-nm space, especially since the space also includes the cortical cytoskeleton. Furthermore, the theory cannot explain the shape changes in cells inflated to form spheres or the irregular deformations observed after reduction of intracellular pressure (Holley and Ashmore 1988a). More conclusively, it cannot account for displacements in individual membrane patches, and the lateral cisternae can be stripped from the inner surface of the cortex without inhibiting motility (Kalinec et al. 1992; see below).

Interestingly, the lateral cisternae vesiculate after about 20 minutes in 5 to 10 mM sodium salicylate, and this is associated with a substantial inhibition of cell length changes (Dieler, Shehata-Dieler, and Brownell 1991). Although the effects of sodium salicylate on the motors are worth further study, those on the lateral cisternae are probably different, and we cannot conclude from the experiment that the cisternae are part of the high-frequency mechanism.

2.4.2 The Cortical Lattice

Closely associated with the outermost surface of the lateral cisternae about 25 nm beneath the plasma membrane there is an unusual protein skeleton called the cortical lattice. Although some of its filamentous components can be seen in thin sections for electron microscopy (Bannister et al. 1988), the structure does not contrast well against the adjacent cisternal membranes. However, if the membranes are removed with detergent, the cortical lattice is clearly revealed (Holley and Ashmore 1990; Arima et al. 1991; Fig. 7.8).

If the lattice is observed in the electron microscope after thin sectioning of embedded cells, negative staining of ruptured cells, or freeze-etching of freeze-fractured cells, it appears to be composed of two distinct types of filament (Holley, Kalinec, and Kachar 1992). The most obvious of these is about 5 nm in diameter and follows a circumferential path around the cell. Adjacent circumferential filaments are from 30 to 80 nm apart, and they are held in parallel arrays by thinner cross-links that are only 2-3 nm in diameter. These arrays form discrete domains that vary from just a few parallel filaments only 200 nm long to at least 10 filaments of up to 1 μm long. Their mean angle to the transverse axis of the cell is about 9°-15°, but the variability is great. Across areas of only 7 μm^2, measurements from individual filaments can cover a range of more than 130°.

The protein composition of the lattice is not particularly well defined. Labeling with the fungal toxin phalloidin shows that the protein actin is present. Destruction of the lattice by the enzyme DNAse1, which binds to actin specifically with a very high affinity, suggests that actin is an essential

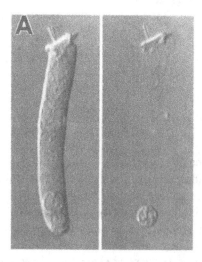

FIGURE 7.8 (A) Guinea pig OHC before (*left*) and after (*right*) incubation with 2.5% Triton X-100, a detergent that removes the cell membranes and soluble cytoplasmic constituents. The only remaining structure in the lateral cortex is the cortical lattice. (From Holley and Ashmore 1988b.)

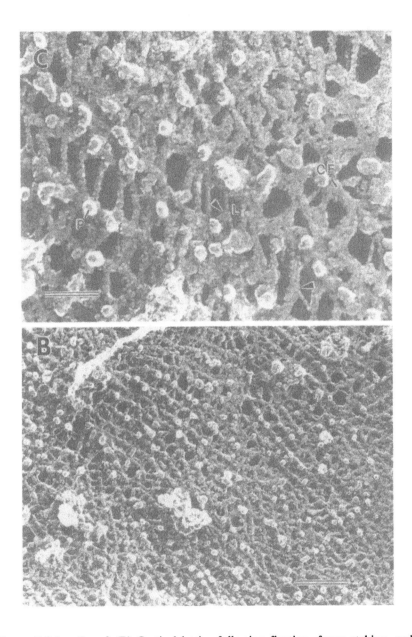

FIGURE 7.8 (*continued*) (B) Cortical lattice following fixation, freeze-etching, and coating with carbon and platinum for electron microscopy. Scale bar = 200 nm. (C) Higher magnification of B showing circumferential filaments (CF) held together by cross-links (L) that may appear single or branched (*arrow heads*). Pillars (P) connect the circumferential filaments to the plasma membrane. Scale bar = 50 nm. (From Holley, Kalinec, and Kachar 1992.)

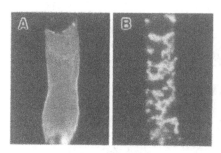

FIGURE 7.9 The structure of the cortical lattice depends upon the protein actin. (A) Preparation of the cortical lattice (see Fig. 7.8B), without the cuticular plate, labeled with a fluorescent probe that is specific for actin. (B) The lattice is destroyed by the enzyme DNase 1. (From Holley, Kalinec, and Kachar 1992.)

part of the structure (Fig. 7.9). The second component is probably some form of the elastic, filamentous protein spectrin, one form of which was first discovered beneath the plasma membrane of erythrocytes (Steck 1974). Antibodies to both blood and brain spectrins label the extracted lattice in the light microscope (Holley and Ashmore 1990), and the thin cross-links can be labeled with similar antibodies conjugated with colloidal gold for electron microscopy (Nishida et al. 1993). The cross-links are too thin to be actin and have similar dimensions to those of spectrin. The circumferential filaments must, therefore, contain actin even if they are not composed of continuous actin polymers.

Despite the similarity between the components, the spectrin skeleton in OHCs is quite different from that in erythrocytes, because it takes the form of an orthotopic array of distinct circumferential and longitudinal elements rather than an isotropic mesh resembling chicken wire (Byers and Branton 1985; Shen, Josephs, and Steck 1986; Fig. 7.10). The attachments to the plasma membrane also differ. The cortical lattice is connected to the plasma membrane by electron-dense "pillars" about 10 nm in diameter and 25 nm long (Flock, Flock, and Ulfendahl 1986; Arima et al. 1991) that are attached specifically to the circumferential filaments at intervals of about 30 nm (Holley, Kalinec, and Kachar 1992). In erythrocytes, the membrane links are associated with sites along the spectrin molecules. In OHCs, the fact that the pillars are not attached to the cross-links has important implications for cell mechanics (Section 2.6).

FIGURE 7.10 Relationship between actin (*filled cylinders*) and spectrin (*paired filaments*) in mammalian red blood cells. The pattern is quite different to that observed in OHCs.

There is little morphological evidence for the attachment sites of the pillars to the plasma membrane (Gulley and Reese 1977; Saito 1983; Forge 1991), thus suggesting either that they do not penetrate the membrane bilayer or that their attachments are obscured by the very high density of membrane protein (see Arima et al. 1991). Nevertheless, the pillars not only anchor the lattice to the plasma membrane but also define a precise, enclosed space of 25–30 nm between the plasma membrane and the outermost cisternal membrane (Fig. 7.11).

2.4.3 The Cortical Lattice as a Force Generator

The molecular properties of spectrin suggest a mechanism for the elastic properties of the cortical lattice. Each molecule is thought to be structured as a flexible chain of homologous, rigid segments each containing 106 amino acids arranged as a triple helix (Speicher 1986). Analysis of individual molecules in partially expanded red blood cell skeletons shows that the a and b subunits form a two-start helix that can accommodate axial length changes by alterations in pitch and diameter, that is, they behave like weak, two-stranded springs (McGough and Josephs 1990). It is this property that gives the erythrocyte skeleton its capacity for elastic deformations, and the same may be true for OHCs. But the molecule may also harbor a mechanism for changing cell shape. Its flexibility and highly acidic amino acid composition means that it can change shape reversibly with ionic strength (Amos and Amos 1991). The length of the tetramer changes from as little as 50 nm in 10 mM sodium chloride to 200 nm in 150 mM sodium chloride (Vertessy and Steck 1989). If this property applies to the cross-links in the cortical lattice, they may be capable of quite large charge-dependent length changes (Holley and Ashmore 1990). Actin might also be able to modulate the balance of forces in the cortical lattice. The mechanical properties of actin filaments can be changed dramatically by bound metals or nucleotides, and modulation of their flexibility, for example by calcium or magnesium, may have important physiological consequences (Orlova and Egelman 1993; Egelman 1994).

In the enclosed 25-nm space beneath the plasma membrane, the cell may be able to generate rapid changes in the ionic environment of the lattice. But there are several problems with this hypothesis. The first is that the lattice extends below the level of the nucleus, where there is apparently no force generation. The second, more important, problem is how the sensors in the plasma membrane might communicate with the cross-links. The mechanism is not dependent on ionic current, and there is no obvious way in which membrane potential might be coupled to the cross-links.

If the motors depend on either the lateral cisternae or the cortical lattice, the high-frequency response should be inhibited when these structures are removed. At present, only one experiment relevant to this question has been done, which depends on the perfusion of proteolytic enzymes into the cell via the patch electrode (Kalinec et al. 1992). The lateral cisternae are inherently unstable and are disrupted by almost any experimental procedure

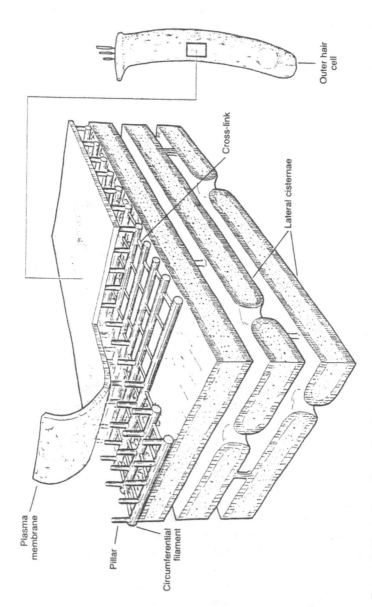

FIGURE 7.11 A three-dimensional diagram of the structure of the lateral cortex of an OHC. (From Holley 1991.)

(Evans 1990). On the other hand, the cortical lattice is a very stable structure, although it cannot be seen in intact cells viewed through the light microscope. The main problem with the enzyme experiment is to be sure that the lattice has been destroyed. The lattice can be isolated by extracting the cell membranes with detergent and it can then be labeled with rhodamine–phalloidin, a fluorescent probe for actin. Digestion of the fluorescent lattice can then be assayed visibly after application of various enzymes. Trypsin at (a concentration of) 0.01 μg/ml is highly effective, and in fact spectrin is known to be very sensitive to this enzyme. Intact cells perfused with 1000 times that concentration lose their cisternae and their nucleus within 5–10 minutes, and they also lose their shape. But 20 minutes later they can still generate voltage-dependent deformations. If the plasma membrane is then removed with detergent, there is no sign of the lattice. The conclusion from this single experiment is that both the sensory and the motor elements of high-frequency length changes are housed within the plasma membrane. Although there is no other direct experimental evidence for this conclusion, it makes very good sense in terms of the physiology (also see Huang and Santos-Sacchi 1993).

2.5 *The Lateral Plasma Membrane*

The plasma membrane has three main components that concern us in the context of cell motility: the glycocalyx, the proteins, and the lipids (Fig. 7.12).

2.5.1 The Glycocalyx

The glycocalyx is a layer of sugar groups that are attached either to membrane proteins or to lipids. This extracellular coat can be analyzed to some extent by labeling with proteins called lectins that bind specifically to different mono- or oligosaccharides. For example, wheat germ agglutinin (WGA) binds to the outer surface of the basolateral membranes of OHCs, thus indicating the presence of sialic acid residues or *N*-acetyl-D-glucosamine (Santi and Anderson 1987; Gil-Loyzaga and Brownell 1988). Another lectin, *Helix pomatia* agglutinin (HPA), binds to the inner surface, indicating

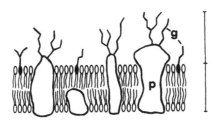

FIGURE 7.12 The three major components of the plasma membranes are the lipids, proteins (p), and glycocalyx (g). The latter is composed of sugars associated with protein (glycoproteins) and lipid (glycolipids). The glycocalyx lies on the outer surface of the plasma membrane.

N-acetyl-D-galactosamine residues. These sugars carry static charges that can serve to separate closely apposed membranes. The negatively charged sialic acid groups on the surfaces of erythrocytes may prevent cells from sticking together as they squeeze through tiny capillaries and blood vessels, and a similar effect may occur between stereocilia within hair bundles (Santi and Anderson 1987). It is thus conceivable that the glycocalyx is important for the high-frequency mechanism in OHCs (see Jen and Steele 1987). However, many sugar groups can be stripped from the cell surface by a variety of enzymes, such as hyaluronidase, neuroaminidase, and trypsin, without any obvious inhibitory effect upon the mechanism (Holley and Ashmore, unpublished).

2.5.2 Membrane Proteins

A high density of integral membrane proteins in the lateral plasma membrane is one of the distinguishing features of OHCs (Fig. 7.13). When the membrane bilayer is fractured so that the two membrane leaflets split apart, a dense array of particles can be seen on the inner surface of the cytoplasmic leaflet (Gulley and Reese 1977). The particles are 8–15 nm in diameter, and the packing density has been estimated to be from 2500 to 3000/μm^2 (Saito 1983; Kalinec et al. 1992) to as much as 6000/μm^2 (Forge 1991). These values are among the highest recorded for any cell, and in the original study the authors noted that outer and inner hair cells must serve different functions in sensory transduction (Gulley and Reese 1977). The cytoplasmic face of the axonal membrane of the Pacinian corpuscle, a mechanoreceptor from vertebrate skin and mesenteries, has a particle density of 2000–4000/μm^2 and similar values for the inner core lamellae (Ide and Hayashi 1987). These figures compare with values of 1300–2500/μm^2 and 300–900/μm^2 for the internodal axolemma of peripheral myelinated fibers and Schwann cells of myelinated axons, respectively. The particles in OHCs must include many of the normal membrane proteins, such as ion channels and structural links with the cytoskeleton, but it is suggested that the motor elements account for a major fraction (Kalinec et al. 1992). Pacinian corpuscles probably operate quite differently from OHCs, because transduction is dependent upon external sodium concentrations and ionic current, as opposed solely to membrane potential.

2.5.3 Membrane Proteins as Motors

Models for the way in which membrane proteins might drive the high-frequency mechanism are drawn entirely from unrelated systems. One early idea was based upon electrophoresis and was derived from experiments in which isolated cells were stimulated by an external electric field (see Kachar et al. 1986). The proposal was that similar voltage gradients within the cochlea act electrophoretically upon charged proteins embedded in the plasma membrane (Jen and Steele 1987). In fact, the charge might also be carried by the glycocalyx or membrane lipids. If one end of the cell is fixed,

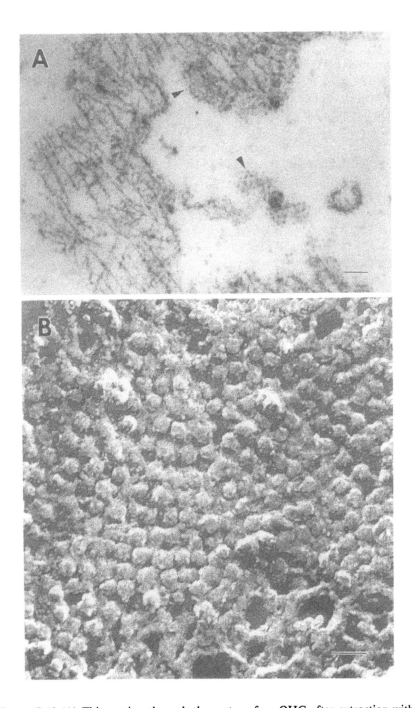

FIGURE 7.13 (A) Thin section through the cortex of an OHC after extraction with detergent. The cortical lattice lies close to an array of electron-dense particles located in the region of the overlying plasma membrane (*arrowheads*). Scale bar = 100 nm. (B) Particles observed in the lateral plasma membrane after the cell was extracted with detergent, chemically fixed, and freeze etched. Scale bar = 20 nm. (From Kalinec et al. 1992.)

changes in the voltage gradient might push or pull the rest of the cell by acting upon the fixed charges. Taking account of the elastic modulus of the cell wall, such a mechanism would require a charge density of 0.08 C/m^2, equivalent to one elementary charge per 200 Å^2, a value that lies within a reasonable physiological range. It is unlikely, however, that hair cells are stimulated in this way in vivo, and the evidence that the high-frequency mechanism is dependent upon transmembrane potential does not favor the theory. Furthermore, the effect might be expected to act on most of the other cells in the organ of Corti.

The best-known membrane proteins that respond to membrane potential are the voltage-gated ion channels. Since they clearly possess a voltage sensor and a mechanism for opening the channel, it is not unreasonable to model the high-frequency motor element as an integral membrane protein that can effect a voltage-dependent shape change. In voltage-dependent sodium channels, the sensor is thought to be a helical transmembrane region of the molecule, known as the S4 region, which carries a regular series of positive charges (Cohen and Barchi 1993). There are homologous structures in voltage-dependent calcium (McCleskey, Womack, and Fieber 1993) and potassium channels (Liman et al. 1991). The function of the S4 region has been explored by constructing channels in which positively charged arginine or lysine residues have been replaced by neutral or negatively charged residues. The effect is to decrease the steepness of the voltage dependence. A sequence of positively charged amino acid residues along the S4 helix sits opposite a sequence of negatively charged residues along another transmembrane helix called S7. A 60° rotation in each helix could lead to a 4.5Å translocation of the S4 helices perpendicular to the plane of the membrane. This involves a displacement of charge, and the activation of the channel is associated with a gating current equivalent to six positive charges moving from the intracellular to the extracellular surface. A slightly different mechanism probably exists in the voltage-dependent gating of gap junctions (Beyer 1993). Here the channel conductance is probably regulated by a tilting or twisting mechanism of the six transmembrane subunits that surround the pore.

Gating currents can also be recorded from OHCs (Ashmore 1992; Santos-Sacchi 1992). Since they do not depend upon the passage of ions through the membrane, they can best be observed when most of the normal ion channels are blocked (Fig. 7.14). The gating current in OHCs follows a very similar time course to that of the cell length change and has a similar potential dependence. It is the only recorded electrical event fast enough to be directly associated with the motor. The transfer of fixed charges is reflected in the membrane capacitance, which depends nonlinearly upon membrane potential and is maximal at a holding potential of about -30 mV (Fig. 7.15). The total charge transferred is, as expected, much greater for larger cells. If the holding potential is changed from -100 to $+100$ mV, the value is about 2 pC for a cell 70 μm long and about 0.25 pC for one only 23 μm long (Ashmore 1992). These values suggest the transfer of one elec-

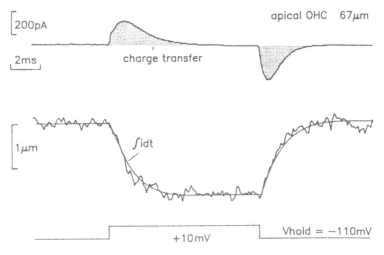

FIGURE 7.14 Recording of nonionic charge movement in an OHC in which ionic channels were blocked with 10 mM tetraethylammonium (TEA) ions. The *middle* trace shows the integrated charge transfer matched to the cell length change. The *bottom* trace shows the command potential. (Reproduced with permission from Ashmore 1990.)

tronic charge per 7–9 nm^2 of the cell membrane, an estimate that is fairly close to the density of integral membrane particles. Furthermore, gating currents can be recorded from membrane patches along the lateral surface of the cell, but not from the basal membrane, where no force generation occurs.

Gating currents and cell length changes are simultaneously inhibited by the external application of 0.5 mM gadolinium ions, thus providing experimental evidence that the two may be linked (Santos-Sacchi 1991). Gadolinium also shifts the voltage dependence of the response, reduces the voltage-dependent capacitance of the plasma membrane, and causes cell elongation. All these effects are reversible. At present, we can only speculate that the positively charged ions bind to functionally important, negatively charged sites on the cell surface.

The elements that determine the nonlinear capacitance of the cell can be mapped using an electrical "guillotine," an elegant combination of the microchamber and patch-clamp techniques (Huang and Santos-Sacchi 1993). The patch electrode can be attached to the portion of the cell that is excluded from the microchamber, and when membrane potential is clamped simultaneously via both systems, the cell capacitance of that portion alone can be measured. The nonlinear capacitance cannot be detected in the top few micrometers of the cell or within about 7 μm of the base. It is thus concluded that the sensors for high-frequency length changes have the same distribution as the motors. From an estimate of the cell surface area that contains sensors, it is also calculated that the charge density may be as high as 7500 $e^-/\mu m^2$, implying that the charge transfer within each elementary

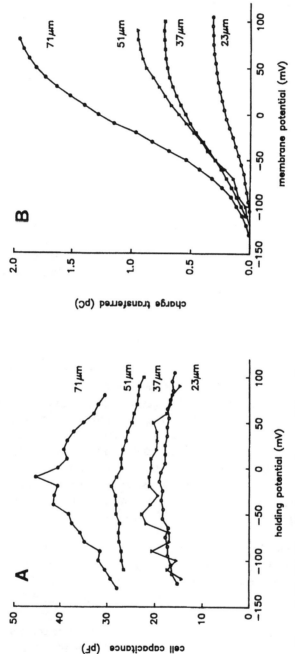

FIGURE 7.15 (A) Plot of the capacitance of the cell membrane as a function of membrane potential for four cells of different lengths. The relationship is nonlinear. (B) Integrated capacitance showing the total charge transfer for each of the four cells. (Reproduced with permission from Ashmore 1992.)

sensor may be equivalent to the displacement of at least two elementary charges across the plasma membrane rather than one. However, this calculation depends upon accurate measurements of the density of membrane particles in electron microscopical images. These measurements vary considerably (see Section 2.5.2), and it is important to obtain much more accurate data from much larger surface areas of the cell before the charge density can be effectively related to protein density.

In terms of the high-frequency mechanism, the existence of gating currents and the nonlinear capacitance of the plasma membrane favor hypotheses based upon protein conformational shape changes rather than electrostatic repulsion between separate elements within the plane of the lipid bilayer.

Gating currents can be generated not only by changes in membrane potential but also by mechanical stimulation (Iwasa 1993; Gale and Ashmore 1994). This suggests that the motor elements can be driven in reverse. If a dissociated OHC is anchored at one end by a patch electrode measuring whole-cell currents and at the other by a piezoelectric bimorph that can stretch the cell by up to 1 μm at velocities of up to 1500 μm/s. then transient inward and outward currents of 30 pA/μm/s can be recorded when the cell is stretched and shortened, respectively (Gale and Ashmore 1994). The current depends not only upon the velocity of the probe but also upon membrane potential. In one cell held at -30 mV, it was 105 pA, but it decreased progressively to 20 pA as the holding potential was either decreased to -120 mV or increased to 50 mV. This bell-shaped relationship with membrane potential clearly corresponds to the relationship between membrane potential and capacitance (Fig. 7.15). The kinetics, velocity dependence, and nonreversibility of the currents all suggest that they are the gating currents associated with the high-frequency motor elements. Capacitance also decreases in response to stretch, further suggesting that the current is nonionic. In the two experimental studies, the results have been used to calculate the area change of one elementary motor as 0.37 nm^2 (Gale and Ashmore 1994) and 2 nm^2 (Iwasa 1993).

2.5.4 The Membrane Protein Band 3

One family of membrane proteins that has attracted much attention recently is that of the anion exchangers (AEs) (Kalinec, Jaeger, and Kachar 1993; Tanner 1993). These proteins are involved in cellular pH and volume regulation, but the reason for looking at them in OHCs is that one member of the family (AE1) is responsible for linking the erythrocyte membrane to the spectrin skeleton. In erythrocytes there is a membrane domain of about 55 kDa that can form dimers or tetramers, and about 10%–15% of the monomers are directly linked to the spectrin skeleton via a cytoplasmic domain of about 45 kDa that is about 25 nm long (Low et al. 1984; Ursitti et al. 1991). The cytoplasmic domain could thus form the pillars that link the cortical lattice to the plasma membrane in OHCs. The membrane domain partitions preferentially to the cytoplasmic face, and in erythrocytes it accounts for a particle density of 2600–3300/μm^2. The majority of these are likely to represent the dimeric form. The membrane domain is composed of 14 transmembrane segments that are protected by the membrane from proteolytic digestion. We have already noted that the motors have similar protection. In reconstituted two-dimensional crystals, each dimer has a diameter of about 12 nm and can exist in a compact or extended conformation (Wang et al. 1993). Could this structural change behave as a motor element?

The initial studies have been encouraging for this theory (Kalinec, Jaeger, and Kachar 1993). The protein (band 4.1) required to link AE1 to the spectrin skeleton is present in OHCs, and polyclonal antibodies to AEs label the cortex as determined by immunofluorescence. Since the motor elements are unlikely to be channels, it is particularly interesting that when cochlear cDNA libraries were probed with a conserved region of the AE structure, a novel molecule was discovered that is apparently structurally altered in the region corresponding to the channel (Kalinec, Jaeger, and Kachar 1993). Nevertheless, there are numerous problems with this model. The pillars do not associate with the cross-links in the cortical lattice, the proposed location of spectrin (Holley, Kalinec, and Kachar 1992). The antibodies to the AEs label many cells in the organ of Corti, and it is no surprise to find AEs in the OHC. Furthermore, the structurally altered AE identified from the cochlear cDNA library could be associated with any cochlear cell, even if one assumes that it is naturally expressed.

2.5.5 Plasma Membrane Lipids

If the high-frequency motor elements change their conformation, they must be mechanically coupled to drive changes of cell shape. In this context, the composition of the lipid in the plasma membrane could be extremely important. One of the best-known membrane proteins is the light-sensitive protein bacteriorhodopsin that is found in the specialized purple membrane of the bacterium *Halobacterium halobium*. Under normal conditions, many copies of this protein are closely packed into a highly ordered two-dimensional hexagonal array, but if the membrane is treated with small amounts of detergent, various unusual orthogonal arrays occur (Sternberg et al. 1992). Thus the normal membrane needs a component that is extracted under these conditions. If either of two particular polar lipids is added to artificial membranes, the normal packing is observed, but in their absence it is not. The membrane particles in the freeze-etched replicas of the OHC plasma membrane appeared to be packed in hexagonal arrays in some areas and in orthogonal arrays in others (Kalinec et al. 1992). Since the preparation involved quite vigorous extraction with the detergent Triton X-100, these patterns are unlikely to represent realistic packing patterns or densities in vivo.

Little is known of the lipid content of the plasma membrane in OHCs, but it seems to include high levels of cholesterol that may reduce its flexibility and fluidity (Forge 1991). This and the high protein content should mean that the plasma membrane is relatively stiff. Lipid–lipid interactions in the plane of the membrane could vary by as much as 5%, which is similar to the movement that is proposed for the protein. This might absorb the forces along the length of the cell. Thus the total density of protein is important. The relatively high stiffness of the plasma membrane may resist shrinkage during fixation for electron microscopy, which is why it appears to have an undulating form in thin section whereas the cisternal membranes are straight (Flock, Flock, and Ulfendahl 1986; Holley, Kalinec, and Kachar 1992; Fig. 7.7).

2.6 The Vectorial Component of the Mechanism

Changing the surface area of the plasma membrane is not enough to drive a cell length change, and we must consider how forces or displacements produced in the plane of the membrane are directed into the longitudinal axis of the cell. The fact that electrically induced changes of both cell diameter and length occur following inflation from the normal cylindrical form to the spherical form must also be considered. The proposed motor elements in the membrane must be dramatically reorganized during these cell shape changes. There are no structural or experimental data to suggest that the motor elements change shape asymmetrically or that they are disposed within the membrane with any specific orientation. The cortical lattice may, however, be important in this respect. The pillars between the lattice and the plasma membrane are specifically located along the thick, circumferential filaments. Under normal conditions, the relatively inextensible circumferential filaments may restrain large changes in cell diameter while the elastic cross-links offer less resistance. The cortical lattice could thus behave as an orthotopic sheet that is stiffer circumferentially than it is longitudinally. This might ensure that the output of the motor elements is expressed along the longitudinal axis (Fig. 7.16), a conclusion supported by the theoretical model discussed in Section 2.7.

This conclusion finds support from mechanical measurements of the stiffness of OHCs. The axial stiffness of isolated cells, measured by longitudinal compression, is about 544×10^{-6} N/m (Holley and Ashmore 1988b). If the stiffness of the cortex is measured by the forces required to pump up the cell hydraulically, it appears to be much greater at about 7000×10^{-6} N/m (Iwasa and Chadwick 1992). In the second estimate, it is assumed that the cortex is isotropic, but if it is modeled as an orthotopic structure, the two measurements can be reconciled, because the cell will appear to be relatively stiff for internal pressure but compliant for axial force (Steele et al. 1993). This is clearly important, because it means that the elastic reactance of the cell is less of an obstacle to the transmission of

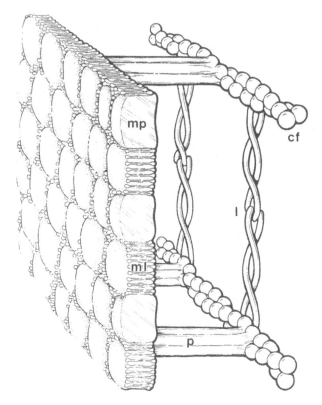

FIGURE 7.16 Three-dimensional construction of a section of the lateral plasma membrane with the proposed motor proteins (mp), the membrane lipid (ml), the pillars (p), and the circumferential filaments (cf) and cross-links (l) of the cortical lattice. Surprisingly little hard information is available to draw such a diagram, but it is useful as a focus for discussion.

forces to the basilar membrane. The orthotropy of the cortex is most likely to be a function of the cortical lattice. The axial stiffness is the same per unit length for both short and long cells (Fig. 7.17). This also applies to the isolated lattice, although its axial stiffness is only about 34×10^{-6} N/m (Fig. 7.17). The stiffness of the structurally irregular lateral cisternae is very unlikely to contribute significantly to that of the intact cell.

Morphological measurements suggest that the discrete structural domains of the cortical lattice may be able to move relative to each other while broadly retaining their orientation with respect to the cell axis (Holley, Kalinec, and Kachar 1992). Such measurements cannot at present be made from images of the membrane particles, but if we assume that each domain of the lattice is associated with a corresponding domain of motor elements in the plasma membrane, this could explain the ability of a cell to retain the function of the high-frequency mechanism despite substantial shape changes. Each domain could then be considered as a high-frequency "motor unit" composed of between 20 and 2000 motor elements (Fig. 7.18).

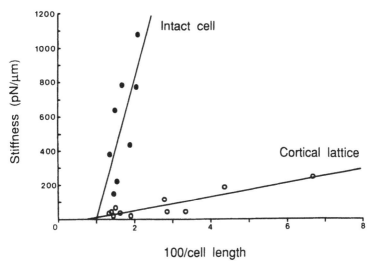

FIGURE 7.17 Stiffness of intact OHCs and extracted cortical lattices to longitudinal compression. Measurements were made with a fine, vibrating glass probe of known stiffness (see Holley and Ashmore 1988b.)

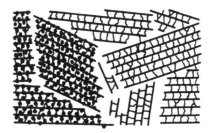

FIGURE 7.18 Diagram of the domain structure of the cortical lattice and the protein arrays (partially shown) in the lateral plasma membrane. It is not possible from present data to confirm this arrangement for the membrane proteins, but the data are relatively easy to obtain from large surface areas of the cortical lattice viewed in whole, extracted cells negatively stained for electron microscopy.

2.7 Summary Model of the Mechanism

One detailed model of the mechanism based primarily upon the microchamber experiments successfully predicts many of the observed cellular responses (Dallos, Hallworth, and Evans 1993). The theory is that each elementary motor, possibly a membrane protein, is a displacement generator that exists in two distinct, voltage-dependent states, long or short. The displacement is anisotropic, so that in the short state the element is fatter than it is in the long state. Conversion between the two states with respect to membrane potential is probabilistic, in the same way that it is for voltage-sensitive ion channels, and the transition involves the transfer of one electron charge across the membrane. Many elements are packed uniformly throughout the lateral surface of the cell to generate mechanical displacements that sum to generate a change of cell shape (Fig. 7.16).

In the model, each element is equipped with springs to account for the stiffness of the cell both axially and circumferentially and to link the elements in the two planes. To a large extent, these springs represent the mechanical properties of the cortical lattice. The model defines the "motor unit" as a circumferential ring of motor elements wound around the cell like a pearl necklace. This construction is directly related to the type of measurement that can be obtained from the microchamber preparation, and it appears to be a justifiable approximation. It should not be confused with the same term applied to the cytoskeletal and motor element domains described earlier. Finally, it is assumed that the cell operates at constant volume with a small, positive, internal hydrostatic pressure.

When quantified mathematically, the model describes the mechanical output of the cell, including the nonlinear changes in length and diameter that are expressed by experimental cells in the microchamber. The best results from the model are dependent upon the motor elements being tilted from their axes by about 20°, a value that agrees remarkably well with the mean angle of the circumferential filaments of the cortical lattice.

2.8 *The Stimulus for the High-Frequency Mechanism*

Any stimulus that drives a change in plasma membrane potential should activate the motors. Deflection of the stereociliary bundle toward the tallest row of stereocilia leads to depolarization and cell shortening, whereas deflection in the opposite direction leads to hyperpolarization and cell elongation (Evans and Dallos 1993). This conclusion comes from experiments with cells inserted up to their cuticular plates in the microchamber. Length changes are abolished reversibly if mechanoelectrical transduction is blocked by removing extracellular calcium or by applying aminoglycosides. Thus the normal stimulus must include sound stimulation to the stereocilia.

2.8.1 Stretch Activation

There are two other ways of changing membrane potential in OHCs. The first is via stretch-activated ion channels in the basolateral membrane. The experimental evidence here is slightly confusing, largely because it is very difficult to control the stimulus. In one set of data, there appear to be stretch-sensitive, nonselective cation channels with a reversal potential of about -12 mV (Ding, Salvi, and Sachs 1991). In another report, the data indicate that the main stretch-activated channel is potassium-selective (Iwasa et al. 1991). Direct mechanical stimuli to the cell bodies might thus stimulate the motors. In practice, the sensitivity of the stretch-activated channels is unlikely to match that of the stereociliary transducer, and the channels are more likely to be involved in mechanisms of volume regulation.

The second method of changing membrane potential comes from evidence that the high-frequency motors can be driven in reverse. The obvious problem here is that such a mechanism cannot draw additional energy for amplification, unlike that of the stereocilia, which can release some of the electrochemical energy provided by the potassium gradient between the endolymph and the perilymph.

It is not clear what kind of mechanical stimulus might be delivered to the cell body in vivo, but it seems unlikely that stretch activation of either ionic current or gating current is involved in driving cell length changes.

2.9 Expression of High-Frequency Displacements Within the Organ of Corti

The most striking thing about an OHC is the regular cylindrical shape of the cell body, which in the guinea pig is usually about 10 μm in diameter. The cuticular plate, supporting the stereocilia, sits in the apex like a stopper in the top of a test tube. The way that the cell is mounted into the sensory epithelium represents one of the most important specializations of the mammalian cochlea. All other hair cells that have been studied in the animal kingdom are integrated into the sensory epithelium like ordinary epithelial cells. But each OHC is attached to its adjacent supporting cells only at the apex and the base. The lateral surfaces of the cell body are free from cell contacts and are bathed in perilymphatic fluid (Slepecky, Chapter 2; Fig. 7.19).

The apical cuticular plates of both hair cells and supporting cells are tightly bound to form the reticular lamina. This has been presumed by many authors to provide a rigid base for interactions between the stereocilia and the tectorial membrane (eg. Gulley and Reese 1976). Indeed, it seems likely to be so, given that the stereocilia are sensitive to angular displacements of only $10^{-2°}$. The apical intercellular junctions are tight junctions that form complexes that extend up to 3 μm below the surface of the reticular lamina (Gulley and Reese 1976). These junctions are important not only for cell attachment but also for establishing a diffusion barrier between the perilymph and the endolymph. They are relatively well known, because similar junctions occur between many other cell types and they are fairly easy to prepare for standard electron microscopy. Although the junctions appear to be strong, structures observed in the microscope tell us very little about their mechanics. Nevertheless, although their mechanical strength has not been directly measured they are the last junctions to break during mechanical dissociation of cells, and they remain intact following total membrane extraction with detergent (Fig. 7.20).

The base of every OHC is cradled by a Deiters cell, but the structure of this junction shows no particular specialization and it is less well known. The key structure in the Deiters cell is the cytoskeletal bundle, composed mainly of microtubules and actin filaments. At its base the bundle is attached to the basilar membrane, but as it extends toward the reticular

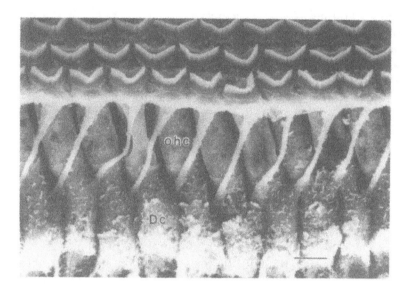

FIGURE 7.19 Scanning electron micrograph of the organ of Corti looking onto the third, outer row of outer hair cells after removal of the Hensens' cells. Note the relationship between the angles of the outer hair cells (ohc) and the processes of the Deiters' cells (Dc). Scale bar = 10 μm. (Courtesy of Dr. Andrew Forge.)

lamina it bifurcates (Angelborg and Engstrom 1973; Slepecky and Chamberlain 1983). One half goes to the base of the OHC, and the other projects up to the apex of the next OHC but one toward the apex of the cochlea (Fig. 7.19). The stiffness of these bundles is likely to be much less than that of the pillar cells, because they possess less than 20% of the number of microtubules, but it is probably an important factor in the context of OHC length changes. When OHCs change length, they should be expected to change the

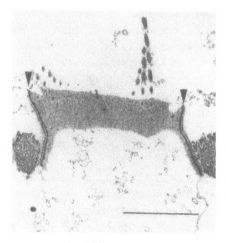

FIGURE 7.20 Thin section of the apical region of a hair cell after extraction with 2.5% Triton X-100. Nearly all the cell membranes were extracted with many soluble cellular constituents, but the structural integrity of the junctions (*arrowheads*) around the cuticular plate remained. Scale bar = 5 μm.

distance between the basilar membrane and the reticular lamina. The hair cell, reticular lamina, and Deiters cell process form a triangular structure, and the angle between the Deiters cell process and the hair cell means that length changes can indeed be expressed within the epithelium.

2.10 Can Outer Hair Cells Produce Enough Force to Influence the Basilar Membrane?

This question is central to the theory that OHCs are "cochlear amplifiers." The cell can generate a physiological response to a physiological stimulus, but does this mean anything to the cochlea or is it an experimental artifact? Calculations based upon the energy available and the cellular and fluid forces that must be overcome suggest that OHCs can produce sufficient force to displace the basilar membrane in vivo (Ashmore 1992; Iwasa and Chadwick 1992). But such calculations cannot entirely account for all the mechanical constraints that must exist within the complex structure of the organ of Corti. Indirect evidence from studies on otoacoustic emissions and distortion products already imply that forces produced by OHCs are important for normal cochlear mechanics (see Patuzzi, Chapter 4), but more direct evidence is needed. Recent experiments show that electrical stimulation across the organ of Corti, designed to drive length changes in OHCs, can generate basilar membrane responses that are very similar to those generated by normal acoustic stimuli (Nuttall and Dolan 1993; Xue, Mountain, and Hubbard 1993) and that are also associated with tuned responses in the cochlear nerve (Yates and Kirk 1993). Furthermore, simultaneous displacements of the basilar membrane the reticular lamina can be measured during electrical stimulation across the organ of Corti (Mammano and Ashmore 1993). The two structures move in opposite phase, which suggests that they are pushed apart and pulled together as the OHCs change length. The main problem with all these experiments is that they involve opening the cochlea, which inevitably alters the mechanical behavior of the cochlear fluids. Nevertheless, we may conclude that high-frequency length changes in OHCs really are a realistic component of cochlear function.

3. Slow Somatic Motility

Several different types of stimulus cause dissociated OHCs to change shape on a time scale of seconds or minutes rather than microseconds. These responses are quite distinct from the fast response described above, but they are much more difficult to characterize. In this section, several different stimuli are discussed, although they may in fact stimulate different parts of the same mechanism.

3.1 Effects of Elevated External Potassium

Dissociated OHCs depolarize and contract when the extracellular potassium concentration is raised to 25-125 mM by adding potassium chloride or potassium gluconate (Zenner, Zimmermann, and Schmitt 1985). They then relax within about 60 seconds when returned to about 5 mM potassium. Most experimental data suggest that the response is not dependent upon calcium. No inhibition is observed after removal of extracellular calcium with 1 mM EGTA (Zenner, Zimmerman, and Schmitt 1985; Dulon, Aran, and Schacht 1988) or when inward calcium currents are blocked by 50 μM methoxyverapamil (Dulon, Aran, and Schacht 1988). Intracellular calcium release can also be blocked by sodium dantrolene without effect (Slepecky, Ulfendahl, and Flock 1988). The potassium ionophore valinomycin has no effect, and neither does ouabain, which blocks the sodium-potassium ATPase, so it seems that shortening may be caused by osmotic uptake of water as a result of sustained depolarization (Dulon, Aran, and Schacht 1988). Since we have so little knowledge of the effects of potassium or of the relevance of the response, except perhaps in the context of Menière's disease (Dulon, Aran, and Schacht 1987), it is possibly better not to use it as a stimulus to evaluate slow motility.

3.2 Effects of ATP

When 1 mM ATP and 5 mM calcium are applied to dissociated OHCs that have been permeabilized with the detergent Triton X-100, the cells contract by about 14% at a maximum velocity of 3-24 nm/ms (Flock, Flock, and Ulfendahl 1986; Zenner 1988). The degree and rate of contraction vary with the concentration of ATP, and both ATP and calcium are required. This can be shown by reducing the calcium level with EGTA and by blocking ATP binding with adenylylimidodiphosphate, a nonhydrolyzable analog of ATP (Zenner 1988). During contraction, cell length and diameter decrease and the nucleus also becomes smaller (Flock, Flock, and Ulfendahl 1986). These observations indicate a general shrinkage of tissue, and the behavior of the nucleus implies a relatively nonspecific response from cellular components.

The addition of ATP and calcium to permeabilized cells does not cause cell elongation. This may be due to the absence of the usual cellular control mechanisms that are used to regulate ATP and calcium locally, or to the loss of cytoplasmic fluid pressure that allows mechanical forces to be transmitted through the cell cortex. The skinned muscle fiber has been a valuable preparation in muscle research. The highly structured contractile machinery remains intact following glycerine extraction, and the direct application of ATP and calcium merely bypasses earlier stages of signal transmission. But the motile mechanisms in OHCs are almost certainly damaged during detergent extraction, and extracted models may not yield coherent results.

3.3 Effects of Calcium

The concentration of ATP in most cells is 3–5 mM, so it is possible to get responses in intact cells by raising intracellular calcium, which is normally buffered at concentrations of 20–50 μM. The ionophores A23187 and ionomycin both selectively permeabilize the membrane to calcium so that it then equilibrates with the external concentration. With 125 mM extracellular calcium, the intracellular calcium takes some 30–60 seconds to reach 120 mM following addition of the ionophores (Dulon, Zajic, and Schacht 1990; Fig.7.21). The influx is apparently uniform across the cell surface. In contrast to the permeabilized cells, however, the influx leads to an increase of 1–2 μm in cell length with a simultaneous decrease in diameter. The change in diameter was greatest halfway along the cell and least at the basal

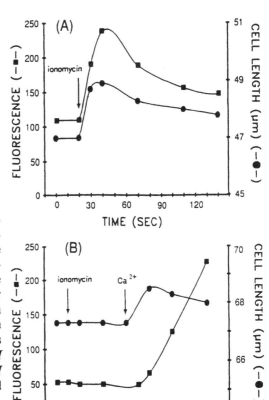

FIGURE 7.21 Effects of ionomycin on cell length and calcium-dependent fluorescence in the absence of extracellular calcium. OHCs loaded with the fluorescent calcium indicator were washed and incubated in calcium-free saline for 5 min (A) or 45 min (B). At the times indicated by the arrows, 10 μM ionomycin (A) or 1.25 mM calcium chloride (B) was added to the extracellular medium. (Reproduced with permission from Dulon, Zajic, and Schacht 1990.)

and apical regions, although in another report it seemed to be greatest in the basal region (Zajic and Schacht 1991). There is also an apparent increase of about 5% in the volume of the cell. The response is particularly interesting, because it involves cell elongation rather than shortening. Hypoosmotic solutions cause an apparently controlled shortening, but hyperosmotic solutions cause the cell to collapse. OHCs can only lengthen effectively by generating coherent forces in the cell cortex. The length change is almost completely reversible if the extracellular calcium is subsequently decreased with 2 mM EGTA. The ionophores can also release intracellular calcium, since they work, if only transiently, when the extracellular calcium has been removed.

There is evidence that the effects of calcium are modulated by calmodulin, an intracellular calcium-binding protein. Calmodulin is one of the most common mediators of calcium regulation, and it is present throughout the OHC (Slepecky and Ulfendahl 1993). It occurs in all cells from animals to plants and is highly conserved, with a molecular weight of about 17 kDa. It operates by binding to other functionally important molecules, such as myosin, the muscle ATPase. Both trifluoperazine and pimozide, which inhibit calmodulin, inhibit cell length changes but not the intracellular increase of calcium following the application of ionophores (Dulon, Zajic, and Schacht 1990; Fig. 7.22). Intracellular calcium can also be released by caffeine, and this generates a response similar to that obtained with high potassium (Slepecky, Ulfendahl, and Flock 1988). The effect is antagonized by tetracaine, which in other cells blocks caffeine-induced calcium release.

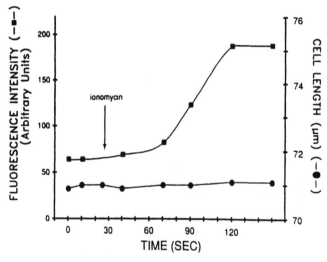

FIGURE 7.22 Effect of trifluoperazine on ionomycin-induced calcium increase and cell elongation. Hair cells were incubated with the inhibitor of calmodulin before the application of ionophore. The rise in intracellular calcium was not accompanied by a cell length change. (Reproduced with permission from Dulon, Zajic, and Schacht, 1990.)

Like the high-frequency mechanism, these calcium-dependent length changes are probably mediated within the cell cortex (Dulon, Zajic, and Schacht 1990). Spectrin possesses a calmodulin-binding site and we can speculate that changes in the geometry of the cortical lattice might provide the necessary forces (Section 2.4.3). But there are plenty of other possibilities. Among the lateral cisternae, there are significant concentrations of relatively unstable actin filaments that may also be important for force generation. Given the dependence upon calcium and ATP, the mechanism is unlikely to reside in the plasma membrane, but there are as yet no experimental leads to the problem.

3.4 The Function of Slow Somatic Motility

Slow cell length changes should have measurable effects on the geometry of the organ of Corti. Large-scale reductions in the distance between the basilar membrane and the reticular lamina are observed following reversible noise trauma (Harding, Baggot, and Bohne 1992). Such responses may be expected to inhibit the effect of the high-frequency mechanism and should also influence activation of the stereocilia. Thus it is proposed that slow length changes may account for the temporary threshold shifts that are observed during prolonged noise exposure (Johnstone, Patuzzi, and Yates 1986; Decory, Hiel, and Aran 1991).

4. Movement of the Cuticular Plate

All stimuli that generate cell length changes in dissociated cells also generate tilting in the cuticular plate with the same response characteristics (Zenner, Zimmerman, and Gitter 1988). The close association with both fast and slow cell length changes suggests that the response is most likely to be a passive property of dissociated cells. If the response is physiological, then the stimulus may be any that generates a cell shape change, but it is unlikely to occur in vivo. Any motion of the cuticular plate should be resisted by the deep belt of very well-developed intercellular junctions that lock the apex of the cell into the reticular lamina. Tight junctions are formed by close interactions between adjacent cell membranes, and in OHCs they are associated with a layer of densely stained cytoskeletal material on the cytoplasmic surface that may project some way below the cuticular plate (Fig. 7.20). It is extremely unlikely that the cuticular plate can tilt relative to the rest of the reticular lamina, and until this can be shown it would be bold to build the mechanism into a cochlear model.

5. Cell Length Changes to "Sound" Stimuli

"Sound-induced" length changes have been reported from OHCs (Canlon, Brundin, and Flock 1988; Brundin, Flock, and Canlon 1989; Brundin and

Russell 1993). In fact, the stimulus is not the same as that caused by sound pressure via the stapes, and it is not transmitted through the stereocilia. It is delivered to the basolateral membrane as sinusoidal pressure changes or fluid displacements via a fluid-filled micropipette driven by a piezoelectric vibrator. The cell responds with both tonic and phasic length changes.

The earlier experiments described the tonic response. When a 200-Hz fluid pressure stimulus is applied for about 1 second some 30 μm from the lateral cortex, the cell changes length unidirectionally by up to 4%. The long apical cells tend to elongate, but the short basal cells tend to shorten, and there is no response from the inner hair cells (IHCs). Although longer cells are generally more sensitive, each cell is best stimulated at frequencies associated with its location along the cochlea. For example, the best frequency for stimulating apical cells 80–90 μm long is about 0.2–0.6 kHz, whereas that for basal cells about 30 μm long is 3–10 kHz (Fig. 7.23).

More recent experiments describe faster length changes that follow the phase of the stimulus at frequencies of up to 3 kHz (Brundin and Russell 1993), the maximum frequency being limited by the apparatus. The stimulus causes lateral displacements of the cell cortex, so that the cell appears to become thinner and fatter as it lengthens and shortens, respectively (Fig. 7.24A). Thus it resembles high-frequency length changes. Both the phasic and tonic length changes are tuned to the same particular stimulus frequency for a particular cell (Fig. 7.24B).

The two responses are driven by mechanical deformations of any part of

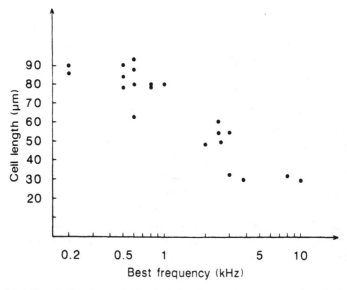

FIGURE 7.23 Correlation between the length of an OHC and the best frequency of fluid stimulation to elicit a change in cell length. Longer cells respond best to lower-frequency stimulation, and shorter cells respond best to higher-frequency stimulation. (Reproduced with permission from Brundin, Flock, and Canlon 1989.)

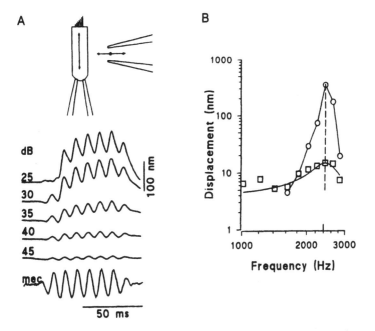

FIGURE 7.24 (A) Diagram of a guinea pig OHC held at the base by a pipette and stimulated from the side. The *bottom* trace shows the sinusoidal fluid displacement in the stimulating pipette. The other traces record cell length changes to stimuli of increasing intensity. The figures on the *left* show the relative intensity as attenuation of the stimulus in 5-dB steps. The absolute intensity cannot be determined accurately. At the maximum stimulus with an attenuation of 25-dB, the phasic response is clearly superimposed on the tonic response. (B) Tuned response of both the phasic (□) and tonic (○) components to a stimulus at 2400 Hz. (Reproduced with permission from Brundin and Russell 1993.)

the lateral cortex, but they probably depend upon different mechanisms. The tonic response is activated within milliseconds, but it is sensitive to pH changes and is blocked within about 30 minutes following application of sodium cyanide or dinitrophenol, both of which block the aerobic production of ATP (Canlon and Brundin 1991). In this respect, it is more like the slow, ATP-dependent length changes. It is also inhibited by external application of 3 μM poly-L-lysine, a positively charged molecule that binds to the plasma membrane. Poly-L-lysine inhibits smooth muscle contraction, possibly by inhibiting the formation of inositol triphosphate (IP_3) and calcium release from the sarcoplasmic reticulum. It has thus been argued that the relationship between the OHC plasma membrane and the lateral cisternae is analogous to that between the muscle cell plasma membrane and the sarcoplasmic reticulum (Brundin et al. 1992). In both cases, the two membrane systems are connected by a regular array of protein "pillars." In muscle cells, the pillars have a similar spacing, but they are only 15 nm long, half the length of the pillars in the OHCs. The strength of this argument, however, must depend upon physiological experiments, because although skeletal muscle shares these structures, it is not inhibited by poly-L-lysine.

The mechanism of the phasic response may be based on the same molecular structures as those responsible for the electrically stimulated, high-frequency length changes. Measurements of membrane potential, ionic current, and capacitance will be required before informed conclusions can be drawn. The normal stimulus might be via direct mechanical stimuli from the basilar membrane or from pressure changes in the surrounding perilymph. Stretch-activated ion channels are not thought to be involved, because they appear to require a much greater mechanical stimulus than that provided by the fluid displacements (Brundin and Russell 1993).

6. Motile Mechanisms in Nonmammalian Hair Cells

Motile mechanisms discovered in other hair cell types may be relevant to those in OHCs. These refer exclusively to the stereociliary bundle.

6.1 The Tip Link Motor

The work relating to the tip link motor revolves around a very clear question. The transducer mechanism is sensitive to displacements of less than 1 nm, but does it have to be precisely reset every time it is stimulated? Is it possible or necessary to maintain what would have to be an extraordinary level of mechanical stability in the entire sensory epithelium? The answer may lie in the process of adaptation, that is, the decrease in the depolarizing transducer current following sustained deflection of the hair bundle in the excitatory direction. Adaptation has been observed in amphibian (Shepherd and Corey 1994), reptilian (Crawford, Evans, and Fettiplace 1989), avian (Kimitsuka and Ohmori 1992), and mammalian hair cells (Kros, Rusch, and Richardson 1992), and the following mechanism may be generally applicable. It can be explained by a relaxation of the tip link tension, thus allowing the transducer channels to close. The model that has been most thoroughly tested involves an actin–myosin interaction at the top end of the tip link. The core of each stereocilium is composed of actin, which can act as a substrate for the translocation of myosin (Shepherd, Corey, and Block 1990). The tip link could be connected to a cytoplasmic form of myosin via some sort of integral membrane protein.

It is proposed that the interaction between the actin and myosin depends upon ATP and calcium, as in normal muscle. This means that there is a series of bound and unbound steps. In the absence of ATP, actin and myosin are strongly bound, but myosin detaches when it binds ATP. The myosin then hydrolyzes the ATP and becomes weakly bound to the actin. With the release of the free phosphate, the myosin binds tightly to actin and

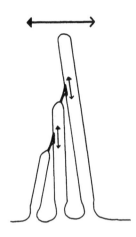

FIGURE 7.25 Diagram of the action of the tip links during displacement of the hair bundle. When the bundle is displaced in the inhibitory direction (*large, filled arrow*), the tip links are translocated toward the tips of the stereocilia (*small, filled arrows*). The opposite occurs when the bundle is displaced in the excitatory direction (*unfilled arrows*).

at the same time generates a power stroke. Calcium stimulates the binding of ATP. Thus, when the bundle is deflected in the excitatory direction, calcium should enter and promote a release in the tension of the crosslink by slippage (Fig. 7.25).

The attraction of this model is that it allows several experimental predictions (Gillespie and Hudspeth 1993). In the absence of ATP, there should be no adaptation. In fact, this proves to be so. When 5 to 25 mM ADP is dialyzed into the cell, adaptation is reduced, although it is not blocked. This may be because some of the ADP is converted to ATP by adenylate cyclase. Dialysis with ADP(βS), which cannot be converted, leads to a convincing block in most cells, and the effect is reversed by adding 2 mM ATP. These experiments show that adaptation is dependent upon ATP.

The search for myosin in stereocilia has produced conflicting conclusions over the years, but there is now strong evidence for a myosin of 120 kDa that is located near to stereociliary tips of hair cells from the bullfrog sacculus (Gillespie, Wagner, and Hudspeth 1993).

In muscle the effects of calcium are modulated by calcium-binding proteins such as troponin C and calmodulin. The stereocilia contain numerous calmodulin-binding proteins (Walker, Hudspeth, and Gillespie 1993). The key to this work is the ability to concentrate the protein from thousands of hair bundles and to detect calmodulin-binding proteins using a calmodulin–alkaline phosphatase probe with a highly sensitive chemiluminescent detection system. It is assumed that the proteins are located at the tip of the stereocilia where they would be optimally located to regulate calcium.

6.2 Voltage-Dependent Motility of Stereocilia

Forced step displacements of the stereociliary bundle in turtle hair cells cause damped mechanical oscillations that follow oscillations of the receptor potential (Crawford and Fettiplace 1985 Fig. 7.26). The response is probably driven by membrane potential, because the oscillations can be abolished by attenuating the receptor potentials with large depolarizing currents. Furthermore, a fine mechanical probe placed against the bundle can be displaced sinusoidally, with total deflections of as much as 30 nm. The mechanism of these movements is unknown, but their operational frequency range is only 20–320 Hz. It could thus be based upon a conventional actin–myosin interaction, and there is still a chance that it relates to the mechanism of adaptation described above. Although the mechanism is unknown, the phenomenon is convincing, and the conse-

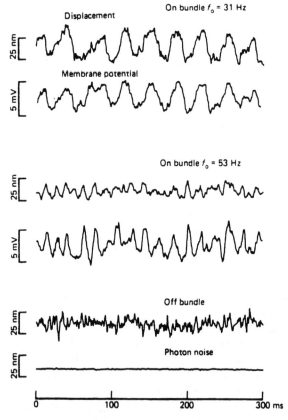

FIGURE 7.26 Spontaneous mechanical and electrical activity in two hair cells. The *top two* pairs of traces are single sweeps of simultaneously recorded bundle displacement and membrane potential. Displacement signals were obtained from the motion of a flexible fiber attached to the ciliary bundle. The *bottom* pair of traces shows the spontaneous motion of the fiber when it was detached from the bundle and an example of the photon noise expressed as an equivalent displacement with the fiber not in the light beam. (Reproduced with permission from Crawford and Fettiplace 1985.)

quence could well be to boost displacement of the basilar membrane near to its threshold (Crawford and Fettiplace 1985).

Displacements of the stereociliary bundles can also be recorded from eel (*Anguilla anguilla* L.) vestibular hair cells (Rüsch and Thurm 1990). Evidence for a motor mechanism comes from deteriorating isolated cells in which the kinocilium produces spontaneous oscillations. Step displacements of the bundle can be driven electrically, and both the kinocilium and the stereociliary bundle can be activated independently. Since the internal structures of the kinocilium and the stereocilia are quite different, the response might be based within the plasma membrane or derived from movements of the whole cell. The mechanism is probably different from that observed during adaptation, because it is independent of calcium and ATP.

7. Discrepancies, Problems, and Suggestions for Future Research

The behavioral repertoire of OHCs is indeed impressive, and the intense interest in high-frequency cell length changes has led to substantial progress. The normal stimulus is almost certainly delivered via the stereociliary transducer. The present challenge is to identify the motor element of the mechanism. This problem is being tackled in several different ways. One is to probe cochlear cDNA libraries for likely peptide sequences. Without a good lead, this can be an extremely difficult strategy. For example, one might search for a voltage-sensitive sequence based upon those found in voltage-dependent ion channels. But the key property here is the sequence of charges along the molecule, and this can be achieved by many different amino acid sequences. Thus, it is difficult to design an accurate probe, and successful results have not yet been obtained by this method. Another approach is to make an educated guess, select a likely family of molecules, generate antibodies to a conserved sequence within that family, check that it reacts with the cell, and then probe a cDNA library. This is the philosophy behind the work with the anion exchanger, band 3 (Kalinec, Jaeger, and Kachar 1993). Although this is a very constructive approach, many features of the motor elements cannot be accounted for by the known properties of band 3. A different approach is to assume nothing about the motor elements but to raise antibodies to the plasma membrane particles seen in freeze-etched hair cells. Appropriate antibodies could then be used to probe cDNA libraries, and they may also be useful for both structural and functional studies of the membrane. There is now a relatively simple method for making antibodies to OHCs in vitro (Holley and Richardson 1994), although it will still be important to prepare an effective antigen in which the plasma membrane fraction is enriched. Unfortunately, although the limited number of cells that can be taken from a single ear does not pose a significant problem for the physiologists who work with single cells, it is still a major problem for those who wish to do cellular biochemistry. Nevertheless, this challenge has been met with considerable success when there is some knowledge of the molecule involved (Gillespie, Wagner, and Hudspeth 1993; Walker, Hudspeth, and Gillespie 1993), and it will probably not be long before the high-frequency motor molecules are identified.

Attempts to draw a three-dimensional diagram of the lateral plasma membrane and cortical lattice reveal numerous weaknesses in our knowledge of the structure of this important complex. In particular, it will be essential to produce more reliable estimates of the density and organization of the membrane proteins. The published physiological measurements of the number of charges transferred per motor molecule depend upon such data. The present structural studies offer images over very small surface areas of membrane, and we cannot tell how many of the observed structures are likely to be motors. This kind of information will probably have to wait until the motors have been identified and can be labeled specifically.

Mutant animal strains that have a defective high-frequency mechanism have not yet been found. The waltzing guinea pig seems to have perfectly good high-frequency responses, but the ATP-dependent slow motility is inhibited (Canlon and Dulon 1993). This may help in our understanding of the slow mechanisms, and it would be particularly interesting to see whether the strain also expresses sound-induced responses. We certainly need more information about the slow, ATP-dependent responses and their control in vivo. At present it would be simplest to assume that they are all expressions of the same mechanism.

The tonic and phasic cell length changes in response to fluid pressure changes (Brundin and Russell 1993) are puzzling, and their significance should become more obvious with further physiological measurement. They are unlikely to effect any mechanical amplification unless they can tap the energy of the potassium gradient between the endolymph and the perilymph. They might be expressions of the other mechanisms in response to an unusual stimulus, hence the importance of establishing what kinds of stimuli can be expected in vivo.

The responses observed in hair bundles from amphibians and reptiles may be equally relevant to mammalian hair cells, even though the organ of Corti has such a distinctive cellular structure. This applies particularly to the mechanism of adaptation in the tip link, since the problem of "setting zero" for the stereociliary transduction mechanism must apply to most hair cells. Mammalian hair cells do not readily lend themselves to such studies, so these problems may take some time to solve.

The functional consequences of all the mechanisms discussed in this chapter are already under intense study. Numerous research groups have tackled the problems of how OHCs influence the mechanics of the basilar membrane. The evidence that OHCs can displace the basilar membrane in the absence of sound is a positive start, but the complex mechanical interactions during sound stimulation, especially at high frequencies, will need further patient and inventive experimentation. One of the greatest attractions of hearing research is the interaction between the many different aspects of the problem, from molecular biology and cell physiology in vitro to the mechanics and performance of the whole organ in vivo, and it seems that the field will remain lively and productive for some time.

Acknowledgments I should like to thank the Royal Society and the Wellcome Trust for their support. I am also very grateful to Jonathan Ashmore, Mike Evans, Jonathan Gale, Paul Kolston, and Fabio Mammano for many valuable discussions, and to Melissa Watson for helping to prepare the manuscript. The work was completed within the tenure of a Royal Society University Research Fellowship.

References

Alberts B, Bray D, Lewis J, Raff M, Roberts K, Watson JD (1989a) Molecular Biology of the Cell. 2nd ed. Chapter 11: The cytoskeleton. New York and London: Garland Publishing.

Alberts B, Bray D, Lewis J, Raff M, Roberts K, Watson JD (1989b) Molecular Biology of the Cell. 2nd ed. New York and London: Garland Publishing, p. 88.

Amos LA, Amos BA (1991) Molecules of the Cytoskeleton. London: Macmillan Education Ltd, p. 69.

Angelborg C, Engstrom H (1973) The normal organ of Corti. In: Moller AR (ed) Basic Mechanisms of Hearing. New York: Academic Press.

Arima T, Kuraoka A, Toriya R, Shibata Y, Uemura T (1991) Quick-freeze, deep-etch visualisation of the "cytoskeletal spring" of cochlear outer hair cells. Cell Tissue Res 263:91-97.

Ashmore JF (1987) A fast motile response in guinea pig outer hair cells: the cellular basis of the cochlear amplifier. J Physiol 388:323-347.

Ashmore JF (1990) Forward and reverse transduction in the mammalian cochlea. Neurosci Res Suppl. 12:S39-S50.

Ashmore JF (1992) Mammalian hearing and the cellular mechanisms of the cochlear amplifier. In: Corey DP, Roper SD (eds) Sensory Transduction. New York: Rockefeller University Press, pp. 396-412.

Ashmore JF, Holley MC (1988) Temperature dependence of a fast motile response in isolated outer hair cells of the guinea pig cochlea. Q J Exp Physiol 73:143-145.

Bannister LH, Dodson HC, Astbury AF, Douek EE (1988) The cortical lattice: a highly ordered system of subsurface filaments in guinea pig outer hair cells. Prog Brain Res 74:213-219.

Beyer EC (1993) Gap junctions. Int Rev Cytol 137c:1-37.

Brownell WE, Bader CR, Bertrand D, de Ribaupierre Y (1985) Evoked mechanical responses of isolated cochlear hair cells. Science 227:194-196.

Brundin L, Russell I (1993) Sound-induced movements and frequency tuning in outer hair cells isolated from the guinea pig cochlea. In: Duifhuis H, Horst JW, van Dijk P, van Netten SM (eds) Biophysics of Hair Cell Sensory Systems. Singapore: World Scientific, pp. 182-191.

Brundin L, Flock A, Canlon B (1989) Sound-induced motility of isolated cochlear outer hair cells is frequency-specific. Nature 342:814-816.

Brundin L, Wiklund NP, Gustafsson LE, Flock Å (1992) Functional and morphological comparisons between cochlear outer hair cells and muscle tissue in the guinea pig. Acta Physiol Scand 144:379-386.

Byers TJ, Branton D (1985) Visualisation of the protein associations in the erythrocyte membrane skeleton. Proc Natl Acad Sci USA 82:6153-6157.

Canlon B, Brundin L (1991) Mechanically induced length changes of isolated outer hair cells are metabolically dependent. Hear Res 53:7-16.

Canlon B, Dulon D (1993) Dissociation between the calcium-induced and voltage-driven motility in cochlear outer hair cells from the waltzing guinea pig. J Cell Sci 104:1137-1143.

Canlon B, Brundin L, Flock Å (1988) Acoustic stimulation causes tonotropic alterations in the length of isolated outer hair cells from the guinea pig hearing organ. Proc Natl Acad Sci USA 85:7033-7035.

Cohen S, Barchi RL (1993) Voltage-dependent sodium channels. Int Rev Cytol 137c:55-103.

Crawford AC, Fettiplace R (1985) The mechanical properties of ciliary bundles of turtle cochlear hair cells. J Physiol 364:359-379.

Crawford AC, Evans MG, Fettiplace R (1989) Activation and adaptation of transducer currents in turtle hair cells. J Physiol 419:405-434.

Dallos P, Evans BN, Hallworth R (1991) Nature of the motor element in electrokinetic shape changes of cochlear outer hair cells. Nature 350:155-157.

Dallos P, Hallworth R, Evans BN (1993) Theory of electrically-driven shape changes of cochlear outer hair cells. J Neurophysiol 70:299-323.

Decory L, Hiel H, Aran J-M (1991) In vivo noise exposure alters the in vitro motility and viability of outer hair cells. Hear Res 52:81-88.

Dieler R, Shehata-Dieler WE, Brownell WE (1991) Concomitant salicylate-induced alterations of outer hair cell subsurface cisternae and electromotility. J Neurocytol 20:637-653.

Ding JP, Salvi RJ, Sachs F (1991) Stretch-activated ion channels in guinea pig outer hair cells. Hear Res 56:19-28.

Dulon D, Aran JM, Schacht J (1987) Osmotically-induced motility of outer hair cells: implications for Menière's disease. Arch Otorhinolaryngol 244:104-107.

Dulon D, Aran JM, Schacht J (1988) Potassium-depolarisation induces motility in isolated outer hair cells by an osmotic mechanism. Hear Res 32:123-130.

Dulon D, Zajic G, Schacht J (1990) Increasing intracellular free calcium induces circumferential contractions in isolated cochlear outer hair cells. J Neurosci 10:1388-1397.

Egelman EH (1994) The ghost of ribbons past. Curr Biol 4:79-81.

Evans BN (1990) Fatal contractions: ultrastructural and electromechanical changes in outer hair cells following transmembranous electrical stimulation. Hear Res 45:265-282.

Evans BN, Dallos P (1993) Stereocilia displacement induced somatic motility of cochlear outer hair cells. Proc Natl Acad Sci USA 90:8347-8351.

Evans BN, Hallworth R, Dallos P (1991) Outer hair cell electromotility: the sensitivity and vulnerability of the DC component. Hear Res 52:288-304.

Flock Å, Flock B, Ulfendahl M (1986) Mechanisms of movement in outer hair cells

and a possible structural basis. Arch Otorhinolaryngol 243:83–90.
Forge A (1991) Structural features of the lateral walls in mammalian cochlear outer hair cells. Cell Tissue Res 265:473–483.
Forge A, Zajic G, Li L, Nevill G, Schacht J (1993) Structural variability of the sub-surface cisternae in intact, isolated outer hair cells shown by fluorescent labelling of intracellular membranes and freeze-fracture. Hear Res 64:175–183.
Furness DN, Hackney CM (1990) Comparative ultrastructure of subsurface cisternae in inner and outer hair cells of the guinea pig cochlea. Eur Arch Otorhinolaryngol 247:12–15.
Gale JE, Ashmore JF (1994) Charge displacement induced by rapid stretch in the basolateral membrane of the guinea pig outer hair cell. Proc R Soc Lond B 255:243–249.
Gillespie PG, Hudspeth AJ (1993) Adenine nucleotide diphosphates block adaptation of mechanoelectrical transduction in hair cells. Proc Natl Acad Sci USA 90:2710–2714.
Gillespie PG, Wagner MC, Hudspeth AJ (1993) Identification of a 120kD hair-bundle myosin located near stereociliary tips. Neuron 11:581–594.
Gil-Loyzaga P, Brownell WE (1988) Wheat germ agglutinin and *Helix pomatia* agglutinin lectin binding on cochlear hair cells. Hear Res 34:149–156.
Gulley RL, Reese TS (1976) Intercellular junctions in the reticular lamina of the organ of Corti. J Neurocytol 5:479–507.
Gulley RL, Reese TS (1977) Regional specialisation of the hair cell plasmalemma in the organ of Corti. Anat Rec 189:109–124.
Harding GW, Baggot PJ, Bohne BA (1992) Height changes in the organ of Corti after noise exposure. Hear Res 63:26–36.
Holley MC (1991) High frequency force generation in outer hair cells from the mammalian ear. Bioessays 13:1–6.
Holley MC, Ashmore JF (1988a) On the mechanism of a high-frequency force generator in outer hair cells isolated from the guinea pig cochlea. Proc R Soc Lond B 232:413–429.
Holley MC, Ashmore JF (1988b) A cytoskeletal spring in cochlear outer hair cells. Nature 335:635–637.
Holley MC, Ashmore JF (1990) Spectrin, actin and the structure of the cortical lattice in mammalian cochlear outer hair cells. J Cell Sci 96:283–291.
Holley MC, Richardson GP (1994) Monoclonal antibodies specific for endoplasmic membranes of mammalian cochlear outer hair cells. J Neurocytol 23:87–96.
Holley MC, Kalinec F, Kachar B (1992) Structure of the cortical cytoskeleton in mammalian outer hair cells. J Cell Sci 102:569–580.
Huang G, Santos-Sacchi J (1993) Mapping the distribution of the outer hair cell motility voltage sensor by electrical amputation. Biophys J 65:2228–2236.
Ide C, Hayashi S (1987) Specialisations of plasma membranes in Pacinian corpuscles: implications for mechano-electrical transduction. J Neurocytol 16:759–773.
Iwasa KH (1993) Effect of stress on the membrane capacitance of the auditory outer hair cell. Biophys J 65:492–498.
Iwasa KH, Chadwick RS (1992) Elasticity and active force generation of cochlear outer hair cells. J Acoust Soc Am 92:3169–3173.
Iwasa KH, Minxu L, Jia M, Kachar B (1991) Stretch sensitivity of the lateral wall of the auditory outer hair cell from the guinea pig. Neurosci Lett 133:171–174.
Jen DH, Steele CR (1987) Electrokinetic model of cochlear hair cell motility. J Acoust Soc Am 82:1667–1678.

Johnstone BM, Patuzzi R, Yates GK (1986) Basilar membrane measurements and the travelling wave. Hear Res 22:147–153.

Kachar B, Brownell WE, Altschuler R, Fex J (1986) Electrokinetic shape changes of cochlear outer hair cells. Nature 322:365–368.

Kalinec F, Holley MC, Iwasa K, Lim DJ, Kachar B (1992) A membrane-based force generation mechanism in auditory sensory cells. Proc Natl Acad Sci USA 89:8671–8675.

Kalinec F, Jaeger RG, Kachar B (1993) Mechanical coupling of the outer hair cell plasma membrane to the cortical cytoskeleton by anion exchanger and 4.1 proteins. In: Duifhuis H, Horst JW, van Dijk P, van Netten SM (eds) Biophysics of Hair Cell Sensory Systems. Singapore: World Scientific, pp. 175–181.

Kimitsuka T, Ohmori H (1992) The effect of caged calcium release on the adaptation of the transduction current in chick hair cells. J Physiol 458:27–40.

Kimura R (1975) The ultrastructure of the organ of Corti. Int Rev Cytol 42:173–222.

Kros CJ, Rusch A, Richardson GP (1992) Mechano-electrical transducer currents in hair cells of the cultured mouse cochlea. Proc R Soc Lond B 249:185–193.

Lee C, Chen LB (1988) Dynamic behaviour of endoplasmic reticulum in living cells. Cell 54:37–46.

Liman ER, Hess P, Weaver FW, Koren G (1991) Voltage-sensing residues in the S4 region of a mammalian K^+ channel. Nature 353:752–756.

Low PS, Westfall MA, Allen DP, Appell KC (1984) Characterisation of the reversible conformational equilibrium of the cytoplasmic domain of erythrocyte membrane band 3. J Biol Chem 259:13070–13076.

Mammano F, Ashmore JF (1993) Reverse transduction measured in the isolated cochlea by laser Michelson interferometry. Nature 365:838–841.

McCleskey EW, Womack MD, Fieber LA (1993) Structural properties of voltage-dependent calcium channels. Int Rev Cytol 137c:39–54.

McGough AM, Josephs R (1990) On the structure of erythrocyte spectrin in partially expanded membrane skeletons. Proc Natl Acad Sci USA 87:5208–5212.

Nishida Y, Fujimotor T, Takagi A, Honjo I, Ogawa K (1993) Fodrin is a constituent of the cortical lattice in outer hair cells of the guinea pig cochlea: immunocytochemical evidence. Hear Res 65:274–280.

Nuttall AL, Dolan DF (1993) Basilar membrane velocity responses to acoustic and intracochlear electrical stimuli. In: Duifhuis H, Horst JW, van Dijk P, van Netten SM (eds) Biophysics of Hair Cell Sensory Systems. Singapore: World Scientific, pp. 288–295.

Orlova A, Egelman EH (1993) A conformational change in the actin subunit can change the flexibility of the actin filament. J Mol Biol 232:334–341.

Raphael Y, Wroblewski R (1986) Linkage of sub-membrane-cisterns with the cytoskeleton and the plasma membrane in cochlear outer hair cells. J Submicrosc Cytol 18:731–737.

Rüsch A, Thurm U (1990) Spontaneous and electrically induced movements of ampullary kinocilia and stereovilli. Hear Res 48:247–264.

Saito K (1983) Fine structure of the sensory epithelium of guinea pig organ of Corti: subsurface cisternae and lamellar bodies in the outer hair cells. Cell Tissue Res 229:467–481.

Santi PA, Anderson CB (1987) A newly identified surface coat on cochlear hair cells. Hear Res 27:47–65.

Santos-Sacchi J (1989) Asymmetry in voltage-dependent movements of isolated hair cells from the organ of Corti. J Neurosci 9:2954–2962.

Santos-Sacchi J (1991) Reversible inhibition of voltage-dependent outer hair cell motility and capacitance. J Neurosci 11:3096–3110.

Santos-Sacchi J (1992) On the frequency limit and phase of outer hair cell motility: effects of the membrane filter. J Neurosci 12:1906–1916.

Santos-Sacchi J, Dilger JP (1988) Whole cell currents and mechanical responses of outer hair cells. Hear Res 35:143–150.

Schulte BA (1993) Immunohistochemical localisation of intracellular Ca-ATPase in outer hair cells, neurons and fibrocytes in the adult and developing inner ear. Hear Res 65:262–273.

Shen BW, Josephs R, Steck TL (1986) Ultrastructure of the intact skeleton of the human erythrocyte membrane. J Cell Biol 102:997–1006.

Shepherd GMG, Corey DP (1994) The extent of adaptation in bullfrog saccular hair cells. J Neurosci 14:6217–6229.

Shepherd GMG, Corey DP, Block SM (1990) Actin cores of hair-cell stereocilia support myosin motility. Proc Natl Acad Sci USA 87:8627–8631.

Slepecky N, Chamberlain SC (1983) Distribution and polarity of actin in inner ear supporting cells. Hear Res 10:359–370.

Slepecky NB, Ulfendahl M (1993) Evidence for calcium-binding proteins and calcium-dependent regulatory proteins in sensory cells of the organ of Corti. Hear Res 70:73–84.

Slepecky N, Ulfendahl M, Flock Å (1988) Effects of caffeine and tetracaine on outer hair cell shortening suggest intracellular calcium involvement. Hear Res 32:11–22.

Speicher DW (1986) The present status of erythrocyte spectrin structure: the 106-residue repetitive structure is a basic feature of an entire class of proteins. J Cell Biochem 30:245–258.

Steck TL (1974) The organisation of proteins in the human red blood cell membrane. J Cell Biol 62:1–19.

Steele CR, Baker G, Tolomeo J, Zetes D (1993) Electro-mechanical models of the outer hair cell. In: Duifuis H, Horst JW, van Dijk P, van Netten SM (eds) Biophysics of Hair Cell Sensory Systems. Singapore: World Scientific, pp. 207–215.

Sternberg B, L'Hostis C, Whiteway CA, Watts A (1992) The essential role of specific *Halobacterium halobium* polar lipids in 2D-array formation of bacteriorhodopsin. Biochim Biophys Acta 1108:21–30.

Tanner M (1993) Molecular and cellular biology of the erythrocyte anion exchanger (AE1). Semin Hematol 30:34–57.

Ursitti JA, Pumplin DW, Wade JB, Bloch RJ (1991) Ultrastructure of the human erythrocyte cytoskeleton and its attachment to the membrane. Cell Motil Cytoskel 19:227–243.

Vertessy BG, Steck TL (1989) Elasticity of the human red cell membrane skeleton. Effects of temperature and denaturants. Biophys J 55:255–262.

von Lubitz E (1981) Sub-surface tubular system in the outer sensory cells of the rat cochlea. Cell Tissue Res 220:787–795.

Walker RG, Hudspeth AJ, Gillespie PG (1993) Calmodulin and calmodulin-binding proteins in hair bundles. Proc Natl Acad Sci USA 90:2807–2811.

Wang DN, Kühlbrandt W, Sarabia VE, Reithmeier RAF (1993) Two-dimensional structure of the membrane domain of human band 3, the anion transport protein of the erythrocyte membrane. EMBO J 12:2233–2239.

Xue S, Mountain DC, Hubbard AE (1993) Direct measurement of electrically-evoked basilar membrane motion. In: Duifuis H, Horst JW, van Dijk P, van Netten SM (eds) Biophysics of Hair Cell Sensory Systems. Singapore: World Scientific, pp. 361-369.

Yates GK, Kirk DL (1993) Electrically evoked travelling waves in the guinea pig cochlea. In: Duifhuis H, Horst JW, van Dijk P, van Netten SM (eds) Biophysics of Hair Cell Sensory Systems. Singapore: World Scientific, pp. 352-360.

Zajic G, Schacht J (1991) Shape changes in isolated outer hair cells: measurements with attached microspheres. Hear Res 52:407-410.

Zenner H-P (1988) Motility of outer hair cells as an active, actin-mediated process. Acta Otorhinolaryngol 105:39-44.

Zenner H-P, Zimmermann U, Schmitt U (1985) Reversible contraction of isolated mammalian cochlear hair cells. Hear Res 18:127-133.

Zenner H-P, Zimmermann R, Gitter AH (1988) Active movements of the cuticular plate induce sensory hair motion in mammalian outer hair cells. Hear Res 34:233-240.

8
Physiology of Olivocochlear Efferents

JOHN J. GUINAN JR.

1. Introduction

Olivocochlear efferent neurons originate in the brain stem and terminate in the organ of Corti, thereby allowing the central nervous system to influence the operation of the cochlea. This chapter reviews the physiology and possible functional utility of mammalian cochlear efferents. Efferent physiology in a few other hair cell systems is also reviewed for the insight provided into mammalian efferent physiology. First a historical overview is presented, and then specific topics are considered in detail.

1.1 Historical Overview

Work on olivocochlear efferents is usefully considered in two periods. Initially, efferents were divided into crossed and uncrossed efferents. Later, it was found that a more basic division was into medial and lateral efferents. Many of the important physiological effects of efferents were discovered during the early period, but in the original publications some interpretations were clouded by the lack of understanding of the underlying anatomy.

1.1.1 Overview of Early Work on Efferents

The division of efferents into crossed and uncrossed groups came from the pioneering work of Rasmussen (1946, 1960). Rasmussen (1946) described fibers that originated in the medial part of the superior olivary complex, formed a compact bundle that crossed the midline near the floor of the fourth ventricle (forming the crossed olivocochlear bundle or COCB) and innervated the cochlea (Fig. 8.1, *top*). Later, he described uncrossed efferent fibers (the uncrossed olivocochlear bundle or UOCB) that also innervated the cochlea (Rasmussen 1960).

In the cochlea (Fig. 8.1, *bottom*), efferent fibers were found to synapse directly on outer hair cells (OHCs) and on radial auditory nerve fibers beneath inner hair cells (IHCs) (Smith 1961; Kimura and Wersäll 1962;

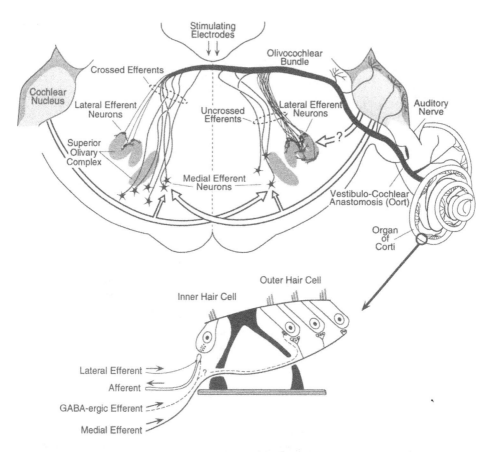

FIGURE 8.1. Schematic of the olivocochlear efferents to the right cochlea of a cat. *Top*: Transverse section through the brain stem showing the locations of olivocochlear neurons, the course of their axons to the cochlea, and the pathways of their known inputs. The *open arrows* indicate input pathways to efferent groups. Hypothetical inputs to lateral efferent neurons are indicated by (?). Stimulating electrodes are shown at their typical location in the floor of the fourth ventricle near the crossed olivocochlear bundle. *Bottom*: The main efferent innervation of the cochlea. Omitted are synapses between lateral and medial efferents in the tunnel of Corti, and efferent synapses onto supporting cells (see Liberman 1980; Liberman, Dodds, and Pierce 1990). The afferent shown is a myelinated, radial auditory nerve fiber; unmyelinated, spiral auditory nerve fibers are not shown. For GABAergic efferents, a subclass of lateral efferents, the question mark indicates uncertainty as to whether the same fiber that innervates radial auditory nerve fibers also innervates OHCs. (Top panel adapted from Liberman 1990.)

Spoendlin 1966). Although during this early period the principal targets of efferents in the cochlea were identified, the correspondence between the origins of the efferents and their targets was poorly understood.

Galambos (1956) excited efferent fibers by shocks at the floor of the fourth ventricle and showed that efferent activity inhibits click-evoked compound action potentials of the auditory nerve (N_1). With this paradigm, the greatest percentage reduction of N_1 was at low sound levels (Fig. 8.2). Desmedt (1962) noted that at low sound levels, the efferent suppression of N_1 was equivalent to shifting the sound-level function a constant amount toward higher sound levels. This efferent-induced "level shift" (Fig. 8.2) has become the standard index of the strength of efferent inhibition. Wiederhold (1970) showed that COCB stimulation produced a similar "level shift" in rate-versus-level functions from single auditory nerve fibers, with the biggest level shifts for auditory-nerve fibers with mid-range (6–15 kHz) characteristic frequencies (CFs). Finally, Nieder and Nieder (1970) showed that in a noisy background, efferent shocks can enhance the N_1 evoked by moderate-level clicks. Thus, efferents may have a role in reducing the effects of masking.

Efferents were also found to affect nonneural cochlear potentials. Fex (1962, 1967) showed that efferent stimulation increased the amplitude of the cochlear microphonic (CM) and evoked a slow potential that could be recorded at the round window or in the cochlea. These changes in cochlear potentials are due to efferents increasing the conductance of OHCs and changing current flows within the cochlea (Fex 1967). It was also found that shock-evoked efferent effects were largest for shocks at high rates (400/s) and that these effects take several hundred milliseconds to begin and to decay (see Desmedt 1962).

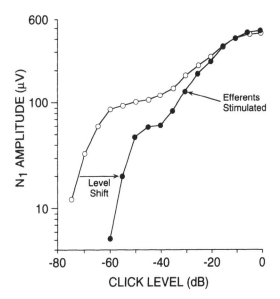

FIGURE 8.2 Effects of efferent stimulation on the compound action potential of the auditory nerve, N_1. The curves show N_1 amplitude without medial efferent stimulation (○) and following medial efferent stimulation (●). The level shift is the amount by which efferent stimulation shifts the response curve to higher sound levels. N_1s were evoked by clicks. Efferents were excited by midline OCB stimulation. (Adapted from Wiederhold 1970.)

Another important finding was that efferent fibers can be excited by contralateral sound and probably also by ipsilateral sound (Fex 1962). Thus, efferents provide feedback control of the cochlea that could function to change the dynamic range of hearing. Several papers suggested that efferents can be activated by the central nervous system in addition to being activated by sound, and that they may play a role in selective attention (Oatman 1971, 1976).

Finally, several papers reported that electrical stimulation of uncrossed efferents depressed N_1 without causing an increase in CM and interpreted these effects as being due to efferent synapses on auditory nerve fibers beneath IHCs (Sohmer 1966; Fex 1967). As will be seen later, this interpretation was based on an incorrect application of efferent anatomy.

The first step toward the new view of efferent anatomy came from Warr (1975). He injected the tracer horseradish peroxidase into the cochleae of cats and identified the retrogradely labeled efferent cell bodies in the brain stem. This showed that there were far more efferent neurons than previously thought and that there were more uncrossed efferents than crossed efferents.

1.1.2 Overview of Recent Work on Efferents

A turning point in our understanding of olivocochlear efferents was the classification of efferents into lateral and medial groups (Fig. 8.1) (Warr and Guinan 1979; Guinan, Warr, and Norris 1983). This led to the demonstration that all of the then-known effects of olivocochlear efferents are produced by the medial group (Gifford and Guinan 1987; Guinan and Gifford 1988a,b,c). In addition, single-fiber recording and labeling indicated that all recordings from olivocochlear efferents were from medial efferents (Robertson and Gummer 1985; Liberman and Brown 1986; Brown 1989).

During this period, efferent effects provided important clues for understanding cochlear mechanisms, which, in turn, helped in understanding efferent mechanisms. A revolution in thinking about cochlear function was brought about by the discoveries of (1) otoacoustic emissions (sounds detectable in the ear canal that are produced by the cochlea; Kemp 1978), (2) changes in otoacoustic emissions produced by efferent synapses on OHCs (Mountain 1980; Siegel and Kim 1982), and (3) OHC motility (Brownell et al. 1985). It is currently thought that OHCs provide the motile element in a "cochlear amplifier" that enhances the sensitivity and frequency selectivity of the cochlea (Davis 1983; Dallos and Evans 1995). Since medial efferents synapse on OHCs, it seems likely that medial efferents influence the cochlear amplifier and act to control the sensitivity of the cochlea. The best-known efferent effect, the shifting of sound-level functions to higher sound levels, would then be due to a reduction of the gain of the cochlear amplifier.

Although the early literature emphasizes efferent effects at low sound levels, recent work suggests that the most significant effects of medial efferents may be at moderate and high sound levels. At very high sound levels, medial efferents inhibit auditory nerve fibers with low spontaneous rates (SRs) (the group that can signal changes in level at high sound levels) by at least as much as the 20-dB level shifts seen at low sound levels (Guinan and Stankovic 1995). New data confirm and extend the older observations suggesting that medial efferents play a significant role in reducing the masking of moderate-level transient sounds by low-level background noise (Kawase, Delgutte, and Liberman 1993). Other recent data suggest a role of efferents in preventing damage due to intense sounds, but this may be a limited role (Rajan and Johnstone 1983; Rajan 1988b; Reiter and Liberman 1995).

Efferent inhibition at high sound levels cannot be fully explained by a reduction of the gain of the cochlear amplifier. The additional mechanism(s) involved are poorly understood, but recent work showing differences in the time courses of mechanical versus electrical effects of efferents provides an indication that electrical effects may have an important role at high sound levels (Guinan and Stankovic 1995).

One recent line of work has been on the "olivocochlear acoustic reflex." Medial efferents respond to ipsilateral and contralateral sound with sharp tuning and feed back to a cochlear place tuned to the same frequency as the efferent fiber (Robertson 1984; Liberman and Brown 1986). Contralateral sound has been used to evoke medial efferent activity and to study the effects produced on N_1, single auditory nerve fibers and otoacoustic emissions (Buño 1978; Warren and Liberman 1989a; Collet et al. 1990; Puel and Rebillard 1990). Since measurements of otoacoustic emissions are noninvasive, they can be used to measure efferent effects in humans. In addition to indicating that efferent effects in humans are like those in animals, such studies may help to reveal the functional significance of efferents, such as a possible role in selective attention (e.g., Puel, Bonfils, and Pujol 1988).

Since all known efferent effects and all efferent recordings appear to be from medial efferents, this leaves us knowing very little about lateral efferents, the most numerous efferent group. Most efferents are cholinergic (use acetylcholine as a neurotransmitter), but lateral efferents release a variety of potentially neuroactive substances (see Sewell, Chapter 9). What these neuroactive substances do is another little-understood aspect of lateral efferent physiology.

1.2 Efferent Anatomy

The following is a brief review of the anatomy of olivocochlear efferents, applicable to most mammals. For more complete reviews see Warr, Guinan, and White (1986) and Warr (1992).

1.2.1 The Lateral and Medial Efferent Groups

By anterograde tracing, Warr and Guinan (1979) and Guinan, Warr, and Norris (1983) showed that there are two main groups of olivocochlear efferents: lateral and medial. Lateral olivocochlear (LOC) efferents originate from small neurons in and near the lateral superior olivary nucleus (LSO), have unmyelinated axons, and terminate on the dendrites of auditory nerve radial afferent fibers in the region beneath IHCs (Fig. 8.1). Medial olivocochlear (MOC) efferents originate from larger neurons located medial, ventral, and anterior to the medial superior olivary nucleus (MSO), have myelinated axons, and terminate directly on OHCs (Fig. 8.1). The exact locations of the neurons of origin vary with species, but the principal cochlear termination of each group is consistent across species. In most species, lateral efferents are the most numerous and project predominantly to the ipsilateral cochlea, whereas medial efferents project predominantly to the contralateral cochlea.

Lateral and medial efferents have different patterns of innervation along the length of the cochlea, as shown for the cat in Fig. 8.3. This figure was obtained by combining data from the two quantitative studies of these patterns in the cat (Guinan, Warr, and Norris 1984; Liberman, Dodds, and Pierce 1990). Medial efferent innervation is largest near the center of the cochlea, with the crossed innervation biased toward the base compared to the uncrossed innervation (Fig. 8.3A,B). In contrast, lateral efferent innervation is relatively constant in the center and base of the cochlea, but the crossed lateral innervation shows a strong bias toward the apical, low-frequency end of the cochlea (Fig. 8.3C,D).

One of the most important differences between medial and lateral efferents is that medial efferents are myelinated but lateral efferents are unmyelinated (evidence reviewed by Guinan, Warr, and Norris 1983; Warr 1992). Myelinated fibers are far more readily stimulated by extracellular currents than are unmyelinated fibers (Hallin and Torebjork 1973; Fitzgerald and Woolf 1981). Because of this, Guinan, Warr, and Norris (1983) hypothesized that (1) electrical stimulation excites medial efferents but not lateral efferents, and (2) the efferent effects that can be demonstrated by electrical stimulation (i.e., almost all known efferent effects) are due to medial efferents. Myelination is also important, because myelinated fibers are readily impaled and recorded from using high-impedance pipet electrodes, whereas unmyelinated fibers are almost impossible to record from with such electrodes. Together, these observations imply that almost all physiological knowledge is about medial efferents, and that almost nothing is known about the physiology of lateral efferents. Throughout this chapter, the involvement of medial versus lateral efferents will be reconsidered wherever it is critical for the topic being discussed.

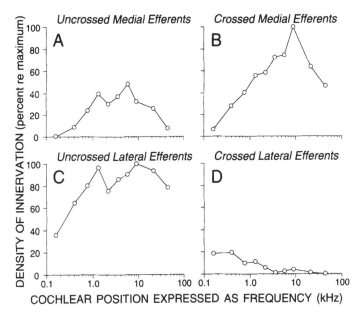

FIGURE 8.3 Efferent innervation density as a function of cochlear frequency in the cat. These plots were obtained by combining the anterograde-transport data of Guinan, Warr, and Norris (1984), which separate crossed and uncrossed innervation but use a less accurate method to determine position along the length of the cochlea, with the electron microscopic (EM) synaptic counts (for lateral efferents) and synaptophysin immunostaining (for medial efferents) of Liberman, Dodds, and Pierce (1990). The distances along the length of the cochlea of all the anterograde-transport data were adjusted to position the peak of the total anterograde OHC innervation at 9 kHz, which is close to the peak of the synaptophysin immunostaining. The medial efferent densities (panels A and B) are the anterograde data with the adjusted frequency axis. The lateral efferent densities (panels C and D) were obtained by averaging the combined ipsilateral and contralateral anterograde transport data with the EM synaptic counts to get the total lateral efferent innervation density along the length of the cochlea (done visually because the plots had different point spacings). This was then divided into uncrossed and crossed using the ratios obtained from the anterograde-transport data. Cochlear position is expressed in terms of frequency using the map of Liberman (1982). Density of innervation is expressed relative to the maximum point with the medial and lateral efferents considered separately.

1.2.2 Cholinergic and GABAergic Efferents

Recent work has shown that efferents are neurochemically complex. Medial efferents and most lateral efferents use the neurotransmitter acetylcholine (ACh), i.e., they are cholinergic (see Sewell, Chapter 9). The remaining lateral efferents appear to use the neurotransmitter γ-aminobutyric acid (GABA). In addition, lateral efferent neurons may contain other neurotransmitters or neuromodulators, including enkephalins and calcitonin gene-related peptide (CGRP) (see Sewell, Chapter 9).

Neurochemical staining indicates that there are two subgroups of lateral efferents but only one group of medial efferents. In rodents, cholinergic and GABAergic lateral efferents are distinct populations, with cholinergic lateral efferents also containing CGRP whereas GABAergic lateral efferents do not (Eybalin and Pujol 1989; Vetter, Adams, and Mugnaini 1991). These lateral efferents innervate the dendrites of radial auditory nerve fibers throughout the length of the cochlea, and OHCs primarily in the apical half of the cochlea (Altschuler and Fex 1986; Eybalin et al. 1988; Vetter, Adams, and Mugnaini 1991). Although they innervate OHCs, these are lateral efferents, not medial efferents, because they originate in the lateral part of the superior olivary complex (Schwarz et al. 1988; Vetter, Adams, and Mugnaini 1991). In cats, there are few, if any GABAergic efferents (Vetter, Adams, and Mugnaini 1991) and, correspondingly, only a minor projection from the lateral region to the OHCs (Guinan, Warr, and Norris 1983). In both cats and rodents, medial efferents appear to be cholinergic but not GABAergic or containing CGRP. Although there is little evidence that medial efferents have different branching patterns or neurochemistry in the apex versus the base, it is interesting that there are high-frequency versus low-frequency differences in their responses to sound (Section 6.1.1) and their inhibition of the tuning curves of auditory nerve fibers (Section 3.2.1). Thus, both lateral and medial efferents are different in some way in the apex versus the base.

In guinea pigs and rats, two classes of lateral efferents have been distinguished by their morphology in the cochlea (Brown 1987; Warr and Beck 1995). It is not known whether there is an association between these two classes and the cholinergic-GABAergic division of lateral efferents.

2. Medial Efferent Effects and Mechanisms: I. Cochlear Gross Potentials

The following three sections review medial efferent effects in detail. This section reviews effects on gross cochlear potentials, the identification of these effects as being due to medial efferents, and the mechanisms involved in producing these effects. The next section (Section 3), reviews primarily efferent effects on single auditory nerve fibers at low sound levels and the mechanisms involved in producing these effects (i.e., medial efferent control of the cochlear amplifier). The following section (Section 4) reviews a variety of efferent effects on auditory nerve fibers, particularly effects that are not completely explained by medial efferent control of the cochlear amplifier.

In most studies, efferents were activated by shocks in the brain stem because such shocks provide strong activation of medial efferents in which firing rate and timing are readily controlled. Of course, these same factors make the activation pattern unnatural, in that all activated efferents fire synchronously and at the same rate. There is, however, no evidence that synchrony or rate constancy across fibers is important. Recent studies using contralateral sound to activate efferents have found no qualitative differences between efferent effects produced by shocks versus those produced by sound. Perhaps more important than efferent synchrony or rate constancy, the percentage of medial efferents that are activated by brain stem shocks is unknown.

2.1 Efferent Stimulation Inhibits N_1

Perhaps the best-known effect of efferent activity is to depress the compound action potential of the auditory nerve, N_1 (Fig. 8.2) (Galambos 1956; Desmedt 1962; Wiederhold and Peake 1966). This inhibition can be seen with N_1s evoked by clicks or by tone pips. At low sound levels, the efferent effect is to shift the level function to higher sound levels, an effect which is like turning down the gain of the cochlea (Fig. 8.2). As sound level is increased, there is less N_1 depression, and for high-level clicks, the depression is close to zero or even reversed.

Efferent inhibition of N_1 changes with variations in shock rate in a characteristic way (Desmedt 1962; Gifford and Guinan 1987). The inhibition is greatest at shock rates of 200–400/s and falls off at lower or higher rates (Fig. 8.4). For instance, the inhibition is about one quarter of maximum at 60/s, a rate close to the highest medial efferent firing rate evoked by moderate-level monaural sound in a quiet background. Aside from the inhibition of N_1, the dependence of efferent effects on efferent

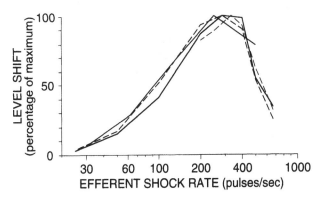

FIGURE 8.4 Medial efferent inhibition as a function of shock rate. Each curve is from a different animal and shows the efferent-induced level shift of click-evoked N_1, normalized so that the largest level shift in each is 100%. (Adapted from Gifford and Guinan 1987.)

firing rate has been studied very little. The data that exist suggest that all medial efferent effects have a similar dependence on shock rate (Desmedt 1962; Konishi and Slepian 1971; Gifford and Guinan 1987; Rajan 1988a).

2.2 Which Efferent Fibers Produce the Inhibition of N_1?

The inhibition of N_1 seen with brain stem shocks is almost certainly due to medial efferents. The most important evidence for this comes from stimulation with an electrode near the origin of medial efferents in cats, using shock levels expected to stimulate only fibers near the electrode tip (Gifford and Guinan 1987). This "focal MOC stimulation" depressed N_1 in both the ipsilateral (referenced to the side of stimulation) and contralateral cochleae. On the average, the N_1 level shift was 2.6 times larger on the contralateral side than on the ipsilateral side (14.4 dB contralateral; 5.6 dB ipsilateral). Correspondingly, in the cat, 2.4 times as many medial efferent fibers project to the contralateral cochlea as project to the ipsilateral cochlea. In contrast, the projection ratio for lateral efferents is 0.25 (Warr, White, and Nyffeler 1982). These data are consistent with the interpretation that the inhibition of N_1 is due to medial efferents and that individual crossed and uncrossed medial efferent fibers produce similar effects.

Experiments with stimulation near the origins of lateral efferents (focal LOC stimulation) did not reveal any effects attributable to lateral efferents (Gifford and Guinan 1987; Guinan and Gifford, unpublished). At shock levels comparable to those used with focal MOC stimulation, focal LOC stimulation produced little inhibition of N_1 (zero to a few decibels of level shift bilaterally). Since myelinated axons have substantially lower thresholds for electrical stimulation than unmyelinated axons (Hallin and Torebjork 1973; Fitzgerald and Woolf 1981), at the standard shock levels used by Gifford and Guinan very few lateral efferent fibers were probably stimulated. In addition, at the shock rates of 200–400/s used for stimulating medial efferent fibers, the unmyelinated lateral efferent fibers would not be expected to fire following each shock and might block completely, since unmyelinated axons do not fire to each shock at high rates (over 10–100/s) (Hallin and Torebjork 1973; Fitzgerald and Woolf 1981). Finally, with high-level focal LOC stimulation, the observed N_1 shifts were similar in both cochleae, in sharp contrast to the pattern of lateral efferent innervation, which is predominantly to the ipsilateral cochlea. The most parsimonious explanation for the data from focal LOC stimulation is that the effects were produced by exciting distant medial efferent fibers.

The finding that medial efferents can inhibit N_1 has important implications for understanding cochlear function. N_1 is produced by the synchronous firing of the myelinated, radial auditory nerve fibers that innervate IHCs, with no contribution from the unmyelinated, spiral auditory nerve fibers that innervate OHCs (Kiang, Moxon, and Kahn 1976). Thus, medial efferents that synapse on OHCs can affect the firing of radial auditory nerve fibers that innervate IHCs. This is also shown by efferent effects on IHC receptor potentials (Section 3.1) and by direct measurements of efferent effects on radial auditory nerve fibers (Sections 3 and 4). Since there are no known neural connections between OHCs and IHCs, or between medial efferent fibers and either IHCs or radial auditory nerve fibers, the mechanism for medial efferent inhibition was a puzzle for many years. As will be explained later, it now seems likely that, at least at low sound levels, medial efferents exert their effect by depressing basilar membrane motion in response to sound.

2.3 Efferent Stimulation with a Midline Electrode Excites Medial Efferents

Much of the existing data on efferent effects has been obtained by exciting the olivocochlear bundle (OCB) with shocks from an electrode at the midline of the floor of the fourth ventricle (midline OCB stimulation). This location is used because efferent fibers are close to the surface and easy to access (Galambos 1956). However, the anatomy of this region is complex. Near the midline in the cat (a similar pattern is found in rodents; Brown et al. 1991), there are many crossing medial efferent fibers, a smaller number of crossing lateral efferent fibers, uncrossed medial efferent fibers that loop close to the midline, and uncrossed lateral efferent fibers that loop slightly lateral to the medial efferent fibers (see Fig. 8.1). By recording from medial efferent fibers near the vestibulocochlear anastomosis in cats, McCue and Guinan (unpublished) found that both crossed and uncrossed medial efferent fibers can be stimulated by a midline electrode. Interestingly enough, in the study by Gifford and Guinan (1987), the average N_1 shift produced by midline stimulation was 19.9 dB, which is almost exactly the sum of the effects found with crossed and uncrossed focal MOC stimulation. These findings are consistent with the interpretation that midline OCB stimulation can activate both crossed and uncrossed medial efferent fibers, at least in the cat. Finally, midline OCB stimulation probably does not activate lateral efferent fibers (for most of the same reasons cited in considering focal LOC stimulation; Section 2.2). Thus, it is likely that both focal MOC stimulation and midline OCB stimulation can be used to activate medial efferents with little, if any, excitation of lateral efferents.

2.4 Medial Efferent Stimulation Increases CM and Evokes the MOC Potential

In addition to inhibiting N_1, medial efferents also affect nonneural cochlear potentials. Efferent stimulation increases the amplitude of CM and evokes a slow potential, the "MOC potential." The increase in CM is typically larger at high sound levels than at low sound levels and can be as large as 4 dB (Fex 1959; Kittrell and Dalland 1969; Konishi and Slepian 1971; Teas, Konishi, and Nielsen 1972; Mountain, Geisler, and Hubbard 1980; Gifford and Guinan 1987). The increase in CM is largest at low frequencies (e.g., 1 kHz) and decreases at higher frequencies, with a higher cutoff frequency for more basal intracochlear electrodes (Kittrell and Dalland 1969; Konishi and Slepian 1971). The "MOC potential" can be recorded as a negative potential equivalent to a few millivolts decrease in endocochlear potential (EP) or as a positive potential of a few millivolts in the organ of Corti and a few hundred microvolts at the round window (Fig. 8.5) (Fex 1959, 1967; Konishi and Slepian 1971; Teas, Konishi, and Nielsen 1972; Brown and Nuttall 1984; Gifford and Guinan 1987). Fex called the MOC potential the "crossed olivocochlear" (COC) potential. However, as shown by Gifford and Guinan (1987), both crossed and uncrossed medial olivocochlear fibers produce this potential, so it will be called the "MOC potential."

2.5 Mechanisms of Medial Efferent Effects: I. CM and the MOC Potential

Although an efferent *increase* in the CM might seem anomalous when other efferent effects are inhibitory, this increase, and the MOC potential, can be understood from the electrical properties of the cochlea (Fig. 8.5A). Activation of medial efferent synapses increases an OHC conductance in series with a hyperpolarizing battery (i.e., activates inhibitory synaptic channels). The CM is produced by sound-frequency motion that changes the resistance of the OHC stereocilia (R_o in Fig. 5A), thereby creating a sound-frequency variation in the current flow through the OHCs and throughout the cochlea (Davis 1965; Dallos and Cheatham 1976). Efferent stimulation, by increasing OHC basolateral conductance and hyperpolarizing the OHCs, increases this current flow and thereby increases the CM. As will be seen in Section 3.4, this increase in OHC conductance also leads to a decrease in basilar membrane motion and to the inhibition of N_1. Thus, the same action of the medial efferents leads to an increase in the CM and a decrease in N_1. In addition to the above, the increased conductance and hyperpolarization of the OHCs causes an increased DC current flow through the OHC stereocilia, thereby decreasing the large, positive endocochlear potential (Fig. 8.5C). The MOC potential is the decrease in the EP and the potentials caused throughout the cochlea (including at the round window) by the increased DC current flow.

Although an efferent-evoked hyperpolarization of the OHCs has never been directly measured, the indirect evidence for this is very strong. In several nonmammalian hair cell systems, efferent stimulation increases the conductance and hyperpolarizes the hair cells (Flock and Russell 1976; Art, Fettiplace, and Fuchs 1984). In isolated OHCs, ACh produces an increased conductance and an outward membrane current that is due to Ca^{2+}-

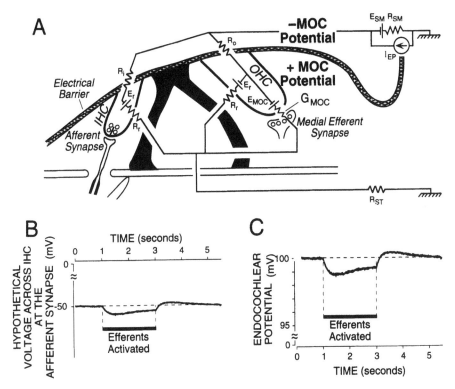

FIGURE 8.5 (A) Cochlear schematic showing the MOC potential and the main elements involved in generating it. The *hatched line* represents the electrical barrier formed by the tight junctions surrounding the scala media. The MOC potential is negative (a reduction of EP) on the scala media side of the barrier and positive on the scala tympani side. The lumped elements approximate distributed components and illustrate the production of the MOC potential. An increase in G_{MOC}, the OHC conductance activated by the medial efferent synapse, increases current flow from the scala media through the OHCs to the scala tympani, with a return path through distant tissue represented by R_{ST} and R_{SM}. The MOC potentials can be thought of as the voltage drops produced by this current in the resistors R_{ST} and R_{SM}. Some fraction of the difference between the negative MOC potential in the scala media and the positive MOC potential in the scala tympani will appear across the IHC–afferent synapse and produce a voltage such as that shown in panel (B) (see Section 4.4.1). Panel (C) shows a measurement of the MOC potential in the scala media (adapted from Guinan and Stankovic 1995). In panel A, R_i and R_o represent the apical resistances of the IHCs and OHCs. R_r and E_r represent the rest resistances and equilibrium potentials of the hair cells. E_{SM} and E_{MOC} represent the scala media voltage and the equilibrium potential of the OHC efferent synapse. I_{EP} is a current source that is hypothesized to regulate EP and cause the decay and overshoot of the MOC potential (see Section 4.3.2). Capacitance is not shown because its effects are negligible at the frequencies contained in the MOC potential.

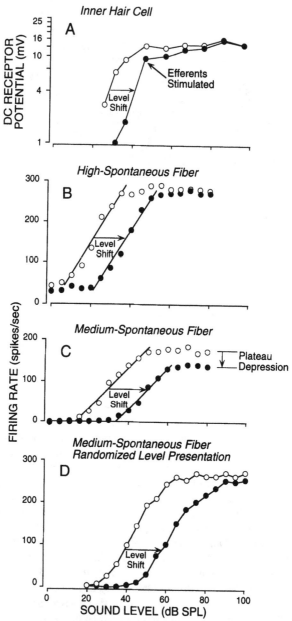

FIGURE 8.6 Sound-level functions of IHC receptor potential (panel A) and auditory nerve firing rate (panels B–D), without (○) and with (●) medial efferent stimulation. The level shift is the amount by which efferent stimulation shifts the responses to higher sound levels. Panel A was from a guinea-pig IHC with a best frequency of 20 kHz and used midline OCB stimulation. (Adapted from Brown and Nuttall 1984.) Panels B and C were from cat auditory nerve fibers with CFs of 1.29 kHz (B) and 2.8 kHz (C), and spontaneous rates (SRs) of 27 spikes/s (B) and 1.7 spikes/s (C)

(*Continued*)

activated K^+ channels (Housley and Ashmore 1991; Doi and Ohmori 1993; Kakehata et al. 1994; see also Fuchs and Murrow 1992). Finally, the MOC potential and the efferent-evoked increase of the CM are consistent with the hypothesis that activation of the medial efferents increases the membrane conductance and hyperpolarizes the OHCs.

The MOC potential and the efferent-evoked increase of the CM provide another line of argument that midline OCB stimulation and focal MOC stimulation excite the efferents to the OHCs (medial efferents) rather than the efferents to the dendrites of the auditory nerve fibers (lateral efferents). First, the above model provides a ready explanation of how these effects could be produced by synapses that are directly on the hair cells, but no comparable explanation exists for how they could be produced by lateral efferent synapses on the dendrites of the auditory nerve fibers. Second, the laterality of the effects (weak but present with ipsilateral medial efferent stimulation, stronger with contralateral medial efferent stimulation, and strongest with midline OCB stimulation; Gifford and Guinan 1987) and the dependence on shock rate (greatest effect at 200–400/s) (Desmedt 1962; Gifford and Guinan 1987) are again consistent with effects produced by myelinated, mostly crossed, medial efferent fibers rather than unmyelinated, mostly uncrossed lateral efferent fibers. Some older work reported that when uncrossed efferent fibers were stimulated, N_1 was inhibited but there was no increase in the CM (Sohmer 1966; Fex 1967). Using modern averaging techniques, Gifford and Guinan (1897) found that there was an approximately constant ratio of N_1 decrease (expressed as decibel shift) to CM increase with uncrossed MOC stimulation, crossed MOC stimulation, or midline OCB stimulation.

3. Medial-Efferent Effects and Mechanisms: II. Control of the Cochlear Amplifier

This section reviews medial efferent effects on responses of IHCs and single auditory nerve fibers, primarily for low-level sounds. The effects reviewed appear to be explained by medial efferent control of the cochlear amplifier. First, the effects are reviewed, then an explanation is presented of the mechanisms that may account for these effects.

FIGURE 8.6 (*Continued*) using crossed focal MOC stimulation. The lines in panels B and C were constrained to be parallel and were fit to the points in the rising phase of the level functions. (Adapted from Guinan and Gifford 1988a.) In panels B and C, the level functions were run in sequence from low to high sound levels, and at each level, efferents-stimulated and efferents-not-stimulated conditions were alternated several times. In panel D, the level functions were obtained with a multiple randomized presentation of both sound level and the presence or absence of shocks. Panel D was from a cat auditory nerve fiber with a CF of 3.4 kHz and an SR of 1.6 spikes/s using midline OCB stimulation. (Adapted from Guinan and Stankovic 1995.)

3.1 Medial Efferent Effects on IHC Potentials

An important step in understanding how efferents that synapse on OHCs affect the firing of auditory nerve fibers that innervate IHCs was provided by intracellular recordings from IHCs. Brown and co-workers (Brown, Nuttall, and Masta 1983; Brown and Nuttall 1984) stimulated medial efferents in guinea pigs while recording intracellularly from IHCs. Efferent stimulation reduced both AC and DC receptor potentials in the IHCs (Fig. 8.6A) without producing a conductance change in the IHCs. The lack of a conductance change indicates that the efferent effect was *not* produced by efferent synapses that were directly on the IHCs, and therefore that the site of medial efferent inhibition is functionally peripheral to the IHCs. In addition, for tone bursts at the most sensitive frequency of the IHCs, efferent-induced level shifts in IHC receptor potentials were approximately equal to efferent-induced level shifts of N_1. Thus, the efferent inhibition of N_1 could be accounted for by the efferent-induced decrease in the IHC receptor potential, and the agent that produced this change acted peripherally to the IHCs.

3.2 Medial Efferent Effects on Auditory Nerve Fibers at Low to Moderate Sound Levels

3.2.1 Medial Efferent Effects on Tuning Curves

Medial efferent stimulation shifts the thresholds of auditory nerve fibers to higher sound levels throughout the tuning curves (Guinan and Gifford 1988c). The largest threshold shifts are in the tip region of the tuning curve (the most sensitive part) (Fig. 8.7). For auditory nerve fibers with high (>3 kHz) characteristic frequencies (CFs), the threshold shift is usually greatest at the fiber's CF and decreases for higher and lower frequencies (Fig. 8.7B). The result is that efferent stimulation makes tuning curves wider (Wiederhold 1970; Guinan and Gifford 1988c). For auditory nerve fibers with high to medium spontaneous rates (SRs) and midrange CFs (CFs of 2–8 kHz), medial efferent stimulation typically produced 30%–40% increases in tuning curve width as measured by the change in Q_{20} (Q_{20} is the CF divided by the bandwidth of the tuning curve measured at 20 dB above the threshold). However, the largest efferent-induced threshold change is not always at the CF of an auditory nerve fiber. For some fibers, particularly low-SR and medium-SR fibers with CFs of 1–2 kHz, the largest increase in threshold was at the low-frequency edge of the tuning curve tip (e.g., Fig. 8.7A) (Guinan and Gifford 1988c). In such cases, efferent stimulation made the tuning curve more narrow, opposite to the typical efferent effect on high-CF auditory nerve fibers.

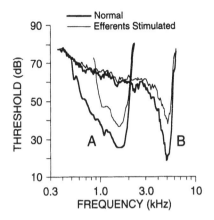

FIGURE 8.7 Tuning curves without (*thick lines*) and with (*thin lines*) medial efferent stimulation. Data are from cat auditory nerve fibers with spontaneous rates of 2.4 spikes/s (A) and 0.0 spikes/s (B) using midline OCB stimulation. (Adapted from Guinan and Gifford 1988c.)

Medial efferent stimulation has little effect on tuning curve tails (the insensitive, low-frequency regions). For high-CF fibers in which 1 kHz was in the tail, threshold shifts at 1 kHz averaged less than 1 dB (Guinan and Gifford 1988c). This lack of efferent effect in the tail is consistent with the observations of Brown and co-workers (Brown, Nuttall, and Masta 1983; Brown and Nuttall 1984) who found little medial efferent-induced change in IHC receptor potentials at frequencies well below the most sensitive frequency of the IHCs.

3.2.2 Medial Efferent-Induced Threshold Shifts as a Function of CF

The strength of medial efferent inhibition varies dramatically across auditory nerve fibers. For stimulation with tones at the CF in the cat, efferent-induced threshold shifts are largest (e.g., 20 dB) for fibers with mid-to-high CFs (CFs of 2-10 kHz) and they decrease for fibers of higher or lower CF (Fig. 8.8) (Wiederhold 1970; Guinan and Gifford 1988c). Guinea pigs show a similar pattern, except that the CF range with the largest shifts is slightly higher, 7-10 kHz (Teas, Konishi, and Nielsen 1972). Similar frequency patterns are also obtained from plots of the efferent inhibition of N_1 evoked by tone pips versus tone-pip frequency (Patuzzi and Rajan 1990; Liberman 1991).

The distribution of threshold shifts as a function of CF for high-SR and medium-SR fibers is similar to the distribution of medial efferent endings along the length of the cochlea (compare Figs. 8.8 and 8.3A,B). This correspondence is consistent with evidence that medial efferents produce these threshold shifts. The close correspondence of the distribution of medial efferent endings and medial-efferent inhibition suggests that this effect of medial efferent synapses is relatively local and does not spread a great distance along the length of the cochlea.

In contrast, the pattern of efferent-induced threshold shifts in low-SR fibers appears to be different from the patterns for high-SR and medium-

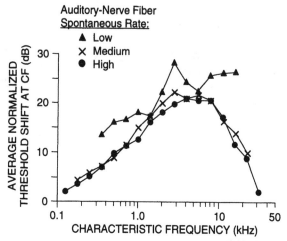

FIGURE 8.8 Medial-efferent-induced threshold shift as a function of auditory nerve fiber CF for fibers in each SR class. Threshold shifts were measured at each fiber's CF. Each point shows the average over an octave band (adjacent points have half-octave overlaps). Data are from 406 auditory nerve fibers (255 with midline-OCB stimulation, 151 fibers with crossed-focal-MOC stimulation). To weight all cats evenly before averaging, data from each cat were normalized to an N_1 shift of 20 dB, the average found in a series of experiments. (Adapted from Guinan and Gifford 1988c.)

SR fibers (Fig. 8.8) or for medial efferent endings (Fig. 8.3A,B). In Figure 8.8, there were fewer low-SR fibers averaged than medium-SR or high-SR fibers, which may account for some of the irregularity in the low-SR data. However, the large differences at the highest frequencies suggest that the factors involved in efferent effects on low-SR fibers are different (at least in part) from those involved in efferent effects on other auditory nerve fibers (see Section 4.4).

3.2.3 Medial Efferent Inhibition in the Rising Phase of Rate Level Functions

Medial efferent stimulation shifts the rising phase of auditory nerve rate versus sound-level functions to higher sound levels (Fig. 8.6B,C,D) (Wiederhold 1970; Teas, Konishi, and Nielsen 1972; Gifford and Guinan 1983; Winslow and Sachs 1987; Guinan and Gifford 1988c; Warren and Liberman 1989b; Guinan and Stankovic 1995). Medial efferent stimulation also shifts the rising phase of synchrony versus sound-level functions (synchrony is the tendency of firing to be locked to sound phase) by approximately the same amount as the shift of the rate-level function in the same fiber. In contrast, efferents produce little change in phase versus sound-level functions (Gifford and Guinan 1983; Warren and Liberman 1989b). These level-function changes are consistent with medial efferents inserting an attenuation peripheral to the place at which spike rate, synchrony, and phase are determined from an analog signal. Thus, efferent-induced effects in the rising phases of both IHC receptor potentials (Fig. 8.6A) and auditory nerve fiber responses (Fig. 8.6B,C,D) indicate that the medial efferents produce a signal attenuation that is functionally peripheral to the IHCs. In addition, the small change in phase with efferent stimulation is consistent with efferent stimulation's having little effect on the speed of the traveling wave along the basilar membrane.

Efferent-induced shifts in rate-level functions depend on fiber CF and sound frequency relative to CF, and have patterns that are similar to the patterns for efferent-induced threshold shifts. This, of course, is expected because the threshold change is just the lowest-level part of the overall sound-level function, and efferents produce approximately parallel shifts in sound-level functions (Fig. 8.6). For CF tones, the level-function shifts are similar to the pattern in Fig. 8.8 for threshold shifts (Wiederhold 1970; Guinan and Gifford 1988a). For tones not at CF, the few data available are mostly consistent with a decrease in the shift as the test frequency to CF difference increases. However, a few low-CF, low-SR fibers have particularly large shifts near the lower edges of their tuning curves (Fig. 8.7A) (Gifford and Guinan 1983).

3.3 Medial Efferent Effects on Otoacoustic Emissions

A variety of experiments indicate that medial efferent activity influences otoacoustic emissions (OAEs). Such effects are of interest because OAEs are thought to reflect aspects of basilar membrane motion. Measurements of OAEs, however, are much easier to make than measurements of basilar membrane motion, and in one animal they can provide information about a wide range of cochlear places. Furthermore OAE measurements are noninvasive and can be done in humans. Measurements of efferent effects on OAEs have been made with efferent activity evoked by brain stem shocks or by sound in the opposite ear. This section will concentrate on the kinds of changes produced in the OAEs and data that might provide insight into the cochlear mechanisms that underlie these changes. Data that primarily concern how efferent effects on OAEs depend on the sounds used to evoke the efferent activity will be dealt with in Section 6.1.3. See Whitehead et al. (1995) for an overview of OAEs.

3.3.1 Medial-Efferent Effects on DPOAEs

Distortion product otoacoustic emissions (DPOAEs) are tones produced by distortion in the cochlea in response to two externally supplied tones, the primary tones ($f_1 < f_2$). Energy at the distortion frequency is generated by cochlear nonlinearities (which are probably in OHCs), and this energy travels in both directions away from its origin near the peak of the f_2

response (Kim, Molnar, and Matthews 1980). For low-level primary tones, the $2f_1 - f_2$ distortion product has the highest amplitude and has been the most studied. The $f_2 - f_1$ DPOAE is also interesting because it may depend on different processes than the $2f_1 - f_2$ DPOAE and may provide additional information about the cochlea (Brown 1988).

Efferent stimulation usually decreases DPOAEs, but sometimes it increases them (Mountain 1980; Siegel and Kim 1982). On average, efferent inhibition of DPOAEs is greatest for low-level primaries and decreases as primary-tone level is increased (Mountain 1980; Moulin, Collet, and Duclaux 1993). The pharmacologic properties of the medial efferent inhibition of DPOAEs show that the inhibition is mediated by a nicotinic-like cholinergic receptor, which is consistent with other pharmacologic studies of medial efferent effects (Kujawa et al. 1993, 1994; see Sewell, Chapter 9).

Medial-efferent effects on OAEs confirm that medial efferents produce a mechanical change in the cochlea. However, it is not possible to translate from changes in DPOAEs to changes in basilar membrane motion. Efferent suppression of $2f_1 - f_2$ DPOAEs in cats was sometimes equal to and sometimes less than the suppression of N_1 evoked by tone pips at the f_2 frequency (Puria, Guinan, and Liberman 1996). Since suppression of N_1 is reported to match closely to suppression of basilar membrane motion (Dolan and Nuttall 1994), the suppression of DPOAEs does not always accurately reflect the amount of suppression of basilar membrane motion.

3.3.2 Medial Efferent Effects on CEOAEs, TBOAEs, and SFEs

Click-evoked OAEs (CEOAEs), tone burst–evoked OAEs (TBOAEs), and stimulus-frequency otoacoustic emissions (SFEs) are ear canal sounds due primarily to cochlear re-radiation of energy at a frequency which is present in the sound stimulus (Kemp 1978; Kemp and Chum 1980; Norton and Neely 1987). These emissions are fundamentally different from DPOAEs in that they do not require a nonlinearity in the cochlea for their production (although their growth may be nonlinear due to the saturation of the mechanisms that produce them). It seems possible that these emissions are a direct consequence of the injection of energy into the basilar membrane during the action of the cochlear amplifier (Fig. 8.9A,B; see Section 3.4). If this is true, efferent effects on these emissions might be more closely tied to efferent effects on basilar membrane motion than are efferent effects on DPOAEs.

Activity in medial efferents affects CEOAEs, TBOAEs, and SFEs. The usual effect is to inhibit with the greatest inhibition for responses to low-level sounds (Guinan 1986, 1991; Collet et al. 1990; Ryan, Kemp, and Hinchcliffe 1991; Norman and Thornton 1993). However, at the frequency of a sharp dip in the amplitude versus frequency function of the CEOAE, TBOAE, or SFE, efferent stimulation might increase the OAE response by

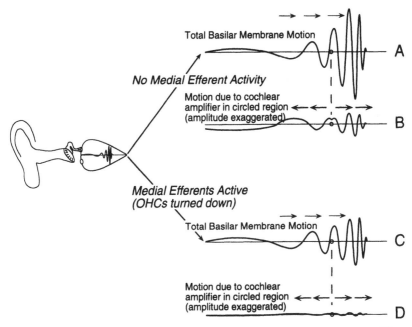

FIGURE 8.9 Schematic illustrating the presumed action of the cochlear amplifier at one place along the basilar membrane, and medial efferent inhibition of the cochlear amplifier. Trace A is a snapshot of normal basilar membrane motion in response to a tone. The cochlear amplifier at the circled place moves the basilar membrane at that place and creates waves that travel away in both directions (trace B). The forward-traveling wave (trace B →→) adds to the sound-driven wave and amplifies it. Note that the amplitude scale is exaggerated in traces B and D relative to traces A and C. For traces C and D, medial efferents are activated and reduce the gain of the cochlear amplifier so that the motion created by it is smaller (the amplitude is smaller in trace D than in trace B). Since there is less amplification at many places along the basilar membrane, the resulting traveling wave is less (the amplitude is smaller in trace C than in trace A).

producing a small change in the frequency of the dip (Long, Talmadge, and Shaffer 1994). If CEOAEs, TBOAEs, and SFEs are a direct consequence of the action of the cochlear amplifier, then their inhibition by efferent activity is a demonstration of efferent suppression of the cochlear amplifier (Guinan 1991).

3.3.3 Medial Efferent Effects on SOAEs

Spontaneous otoacoustic emissions (SOAEs) are almost-sinusoidal tones emitted by the cochlea in the absence of external stimulation. SOAEs are present in a large fraction of normal human ears, but are much less prevalent in the ears of lower animals (Zurek 1981; Martin et al. 1988). SOAEs appear to be due to an oscillation originating within the cochlea (van Dijk and Wit 1990) that is likely to depend on OHC motility as a source of mechanical energy.

Medial efferents produce small changes in SOAEs. In humans, efferent activity evoked by contralateral tones or broadband noise produces SOAE frequency and amplitude changes in some subjects (Mott et al. 1989; Harrison and Burns 1993). The SOAE frequency shifts were always to higher frequencies, but the amplitude could change in either direction. For tones, the largest frequency shifts occurred when the tone was approximately one half octave below the SOAE. It is presently impossible to interpret the mechanisms by which efferents change SOAEs, because the production of SOAEs is so poorly understood. However, the fact that efferents change SOAEs, and medial efferent synapse on OHCs, is consistent with the hypothesis that OHC motility is directly involved in the production of SOAEs.

3.4 Mechanisms of Medial Efferent Effects: II. Depression of Basilar Membrane Motion

At low sound levels, medial efferent-induced depression of basilar membrane motion appears to be the dominant mechanism by which efferents that synapse on OHCs change the firing of afferents that synapse on IHCs. Since the discovery of efferent effects on OAEs, it has been clear that efferents produce mechanical changes in the cochlea. The nature of these changes, however, had been unknown. Dolan and Nuttall (1994) have recently reported that in guinea pigs, efferent stimulation depresses basilar membrane motion, and the depression is enough to account for efferent suppression of N_1 at low sound levels. The depression of basilar membrane motion was greatest at frequencies near the CF of the region tested (18–20 kHz) and could be as much as 20–22 dB. The reduction in motion was greatest at the lowest sound levels (30–40 dB SPL) and decreased with increasing sound level. The pattern of these data matches well with medial efferent suppression of auditory nerve fiber responses to low- and moderate-level sounds. However, the abstract by Dolan and Nuttall (1994) does not provide detail on how well, or under what circumstances, the depression of basilar membrane motion matches the level shift in N_1.

3.4.1 Medial-Efferent Mechanisms: Suppression
of the Cochlear Amplifier

The demonstration that efferents depress basilar membrane motion leads to the next issue, the mechanisms involved in this depression. Understanding of these mechanisms requires understanding how the high sensitivity and frequency selectivity of the cochlea are produced. Over the last 15 years it has become increasingly clear that an active process, often called the "cochlear amplifier," is involved. In this chapter, the term "cochlear amplifier" means the mechanisms that increase the amplitude of the traveling wave from the amplitude in a dead or passive cochlea. It is widely thought that OHC motility is a central element in this cochlear amplifier, but many aspects of how OHC motility leads to increased basilar membrane motion are not known. The following paragraphs provide a hypothetical outline of the mechanisms invoked and serve to show how medial efferents might exert their effect. Patuzzi, Chapter 4, and de Boer, Chapter 5, provide more information on the mechanisms involved in shaping basilar membrane motion.

In a normal cochlea, the basilar membrane response to sound is due to a combination of passive mechanical factors and active processes. The active process is produced, most likely, by basilar membrane motion causing radial shearing of the reticular lamina relative to the tectorial membrane, which bends OHC stereocilia. This bending opens ion channels in the stereocilia so that current flows into the OHC and creates a sound frequency change in the OHC membrane voltage. This OHC receptor potential causes the OHC to shorten and elongate at the sound frequency, and somehow this vibration is coupled back to the basilar membrane. When the OHC-induced basilar membrane motion is in phase with the forward traveling wave, it amplifies the traveling wave. Components of the OHC-induced basilar membrane motion traveling in the backward direction may cause OAEs.

The process by which OHC-induced motion of the basilar membrane produces amplification is illustrated in Figure 8.9. Sound energy enters the cochlea through the middle ear and creates a traveling wave along the basilar membrane (Fig. 8.9A). Considering just one place along the cochlea (*circled* in Fig. 8.9A), the OHC response creates a small basilar membrane motion. The basilar membrane acts as a transmission line and this motion flows away from its origin traveling both apically and basally (Fig. 8.9B). If the apically directed motion created by OHCs has the same phase as the sound-driven traveling wave (as shown in Fig. 8.9), then these waves add and the traveling wave is amplified. The basally directed motion created by the OHCs produces motion of the ossicles and OAEs in the ear canal.

Efferent suppression of basilar membrane motion is illustrated in the bottom half of Fig. 8.9. In this case, medial efferent synapses on OHCs are assumed to be active and to reduce the motion response of the OHCs. The OHCs then create less motion on the basilar membrane (Fig. 8.9D), resulting in less amplification and a smaller total basilar membrane motion (Fig. 8.9C).

Medial efferents have large synapses on OHCs and seem ideally situated to affect the operation of OHCs. However, the mechanisms by which these synapses change OHC properties are not known. One attractive possibility is that medial efferent synapses produce a large increase in conductance at the base of the OHCs (Sections 2.5 and 7.1), and the increased conductance reduces the AC receptor potential produced by the stereocilia receptor current. Since OHC motility is controlled by the OHC voltage (Santos-Sacchi and Dilger 1988), this reduction in OHC receptor potential would reduce OHC motion. An additional mechanism, hypothesized by Roddy et al. (1994), is that an efferent hyperpolarization of the OHCs moves the membrane potential away from the optimum voltage for voltage-to-length transduction. Another possibility is that efferent-induced contractions of the OHCs distort the organ of Corti, thereby lowering the gain of the cochlear amplifier. However, it seems likely that the efferents do not produce substantial contractions of the OHCs (Bobbin et al. 1990; Patuzzi and Rajan 1990). Yet another possibility is that the activation of medial-efferent synapses does something that directly interferes with OHC fast motility. However, electrically evoked OAEs (which are presumably due to OHC fast motility) are changed very little during activation of medial efferents (Murata et al. 1991). Finally, the medial efferent reduction of the EP will lower the gain of the cochlear amplifier by lowering the OHC receptor current, but this accounts for at most a few decibels of efferent inhibition (Sewell 1984b). Of all these possible mechanisms, the most attractive is that medial efferent synapses shunt the receptor currents of the OHCs. The main elements of this hypothesis seem almost certain to be true. It is not known, however, whether the efferent-induced conductance change is large enough to reduce the gain of the cochlear amplifier and account for efferent inhibition at low sound levels.

3.4.2 Medial-Efferent Mechanisms: Otoacoustic Emissions

There are a wide variety of mechanisms by which medial efferents might affect OAEs. First, an efferent-induced depression of basilar membrane motion in the forward traveling wave will automatically reduce the energy "reflected" into the backward-traveling wave. Efferent action on OHCs might also change the processes by which this energy is "reflected" and thereby affect OAEs. Medial efferent effects on DPOAEs are particularly complicated: efferents may affect the response of the basilar membrane to each of the two primary tones and to the distortion product; efferents may affect the "reflection process"; and additionally, efferents may affect the nonlinear process that creates the distortion product.

Despite the many ways by which medial efferents might inhibit OAEs, a direct comparison has shown that efferent inhibition of DPOAEs is usually *less than,* or equal to, efferent inhibition of neural responses (Puria, Guinan, and Liberman 1996). One explanation is that DPOAEs are generated mostly by a region that is basal to the peak of the traveling wave (Avan et al. 1995) where basilar membrane motion has received only part of its amplification by the cochlear amplifier. An efferent-induced reduction of cochlear amplifier gain would produce a smaller effect in this region than at the peak of the traveling wave.

3.5 Medial Efferent Effects in Noisy Backgrounds

Although medial efferent activity reduces N_1 evoked by brief sounds in a quiet background, efferent activity can *increase* N_1 when the background is noisy (Nieder and Nieder 1970; Dolan and Nuttall 1988; Kawase and Liberman 1993). This increased N_1 is due to a reduction of masking by the background noise (see below). It is prominent for moderate-level transient sounds (e.g., clicks or tone pips) in moderately noisy backgrounds, and is not present for very low-level or high-level transient sounds. This effect is significant because we normally hear in noisy environments.

Studies on single auditory nerve fibers have confirmed that medial efferent activity increases responses to transient sounds in a noisy background and have provided insight into the mechanisms involved (Winslow and Sachs 1987, 1988; Kawase, Delgutte, and Liberman 1993). When a continuous background sound excites an auditory nerve fiber, the fiber's response to an additional transient sound is compressed compared to its response in a silent background (Fig. 8.10C). This happens because the noise increases the background firing rate, and this rate increase produces

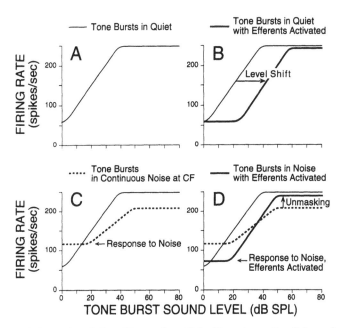

FIGURE 8.10 Schematic of the effects of medial efferents on the firing of auditory nerve fibers. Each panel shows rate versus level functions for a single auditory nerve fiber with different conditions coded by different line styles. Schematized from the results of Wiederhold (1970), Winslow and Sachs (1987), Guinan and Gifford (1988a), and Kawase, Delgutte, and Liberman (1993).

adaptation that reduces the maximum rate that can be evoked by a transient sound. Efferent stimulation reduces the response to the background sound, which reduces the adaptation, and allows a greater response to a transient sound (Fig. 8.10D) (Winslow and Sachs 1987; Kawase, Delgutte, and Liberman 1993). In addition to reducing adaptive masking, efferent stimulation may also reduce suppressive masking by reducing the response to the suppressor. The medial efferent–induced reduction in adaptive masking is present for transient stimuli at frequencies both near and off the CF, because adaptation has the same effect on responses to any sound frequency. However, for there to be an efferent-induced reduction in adaptive masking, the low-level masking sound must have energy near the CF so that it produces adaptation that can be prevented by efferent stimulation. The efferent antimasking effect is greatest for mid- and high-frequency maskers (where the medial efferent innervation is greatest) and for low-level maskers (i.e., within 20 dB of the fiber's threshold to the masker) where medial efferents can substantially reduce the response to the masker (Kawase, Delgutte, and Liberman 1993). Efferent stimulation has little effect on an auditory nerve fiber's masked threshold, because at levels near threshold, the responses to the transient and the masker are both inhibited by approximately the same amount (see Fig. 8.10). However, for transients 10 dB or more above threshold, efferent activity can produce a substantial increase in the discriminability of transient signals in continuous noise (Winslow and Sachs 1988; Kawase, Delgutte, and Liberman 1993).

4. Medial Efferent Effects and Mechanisms: III. Beyond the Cochlear Amplifier

This section reviews a variety of efferent effects on auditory nerve fibers, particularly effects that are not completely explained by medial efferent control of the cochlear amplifier.

4.1 Medial Efferent–Induced Changes in Auditory Nerve Spontaneous Rate

Stimulation of medial efferents reduces the spontaneous activity of auditory nerve fibers (Wiederhold and Kiang 1970; Guinan and Gifford 1988b; Kawase, Delgutte, and Liberman 1993), apparently by two effects. In very sensitive auditory nerve fibers, many, perhaps most, "spontaneous" firings are probably responses to sounds generated by the animal (Wiederhold and Kiang 1970; Guinan and Gifford 1988b; Kawase, Delgutte, and Liberman 1993; see also Lewis and Henry 1992). For these fibers, efferent inhibition of "spontaneous" firing is mostly an inhibition of responses to sound. However, even in very insensitive fibers, medial efferent stimulation still produces a small decrease in spontaneous firing (Guinan and Gifford 1988b; Guinan and Stankovic 1995), and in these fibers, medial efferent stimulation appears to be inhibiting "true spontaneous" activity (i.e., activity that is not in response to internal or external sound).

Properties of the medial efferent inhibition of "true spontaneous" activity have been studied by attempting to exclude from consideration auditory nerve fibers in which part of the "spontaneous" activity is a response to sound. Guinan and Gifford (1988b) did this by excluding fibers from one very sensitive cat. In the remaining animals, efferent inhibition of spontaneous rate (SR) was greater the lower the SR (the opposite of what is expected if part of the SR is a response to sound), and had a maximum for fibers with CFs of about 10 kHz (which is higher than the CF with the maximum medial efferent inhibition of responses to sound). These observations are consistent with the conclusion that medial efferent inhibition of "true spontaneous" activity is not due to an efferent-induced reduction of cochlear amplifier gain, which reduces the basilar membrane response to sound. This inhibition of spontaneous rate is highly likely to be due to medial efferent synapses on OHCs rather than lateral efferent synapses on auditory nerve fibers, because this inhibition can be produced by focal MOC stimulation, it is greatest on the side contralateral to the focal stimulation, and its distribution as a function of CF is more similar to the distribution of medial efferent endings than the distribution of lateral efferent endings (Guinan and Gifford 1988b).

Guinan and Gifford (1988b) hypothesized that the medial efferent effect on spontaneous activity is due to the MOC potential (see Section 4.4 for a discussion of the mechanisms involved). To test this hypothesis, Guinan and Stankovic (1995) compared the time courses of the efferent-induced change in spontaneous activity and the MOC potential. At the onset of a 1- to 2-s shock burst, the inhibition of spontaneous activity rose to a peak in a few hundred milliseconds, then decayed with a time constant of 1–3 s, and after termination of the shock burst, the efferent effect on spontaneous activity had a prominent overshoot reversal. This time course matched well with that of the MOC potential (Fig. 8.5C), as predicted by the hypothesis. In contrast, efferent effects on the CM and otoacoustic emissions (SFEs) did not show much decay during the shocks or much of an overshoot after the termination of the shock burst. The results are consistent with the hypothesis that medial efferent inhibition of spontaneous activity is due to the MOC potential.

4.2 Medial-Efferent Inhibition of Auditory Nerve Responses to High-Level Sounds

In addition to the well-known changes for low-level sounds, medial efferents also decrease auditory nerve fiber responses to high-level sounds. In most studies of efferent effects as a function of sound level (Wiederhold

1970; Gifford and Guinan 1983; Guinan and Gifford 1988a; Warren and Liberman 1989b), rate versus sound-level functions were obtained by presenting low-level to high-level sounds in sequence, with efferent-stimulation and no-efferent-stimulation conditions alternating at each sound level. Such level functions show a region in which the rate increases rapidly with the sound level, and at higher sound levels, a "saturated" or "plateau" region in which the rate changes little, if at all, with level (Fig. 8.6B,C). In these level functions, efferent stimulation produced small depressions in the plateau region (Fig. 8.6B,C). For midfrequency auditory nerve fibers, efferent stimulation reduced plateau rates in high-SR fibers by about 5% and in low-SR fibers by more, often 15%–20% (Guinan and Gifford 1988b; also see Warren and Liberman 1989b). Even greater depressions of the plateau rate were found with off-CF tones in low-frequency, low-SR auditory nerve fibers (Gifford and Guinan 1983). Although these data demonstrate that efferents have substantial effects at high sound levels, the interpretation of these results is clouded, because with the sequential paradigm, the firing rates in response to high-level sounds are strongly influenced by adaptation.

Recent experiments show that medial efferents produce large effects on responses to high-level sounds, even when the influence of adaptation is minimized by randomizing the sound level presentation (Guinan and Stankovic 1995). With randomized level functions, high-SR fibers still show little effect of efferent stimulation at high sound levels. However, for low-SR and medium-SR fibers in which the rate grows with sound level up to very high sound levels, efferent stimulation has a strong effect, even at 100 dB SPL (Fig. 8.6D). At 100 dB SPL, efferent inhibition can reduce the rate by an amount equivalent to more than a 20-dB reduction in sound level. At these high sound levels, 20 dB of efferent inhibition is unlikely to be produced by an efferent-induced reduction in basilar membrane motion (see Dolan and Nuttall 1994). Inhibition at high sound levels is likely to be due to medial efferents rather than lateral efferents, because such inhibition can be produced by focal MOC stimulation, it is greatest on the side contralateral to the focal stimulation, and is still stronger for midline OCB stimulation (Guinan and Gifford 1988a); in contrast, lateral efferent neurons are far more numerous on the ipsilateral side than on the contralateral side.

There is one aspect of auditory nerve fiber responses to high-level sounds that medial efferents do not appear to inhibit. In some sound level functions, at high sound levels (80–100 dB SPL) there are abrupt changes in response phase, and dips in rate and synchronization index (Gifford and Guinan 1983). With medial efferent stimulation, the firing rate is depressed at sound levels below the dip, and the abrupt phase changes and dips are shifted to lower sound levels, but there is little effect at sound levels above the dip (Gifford and Guinan 1983; Warren and Liberman 1989b). These phenomena have been explained by postulating that the excitation of an auditory nerve fiber is the result of two factors that are out of phase and have different thresholds and growth functions. The two factors cancel when they are equal in amplitude, producing the dips and phase changes. Presumably, medial efferent activity reduces the more sensitive factor but does not change the other factor, so that with medial efferent stimulation there is inhibition at sound levels below the rate dip but not above, and the two factors cancel at a lower sound level (Gifford and Guinan 1983).

4.3 Time Courses of Medial Efferent Effects

4.3.1 Rise and Decay Times of Medial Efferent Effects

Most medial efferent effects begin and end with time constants in the hundred-millisecond range (Galambos 1956; Desmedt 1962; Fex 1967; Wiederhold and Kiang 1970; Gifford and Guinan 1987; Guinan and Stankovic 1995). Since one shock evokes very little inhibition, trains of shocks are required to produce a strong efferent inhibition. At the onset of a train of shocks, efferent effects take one hundred to a few hundred milliseconds to build up, and after the termination of the shock train, the decay times, although not usually identical to the rise times, are in the same range. The rise time most likely includes time for a buildup of transmitter release due to presynaptic facilitation (Art, Fettiplace, and Fuchs 1984), plus time for the transmitter to produce an increase in Ca^{2+}-activated K^+ conductance in the OHC (see Section 2.5 and Sewell, Chapter 9). The decay time may be mainly the decay of the Ca^{2+}-activated K^+ conductance, since the abundant supply of acetylcholinesterase around the OHCs would be expected to remove residual ACh quickly.

Kemp and Souter (1988) suggested that efferents can produce effects with rise times in the tens of milliseconds range, but their method of measurement leaves this interpretation open to question. They used paired shock bursts and measured the difference between the first and second CM (or SFE) response. This difference sometimes rose to a maximum in as little as 20 ms. However, this result may be due to the first and second responses having different time courses, with their differences being greatest 20 ms after onset. Fast components comparable to those reported by Kemp and Souter have not been seen with direct measurements of efferent effects on CM or SFEs (Guinan 1986, 1990; Guinan and Stankovic 1995; also see Mountain, Geisler, and Hubbard 1980).

4.3.2 Medial Efferent Time Courses over Seconds

With excitation by shock bursts at 200/s or less, most efferent effects have a relatively simple time course, but for shock rates over 200/s, efferent effects often have complex time courses. In a study on auditory nerve fibers, Wiederhold and Kiang (1970) found that with shocks at rates of 200/s or less, efferent inhibition increased to a maximum within a few hundred milliseconds of onset and then adapted by varying amounts with no additional peak of inhibition. However, for shock rates over 200/s, after the initial peak and adaptation of inhibition, some fibers had a second, smaller increase and decrease of inhibition. Brown and Nuttall (1984) also saw more complex time patterns in IHC receptor potentials with 400/s shocks than with 200/s shocks. The origin of the complex time variation in efferent effects at high-rate shocks is not known, but with high-rate shocks, there could be depletion followed by mobilization of efferent transmitter, or intermittent failure of efferent fibers to follow every shock in a train of shocks.

Even with shocks at 200/s, the MOC potential has a complex time course, which, surprisingly, is different from the time course of efferent-induced changes in CM or OAEs (Guinan and Stankovic 1995; also see Konishi and Slepian 1971). Efferent-induced changes in CM and SFEs peak a few hundred milliseconds after shock onset and decay relatively little while shocks continue. In contrast, the MOC potential peaks in a few hundred milliseconds, then decays with a time constant of a few seconds while the shocks continue (Fig. 8.5C). After termination of a shock burst, the MOC potential has a prominent overshoot reversal, whereas CM and SFEs do not. As described in Section 2.5, the MOC potential and the efferent-induced increase in CM are both produced by an efferent-induced increase in OHC conductance (G_{MOC} in Fig. 8.5A). However, it is difficult to see how a change in G_{MOC} can simultaneously produce the two different time courses observed for CM and the MOC potential. Two time courses could be produced if a second electrical element (in addition to G_{MOC}) varied with time during efferent stimulation. One hypothesis is that the efferent-induced changes in CM and SFEs follow the time course of the change in G_{MOC}, and that the deviation of the MOC potential from this time course is due to a homeostatic regulation of the EP (represented by I_{EP} in Fig. 8.5A) that has a time constant of a few seconds. Thus, after an efferent increase in G_{MOC} lowers the EP, I_{EP} increases and brings the EP approximately back to its original value. Similarly, after efferent shocks are turned off, G_{MOC} decreases in a few hundred milliseconds but I_{EP} takes much longer to return to its original value and produces the overshoot of the EP. With this hypothesis, G_{MOC} and I_{EP} together produce the time course of the MOC potential.

4.3.3 Medial Efferent Time Courses over Minutes

In addition to the "fast component" of efferent inhibition, there is a "slow transient component" that builds up and decays on a time scale of minutes (Sridhar et al. 1995). During continuing efferent shocks, the slow transient component is manifested as an increased inhibition of N_1 and a parallel increase in the CM that build up over 1 minute or so. With efferent shocks lasting over several minutes, this slow transient component then decays, leaving only the normal "fast" component of efferent inhibition (Reiter and Liberman 1995; Sridhar et al. 1995). Slow and fast effects are probably due to the same neurotransmitter and receptor on OHCs, because they have the same pharmacologic properties (Sridhar et al. 1995).

4.4 Mechanisms of Medial Efferent Effects: III. Other Effects

4.4.1 Medial Efferent Mechanisms: Spontaneous Rate

The medial efferent-induced reduction, and rebound increase, of "true spontaneous" activity seem likely to be due to the MOC potential (Section 4.1), but the processes involved are not fully understood. Before considering these processes, note that efferent effects on "true spontaneous" activity are not likely to be due to an efferent-induced reduction of cochlear amplifier gain leading to reduced basilar membrane motion, because this activity is not due to basilar membrane motion. This conclusion is supported by the dissimilarity of the time course of the medial efferent effects on "true spontaneous" activity and the time courses of efferent-induced changes in the CM and SFEs. In contrast, the medial efferent influence on "true spontaneous" activity is strongly coupled in time with the MOC potential, so that a causal relationship between them seems reasonable.

There are two main ways in which the MOC potential might affect the firing of auditory nerve fibers. First, the MOC potential might change the effective IHC membrane potential at the afferent synapse (Fig. 8.5B) and modulate the release of transmitter by the IHC. Brown and Nuttall (1984) found MOC potentials of 0 to +1.5 mV in IHCs, +0.5 to +2 mV in the organ of Corti near the IHCs, and −2 to −4 mV in the scala media. This suggests that the net effect of efferent stimulation is to hyperpolarize the IHC presynaptic area by 0.5 mV. However, the penetration of the IHC by the microelectrode is likely to have made the MOC potential in the IHC artificially close to the MOC potential of the organ of Corti, so that actual hyperpolarization of the IHC presynaptic area by the MOC potential may be greater than 0.5 mV. Second, the MOC potential might change the number of auditory nerve firings produced by a given IHC transmitter release by changing the transmembrane potentials of the afferent dendrites (Fig. 8.5A). In favor of the first hypothesis, when the EP was changed by the drug furosemide, the firing rate of the auditory nerve fiber varied exponentially with changes in the EP, a result that fits with the established relationship between presynaptic voltage and transmitter release (Sewell 1984a). The change in the EP produced by the MOC potential might be expected to produce a similar result.

4.4.2 Medial Efferent Mechanisms: Responses to High-Level Sounds

An efferent-induced reduction of cochlear amplifier gain might account for some but not all of the inhibition of auditory nerve responses to high-level sounds. To account for this inhibition, the cochlear amplifier must produce a substantial increase (20 dB or more) in basilar membrane motion at sound levels as high as 100 dB SPL, and efferent activity must reduce the amplification enough to produce a 20-dB or greater effect. Some measurements of basilar membrane motion suggests that the cochlear amplifier may influence the peak of the traveling wave at 80–110 dB SPL. In particular, at these high sound levels, there is less than linear growth of basilar membrane motion with increasing sound level, and basilar membrane motion changes soon after death or from acoustic trauma (Rhode 1971, 1973; Robles, Ruggero, and Rich 1986; Cooper and Rhode 1992). The one report on efferent inhibition of basilar membrane motion (Dolan and Nuttall 1994) found motion reductions as much as 20–22 dB at 30–40 dB SPL but smaller reductions at high sound levels. However, even substantial efferent inhibition of basilar membrane motion does not appear adequate to explain the efferent inhibition of the plateau rate that is present in some fibers with randomized presentation of sound levels (Fig. 8.6D). It appears that medial efferent inhibition of responses to high-level sound may be due, in part, to efferent inhibition of basilar membrane motion, but must also include inhibition from some other mechanism.

In addition to the efferent inhibition of basilar membrane motion, there are several possible mechanisms by which medial efferents might suppress responses of radial auditory nerve fibers. A possibility suggested by de Boer (1990) is that efferents produce a mechanical change (e.g., a distortion of the organ of Corti) that reduces the mechanical coupling of basilar membrane motion to sound-frequency bending of IHC stereocilia. Another possibility is that mechanical rectification in OHCs leads to a slow bending of IHC stereocilia (this might be the main method of exciting auditory nerve fibers for high-frequency sounds; Evans, Hallworth, and Dallos 1991), and efferent activity may inhibit by reducing the OHC rectification or the coupling of the resulting motion to IHCs. A third possibility is that medial efferents change the firing of afferents by an electrical effect of the MOC potential (Fex 1967; Geisler 1974a; Guinan and Gifford 1988b). The MOC potential might lead to less transmitter being released by IHCs (by slightly hyperpolarizing the IHC, Fig. 8.5B). Alternatively, the MOC potential in the region of the dendrites of radial auditory nerve fibers may cause there to be fewer action potentials in response to a given transmitter release (see Fig. 8.5A).

A key element in judging the above hypotheses is how well the efferent inhibition at low and moderate sound levels is accounted for by the efferent-induced reduction of basilar membrane motion. Some of the above mechanisms, particularly the efferent-induced reduction in mechanical coupling between basilar membrane motion and sound-frequency bending of IHC stereocilia, would reduce IHC drive by the same ratio at all sound levels and affect responses at both high and low sound levels approximately equally. If the efferent inhibition of basilar membrane motion fully accounted for the efferent inhibition of responses at low sound levels (as indicated by Dolan and Nuttall 1994), then there would be little room for an additional reduction due to another mechanism. If, as sound level is increased, the efferent inhibition of basilar membrane motion becomes less (Dolan and Nuttall 1994) but the efferent inhibition of auditory nerve firing stays the same or becomes more (Fig. 8.6D), then an efferent effect that is greatest at high sound levels appears to be needed. Alternatively, an efferent effect that reduces auditory nerve firing by a certain percentage might account for the data because of the reduced slope of the rate-versus-level function at high sound levels (Fig. 8.6D).

The reduction in the plateau rate of Figure 8.6D appears to require a mechanism that reduces auditory nerve firing at, or after, the stage responsible for the saturation of rate at high levels (which may be a saturation in the amount of transmitter released by the IHC). The similarity of the time courses of the MOC potential and the efferent inhibition of spontaneous activity indicates that the MOC potential can significantly influence auditory nerve firing (Section 4.3.2). Because of this, it seems likely that the MOC potential is also involved in producing efferent effects at high sound levels. One attractive aspect of the hypothesis that the MOC potential reduces the number of spikes evoked by a given IHC transmitter release is that this effect would reduce auditory nerve firing after the IHC synapse and account for medial efferent reduction of the plateau rate. Further experimental results are required to determine the actual combination of factors by which medial efferents change auditory nerve responses to high-level sounds.

4.4.3 Medial Efferent Mechanisms: Slow Transient Effects

Existing data are consistent with the hypothesis that slow transient medial efferent effects are produced by the same increase in OHC conductance responsible for fast medial efferent effects, plus a mechanism that slowly and transiently exaggerates the OHC conductance increase (Sridhar, Brown, and Sewell 1995; Sridhar et al. 1995). The exaggeration of the efferent-induced OHC conductance increase may be due to Ca^{2+} released during prolonged efferent stimulation, which overwhelms the Ca^{2+}-buffering ability of the OHC region near the efferent synapse, thereby increasing the Ca^{2+} concentration and causing an increased Ca^{2+}-activated K^+ increase.

5. Effects of Lateral Efferents

Although there is no direct evidence of effects of lateral efferents, there are data that provide hints of what lateral efferents may do. There are no single-unit recordings attributable to lateral efferents, and no measured effects attributable to stimulation of lateral efferents. Nonetheless, since lateral efferents contain ACh, calcitonin gene-related peptide (CGRP), enkephalins, and/or GABA (see Sewell, Chapter 9), the application of these neurochemicals in the cochlea may provide insight into lateral efferent effects.

Iontophoretic application of ACh in the region under the IHCs suggests that ACh, and therefore lateral efferents that contain ACh, may *excite* radial auditory nerve fibers. In guinea pigs, application of ACh from a multibarrel pipet increased the rate of extracellular action potentials presumed to be recorded from the dendrites of auditory nerve fibers (Felix and Ehrenberger 1992). Furthermore, ACh also increased responses to glutamate, an amino acid that is thought to excite the same postsynaptic receptors as the endogenous transmitter of the IHCs (see Sewell, Chapter 9). Both results suggest that lateral efferents that contain ACh may excite auditory nerve fibers.

Iontophoretic application of GABA in the region under the IHCs suggests that GABA, and therefore lateral efferents that contain GABA, may *inhibit* radial auditory nerve fibers. In guinea pigs, application of GABA from a multibarrel pipet produced a slow, long-lasting decrease in the spontaneous firing of 3 of 11 recordings presumed to be from radial auditory nerve fibers (Felix and Ehrenberger 1992). GABA also decreased activity evoked by glutamate (which presumably mimics the action of the normal IHC transmitter; see Sewell, Chapter 9) and activity evoked by ACh (which presumably is released by lateral efferents) (Felix and Ehrenberger 1992). Thus, GABAergic lateral efferents appear to produce distinctly different responses from those of cholinergic lateral efferents.

Although a slow decrease in the $f_2 - f_1$ distortion product otoacoustic emission has been suggested to be due to efferents, possibly GABAergic efferents, recent evidence rules out an efferent origin of this effect. Several investigations found that in guinea pig, gerbil, and rabbit, sounds that remained on for many minutes produced a decrease in the $f_2 - f_1$ DPOAE that was not accompanied by a decrease in the more commonly measured $2f_1 - f_2$ DPOAE (Brown 1988; Whitehead, Lonsbury-Martin, and Martin 1991; Kirk and Johnstone 1993). Based on a blocking of the slow change in $f_2 - f_1$ by bicuculline, a GABAergic antagonist, the change in $f_2 - f_1$ has been attributed to GABAergic synapses on OHCs (Kirk and Johnstone 1993). However, after all of the olivocochlear efferents were cut in the brain stem, the slow change in $f_2 - f_1$ was present and little different (Kujawa, Fallon, and Bobbin 1995; Lowe and Robertson 1995). Thus, this effect is not due to olivocochlear efferents.

Sahley and colleagues (Sahley et al. 1991; Sahley and Nodar 1994) have suggested that enkephalins found in lateral efferents may increase auditory sensitivity near threshold. In anesthetized chinchillas, they intravenously administered pentazocine, a drug that is presumed to activate receptors normally activated by enkephalins released from lateral efferents. Pentazocine increased N_1 amplitude for sound levels near threshold, but significantly, this increase was *not* accompanied by a change in CM (Sahley et al. 1991; Sahley and Nodar 1994). If the effect had been produced by medial efferent synapses on OHCs, a change in CM would have been expected. Although this result suggests that lateral efferent enkephalins may change auditory thresholds, this is not the only possibility, since pentazocine was administered intravenously and might have changed many things, such as cochlear blood flow.

A different kind of experiment, chronic cuts of efferents, indicates that lateral efferents may excite auditory nerve fibers (Liberman 1990). The basic response properties of auditory nerve fibers were recorded in normal cats and in cats in which all of the efferents had been cut 3 to 30 weeks earlier. The animals with cut efferents had normal thresholds, tuning curves, and rate-level functions, but only about two-thirds as much spontaneous activity. These data are consistent with the hypothesis that ongoing activity in lateral efferents excites radial auditory nerve fibers, and that cutting the efferents removed this activity and thereby reduced spontaneous activity. However, here again, there are other possible ways to explain the results (see Liberman 1990).

Altogether, the data provide tantalizing hints, but no unequivocal demonstration, that cholinergic lateral efferents excite radial auditory nerve afferents. In addition to the above evidence, excitation is the dominant efferent effect in mammalian vestibular afferent fibers (see Section 7.2), and CGRP, a neuropeptide found in lateral efferents, excites frog lateral line afferents (Adams, Mroz, and Sewell 1987; Sewell and Starr 1991). Compared with cholinergic efferents, even less is known about GABAergic lateral efferents and about the effects of the other neurotransmitters or neuromodulators found in lateral efferents.

6. Efferent Responses to Sound and Efferent Acoustic Reflexes

Most likely, all reported recordings from single efferent fibers have been from medial efferents (Fex 1962, 1965; Cody and Johnstone 1982a; Robertson 1984; Robertson and Gummer 1985; Liberman and Brown 1986; Gummer, Yates, and Johnstone 1988; Liberman 1988a,b; Brown 1989). The best evidence for this is that adequately labeled, physiologically identified efferent fibers have all innervated OHCs, and the few fibers that were traced centrally originated from the region of medial efferents (Robertson 1984; Robertson and Gummer 1985; Liberman and Brown 1986; Brown 1989). Since these studies used high-impedance pipet electrodes, it is not surprising that all recordings would be from the large, myelinated medial efferents and not the small, unmyelinated lateral efferents. As will be seen in the rest of this section, this recording bias has led to our having considerable knowledge of the response properties of medial efferents, but no direct knowledge of the response properties of lateral efferents.

6.1 The Medial Efferent Acoustic Reflex

6.1.1 Response Properties of Medial Efferents

The data obtained from medial efferent recordings depend on a number of methodological variables, including the method of anesthesia. All investigators used anesthesia, except Fex (1962, 1965) who used decerebrate animals, and all, including Fex, found qualitatively similar results. There are, however, small but statistically significant quantitative differences attributable to the type of anesthesia, and strong anecdotal evidence that efferent firing diminishes following a booster dose of anesthesia (Robertson and Gummer 1985; Liberman and Brown 1986; Brown 1989). These results, and measurements of efferent effects in humans in waking versus sleeping states (Froehlich et al. 1993), indicate that the magnitude of efferent effects varies with the state of the animal, but the qualitative pattern of efferent effects appears to be maintained.

Recordings from single medial efferents have been made in only two species: cats (Fex 1962, 1965; Liberman and Brown 1986; Liberman 1988a,b) and guinea pigs (Cody and Johnstone 1982a; Robertson 1984; Robertson and Gummer 1985; Gummer, Yates, and Johnstone 1988; Brown 1989), with different methods used for each species. Except for the experiments of Fex, the recordings in cats were at the vestibular-cochlear anastomosis, also known as the bundle of Oort, where efferent fibers course from the vestibular to the cochlear nerve. At this recording location, an electrode can sample efferents that innervate all parts of the cochlea, but there is some possibility that cerebellar retraction and drilling cause damage that affects ipsilateral responses. In contrast, the recordings in guinea pigs were near the intraganglionic spiral bundle in the midbasal turn of the cochlea and were biased toward sampling fibers that go to nearby parts of the cochlea and more apically, that is, they missed the most basally directed efferents (Brown 1987). In this approach, the cochlea is opened and the probability of damage that affects ipsilateral responses is much greater than in the approach used in the cat.

In the preparations used so far, most medial efferents (50%-89%) had no spontaneous activity, and the few with spontaneous activity fired at relatively low rates (0-20 spikes/s) (Robertson and Gummer 1985; Liberman and Brown 1986). However, medial efferent "spontaneous" rates are higher for minutes after stimulation with high-level sound (Liberman 1988a), and thus "spontaneous" rates depend on the stimulation history.

Furthermore, as noted above, "spontaneous" rates appear to depend on the level and kind of anesthesia.

Medial efferents have regular firing patterns, a property that has been used to distinguish medial efferents from the other nearby fibers. In both spontaneous activity and responses to sound, spike trains from medial efferents have interval histograms with symmetric shapes and few, if any, short intervals (<5 ms) (Robertson and Gummer 1985; Liberman and Brown 1986). In both cat and guinea pig preparations, the regular firing of medial efferents has been important in allowing efferents to be distinguished from nearby cochlear afferents that have irregular firing patterns with many short intervals. Near the vestibular-cochlear anastomosis, medial efferents can also be intermixed with vestibular efferents and afferents. Since some vestibular afferents have regular firing and some respond to sound (see McCue and Guinan 1994a), care must be taken in distinguishing medial efferents. Medial efferents are the only fibers with regular firing and long-latency (>5 ms) responses to sound. Whether *all* medial efferents respond to sound, as is generally assumed, is an open question. In all reported experiments, medial efferents would have been missed if they did not have spontaneous activity or respond at the sound levels tested.

Medial efferents have tuning curves that are similar to, or slightly wider than, those of auditory nerve fibers (Fig. 8.11) (Cody and Johnstone 1982a; Robertson 1984; Liberman and Brown 1986). Contrary to the initial report of Fex (1965), there is normally a well-defined best frequency (BF: the frequency with the lowest threshold). Medial efferent BFs span the range of hearing with a distribution across frequency similar to the overall pattern of efferent innervation of OHCs (see Fig. 8.2A,B) (Liberman 1988a).

Based on their response laterality, medial efferents have been divided into three types: ipsi, contra, and either-ear (either-ear efferents have been called

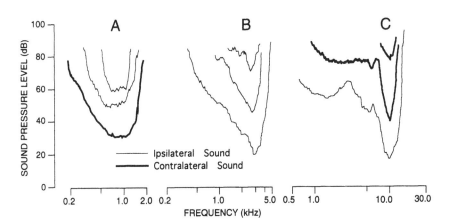

FIGURE 8.11 Tuning curves from cat medial efferents chosen to illustrate fibers with different thresholds in each BF range. (Adapted from Liberman and Brown 1986.)

"binaural efferents," but this is confusing because almost all efferents are binaural) (Robertson and Gummer 1985; Liberman and Brown 1986; Brown 1989). For monaural sounds, ipsi efferents respond only to ipsilateral (referenced to the ear innervated by the efferent) sound, contra efferents respond only to contralateral sound, and either-ear efferents respond to sound in either ear. Approximately two-third of medial efferents are ipsi, one-third are contra, and a small fraction (4%-11%) are either-ear. Although with monaural sound, ipsi and contra efferents respond only to one ear, once they are excited by sound in the primary ear, most efferents can be influenced by sound in the secondary ear (Liberman 1988a; Robertson and Gummer 1988). Most medial efferents, therefore, are binaural. Usually sound in the second ear produces additional excitation, but sometimes it produces inhibition; this aspect of medial efferent responses to sound has not been studied extensively.

The firing rates of medial efferents depend on many factors, including sound laterality, sound level, and sound stimulation history. For monaural sounds at the highest levels normally used (approximately 90 dB SPL), the fastest-firing medial efferent has a maximum rate of approximately 60/s and the average medial efferent fires half to two-thirds as fast (Robertson and Gummer 1985; Liberman 1988a). With binaural sound or a history of high-level sound, rates as high as 134/s have been found (Liberman 1988a), but even under these conditions, the maximum rate of most efferents is under 100/s (Fig. 8.12). Medial efferent rates increase with sound level over a wide range of sound levels, and in many medial efferents (particularly those with BF > 2 kHz), the rate was still increasing with sound level at the highest levels

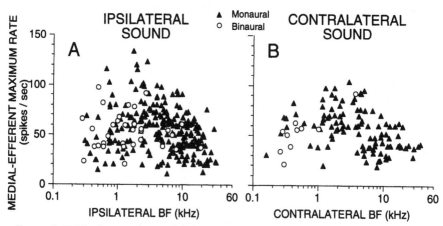

FIGURE 8.12 Maximum observed firing rates of medial efferent fibers versus fiber BFs. BF tone bursts were presented at levels between 0 and 90 dB SPL. For monaural fibers, the maximum rates were for stimulation in the primary ear. For many fibers, especially those with high BFs, the rate-level functions had not saturated at 90 dB SPL so the true maximum may be higher. (Adapted from Liberman 1988a; fibers classed as "uncertain" by Liberman are not distinguished from those that were "certain.")

tested. This suggests that medial efferents might have even higher rates at higher (perhaps traumatic) sound levels. The maximum rates of medial efferents can also depend on the anesthesia used (Brown 1989).

The question of how fast medial efferents fire in response to sound is important for understanding the medial efferent acoustic reflex. With electrical stimulation, the effects of medial efferents are highly dependent on the stimulation rate. The largest effects are for high shock rates (200–400/s); much lower effects are produced by stimulation rates less than 100/s (Fig. 8.4). When the highest reported efferent rate was 60/s, Pfalz (1969) argued that efferents did not fire fast enough to produce substantial effects in the cochlea. This problem is ameliorated by the higher rates that have been found with binaural stimulation (Liberman 1988a). However, the effects seen with sound-evoked efferent activity (see the next section) are surprisingly large considering the relatively low medial efferent firing rates usually evoked by such sounds and the high shock rates required to produce similar effects. A possible explanation is that with electrical stimulation, artificially high shock rates are needed to produce large efferent effects because electrical stimulation only excites a fraction of medial efferents. With this line of reasoning, if sound evokes firing in all, or almost all, medial efferents, then strong effects can be produced at lower rates than are required by shocks. It might be argued that shocks excite all medial efferents because Wiederhold and Kiang (1970) observed that with 400/s shocks, efferent inhibition saturated at the highest shock levels. This, however, need not imply that all medial efferents were excited. The saturation could have occurred at many different places (e.g., there may be a maximum OHC conductance change, limited by the number of Ca^{2+}-activated K^+ channels; see Section 2.5 and Sewell, Chapter 9).

The response properties of efferents vary across BFs. Low-BF efferents have lower thresholds, higher maximum rates, and shorter latencies than high-BF efferents (Liberman and Brown 1986). In addition, binaural facilitation by secondary-ear stimulation with tones is most pronounced among low-BF medial efferents, but facilitation by a background noise is greatest for high-BF medial efferents (Liberman 1988a).

Medial efferents have longer latencies than auditory nerve fibers. In response to tone or noise bursts, most medial efferents have minimum latencies of 10–40 ms, but some have latencies as short as 5 ms (Robertson and Gummer 1985; Liberman and Brown 1986; Brown 1989). In contrast, medial efferent group delays averaged 8.2 ± 1 ms when measured from modulation transfer functions using amplitude-modulated tones in guinea pigs (Gummer, Yates, and Johnstone 1988). This method of measuring latency gives a better estimate of the time it takes information to flow from the periphery to the medial efferent fiber during ongoing firing, that is, it is less influenced by factors that may have to build up when an efferent is first activated. Perhaps because of this, medial efferent group delays are much more tightly clustered than medial efferent latencies (Gummer, Yates, and Johnstone 1988).

6.1.2 Organization of Medial Efferent Acoustic Reflexes

Since medial efferents respond to sound and their activity influences cochlear responses, they are part of a sound-evoked efferent reflex. Single-fiber labeling indicates that medial efferent fibers project approximately to the region of the cochlea that is tuned to the BF of the efferent fiber (Robertson 1984; Robertson and Gummer 1985; Liberman and Brown 1986; Brown 1989). Thus, medial efferent feedback to the organ of Corti is frequency-specific, with the response of a given cochlear region feeding back to inhibit itself. The frequency specificity of this feedback is limited by the spread of medial efferent innervation in the cochlea and by the extent to which OHCs at a specific cochlear place influence cochlear responses over a range of frequencies. In the cat, a single medial efferent fiber can innervate a cochlear region spanning up to an octave (Liberman and Brown 1986), but in the guinea pig, single efferent fibers may innervate a shorter region (Robertson and Gummer 1985; Brown 1989). Such differences suggest the possibility that the length of medial efferent innervation is proportional to the cochlear length involved in amplifying the response to a single-frequency sound.

One potentially confusing aspect of the medial efferent acoustic reflex is that, most likely, crossed efferent fibers mediate the ipsilateral reflex and uncrossed efferent fibers mediate the contralateral reflex (see Fig. 8.1, *top*). The reason for this is that in the medial part of the superior olivary complex, the origin of medial efferents, neurons on one side respond predominantly to sound in the opposite ear (Guinan, Norris, and Guinan 1972). Thus, for the crossed reflex, the signal crosses the midline in the fibers that are inputs to the medial efferent neurons, not in the axons of medial efferent neurons. Furthermore, the ipsilateral reflex is really a double-crossed reflex, that is, the signal crosses the midline twice, once in the inputs to the medial efferents and a second time in the crossed efferent fibers. The clearest evidence for this is a few physiologically identified medial efferent fibers that have been labeled and traced to their origins (Liberman and Brown 1986). Also consistent with this is the fact that the fraction of ipsi medial efferents approximately equals the fraction with crossed axons (both approximately two-thirds). Similarly, the fraction of contra medial efferents approximately equals the fraction with uncrossed axons (both approximately one-third). The origin of either-ear medial efferents is not known. Robertson, Cole, and Corbett (1987) suggested that they correspond to the small fraction of medial efferents that, based on double-labeling experiments, innervate both ears.

Many things are known about the central pathways involved in medial efferent acoustic reflexes, but crucial details are missing. The observation of labeled endings on labeled medial efferents following tracer injections into the cochlea and into the cochlear nucleus (CN) indicates that there are direct connections from the CN to medial efferents (Fig. 8.1, *top*) (Robertson and

Winter 1988; Thompson and Thompson 1991). Both crossed and uncrossed CN inputs were found but, as expected, crossed inputs predominated (Fig. 8.1, *top*). Unfortunately, these studies do not provide a good estimate of the number of direct CN inputs to medial efferents or their importance relative to other inputs to medial efferents. The short latency (5 ms) of a few medial efferents is consistent with a direct input from CN neurons to medial efferents. However, for most medial efferents, the group delays measured from modulation transfer functions were longer (8.2 ± 1 ms), which suggests that the activation of most medial efferents may be mediated by a chain that includes one or more neurons between the CN and medial efferents (Gummer, Yates, and Johnstone 1988). Alternatively, a direct CN connection plus a long delay in medial efferent dendrites might also account for the long group delays.

6.1.3 Measurements of Medial Efferent Acoustic Reflexes in Animals

Since attempts to measure ipsilateral efferent reflexes encounter several difficult problems, efferent reflexes have been studied mainly by evoking efferent activity with contralateral sound and/or by cutting or pharmacologically blocking the efferents. Ipsilateral efferent reflexes are difficult to study, in part, because they are closed loop so that responses to sound already include effects of the ipsilateral medial efferent reflex. Furthermore, separating efferent from nonefferent (e.g., two-tone suppression or adaptation) effects of ipsilateral sound can be difficult. These problems are avoided if contralateral sound is used to evoke efferent activity. However, care must still be taken to rule out effects of the middle ear muscles (these can be activated by ipsilateral or contralateral sound) or of conduction of sound from one ear to the other. Another issue is that monaural, contralateral sound evokes activity principally in the uncrossed medial efferents, which are only one-third of the medial efferent population (see Fig. 8.1 and Section 6.1.2). In contrast, cutting the efferents allows study of the effects of all efferents using both ipsilateral and contralateral stimuli. Cuts, however, are not reversible, and care must be taken to avoid relevant nearby structures (e.g., stapedius motor axons that travel near the olivocochlear efferents in the floor of the fourth ventricle; Guinan, Joseph, and Norris 1989).

Efferent activity evoked by contralateral sound affects cochlear responses, and the similarity between these effects and those produced by shock-evoked efferent activity supports the hypothesis that these effects are produced by medial efferents. Many studies have found that contralateral sound inhibits N_1 and the firing of single auditory nerve fibers (Buño 1978; Murata et al. 1980; Folsom and Owsley 1987; Liberman 1989; Warren and Liberman 1989a,b). Warren and Liberman (1989a,b), in a wide-ranging study on auditory nerve fibers, found that the effects of contralateral sound are qualitatively similar to, but quantitatively less than, the effects produced by shock excitation of efferents. They found that the greatest inhibition occurs at the CF of a fiber. Within a fiber's dynamic range, there are approximately equal shifts in plots of rate and synchrony versus sound level, but little effect on phase. For contralateral tones, the greatest inhibition of the response to an ipsilateral tone is produced when the contralateral tone is approximately the same frequency as the ipsilateral tone. Presumably, this is because each medial efferent fiber projects to a region of the cochlea tuned to the efferent BF. Contralateral sound also depresses the spontaneous rate, plateau rate, and plateau rate at sound levels below, but not above, a rate dip. Finally, contralateral sound depresses otoacoustic emissions (Collet et al. 1990; Puel and Rebillard 1990), which provides additional evidence that medial, not lateral, olivocochlear efferents are involved.

Efferent effects evoked by contralateral sound are smaller than those evoked by shocks, but nonetheless, they are surprisingly large. Presumably, monaural contralateral sound evokes activity primarily in the one-third of medial efferents that are uncrossed, at rates that are well below those evoked by 200–400/s shocks (Fig. 8.12). Although quantitative comparisons are difficult because sound-evoked efferent effects vary in strength across animals and time (possibly because of varying levels of anesthesia), two well-studied cats of Liberman (1989) provide a useful example. In these cats, moderate-level contralateral noise inhibited N_1s evoked by 6-kHz tone pips by amounts ranging from 1 to 8 dB equivalent shift. The animal with the greatest inhibitions (which averaged 6 dB equivalent shift) also had the highest medial efferent firing rates (the average of the maximum rates for sounds up to 90 dB SPL was about 80/s). Contralateral sounds at the levels actually used would have evoked lower firing rates, averaging perhaps 60 spikes/s. Midline OCB shocks at 60/s would produce approximately a 5-dB level shift (Fig. 8.4 with 100% set to 20 dB; see Gifford and Guinan 1987), which compares to a 6-dB shift in Liberman's cat. However, midline OCB shocks are thought to excite most medial efferents (the two-thirds of medial efferents that are crossed plus some fraction of uncrossed medial efferents; Gifford and Guinan 1987), whereas contralateral sound is thought to excite only one-third of medial efferents. The apparent discrepancy might come about because (1) shocks only excite a fraction of medial efferents; (2) more than one-third of medial efferents are excited in the contralateral sound experiments (although the ipsilateral clicks technically make the stimulus "binaural," clicks have relatively little ability to excite medial efferents; Liberman and Brown 1986); and/or (3) the pattern of activity evoked by sound is somehow more efficient in producing cochlear effects than the pattern evoked by shocks.

Efferent effects evoked by contralateral sound are largest in the 1- to 2-kHz region of the cochlea, which contrasts with shock-evoked efferent effects, which are greatest from 2 to 10 kHz. To understand why moderate-level contralateral sound is most effective at 1–2 kHz, recall that the contralateral reflex is mediated by uncrossed medial efferent fibers and that their distribution is peaked in the center of the cochlea (Fig. 8.3A). The maximum rates evoked by contralateral sound are greatest in a broad range around 2 kHz for sounds up to 90 dB SPL (Fig. 8.12B). However, 90 dB SPL is much greater than the levels actually used in contralateral sound experiments. Since the rates of medial efferent fibers with BFs below 2 kHz often saturate at sound levels below 90 dB SPL (Liberman and Brown 1986; Liberman 1988a), lowering the contralateral sound will preferentially lower the rates of fibers with BFs above 2 kHz. The net effect of all of these factors would be to produce large, contralaterally evoked efferent inhibitions primarily in the 1- to 2-kHz region.

6.1.4 Measurements of Medial Efferent Acoustic Reflexes in Humans

Recently, the olivocochlear acoustic reflex has been studied in humans, principally by measuring the effect of contralateral sound of OAEs. To distinguish efferent effects from middle ear muscle (MEM) effects, investigators have (1) shown that the effects are removed by vestibular neurectomy (olivocochlear efferents exit the brain with the vestibular nerve, see Fig. 8.1); (2) used subjects with weak or absent MEM reflexes (note, however, that very weak MEM reflexes may not be detected by typical clinical devices but, nonetheless, can have a substantial effect on OAEs; Burns et al. 1993); and/or (3) demonstrated frequency-specific inhibition (e.g., Fig. 8.13), an effect that MEMs are probably not capable of producing because MEM contractions attenuate broad frequency ranges (Pang and Peake 1986) and binaural facilitation of the MEM acoustic reflex is not frequency dependent (Stelmachowicz and Gorga 1983).

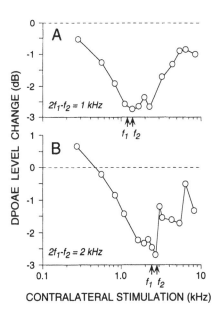

FIGURE 8.13 Frequency specificity of the reduction in DPOAE amplitude produced by contralateral-sound-evoked efferent activity. DPOAEs were from equilevel primary tones at frequencies indicated by the *arrows* and at levels 5 dB above the threshold for DPOAE detection. The contralateral stimulation was with 55 dB SPL, narrow-band noise with ascending and descending slopes of 24 dB/octave. Data were from a subject with large effects. (Adapted from Chéry-Croze, Moulin, and Collet 1993.)

Probably the most important thing learned from human studies of contralateral suppression of OAEs is that the medial efferent acoustic reflex in humans is similar to the reflex in experimental animals. Contralateral sound suppresses CEOAEs, TBOAEs, and DPOAEs (Collet et al. 1990; Ryan, Kemp, and Hinchcliffe 1991; Moulin, Collet, and Duclaux 1993; Norman and Thornton 1993; Williams, Brookes, and Prasher 1994). Little frequency-specific inhibition has been found on CEOAEs, but a considerable degree of frequency specificity has been shown for the inhibition of TBOAEs and DPOAEs (Fig. 8.13) (Ryan, Kemp, and Hinchcliffe 1991; Chéry-Croze, Moulin, and Collet 1993; Norman and Thornton 1993). The inhibition of OAEs is greater as the contralateral sound level is increased or as the ipsilateral sound level is decreased. At the lowest sound level in a recent study, the average equivalent sound level shift is 6 dB (Moulin, Collet, and Duclaux 1993). Although the suppressions typically measured in humans with OEAs are less than those measured in cats with N_1, most of the difference is probably due to differences in experimental conditions (Puria, Guinan, and Liberman 1996). It seems likely that the efferent reflex is as strong in humans as it is in most animals.

Efferent reflexes have also been studied using forward masking. With forward masking, binaural noise has been shown to suppress the click-evoked OAE more than ipsilateral or contralateral noise (Hood et al. 1995). Presumably, this is because binaural noise excites both crossed and uncrossed medial efferents, and the excitation of each group is greater than with monaural noise. Comparisons of efferent reductions of OAEs and psychophysical masking show that very little of psychophysical masking appears to be due to efferent inhibition (Gobsch et al. 1992; Kevanishvili et al. 1992). As noted elsewhere in this chapter, efferents may be more important in producing antimasking (see Sections 3.5 and 8.2) than in producing masking.

6.1.5 Central Influences on Medial Efferents

A variety of evidence indicates that medial efferents receive inputs from central sources; however, very little is known about what these inputs do. Anatomical studies have shown descending projections from higher auditory centers to medial and lateral efferents (Faye-Lund 1986). Chemical activation of the brain stem reticular formation appears to activate medial efferents and inhibit cochlear potentials (Kingsley and Barnes 1973). Electrical stimulation of the inferior colliculus appears to influence efferents to produce a protective effect in the cochlea (Rajan 1990). Sleep can influence an effect produced by medial efferents, the inhibition of OAEs by contralateral sound (Froehlich et al. 1993). Finally, there is considerable evidence that selective attention can modulate cochlear responses through activation of medial efferents (see Section 8.3), again implying central influences on medial efferents.

6.2 The Lateral Efferent Acoustic Reflex

Several aspects of the anatomy of lateral efferents suggest that they respond to sound. Lateral efferent cell bodies are located in the auditory brain stem in and around the LSO, and lateral efferents appear to receive inputs from brain stem auditory neurons (reviewed by Warr 1992). In addition, there is a topographic projection from lateral efferent cell bodies to the cochlea such that regions with similar best frequencies are connected (Guinan, Warr, and Norris 1984; Robertson, Anderson, and Cole 1987). It is difficult to see why lateral efferents would have an apparently tonotopic connectivity unless they respond to sound and have some frequency specificity.

There have been no reported recordings from lateral efferents, but there is evidence that they respond to sound. In cats, sound induced lateral efferents to produce a substance that immunostained for Fos, an early gene product (Adams 1995). In guinea pigs, stimulation with intense noise (110–115 dB SPL) produced a decrease of met-enkephalin (one of the putative neuroactive substances found in lateral efferents) in cochlear tissue and an increase in perilymph of a met-enkephalin-like compound (Drescher, Drescher, and Medina 1983; Eybalin et al. 1987). Presumably this indicates that sound-evoked activity in lateral efferents induced a release of meta-enkephalin, some of which diffused into the perilymph.

Since lateral efferents presumably respond to sound, and they synapse on radial auditory nerve fibers, they form a sound-evoked efferent reflex. If lateral efferent neurons are like most other neurons with cell bodies in the lateral part of the superior olivary complex, they are excited by ipsilateral sound (Guinan, Norris, and Guinan 1972). If, in addition, lateral efferents excite radial auditory nerve fibers (as suggested in Sections 5 and 7.2), then lateral efferents form a positive-feedback loop. However, this is probably an overly simplistic view of lateral efferents in both their central activation and the cochlear effects they produce.

7. Hair Cell Efferents in Preparations Other Than the Mammalian Cochlea

Although our focus is on the mammalian cochlea, it is worth considering data from other vertebrate hair cell systems, particularly when comparable data are not available for the mammalian cochlea. Since hair cell systems are phylogenetically primitive (Roberts and Meredith 1992), many aspects of mammalian efferent physiology can be expected to be shared with other vertebrates.

7.1 Recordings of Efferent Synaptic Potentials in Hair Cells

Although it has not been possible to record from OHCs during efferent stimulation, efferent-evoked synaptic potentials have been recorded in hair cells in both the fish lateral line and the turtle cochlea (Flock and Russell 1973, 1976; Art et al. 1982; Art, Fettiplace, and Fuchs 1984). In these preparations, efferent excitation evoked inhibitory postsynaptic potentials (IPSPs) in the hair cells. These IPSPs were produced by a conductance increase with an equilibrium potential near the K^+ potential. Furthermore, the pharmacology of the IPSPs indicates that they were produced by ACh released from the efferent nerve terminal. This is consistent with data from mammalian and chick preparations in which hair cell conductance changes have been produced by ACh agonists rather than by true IPSPs (Housley and Ashmore 1991; Fuchs and Morrow 1992; Doi and Ohmori 1993; Eróstegui et al. 1994; Kakehata et al. 1994).

Another important insight from the turtle cochlea is that presynaptic facilitation plays a role in increasing efferent synaptic output during a burst of efferent activity (Art, Fettiplace, and Fuchs 1984). With pairs of shocks exciting efferents, the IPSP evoked by the second shock was three times larger than the IPSP evoked by the first shock when the interval between shocks was short (4 ms). This facilitation of the IPSP decreased after the first shock with a time constant of about 100 ms. Fluctuations in the amplitudes of individual IPSPs indicate that each IPSP was composed of a small number of transmitter quanta. As the shock rate was increased, the IPSP amplitudes were larger and the fluctuations were less. The reduced fluctuations indicate that the increased IPSP amplitude is due, at least in part, to an increased probability of quantal release, a presynaptic effect. These data suggest that much of the buildup of efferent effects during a train of shocks is due to increased release of ACh through presynaptic facilitation.

The turtle cochlea shows an interesting efferent effect, a profound detuning without a change in the BF (Art et al. 1985). This detuning (a widening of the tuning curve) appears to be due to the efferent-evoked increase in hair cell membrane conductance. A somewhat similar, although much smaller, efferent-induced "detuning" without a change in CF is found in mammalian high-frequency auditory nerve fibers (Guinan and Gifford 1988c). However, the mechanisms of these effects are likely to be different. In the turtle cochlea, the hair cells are electrically tuned and the efferent-evoked electrical change reduces this electrical tuning (Art et al. 1985). In contrast, in the mammalian cochlea, tuning is primarily determined by mechanical factors, and medial efferent synapses on OHCs change basilar membrane motion and the sharpness of tuning by affecting the gain of the cochlear amplifier (see Section 3.4).

7.2 Effects of Efferent Synapses in the Vestibular System

Studies of efferent effects in the vestibular system may provide insight into one of the biggest gaps in knowledge of the physiology of cochlear efferents, the effects of efferent fibers that synapse directly on afferents (i.e., lateral efferents). However, before describing efferent effects in the vestibular system, basic vestibular anatomy and physiology need to be reviewed. The following picture of the mammalian vestibular system was derived both from classical anatomical techniques and from physiologic studies with single-fiber labeling (e.g., Spoendlin 1970; Baird et al. 1988; Fernández et al. 1988; Goldberg et al. 1990).

The mammalian vestibular system shows many parallels with the cochlear system, including two kinds of hair cells and two kinds of efferent endings (Fig. 8.14A,B). Flask-shaped type I vestibular hair cells are almost surrounded by large-calyx afferent endings, and these afferent endings receive efferent synapses (Fig. 8.14A, *open arrow*), an arrangement that is similar to lateral efferent endings on radial auditory nerve fibers (Fig. 8.14B, *open arrow*). Long, thin type II vestibular hair cells are contacted directly by both afferents and efferents, similar to OHCs (Fig. 8.14A,B). Although there are two kinds of hair cells, there are three classes of vestibular afferent fibers, classed according to whether they innervate type I, type II, or both kinds of hair cells. A major difference between vestibular and cochlear systems is that some vestibular afferents fire regularly and others fire irregularly, whereas radial auditory nerve fibers all fire irregularly. A possible explanation is that regular afferents innervate many hair cells, whereas irregular afferents innervate few.

Efferent stimulation has two effects on vestibular afferents, to increase the overall firing rate (i.e., to excite) and to decrease the within-cycle motion-to-rate "gain" (i.e., to inhibit). "Gain," as used in vestibular physiology, is the variation of the number of spikes per second throughout a motion stimulation cycle, divided by the amplitude of the motion. The mix of efferent excitation and inhibition depends on species and afferent fiber class (Goldberg and Fernández 1980; Boyle and Highstein 1990; McCue and Guinan 1994b). A hypothesis that fits most of the available data is that efferent synapses on hair cells inhibit the within-cycle "gain," and efferent synapses on afferent dendrites excite with little effect on the gain (Goldberg and Fernández 1980). With this hypothesis, the synapses that are comparable to lateral efferents excite the afferents.

In the two mammalian preparations that have been studied in some detail, the dominant effect of efferents was to excite vestibular afferents. In the squirrel monkey (*Saimiri sciureus*), 99% of vestibular afferents were excited, but most also showed a reduction of within-cycle "gain" (Goldberg and Fernández 1980). In the cat, in a smaller sample that was restricted to acoustically responsive afferents in the inferior vestibular nerve, all afferents were excited and there was a corresponding increase in within-cycle "gain" (McCue and Guinan 1994b). In these two preparations, the excitation was similar in showing both fast (~100 ms) and slow (tens of seconds) components.

Efferent effects in cat acoustically responsive vestibular fibers are

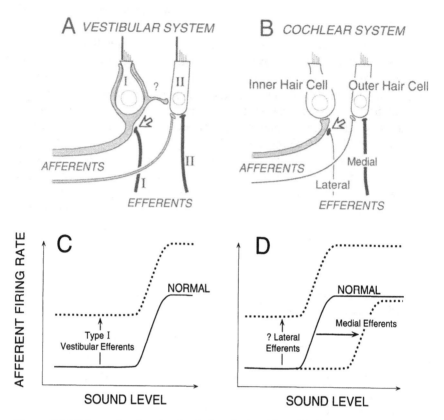

FIGURE 8.14 Schematics showing (*top*) the relationships of hair cells and neurons in the vestibular and cochlear systems, and (bottom) efferent effects on vestibular and cochlear afferents. Each system has flask-shaped cells (type I or IHCs) and cylindrical cells (type II or OHCs). Efferent neurons synapse directly on the afferent fibers under the flask-shaped cells [vestibular efferent I or lateral efferent synapses (*open arrows*)] or on the cylindrical cells (vestibular efferent II or medial efferent synapses). (A) Type I vestibular hair cells are innervated by large myelinated afferent neurons that form a chalice around the cell and that often branch to synapse on type II hair cells. The question mark indicates that not all afferents have these branches. Type II vestibular hair cells are innervated by small myelinated afferents with bouton endings. Separate fibers are shown producing the two kinds of efferent synapses, but a single vestibular efferent neuron may produce both kinds. (B) Each IHC is innervated by many myelinated afferents, each with a small ending. OHCs are innervated by a few unmyelinated afferents. The efferent innervation to the two hair cell regions is mostly separate in the cochlea, with unmyelinated lateral efferents synapsing on radial afferent dendrites and myelinated medial efferents synapsing directly on outer hair cells. *Bottom*: Schematized rate-level functions with efferent stimulation OFF (*solid lines*) and ON (*dashed lines*). (C) Stimulation of type I efferent endings on acoustically responsive vestibular afferent fibers increasing the firing rate of these vestibular afferents without shifting their acoustic thresholds (McCue and Guinan 1994b). (D) Stimulation of medial efferents to OHCs shifts the acoustic threshold of cochlear afferents. The action of lateral efferent synapses on cochlear afferents is unknown, but the analogous neural connections (*top panels*) suggest that lateral efferents may work like type I vestibular efferents, i.e., by raising afferent firing rate without affecting rate threshold. (Adapted from McCue and Guinan 1994b.)

particularly interesting because they appear to be due exclusively to efferent synapses on afferent endings, a synaptic arrangement like that of lateral efferents (Fig. 8.14A,B). In these afferents, efferent activity increased the afferent firing rate without changing the sound threshold or the total number of additional spikes evoked by sound at any sound level (Fig. 8.14C) (McCue and Guinan 1994b). Since efferent endings on hair cells would be expected to change the sound-evoked responses, the lack of a change in the sound-evoked response gives strong support for the hypothesis that the observed excitation is due to efferent synapses on afferent fibers (Fig. 8.14A, *arrow*) and not to efferent synapses on hair cells (McCue and Guinan 1994b). Interestingly, as the firing rate increased, the within-cycle "gain" also increased so that the degree of phase locking to low-frequency tones, as measured by the synchronization index, was not changed by efferent stimulation. These observations place strong constraints on the kinds of mechanisms that might be involved. One possibility is that efferent excitation is produced by an efferent-induced increase in the membrane resistance of the afferent ending. Then, if spontaneous activity and sound-evoked activity are produced by hair cell transmitter release with only a fraction of transmitter releases normally evoking spikes, an efferent-evoked increase in afferent resistance would increase the fraction of releases that evoke spikes and thereby increase the firing rate without changing the synchronization index.

Figure 8.14D pictures a hypothetical effect of lateral efferents, based on the vestibular excitatory effect, and contrasts this to known effects of medial efferents. In this figure, lateral efferents excite auditory nerve fibers and shift rate-level functions to higher rates without changing thresholds or the additional spikes evoked by sound. Note that evidence from other sources also suggests that lateral efferents excite afferents (see Section 5). In contrast, medial efferents shift rate-level functions to higher sound levels, a shift that is orthogonal to the hypothesized shift from lateral efferents. Thus, these two effects would be complementary.

7.3 Responses of Efferents During Motion

Recordings of neural activity from the lateral line, a hair cell mechanoreceptor that detects water movement, provide evidence that efferents are activated during body movement. In fish and frog lateral lines, efferent fibers do not respond to the mechanical stimulation of the lateral line, but they do fire preceding and during animal-initiated body movements (Russell 1971; Roberts and Russell 1972). This efferent activity inhibits the lateral line afferent response, and its function appears to be to suppress afferent responses evoked by the mechanical stimulation from the animal's movement. This inhibition may prevent overstimulation or just prevent unwanted self-stimulation. There is also evidence that vestibular efferents fire prior to and during body movements in goldfish (Klinke and Schmidt 1970).

There are no comparable data from efferents to hair cells in mammalian preparations. However, this activation of efferents prior to motion is similar to activation of the stapedius muscle in humans and bats prior to vocalization (Borg and Zakrisson 1975; Suga and Jen 1975). Thus, it would not be surprising if efferents were found to be activated during body motion or vocalization in mammals, including humans. However, Suga and Jen (1975) found no evidence from changes in N_1 or CM for activation of efferents during vocalization in bats.

8. Role of Olivocochlear Efferents in Hearing

There is evidence for several roles of medial efferents in hearing, but no evidence for a role of lateral efferents. The putative roles of medial efferents and evidence that supports these roles are reviewed below. It is not known, however, whether this list of medial efferent functions is complete, or even whether the most important function of medial efferents is included. For lateral efferents, there is so little information that any suggestion about their function must be highly speculative. Nonetheless, their anatomy puts some constraints on what they might do and leads to a suggested function.

8.1 Efferent Control of the Cochlear Amplifier and the Dynamic Range of Hearing

Medial efferent synapses on OHCs are well situated anatomically to control a cochlear amplifier formed by OHCs, and thus to control the dynamic range of hearing. The primary effect of efferent activity on the cochlear amplifier is to shift thresholds and level functions to higher sound levels and, therefore, to shift the dynamic range of hearing upward (Fig. 8.15). An early view of efferent control of the dynamic range of hearing was that auditory nerve fibers have a 30-dB range of thresholds, each has a 30- to 40-dB dynamic range giving an overall range of 70 dB, and that the 20-dB shift produced by efferents increases this overall range to 90 dB (Geisler 1974b). It is now realized that some auditory nerve fibers have much larger dynamic ranges, but the same concept still applies.

Efferents had been thought to affect responses only at low sound levels; however, current data show that efferent inhibition extends to high sound levels so that an upward extension of the dynamic range of hearing is a realistic role of efferents. Efferents were thought to produce substantial effects only at low sound levels, because at high sound levels efferent stimulation has little effect on N_1 or on the most common auditory nerve fibers, high-SR fibers (Galambos 1956; Wiederhold 1970). However, more recent data show that medial efferent inhibition can be large at very high

FIGURE 8.15 Medial efferent shift of the dynamic rate of hearing, as shown by the effect of medial efferent stimulation on auditory nerve fiber thresholds in one cat. (Δ) Low-SR fibers; (X) medium-SR fibers; (○) high-SR fibers. (Adapted from Guinan and Gifford 1988c.)

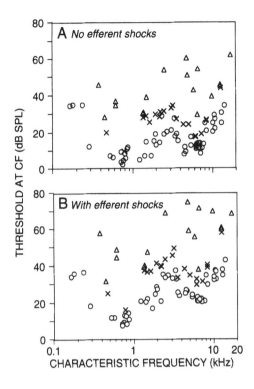

sound levels in low-SR and some medium-SR auditory nerve fibers (Fig. 8.6C,D) (Gifford and Guinan 1983; Guinan and Gifford 1988a; Guinan and Stankovic 1995). Furthermore, at high sound levels, low-SR and medium-SR fibers are the most important to inhibit, because at high sound levels, these fibers are probably the most important carriers of information (Young and Barta 1986; Sachs and Blackburn 1989). In addition, high-rate medial efferent activity is evoked only by high-level sound (see Section 6.1.1). Thus, the medial efferent acoustic reflex may have an important function in extending level functions upwards, thereby allowing a more accurate representation of sounds at high sound levels.

The efferent-induced upward shift of auditory nerve dynamic range may be of value not only in extending the range of response at high sound levels but also in removing the range of response at low sound levels. Although in some situations the loss of responses to low-level sounds is a disadvantage, in many situations it is an advantage, because it prevents adaptive masking (see Sections 3.5 and 8.2). Thus the role of medial efferents in controlling the cochlear amplifier may be to provide just enough amplification for the current acoustic situation, balancing the need to amplify very-low-level sounds against the desirability of having the dynamic range go up to very high sound levels without having too much adaptation and its resultant masking (see below).

8.2 Efferent Reduction of Masking

Physiological experiments have shown that medial efferents can have an antimasking effect on responses to transient sounds in a continuous masker. As explained in Section 3.5, medial efferents produce this antimasking primarily by reducing neural adaptation caused by the response to the masker. Based on the pattern of the physiological results, medial efferent activity should produce a considerable increase in the discriminability of well-above-threshold transient stimuli in continuous background noise, and under some circumstances may lower the threshold in noise (Kawase, Delgutte, and Liberman 1993).

Behavioral experiments have demonstrated an efferent antimasking effect, particularly in the discrimination of above-threshold sounds. Cutting efferents produced no change in the threshold for the detection of a tone in noise in cats and humans (Igarashi et al. 1972; Scharf et al. 1993). However, again in cats, Trahiotis and Elliott (1970) found that cutting the crossed OCB produced a 3- to 4-dB increase in masking at 1 and 2 kHz, although the effect was not statistically significant because of the small number of animals used. Experiments that looked for efferent changes in discrimination ability have had more positive results. Cutting efferents in the floor of the fourth ventricle produced a performance deficit in monkeys in a vowel discrimination task in low-pass noise (Dewson 1968), but the change may have been due to cutting stapedius motor axons (see Kawase, Delgutte, and Liberman 1993). Cats with cut efferents had performance deficits in detecting level changes in a noisy background for 8-kHz tones, but not for 1-kHz tones (McQuone and May 1993; May, McQuone, and Lavoie 1995). Humans with vestibular neurectomies had significantly worse intensity and speech discrimination in a binaurally noisy background (Zeng et al. 1994). In a very different kind of experiment, the detection of 1- or 2-kHz tones in noise was improved by the addition of noise in the contralateral ear in subjects with strong medial efferent feedback (judged from the inhibition of CEOAEs by the same contralateral noise), but contralateral noise did not improve detection in subjects with little or no medial efferent feedback (Micheyl and Collet 1993). This points out the importance of using binaural background noise rather than monaural noise in efferent behavioral experiments, both because binaural noise is more natural and because it evokes more efferent activity.

8.3 Efferent Involvement in Selective Attention

A variety of evidence suggests that medial efferents act to suppress auditory responses when attention is directed at another modality. The principal evidence for this comes from experiments on cats by Oatman (1971, 1976; Oatman and Anderson 1977) who found that during a visual discrimination task, N_1 was suppressed and CM was enhanced, exactly the pattern expected from activation of medial efferents. In humans, a visual detection task suppressed cochlear neural responses (Lukas 1981); however, paying attention to clicks during a discrimination task did not change auditory nerve responses, even though it enhanced cortical responses (Picton et al. 1971; Picton and Hillyard 1974). More recently, experiments on attention in humans have been done with measurements of OAEs. With visual attention tasks, no change was found in DPOAEs and SFEs (Avan and Bonfils 1992), but a change was found in CEOAEs, although only in certain subjects (Puel, Bonfils, and Pujol 1988; Froehlich et al. 1990). Overall, it appears that some aspect of the visual attention task elicits activity in medial efferents. Most likely, medial afferents are only one of many systems that suppress potentially distracting auditory responses.

Scharf and co-workers (1987, 1993, 1994) have presented evidence for what might be frequency-selective attention due to efferents. In their experiments, a subject is primed by an orienting tone, and must detect just-above-threshold tones either at the "priming frequency" or at a nearby, occasionally presented "test frequency." A tone is usually heard when it is at the primer tone frequency but not when it is unexpected and at a test frequency one-half critical band or more away from the priming frequency (the test-frequency tone is heard if it is presented every time). The suppression of unexpected test-frequency responses is thought to be an efferent effect, because after a vestibular neurectomy, the detectability of unexpected test-frequency responses is increased. This interesting effect needs to be explored more to determine its frequency selectivity (see Strickland and Viemeister 1995) and whether it is due to medial or lateral efferents (e.g., by measurements of OAEs).

8.4 Efferent Protection from Damage Due to Intense Sounds

Efferents appear to provide some protection from temporary threshold shift (TTS) due to intense sounds, but only over a narrow range of frequencies and durations. In the last few years, many reports have been published on this phenomenon, with what appeared to be conflicting results: some found a protective effect (Cody and Johnstone 1982b; Handrock and Zeisberg 1982; Rajan and Johnstone 1983, 1989; Puel, Bobbin, and Fallon 1988; Rajan 1988b; Patuzzi and Thompson 1991), some did not (Trahiotis and Elliott 1970; Liberman 1991), and some found that it depended on experimental conditions (Hildesheimer et al. 1990; Takeyama et al. 1992; Reiter and Liberman 1995). A view has emerged that seemingly small differences in experimental conditions can produce big differences in results because the efferent protective effect is present only for a narrow range of conditions. Efferent protection from TTS is strongest for traumatizing stimuli at approximately 10 kHz and for exposure durations of 1–2 minutes, and is weaker for lower or higher frequencies, or for longer durations (Fig. 8.16) (Reiter and Liberman 1995). When it is present, the efferent protective effect lowers the TTS but does not remove it (Fig. 8.16B).

How efferent protection from TTS is produced is not understood, but there are important clues to the mechanisms involved. Peripherally, there appears to be a close tie between the protective effect and the slow, transient increase in efferent inhibition (Reiter and Liberman 1995). There is also evidence that the effect is a protection of mechanoelectric transduction in OHCs (Patuzzi and Thompson 1991). Centrally, activation of the protective effect is complex and interesting. Activation of the protective effect can be produced by direct shocks to the OCB, contralateral sound at the same frequency as the traumatizing tone, ablation or lignocaine treatment of the contralateral cochlea, or electrical stimulation of the contralateral inferior colliculus (Rajan 1988a, 1990; Rajan and Johnstone 1989; Robertson and Anderson 1994). Except for direct OCB shocks, these protection activators can produce their protection without exciting medial efferents (shown by the lack of changes in N_1 and CM). Presumably, these protection activators act centrally to change the efferent acoustic reflex so that more (or different) efferent activity is evoked by the traumatizing stimulus.

It seems unlikely that noise trauma was a sufficiently large problem in primitive mammals that efferents evolved for their protective function. Nonetheless, this might be a useful function that efferents now perform. An important unresolved question is whether efferents help to protect against permanent threshold shifts.

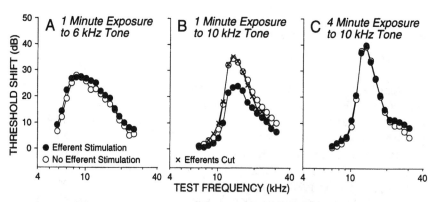

FIGURE 8.16 Reduction in temporary threshold shift (TTS) due to efferent activation for different exposure frequencies (A vs B) and durations (B vs C). Each panel shows the mean threshold shift measured 15 minutes postexposure for sets of animals with the monaural acoustic overexposure parameters listed at the top. The efferent stimulation was 150/s. (A) The 6-kHz tone was 118 dB SPL; (B) the 10-kHz tone was 110 dB SPL; (C) the 10-kHz tone was 104 dB SPL. (Adapted from Reiter and Liberman 1995.)

8.5 Efferent Activity During Speech?

There is no direct evidence, but efferents may be active during speech (see Section 7.3). Presumably, just as with MEM activation during speech, the purpose of such efferent activity would be to prevent self-stimulation with its resultant masking and TTS.

8.6 Possible Roles of Lateral Efferents

One clue to the role of lateral efferents is that they originate in or near the LSO, a brain stem center thought to be involved in the discrimination of interaural intensity differences. Presuming that lateral efferent neurons receive some of the same inputs as LSO neurons (see Warr 1992), they may be able to compare interaural intensity differences, as do LSO neurons (Schwartz 1992). Considering this, a possible function of lateral efferent neurons is to adjust the overall output from each cochlea so as to maintain a balance between the inputs from each ear. Although the output from a given ear may change with time (due to variations in blood flow, etc.), there is a continuing need for the output from the two ears to be balanced so as to maintain proper representation of sounds from different directions. A lateral efferent action of adjusting the firing of radial auditory nerve fibers without changing their incremental response to sound (Fig. 8.14) seems well suited to the function of maintaining the equality in overall firing rates between the two ears. Finally, the observation that vestibular neurectomy (which presumably cuts all efferents) can impair binaural localization ability (Fisch 1970) is consistent with the hypothesis that lateral efferents have a role in maintaining binaural localization abilities.

Since lateral efferents are unmyelinated and conduct action potentials slowly, they are best suited to tasks that do not require fast response and high information capacity. Their speed would, however, be suitable for adjusting the relative excitation from the two ears to compensate for slow drifts.

In contrast to medial efferents, lateral efferents have grossly different patterns of crossed and uncrossed projections. This suggests that the crossed and uncrossed lateral efferents have different functions.

9. Summary and Problems for Future Research

9.1 Medial Efferent Effects

We have considerable knowledge of medial efferent effects on cochlear responses but much less understanding of the mechanisms that produce these effects. The principal medial efferent effect is to shift the response sensitivity of auditory nerve fibers to higher sound levels. Shifts up to 20 dB are found at both low and high sound levels. Similar efferent-induced shifts have been seen in basilar membrane responses to sound. Most likely, this reduction in basilar membrane motion accounts for the level shift in auditory nerve responses at low sound levels and for part of the level shift at high sound levels. Presumably, efferents reduce basilar membrane motion by reducing the gain of the cochlear amplifier, possibly by the OHC synaptic conductance acting to shunt OHC receptor currents. The gross potential produced by activation of medial efferents, the "MOC potential," may also inhibit auditory nerve fibers, particularly at high sound levels.

To extend our understanding of medial efferent mechanisms, both at low and high sound levels, the most pressing need is for data on the efferent reduction of basilar membrane response to sound. We need to know how well the depression of basilar membrane motion matches the level shift in auditory nerve responses and under what circumstances. Discrepancies between efferent-induced depression of basilar membrane motion and shifts in auditory nerve responses would indicate that other efferent mechanisms must be involved. Many additional questions need answers. How does medial efferent activity depress basilar membrane motion? Does medial efferent activity decrease the gain of the cochlear amplifier by shunting receptor current? Why is the medial efferent–induced level shift seldom more than 20 dB when the cochlear amplifier appears to increase hearing sensitivity by much more than that? Presumably, this 20-dB limit is a clue to some aspect of the operation of the cochlear amplifier. Why, for some low-CF, low- and medium-SR auditory nerve fibers, is the greatest effect of medial efferents at the low-frequency edge of the tuning-curve tip, instead of at the CF? Answers to these questions are needed to provide a basic understanding of medial efferent mechanisms at low sound levels.

At high sound levels, the mechanisms of medial efferent inhibition are poorly understood. It seems likely that medial efferent depression of basilar membrane motion cannot fully account for this inhibition. Is this inhibition an effect of the MOC potential? If so, at what site? If not, what mechanism(s) are at work?

9.2 Medial Efferent Acoustic Reflex

We have a good qualitative picture of the medial efferent acoustic reflex. Medial efferents respond to sound and their activity inhibits cochlear responses. This reflex provides frequency-specific negative feedback to both ipsilateral and contralateral cochleae. In contrast to the MEM acoustic reflexes, the medial efferent acoustic reflex has a low threshold and its strength grows slowly with increasing sound level.

Although there is a good qualitative picture of the medial efferent acoustic reflex, there is little quantitative understanding. Medial efferent responses to sound cannot currently be translated into the effects produced in the cochlea. Plots of efferent effect versus efferent shock rate (Fig. 8.4) do not provide a translation, because the fraction of medial efferents excited by the shocks is not known. In addition, relatively little is known about the ipsilateral and binaural medial efferent acoustic reflexes, e.g., about the frequency selectivity of binaural facilitation. Finally, we need to know the response patterns of medial efferents for sounds similar to those for which antimasking effects have been demonstrated.

9.3 Lateral Efferents

Lateral efferents respond to sound, but little else is known about their physiology. Since recording from lateral efferents with microelectrodes seems almost impossible, studying their responses requires another method, perhaps the Fos method (Adams 1995) or something like it. Effects of lateral efferents have been studied by applying lateral efferent neurotransmitters in the cochlea, but such work does not mimic activation of lateral efferents either in the combination of transmitters released or in their spatial specificity. Since electrical stimulation has not worked, another method to selectively stimulate lateral efferents is needed. Perhaps there is a neurochemical method that can selectively stimulate lateral efferents at their origins.

9.4 Roles of Efferents in Hearing

There are several putative roles of medial efferents in hearing, but little evidence for the extent to which efferents actually play these roles. The gradual growth of the medial efferent acoustic reflex with increasing sound level provides continuous adjustment of the dynamic range of hearing and therefore of the adaptive masking produced by low-level sound. Medial efferents may also have a role in preventing damage due to loud sounds, and they may be activated centrally, perhaps to direct attention to another sensory modality. More behavioral work is needed to determine the actual role of these putative efferent functions. For such tests, measurement of OAEs provides a promising assay for medial efferent activation.

Are medial efferents active during speech, swallowing, or body motion? Measurements of OAEs may be of little help in this case, because MEM contractions and potential artifacts due to motion of the ear canal make ear canal sound pressure changes more difficult to interpret. Some other technique is necessary to attack this interesting question.

Acknowledgments I appreciate the comments of Drs. M.C. Brown, S.G. Kujawa, M.C. Liberman, and C.A. Shera on the manuscript, and the help of B.E. Norris in making the figures and in many aspects of preparing the manuscript. Supported by NIH Grant RO1 DC-00235.

References

Adams J (1995) Sound stimulation induces Fos-related antigens in cells with common morphological properties throughout the auditory brainstem. J Comp Neurol 361:645-688.

Adams JC, Mroz EA, Sewell WF (1987) A possible neurotransmitter role for CGRP in a hair-cell sensory organ. Brain Res 419:347-351.

Altschuler RA, Fex J (1986) Efferent neurotransmitters. In: Altschuler RA, Hoffman DW, Bobbin RP (eds) Neurobiology of Hearing: The Cochlea. New York: Raven Press, pp. 383-396.

Art JJ, Crawford AC, Fettiplace R, Fuchs PA (1982) Efferent regulation of hair cells in the turtle cochlea. Proc R Soc Lond B 216:377-384.

Art JJ, Fettiplace R, Fuchs PA (1984) Synaptic hyperpolarization and inhibition of turtle cochlear hair cells. J Physiol 356:525-550.

Art JJ, Crawford AC, Fettiplace R, Fuchs PA (1985) Efferent modulation of hair cell tuning in the cochlea of the turtle. J Physiol 360:397-421.

Avan P, Bonfils P (1992) Analysis of possible interactions of an attentional task with cochlear micromechanics. Hear Res 57:269-275.

Avan P, Bonfils P, Loth D, Elbez M, Erminy M (1995) Transient-evoked otoacoustic emissions and high-frequency acoustic trauma in the guinea pig. J Acoust Soc Am 97:3102-3020.

Baird RA, Desmadryl G, Fernández C, Goldberg JM (1988) The vestibular nerve of the chinchilla. II. Relation between afferent response properties and peripheral innervation patterns in the semicircular canals. J Neurophysiol 60:182-203.

Bobbin RP, Fallon M, Puel JL, Bryant G, Bledsoe SC Jr, Zajic G, Schacht J (1990) Acetylcholine, carbachol, and GABA induce no detectable change in the length of isolated outer hair cells. Hear Res 47:39-52.

Borg E, Zakrisson J-E (1975) The stapedius muscle and speech perception. Symp Zool Soc Lond 37:51-68.

Boyle R, Highstein SM (1990) Efferent vestibular system in the toadfish: action upon horizontal semicircular canal afferents. J Neurosci 10:1570-1582.

Brown AM (1988) Continuous low level sound alters cochlear mechanics: an efferent effect? Hear Res 34:27-38.

Brown MC (1987) Morphology of labeled efferent fibers in the guinea pig cochlea. J Comp Neurol 260:605-618.

Brown MC (1989) Morphology and response properties of single olivocochlear efferents in the guinea pig. Hear Res 40:93-110.

Brown MC, Nuttall AL (1984) Efferent control of cochlear inner hair cell responses in the guinea pig. J Physiol 354:625-646.

Brown MC, Nuttall AL, Masta RI (1983) Intracellular recodings from cochlear inner hair cells: effects of stimulation of the crossed olivocochlear efferents. Science 222:69-72.

Brown MC, Pierce S, Berglund AM (1991) Cochlear-nucleus branches of thick (medial) olivocochlear fibers in the mouse: a cochleotopic projection. J Comp Neurol 303:300-315.

Brownell WE, Bader CR, Bertrand D, de Ribaupierre Y (1985) Evoked mechanical response of isolated cochlear outer hair cells. Science 277:194-196.

Buño W (1978) Auditory nerve fiber activity influenced by contralateral ear sound stimulation. Exp Neurol 59:62-74.

Burns EM, Harrison WA, Bulen JC, Keefe DH (1993) Voluntary contraction of middle ear muscles: effects on input impedance, energy reflectance and spontaneous otoacoustic emissions. Hear Res 67:117-128.

Chéry-Croze A, Moulin A, Collet L (1993) Effect of contralateral sound stimulation on the distortion product $2f_1 - f_2$ in humans: evidence of a frequency specificity. Hear Res 68:53-58.

Cody AR, Johnstone BM (1982a). Acoustically evoked activity of single efferent neurons in the guinea pig cochlea. J Acoust Soc Am 72:280-282.

Cody AR, Johnstone BM (1982b) Temporary threshold shift modified by binaural acoustic stimulation. Hear Res 6:199-205.

Collet L, Kemp DT, Veuillet E, Duclaux R, Moulin A, Morgon A (1990) Effect of contralateral auditory stimuli on active cochlear micro-mechanical properties in human subjects. Hear Res 43:251-262.

Cooper NP, Rhode WS (1992) Basilar membrane mechanics in the hook region of cat and guinea-pig cochleae: sharp tuning and nonlinearity in the absence of baseline position shifts. Hear Res 63:163-190.

Dallos P, Cheatham MA (1976) Production of cochlear potentials by inner and outer hair cells. J Acoust Soc Am 60:510-512.

Dallos P, Evans BN (1995) High-frequency motility of outer hair cells and the cochlear amplifier. Science 267:2006-2009.

Davis H (1965) A model for transducer action in the cochlea. Cold Spring Harbor Symp Quant Biol 30:181-189.

Davis H (1983) An active process in cochlear mechanics. Hear Res 9:79-90.

de Boer E (1990) Wave propagation, activity and frequency selectivity in the cochlea. In: Grandori F, Cianfrone G, Kemp DT (eds) Cochlear Mechanisms and Otoacoustic Emissions. Basel: Karger, pp. 1-12.

Desmedt JE (1962) Auditory-evoked potentials from cochlea to cortex as influenced by activation of the efferent olivocochlear bundle. J Acoust Soc Am 34:1478-1496.

Dewson JH (1968) Efferent olivocochlear bundle: some relationships to stimulus discrimination in noise. J Neurophysiol 31:122-130.

Doi T, Ohmori H (1993) Acetylcholine increases intracellular Ca^{2+} concentration and hyperpolarizes the guinea-pig outer hair cells. Hear Res 67:179-188.

Dolan DF, Nuttall AL (1988) Masked cochlear whole-nerve response intensity functions altered by electrical stimulation of the crossed olivocochlear bundle. J Acoust Soc Am 83:1081-1086.

Dolan DF, Nuttall AL (1994) Basilar membrane movement evoked by sound is altered by electrical stimulation of the crossed olivocochlear bundle. Assoc Res Otolaryngol Abstr 17:89.

Drescher MJ, Drescher DG, Medina JE (1983) Effect of sound stimulation at several levels on concentration of primary amines, including neurotransmitter candidates, in perilymph of the guinea pig inner ear. J Neurochem 41:309-320.

Eróstegui C, Nenov AP, Norris CH, Bobbin RP (1994) Acetylcholine activates a K^+ conductance permeable to Cs^+ in guinea pig outer hair cells. Hear Res 81:119-129.

Evans BN, Hallworth R, Dallos P (1991) Outer hair cell electromotility: the sensitivity and vulnerability of the DC component. Hear Res 52:288-304.

Eybalin M, Pujol R (1989) Cochlear neuroactive substances. Arch Otorhinolaryngol 246:228-234.

Eybalin M, Rebillard G, Jarry T, Cupo A (1987) Effect of noise level on Met-enkephalin content of the guinea pig cochlea. Brain Res 418:189-192.

Eybalin M, Parnaud C, Geffard M, Pujol R (1988) Immunoelectron microscopy identifies several types of GABA-containing efferent synapses in the guinea pig organ of Corti. Neuroscience 24:29-38.

Faye-Lund H (1986) Projection from the inferior colliculus to the superior olivary complex in the albino rat. Anat Embryol 175:35-52.

Felix D, Ehrenberger K (1992) The efferent modulation of mammalian inner hair cell afferents. Hear Res 64:1-5.

Fernández C, Goldberg JM, Baird RA (1988) The vestibular nerve of the chinchilla. III. Peripheral innervation patterns in the utricular macula. J Neurophysiol 63:767-780.

Fex J (1959) Augmentation of cochlear microphonic by stimulation of efferent fibers to the cochlea. Acta Otolaryngol 50:540-541.

Fex J (1962) Auditory activity in centrifugal and centripetal cochlear fibers in cat. Acta Physiol Scand 55:2-68.

Fex J (1965) Auditory activity in the uncrossed centrifugal cochlear fibers in cat. A study of a feedback system, II. Acta Physiol Scand 64:43-57.

Fex J (1967) Efferent inhibition in the cochlea related to hair-cell dc activity: study of postsynaptic activity of the crossed olivo-cochlear fibers in the cat. J Acoust Soc Am 41:666-675.

Fisch U (1970) Transtemporal surgery of the internal auditory canal: report of 92 cases, technique, indications and results. Adv Otorhinolaryngol 17:203-240.

Fitzgerald M, Woolf CJ (1981) Effects of cutaneous nerve and intraspinal conditioning on C-fibre efferent terminal excitability in decerebrate spinal rats. J Physiol (Lond) 318:25-39.

Flock Å, Russell IJ (1973) The post-synaptic action of efferent fibres in the lateral line organ of the burbot *Lota lota*. J Physiol 35:591-605.

Flock Å, Russell IJ (1976) Inhibition by efferent nerve fibres, action on hair cells and afferent synaptic transmission in the lateral line canal organ of the burbot *Lota lota*. J Physiol 257:45-62.

Folsom RC, Owsley RM (1987) N1 action potentials in humans. Influence of simultaneous contralateral stimulation. Acta Otolaryngol (Stockh) 103:262-265.

Froehlich P, Collet L, Chanal JM, Morgon A (1990) Variability of the influence of a visual task on the active micromechanical properties of the cochlea. Brain Res 508:286-288.

Froehlich P, Collet L, Valatx JL, Morgon A (1993) Sleep and active cochlear micromechanical properties in human subjects. Hear Res 66:1-7.

Fuchs PA, Murrow BW (1992) A novel cholinergic receptor mediates inhibition of chick cochlear hair cells. Proc R Soc Lond B 248:35-40.

Galambos R (1956) Suppression of auditory activity by stimulation of efferent fibers to the cochlea. J Neurophysiol 19:424-437.

Geisler CD (1974a) Model of crossed olivocochlear bundle effects. J Acoust Soc Am 56:1910-1912.

Geisler CD (1974b) Hypothesis on the function of the crossed olivocochlear bundle. J Acoust Soc Am 56:1908-1909.

Gifford ML, Guinan JJ Jr (1983) Effects of cross-olivocochlear-bundle stimulation on cat auditory nerve fiber responses to tones. J Acoust Soc Am 74:115-123.

Gifford ML, Guinan JJ Jr (1987) Effects of electrical stimulation of medial

olivocochlear neurons on ipsilateral and contralateral cochlear responses. Hear Res 29:179-194.
Gobsch H, Kevanishvili Z, Gamgebeli Z, Gvelesiani T (1992) Behavior of delayed evoked otoacoustic emission under forward masking paradigm. Scand Audiol 21:143-148.
Goldberg JM, Fernández C (1980) Efferent vestibular system in the squirrel monkey: anatomical location and influence on afferent activity. J Neurophysiol 43:986-1025.
Goldberg JM, Desemadryl G, Fernández C, Baird RA (1990) The vestibular nerve of the chinchilla. V. Relation between afferent discharge properties and peripheral innervation patterns in the utricular macula. J Neurophysiol 63:791-804.
Guinan JJ Jr (1986) Effect of efferent neural activity on cochlear mechanics. Scand Audiol Suppl 25:53-62.
Guinan JJ Jr (1990) Changes in stimulus frequency otoacoustic emissions produced by two-tone suppression and efferent stimulation in cats. In: Dallos P, Geisler CD, Matthews JW, Steele CR (eds) Mechanics and Biophysics of Hearing. Madison, WI: Springer-Verlag, pp. 170-177.
Guinan JJ Jr (1991) Inhibition of stimulus frequency emissions by medial olivocochlear efferent neurons in cats. Assoc Res Otolaryngol Abstr 14:129.
Guinan JJ Jr, Gifford ML (1988a) Effects of electrical stimulation of efferent olivocochlear neurons on cat auditory-nerve fibers. I. Rate-level functions. Hear Res 33:97-114.
Guinan JJ Jr, Gifford ML (1988b) Effects of electrical stimulation of efferent olivocochlear neurons on cat auditory-nerve fibers. II. Spontaneous rate. Hear Res 33:115-128.
Guinan JJ Jr, Gifford ML (1988c) Effects of electrical stimulation of efferent olivocochlear neurons on cat auditory-nerve fibers. III. Tuning curves and thresholds at CF. Hear Res 37:29-46.
Guinan JJ Jr, Stankovic KM (1995) Medial olivocochlear efferent inhibition of auditory-nerve firing mediated by changes in endocochlear potential. Assoc. Res Otolaryngol Abstr 18:172.
Guinan JJ Jr, Norris BE, Guinan SS (1972) Single auditory units in the superior olivary complex II: Locations of unit categories and tonotopic organization. Int J Neurosci 4:147-166.
Guinan JJ Jr, Warr WB, Norris BE (1983) Differential olivocochlear projections from lateral vs. medial zones of the superior olivary complex. J Comp Neurol 221:358-370.
Guinan JJ Jr, Warr WB, Norris BE (1984) Topographic organization of the olivocochlear projections from lateral and medial zones of the superior olivary complex. J Comp Neurol 226:21-27.
Guinan JJ Jr, Joseph MP, Norris BE (1989) Brainstem facial-motor pathways from two distinct groups of stapedius motoneurons in the cat. J Comp Neurol 289:134-144.
Gummer M, Yates GK, Johnstone BM (1988) Modulation transfer function of efferent neurons in the guinea pig cochlea. Hear Res 36:41-52.
Hallin RG, Torebjork HE (1973) Electrically induced A and C fibre responses in intact human skin nerves. Exp Brain Res 16:309-320.
Handrock M, Zeisberg J (1982) The influence of the efferent system on adaptation, temporary and permanent threshold shift. Arch Otorhinolaryngol 234:191-195.

Harrison WA, Burns EM (1993) Effects of contralateral acoustic stimulation on spontaneous otoacoustic emissions. J Acoust Soc Am 94:2649-2658.

Hildesheimer M, Makai E, Muchnik C, Rubinstein M (1990) The influence of the efferent system on acoustic overstimulation. Hear Res 43:263-268.

Hood LJ, Berlin CI, Wakefield L, Hurley A (1995) Noise duration affects bilateral, ipsilateral and contralateral suppression of transient-evoked otoacoustic emissions in humans. Assoc Res Otolaryngol Abstr 18:123.

Housley GD, Ashmore JF (1991) Direct measurement of the action of acetylcholine on isolated outer hair cells of the guinea pig cochlea. Proc R Soc Lond B 244:161-167.

Igarashi M, Alford BR, Nakai Y, Gordon WP (1972) Behavioral auditory function after transection of crossed olivo-cochlear bundle in the cat. I. Pure-tone threshold and perceptual signal-to-noise ratio. Acta Otolaryngol (Stockh) 73:455-466.

Kakehata S, Nakagawa T, Takasaka T, Akaike N (1994) Cellular mechanism of acetylcholine-induced response in dissociated outer hair cells of guinea-pig cochlea. J Physiol 463:227-244.

Kawase T, Liberman MC (1993) Anti-masking effects of the olivocochlear reflex, I: Enhancement of compound action potentials to masked tones. J Neurophysiol 70:2519-2532.

Kawase T, Delgutte B, Liberman MC (1993) Anti-masking effects of the olivocochlear reflex, II: Enhancement of auditory-nerve response to masked tones. J Neurophysiol 70:2533-2549.

Kemp DT (1978) Stimulated acoustic emissions from within the human auditory system. J Acoust Soc Am 64:1386-1391.

Kemp DT, Chum R (1980) Properties of the generator of stimulated acoustic emissions. Hear Res 2:213-232.

Kemp DT, Souter M (1988) A new rapid component in the cochlear response to brief electrical efferent stimulation. Hear Res 34:49-62.

Kevanishvili Z, Gobsch H, Gvelesiani T, Gangebeli Z (1992) Evoked otoacoustic emission: behaviour under the forward masking paradigm. ORL 54:229-234.

Kiang NYS, Moxon EC, Kahn AR (1976) The relationship of gross potentials recorded from the cochlea to single unit activity in the auditory nerve. In: Ruben RJ, Elberling C, Salomon G (eds) Electrocochleography. Baltimore: University Park Press, pp. 95-115.

Kim DO, Molnar CE, Matthews JW (1980) Cochlear mechanics: nonlinear behavior in two-tone responses as reflected in cochlear-nerve-fiber responses and in ear-canal sound pressure. J Acoust Soc Am 67:1704-1721.

Kimura R, Wersäll J (1962) Termination of the olivocochlear bundle in relation to the outer hair cells of the organ of Corti in guinea pig. Acta Otolaryngol (Stockh) 55:11-32.

Kingsley RE, Barnes CD (1973) Olivo-cochlear inhibition during physostigmine-induced activity in the pontine reticular formation in the decerebrate cat. Exp Neurol 40:43-51.

Kirk DL, Johnstone BM (1993) Modulation of $f_2 - f_1$: evidence for a GABA-ergic efferent system in apical cochlea of the guinea pig. Hear Res 67:20-34.

Kittrell BJ, Dalland JI (1969) Frequency dependence of cochlear microphonic augmentation produced by olivo-cochlear bundle stimulation. Laryngoscope 79:228-238.

Klinke E, Schmidt CL (1970) Efferent influence on the vestibular organ during active movements of the body. Pflügers Arch Ges Physiol 318:325-332.

Konishi T, Slepian JZ (1971) Effects of the electrical stimulation of the crossed olivocochlear bundle on cochlear potentials recorded with intracochlear electrodes in guinea pigs. J Acoust Soc Am 49:1762-1769.

Kujawa SG, Glattke TJ, Fallon M, Bobbin RP (1993) Contralateral sound suppresses distortion product otoacoustic emissions through cholinergic mechanisms. Hear Res 68:97-106.

Kujawa SG, Glattke TJ, Fallon M, Bobbin RP (1994) A nicotinic-like receptor mediates suppression of distortion product otoacoustic emissions by contralateral sound. Hear Res 74:122-134.

Kujawa SG, Fallon M, Bobbin RP (1995) Time-varying alterations in the $F_2 - F_1$ DPOAE response to continuous primary stimulation. I. Response characterization and contribution of the olivocochlear efferents. Hear Res 85:142-154.

Lewis ER, Henry KR (1992) Modulation of cochlear nerve spike rate by cardiac activity in the gerbil. Hear Res 63:7-11.

Liberman MC (1980) Efferent synapses in the inner hair cell area of the cat cochlea: an electron microscopic study of serial sections. Hear Res 3:189-204.

Liberman MC (1982) The cochlear frequency map for the cat: labeling auditory-nerve fibers of known characteristic frequency. J Acoust Soc Am 72:1441-1449.

Liberman MC (1988a). Response properties of cochlear efferent neurons: monaural vs. binaural stimulation and the effects of noise. J Neurophysiol 60:1779-1798.

Liberman MC (1988b) Physiology of cochlear efferent and afferent neurons: direct comparisons in the same animal. Hear Res 34:179-192.

Liberman MC (1989) Rapid assessment of sound-evoked olivocochlear feedback: suppression of compound action potentials by contralateral sound. Hear Res 38:47-56.

Liberman MC (1990) Effects of chronic cochlear de-efferentation on auditory-nerve response. Hear Res 49:209-224.

Liberman MC (1991) The olivocochlear efferent bundle and susceptibility of the inner ear to acoustic injury. J Neurophysiol 65:123-132.

Liberman MC, Brown MC (1986) Physiology and anatomy of single olivocochlear neurons in the cat. Hear Res 24:17-36.

Liberman MC, Dodds LW, Pierce S (1990) Afferent and efferent innervation of the cat cochlea: quantitative analysis with light and electron microscopy. J Comp Neurol 301:443-460.

Long G, Talmadge CI, Shaffer L (1994) The effects of contralateral stimulation on synchronous evoked otoacoustic emissions. J Acoust Soc Am 95:2844.

Lowe M, Robertson D (1995) The behaviour of the f_2-f_1 acoustic distortion product: lack of effect of brainstem lesions in anaesthetized guinea pigs. Hear Res 83:133-141.

Lukas JH (1981) The role of efferent inhibition in human auditory attention: an examination of the auditory brainstem potential. Int J Neurosci 12:137-145.

Martin GK, Lonsbury-Martin BL, Probst R, Coats AC (1988) Spontaneous otoacoustic emissions in a nonhuman primate. I. Basic features and relations to other emissions. Hear Res 33:49-68.

May BJ, McQuone SJ, Lavoie A (1995) Effects of olivocochlear lesions on intensity discrimination in cats. Assoc Res Otolaryngol Abstr 18:146.

McCue MP, Guinan JJ Jr (1994a) Acoustically-responsive fibers in the vestibular nerve of the cat. J Neurosci 14:6058-6070.

McCue MP, Guinan JJ Jr (1994b) Influence of efferent stimulation on acoustically-responsive vestibular afferents in the cat. J Neurosci 14:6071-6083.

McQuone SJ, May BJ (1993) Effects of olivocochlear efferent lesions on intensity discrimination in noise. Assoc Res Otolaryngol Abstr 16:51.

Micheyl C, Collet L (1993) Involvement of medial olivocochlear system in detection in noise. J Acoust Soc Am 93:2314.

Mott JB, Norton SJ, Neely ST, Warr WB (1989) Changes in spontaneous otoacoustic emissions produced by acoustic stimulation of the contralateral ear. Hear Res 38:229–242.

Moulin A, Collet L, Duclaux R (1993) Contralateral auditory stimulation alters acoustic distortion products in humans. Hear Res 65:193–210.

Mountain DC (1980) Changes in endolymphatic potential and crossed olivocochlear bundle stimulation alter cochlear mechanics. Science 210:71–72.

Mountain DC, Geisler CD, Hubbard AE (1980) Stimulation of efferents alters the cochlear microphonic and the sound induced resistance changes measured in scala media of the guinea pig. Hear Res 3:231–240.

Murata K, Tanahashi T, Horidawa J, Funai HM (1980) Mechanical and neural interactions between binaurally applied sounds in cat cochlear nerve fibers. Neurosci Lett 18:289–294.

Murata K, Moriyama T, Hosokawa Y, Minami S (1991) Alternating current induced otoacoustic emissions in the guinea pig. Hear Res 55:201–214.

Nieder P, Nieder I (1970) Crossed olivocochlear bundle: electrical stimulation enhances masked neural responses to loud clicks. Brain Res 21:135–137.

Norman M, Thornton ARD (1993) Frequency analysis of the contralateral suppression of evoked otoacoustic emissions by narrow-band noise. Br J Audiol 27:281–289.

Norton SJ, Neely ST (1987) Tone-burst-evoked otoacoustic emissions from normal-hearing subjects. J Acoust Soc Am 81:1860–1872.

Oatman LC (1971) Role of visual attention on auditory evoked potentials in unanesthetized cats. Exp Neurol 32:341–356.

Oatman LC (1976) Effects of visual attention on the intensity of auditory evoked potentials. Exp Neurol 51:41–53.

Oatman LC, Anderson BW (1977) Effects of visual attention on tone burst evoked auditory potentials. Exp Neurol 57:200–211.

Pang XD, Peake WT (1986) How do contractions of the stapedius muscle alter the acoustic properties of the ear? In: Allen JB, Hall JL, Hubbard A, Neely SI, Tubis A (eds) Peripheral Auditory Mechanisms. New York: Springer-Verlag, pp. 36–43.

Patuzzi R, Rajan R (1990) Does electrical stimulation of the crossed olivo-cochlear bundle produce movement of the organ of Corti? Hear Res 45:15–32.

Patuzzi RB, Thompson ML (1991) Cochlear efferent neurones and protection against acoustic trauma: protection of outer hair cell receptor current and interanimal variability. Hear Res 54:45–58.

Pfalz RKJ (1969) Absence of a function for the crossed olivocochlear bundle under physiological conditions. Arch Klin Exp Ohr Nas Kehlk Heilk 193:89–100.

Picton TW, Hillyard SA (1974) Human auditory evoked potentials. II. Effects of attention. Electroenceph Clin Neurophysiol 36:191–199.

Picton TW, Hillyard SA, Galambos R, Schiff M (1971) Human auditory attention: a central or peripheral process? Science 173:351–353.

Puel J-L, Rebillard G (1990) Effect of contralateral sound stimulation on the distortion product 2F1-F2: evidence that the medial efferent system is involved. J Acoust Soc Am 87:1630–1635.

Puel J-L, Bobbin RP, Fallon M (1988) An ipsilateral cochlear efferent loop protects the cochlea during intense sound exposure. Hear Res 37:65–70.

Puel J-L, Bonfils P, Pujol R (1988) Selective attention modifies the active micromechanical properties of the cochlea. Brain Res 447:380–383.

Puria S, Guinan JJ Jr, Liberman MC (1996) Olivocochlear reflex assays: effects of contralateral sound on compound action potentials vs. ear-canal distortion products. J Acoust Soc Am 99:500–507.

Rajan R (1988a) Effect of electrical stimulation of the crossed olivocochlear bundle on temporary threshold shifts in auditory sensitivity. I. Dependence on electrical stimulation parameters. J Neurophysiol 60:549–568.

Rajan R (1988b) Effect of electrical stimulation of the crossed olivocochlear bundle on temporary threshold shifts in auditory sensitivity. II. Dependence on the level of temporary threshold shifts. J Neurophysiol 60:569–579.

Rajan R (1990) Electrical stimulation of the inferior colliculus at low rates protects the cochlea from auditory desensitization. Brain Res 506:192–204.

Rajan R, Johnstone BM (1983) Crossed cochlear influences on monaural temporary threshold shifts. Hear Res 9:279–294.

Rajan R, Johnstone BM (1989) Contralateral cochlear destruction mediates protection from monaural loud sound exposures through the crossed olivocochlear bundle. Hear Res 39:263–278.

Rasmussen GL (1946) The olivary peduncle and other fiber projections of the superior olivary complex. J Comp Neurol 84:141–219.

Rasmussen GL (1960) Efferent fibers of cochlear nerve and cochlear nucleus. In: Rasmussen GL, Windle WF (eds) Neural Mechanisms of the Auditory and Vestibular Systems. Springfield, IL: Thomas, pp. 105–115.

Reiter ER, Liberman MC (1995) Efferent-mediated protection from acoustic overexposure: relation to "slow effects" of olivocochlear stimulation. J Neurophysiol 73:506–514.

Rhode WS (1971) Observations of the vibration of basilar membrane in squirrel monkeys using the Mössbauer technique. J Acoust Soc Am 49:1218–1231.

Rhode WS (1973) An investigation of postmortem cochlear mechanics using the Mössbauer effect. In: Møller AR (ed) Basic Mechanisms of Hearing. New York: Academic Press, pp. 49–67.

Roberts BL, Meredith GE (1992) The efferent innervation of the ear: variations on an enigma. In: Webster DB, Fay RR, Popper AN (eds) The Evolutionary Biology of Hearing. New York: Springer-Verlag, pp. 185–210.

Roberts BL, Russell IJ (1972) The activity of lateral-line efferent neurones in stationary and swimming dogfish. J Exp Biol 57:435–448.

Robertson D (1984) Horseradish peroxidase injection of physiologically characterized afferent and efferent neurones in the guinea pig spiral ganglion. Hear Res 15:113–121.

Robertson D, Anderson C-J (1994) Acute and chronic effects of unilateral elimination of auditory nerve activity on susceptibility to temporary deafness induced by loud sound in the guinea pig. Brain Res 646:37–43.

Robertson D, Gummer M (1985) Physiological and morphological characterization of efferent neurons in the guinea pig cochlea. Hear Res 20:63–77.

Robertson D, Gummer M (1988) Physiology of cochlear efferents in the mammal. In: Syka J, Masterton RB (eds) Auditory Pathways, Structure and Function. New York: Plenum, pp. 269–278.

Robertson D, Winter IM (1988) Cochlear nucleus inputs to olivocochlear neurons revealed by combined anterograde and retrograde labelling in the guinea pig. Brain Res 462:47–55.

Robertson D, Anderson C-J, Cole KS (1987) Segregation of efferent projections to different turns of the guinea pig cochlea. Hear Res 25:69–76.

Robertson D, Cole KS, Cobett K (1987) Quantitative estimate of bilaterally projecting medial olivocochlear neurons in the guinea pig brainstem. Hear Res 27:177–181.

Robles L, Ruggero MA, Rich NC (1986) Basilar membrane mechanics at the base of the chinchilla cochlea. I. Input-output functions, tuning curves, and response phases. J Acoust Soc Am 80:1364–1374.

Roddy J, Hubbard AE, Mountain DC, Xue S (1994) Effects of electrical biasing on electrically-evoked otoacoustic emissions. Hear Res 73:148–154.

Russell IJ (1971) The role of the lateral-line efferent system of *Xenopus laevis*. J Exp Biol 54:621–641.

Ryan S, Kemp DT, Hinchcliffe R (1991) The influence of contralateral acoustic stimulation on click-evoked otoacoustic emission in humans. Br J Audiol 25:391–397.

Sachs MB, Blackburn CC (1989) Processing of complex stimuli in the anteroventral cochlear nucleus. Assoc Res Otolaryngol Abstr 12:5–6.

Sahley TL, Nodar RH (1994) Improvement in auditory function following pentazocine suggests a role for dynorphins in auditory sensitivity. Ear Hear 15:422–431.

Sahley TL, Kalish RB, Musiek FE, Hoffman DW (1991) Effects of opioid drugs on auditory evoked potentials suggest a role of lateral olivocochlear dynorphins in auditory function. Hear Res 55:133–142.

Santos-Sacchi J, Dilger JP (1988) Whole cell currents and mechanical responses of isolated outer hair cells. Hear Res 35:143–150.

Scharf B, Quigley S, Aoki C, Peachey N, Reeves A (1987) Focused auditory attention and frequency selectivity. Percept Psychophys 42:215–223.

Scharf B, Nadol J, Magnan J, Chays A, Marchioni A (1993) Does efferent input improve the detection of tones in monaural noise? In: Verrillo R (ed) Sensory Research: Multimodal Perspectives. Hillsdale, NJ: Erlbaum Press, pp. 299–306.

Scharf B, Magnan J, Collet L, Ulmer E, Chays A (1994) On the role of the olivocochlear bundle in hearing: a case study. Hear Res 75:11–26.

Schwartz IR (1992) The superior olivary complex and lateral lemniscal nuclei. In: Webster DB, Popper AN, Fay RR (eds) The Mammalian Auditory Pathway: Neuroanatomy. New York: Springer-Verlag, pp. 117–167.

Schwarz DWF, Schwarz IE, Hu K, Vincent SR (1988) Retrograde transport of [^{3}H]-GABA by lateral olivocochlear neurons in the rat. Hear Res 32:97–102.

Sewell WF (1984a) The relation between the endocochlear potential and spontaneous activity in auditory nerve fibres of the cat. J Physiol (Lond) 347:685–696.

Sewell WF (1984b) The effects of furosemide on the endocochlear potential and auditory-nerve fiber tuning curves in cats. Hear Res 14:305–314.

Sewell WF, Starr PA (1991) Effects of calcitonin gene-related peptide and efferent nerve stimulation on afferent transmission in the lateral line organ. J Neurophysiol 65:1158–1169.

Siegel JH, Kim DO (1982) Efferent neural control of cochlear mechanics? Olivocochlear bundle stimulation affects cochlear biomechanical nonlinearity. Hear Res 6:171–182.

Smith CA (1961) Innervation pattern of the cochlea. Ann Oto Rhinol Laryngol 70:504–527.
Sohmer H (1966) A comparison of the efferent effects of the homolateral and contralateral olivo-cochlear bundles. Acta Otolaryngol 62:74–87.
Spoendlin H (1966) The organization of the cochlear receptor. Adv Otorhinolaryngol 13:1–227.
Spoendlin H (1970) Auditory, vestibular, olfactory and gustatory organs. In: Bischoff A (ed) Ultrastructure of the Peripheral Nervous System and Sense Organs. St. Louis: C.V. Mosby, pp. 173–338.
Sridhar TS, Brown MC, Sewell WF (1995) Molecular mechanisms involved in olivocochlear efferent slow effects. Assoc Res Otolaryngol Abstr 18:172.
Sridhar TS, Liberman MC, Brown MC, Sewell WF (1995) A novel cholinergic "slow effect" of olivocochlear stimulation on cochlear potentials in the guinea pig. J Neurosci 15:3667–3678.
Stelmachowicz PG, Gorga MP (1983) Investigation of the frequency specificity of acoustic reflex facilitation. Audiology 22:128–135.
Strickland EA, Viemeister NF (1995) An attempt to find psychophysical evidence for efferent action in humans. Assoc Res Otolaryngol Abstr 18:173.
Suga N, Jen PH-S (1975) Peripheral control of acoustic signals in the auditory system of echolocating bats. J Exp Biol 62:277–311.
Takeyama M, Kusakari J, Nishikawa N, Wada T (1992) The effect of crossed olivo-cochlear bundle stimulation on acoustic trauma. Acta Otolaryngol (Stockh) 112:205–209.
Teas DC, Konishi T, Nielsen DW (1972) Electrophysiological studies on the spatial distribution of the crossed olivocochlear bundle along the guinea pig cochlea. J Acoust Soc Am 51:1256–1264.
Thompson AM, Thompson GC (1991) Posteroventral cochlear nucleus projections to olivocochlear neurons. J Comp Neurol 303:267–285.
Trahiotis C, Elliott DN (1970) Behavioral investigation of some possible effects of sectioning the crossed olivocochlear bundle. J Acoust Soc Am 47:592–596.
van Dijk P, Wit HP (1990) Amplitude and frequency fluctuations of spontaneous otoacoustic emissions. J Acoust Soc Am 8:1779–1793.
Vetter DE, Adams JC, Mugnaini E (1991) Chemically distinct rat olivocochlear neurons. Synapse 7:21–43.
Warr WB (1975) Olivocochlear and vestibular efferent neurons of the feline brain stem: their location, morphology and number determined by retrograde axonal transport and acetylcholinesterase histochemistry. J Comp Neurol 161:159–182.
Warr WB (1992) Organization of olivocochlear efferent systems in mammals. In: Webster DB, Popper AN, Fay RR (eds) Mammalian Auditory Pathway: Neuroanatomy. New York: Springer-Verlag, pp. 410–448.
Warr WB, Beck JE (1995) A longitudinal efferent innervation of the inner hair cell region may originate from "shell neurons" surrounding the lateral superior olive in the rat. Assoc Res Otolaryngol Abstr 18:87.
Warr WB, Guinan JJ Jr (1979) Efferent innervation of the organ of Corti: two separate systems. Brain Res 173:152–155.
Warr WB, White JS, Nyffeler MJ (1982) Olivocochlear neurons: quantitative comparison of the lateral and medial efferent systems in adult and newborn cats. Soc Neurosci Abstr 8:346.
Warr WB, Guinan JJ Jr, White JS (1986) Organization of the efferent fibers: the lateral and medial olivocochlear systems. In: Altschuler RA, Hoffman DW,

Bobbin RP (eds) Neurobiology of Hearing: The Cochlea. New York: Raven Press, pp. 333-348.

Warren EH III, Liberman MC (1989a) Effects of contralateral sound on auditory-nerve responses. I. Contributions of cochlear efferents. Hear Res 37:89-104.

Warren EH III, Liberman MC (1989b) Effects of contralateral sound on auditory-nerve responses. II. Dependence on stimulus variables. Hear Res 37:105-122.

Whitehead ML, Lonsbury-Martin BL, Martin GK (1991) Slow variation in the amplitude of acoustic distortion at f_2-f_1 in awake rabbits. Hear Res 51:293-300.

Whitehead ML, Lonsbury-Martin BL, Martin GK, McCoy MJ (1995) Otoacoustic emission: animal models and clinical observations. In: Van De Water T, Popper AN, Fay RR (eds) Clinical Aspects of Hearing. New York: Springer-Verlag, pp. 199-257.

Wiederhold ML (1970) Variations in the effects of electric stimulation of the crossed olivocochlear bundle on cat single auditory-nerve-fiber responses to tone bursts. J Acoust Soc Am 48:966-977.

Wiederhold ML, Kiang NYS (1970) Effects of electric stimulation of the crossed olivocochlear bundle on single auditory-nerve fibers in the cat. J Acoust Soc Am 48:950-965.

Wiederhold ML, Peake WT (1966) Efferent inhibition of auditory nerve responses: dependence on acoustic stimulus parameters. J Acoust Soc Am 40:1427-1430.

Williams EA, Brookes GB, Prasher DK (1994) Effects of olivocochlear bundle section on otoacoustic emissions in humans: efferent effect in comparison with control subjects. Acta Otolaryngol 114:121-129.

Winslow RL, Sachs MB (1987) Effect of electrical stimulation of the crossed olivocochlear bundle on auditory nerve response to tones in noise. J Neurophysiol 57:1002-1021.

Winslow RL, Sachs MB (1988) Single-tone intensity discrimination based on auditory-nerve rate responses in backgrounds of quiet, noise, and with stimulation of crossed olivocochlear bundle. Hear Res 35:165-190.

Young ED, Barta PE (1986) Rate responses of auditory nerve fibers to tones in noise near masked threshold. J Acoust Soc Am 79:426-442.

Zeng FG, Lehmann KM, Soli SD, Linthicum FH (1994) Effects of vestibular neurectomy on intensity discrimination and speech perception in noise. J Acoust Soc Am 95:2993-2994.

Zurek PM (1981) Spontaneous narrowband acoustic signals emitted by human ears. J Acoust Soc Am 69:514-523.

9
Neurotransmitters and Synaptic Transmission

WILLIAM F. SEWELL

1. General Overview

The hair cell transmits information about an acoustic signal by releasing a neurotransmitter to excite afferent nerve fibers. This results in discharge of the auditory nerve. The release of neurotransmitter by the hair cell is triggered by the entry of calcium into the hair cell through voltage-dependent calcium (V_{Ca}) channels. Even in the absence of acoustic stimulation, the small resting current entering through the transduction channels in the stereocilia always slightly depolarizes the hair cell. Because of this slight depolarization, the V_{Ca} current is always activated. Thus, transmitter is always being released, and the auditory nerve fiber is generally always firing. Further depolarization of the hair cell, for example by acoustic stimulation, increases the probability that transmitter will be released. In contrast, hyperpolarization of the hair cell, as during some phases of an acoustic stimulus, decreases the probability of transmitter release. Thus to a first approximation, the release of transmitter from the hair cell reflects the membrane potential of the hair cell and determines the rate of discharge in the auditory nerve fiber.

The hair cell–auditory nerve synapse, although in many ways similar to other chemical synapses that have been studied, has some performance characteristics that are truly remarkable. The synapse is able to successfully transmit timing information about a signal to frequencies of more than 5 kHz (a period of 200 μs) (Johnson 1980). In addition, auditory nerve fibers discharge spontaneously at rates up to 100 spikes per second and can discharge in response to acoustic stimulation at rates of up to several hundred spikes per second. Afferent fiber discharge is graded with voltage in hair cell, and some fibers can respond over a dynamic range of 60 dB in sound level. This is even more extraordinary when one considers that each auditory nerve fiber in the cochlea only makes one synaptic contact, that is, the entire synaptic input to most auditory nerve fibers consists of one synapse (Liberman and Oliver 1984).

Neurotransmitter release by the hair cell is quantal, that is, neurotransmitter is released in packets of constant quantity. These quantal packets could correspond to the amount of transmitter in a synaptic vesicle. It is plausible that in the mammalian cochlea a single quantum might be sufficient to depolarize the fiber to generate an action potential.

The identity of the transmitter released by the hair cell has not been firmly established, although glutamate has been proposed by some investigators. Whatever the endogenous ligand is, it is becoming increasingly evident that its effects may be mediated via glutamate receptors.

The cochlea receives efferent fibers from the brain (Guinan, Chapter 8). Activity in these efferent fibers releases neurotransmitters that alter the way the cochlea responds to sound. The major neurotransmitter released by efferent fibers is acetylcholine. However, other neurotransmitters may be released as well, including opioid peptides (enkephalins and dynorphin), calcitonin gene-related peptide (CARP), and γ-aminobutyric acid (GAB).

The focus of this chapter is on the mammalian cochlea. However, much of our understanding of synaptic transmission and neurotransmitters has come from work in nonmammalian systems, most notably the goldfish (*Carassius auratus*) auditory nerve and the lateral line organ. Thus information gathered from noncochlear systems will be presented as necessary to formulate an understanding of cochlear neurotransmission.

2. Hair Cell Neurotransmitter

2.1 Identity of the Hair Cell Transmitter

The identity of the neurotransmitter released by the hair cell has not been firmly established. A number of investigators have suggested that the transmitter is glutamate (for a recent review, see Eybalin 1993). The arguments are not compelling, but this is a reflection of the extraordinary difficulty associated with clearly establishing a neurotransmitter role for a common metabolite like glutamate. Although attempts have generally been made to establish that glutamate meets the criteria for proof of a neurotransmitter, as articulated by Werman (1966), efforts such as these are perhaps doomed from the beginning by the ubiquity of glutamate in the body. Glutamate is present in hair cells (one of Werman's criteria), but it is also present in all cells of the body. Given that glutamate is present in, and released from, cholinergic and GABAergic synaptosomes (Docherty, Bradford, and Wu 1987), it might not be surprising to find glutamate released (as a metabolite rather than a neurotransmitter) during acoustic stimulation, although that has not yet been demonstrated. Glutamate is released in response to depolarization of the cochlea with potassium (Bobbin, Caesar, and Fallon 1990; Drescher and Drescher 1992), but this phenomenon can be observed even when the hair cells have been destroyed (Bobbin, Caesar, and Fallon 1990), suggesting that the potassium-induced release of glutamate may not necessarily reflect its role as a hair cell transmitter. Glutamate can depolarize the afferent fiber to increase discharge rate, as does the afferent transmitter, although relatively high concentrations (millimolar) of glutamate are required (Bledsoe et al. 1983; Mroz and Sewell 1989; Gleich, Johnstone, and Robertson 1990). And glutamate is no more potent than aspartate and cysteine sulfinic acid (Bledsoe et al. 1983; Morz and Sewell 1989), both of which are present in neural tissue (Collingridge and Lester 1989) and are no less likely candidates to be the hair cell transmitter than glutamate. In addition, other, yet unidentified, neurotransmitter candidates have been extracted from hair cell (and electroreceptor) tissue that can also excite afferent fibers (Umekita et al. 1980; Sewell and Mroz 1987). That the effects of transmitter release from the hair cell are blocked by glutamate receptor antagonists (Littman et al. 1989) suggests that they are acting via a common receptor but does not prove that glutamate is the endogenous ligand released from the hair cell. It only suggests that the transmitter, whatever its identity, is blocked by glutamate antagonists.

However, many of the facets of the physiology and chemistry of glutamate that make it so difficult to prove it is the transmitter also make it difficult to disprove that it is the transmitter. Indeed, unless an alternative candidate is identified and demonstrated to be the transmitter, the debate will likely remain unresolved, if not forgotten.

2.2 Neurotransmitter Receptors on Afferent Fibers

Regardless of the identity of the afferent transmitter, there is compelling evidence for a role of glutamate receptors in afferent transmission. Briefly, the major arguments that a glutamate receptor may mediate transmission at this synapse are as follows: (1) Agonists for the glutamate receptor can produce excitation of afferent nerve fibers innervating hair cells, and some of the agonists, such as domoic and quisqualic acid, can act at very low (micromolar) concentrations at which nonspecific effects are unlikely (Bledsoe et al. 1983; Nakagawa et al. 1991; Sewell 1993). (2) Drugs known to block glutamate receptors can block transmission, although not completely, at the hair cell–afferent fiber synapse (Littman et al. 1989). (3) Glutamate receptors, identified immunohistochemically or with molecular biological techniques, are present in the cochlea (Ryan, Brumm, and Kraft 1991; Doi et al. 1993; Niedzielski and Wenthold 1985).

2.2.1 Pharmacology of Glutamate Agonists and Antagonists

Glutamate receptors can be classified into three major types: NMDA (*N*-methyl-D-aspartate) receptors, non-NMDA receptors, and metabotropic receptors. Much of the extant evidence would suggest that a non-NMDA receptor mediates rapid afferent transmission (see below), although there is evidence, in some hair cell organs, for the presence of NMDA receptors

(Bledsoe et al. 1983; Yamaguchi and Ohmori 1990; Kuriyama et al. 1993; Niedzielski and Wenthold 1995). Metabotropic glutamate receptors have only recently begun to be examined in the inner ear (Devau, Lohouelleur, and Sans 1993), and there is little evidence yet regarding their existence or function in the auditory periphery.

An important means of pharmacologically characterizing a receptor is to determine whether specific antagonists for a given type of receptor block transmission at the synapse. Some of the more potent antagonists specific for the non-NMDA receptor are the quinoxalinediones, 6,7-dinitroquinoxaline-2,3-dione (CNQX) and 6-cyano-7-nitroquinoxaline-2,3-dione (DNQX), which can inhibit binding of glutamate agonists to non-NMDA glutamate receptors at concentrations of 0.3 to 0.5 μM. However, at higher (two orders of magnitude) concentrations, the drugs lose their specificity and can block other transmitter receptors (Honore et al. 1988). These agents have been shown to suppress afferent transmission in the mammalian cochlea, although the suppression was not complete even at applied concentrations of 500 μM (the highest concentration at which the drug could be put in solution; Littman et al. 1989). The antagonists more effectively blocked responses to low-level sounds than to high-level sounds. Although these concentrations are high relative to those required to block glutamate binding, they are within an order of magnitude or so of those reported to block pharmacological effects of glutamate at other putative glutamatergic receptors (Huettner 1990), which is probably indicative of the presence of a non-NMDA receptor in afferent transmission. It is possible that diffusional barriers to the drug account for some of the need for high concentrations of the drug. However, the resistance to antagonism of the responses to high-level sound remains enigmatic.

Additional evidence for a non-NMDA receptor comes from studies of the effects of glutamate agonists, which have been examined in the mammalian cochlea as well as in other hair cell organs. Glutamate agonists that commonly activate non-NMDA receptors include kainic acid, α-amino-3-hydroxy-5-methyl-4-isoxazolepropionic acid (AMPA), quisqualic acid, and domoic acid. NMDA receptors, on the other hand, are specifically activated by NMDA, as the name implies. Afferent fibers innervating hair cells can be activated by non-NMDA agonists (Bledsoe et al. 1983; Bledsoe, Bobbin, and Puel 1988; Sewell 1993), suggesting a role for the non-NMDA glutamate receptor in afferent transmission.

Non-NMDA receptors have been further classified into two major subtypes based upon their affinity for the agonists AMPA and kainate. Given the complexities in receptor organization that are becoming apparent from recent molecular biological studies, this classification may turn out to be too simple. Indeed, the responses of auditory nerve fibers to these agonists do not easily place this receptor into either category. A consistent pattern has emerged as to the relative potencies of these agents in their effects on afferent activity in hair cell organs. The three most commonly studied agents have been kainate, glutamate, and quisqualic acid. In hair cell organs, ranging from the lateral line organ (Bledsoe et al. 1983; Sewell 1993) to the mammalian cochlea (Nakagawa et al. 1991), the order of potencies of these agents is quisqualic acid > kainic acid > > glutamic acid. In general, quisqualic acid is about four times more potent than kainic acid. However, AMPA appears to be equipotent to kainic acid (Sewell 1993), making it difficult to determine if the receptor is an AMPA or a kainic acid receptor.

Some illumination about the receptor characterization has come from molecular biological studies. So far, nine different non-NMDA glutamate receptors have been cloned, sequenced, and identified: there are seven homologous receptors, which include GR_{1-4} (also called GR_{A-D}) and GR_{5-7}, and two other homologous receptors, KA_{1-2}. All have molecular weights near 100 kDa, bind glutamate agonists, and form functional ion channels. These receptors congregate in homomeric or heteromeric clusters, and the binding properties of the receptors can change depending upon the composition of the multiunit receptor (Herb et al. 1992). GR_{1-4} form a subgroup of AMPA-preferring receptors (Boulter et al. 1990), whereas GR_{5-7} and KA_{1-2} form a group of kainate-preferring receptors. In addition, several smaller (40–50 kDA) glutamate receptors have also been identified (Gregor et al. 1989; Wada et al. 1989), and at least one of them is possibly coupled to a functional ion channel (Henley et al. 1992). Of these glutamate receptors, several have been found in abundance in cochlear tissue, including GR_3, GR_4, GR_5, GR_6, and KA_1 (Ryan, Brumm, and Kraft 1991; Safieddine and Eybalin 1992; Doi et al. 1993; Niedzielski and Wenthold 1995).

Evidence also exists for the presence of NMDA receptors on afferent fibers. Yamaguchi and Ohmori (1990) described NMDA receptors in cultured cochlear ganglion neurons of the chick after 5 days in culture. In fact, it is likely that the only glutamate receptors present in those cells were NMDA receptors. Although Nakagawa et al. (1991) did not see NMDA receptors in freshly isolated guinea pig auditory neurons, the enzymatic dissociation procedure might have damaged the NMDA receptors (Allen et al. 1988; Raman and Trussell 1992). The role of NMDA receptors in normal transmission at the hair cell synapse is not clear. Most auditory nerve preparations appear to be relatively insensitive to NMDA. Antagonists for the NMDA receptor can block afferent transmission between the hair cell and afferent fiber, but only at relatively high concentrations where the actions of these antagonists are nonspecific. For example, APV (aminophosphono-valeric acid) does block transmission at the hair cell afferent fiber synapse (Starr and Sewell 1991) but requires higher concentrations than would be expected if the synapse were using NMDA receptor (Watkins and Olverman 1987).

2.3 Excitotoxicity of Glutamate Agonists

Further evidence for the presence of non-NMDA receptors in auditory nerve fibers comes from anatomical studies (Pujol et al. 1985) in which glutamate agonists, such as kainate or AMPA, produced swelling of the radial afferent fiber terminals beneath the inner hair cells (IHCs) and ultimately degeneration of some of the neurons. Such actions probably reflect an excitotoxic action of the glutamate analogs, as has been reported elsewhere in the nervous system (Olney, Ho, and Rhee 1971). Interestingly, the afferent fibers innervating outer hair cells (OHCs) do not appear to be susceptible to this effect. This might be taken to imply that the type II afferent fibers innervating the OHCs differ in the type of receptor used in afferent synaptic transmission and possibly in the neurotransmitter.

3. Efferent Neurotransmitters

3.1 Overview

The cochlea is innervated by efferent nerve fibers that modify the way the cochlea responds to sound. There is abundant evidence that acetylcholine is the major neurotransmitter mediating the effects of electrical stimulation of the efferent fibers. This evidence is covered in several extensive reviews (Guth, Norris, and Bobbin 1976; Bledsoe, Bobbin, and Puel 1988; Eybalin 1993) and will not be detailed here. However, although the major neurotransmitter is acetylcholine, the efferent innervation to the cochlea can be divided into two or three groups that differ from one another in the transmitters they contain. These groups are the medial efferent and two groups of lateral efferents. Detailed descriptions of the anatomy and physiology of the olivocochlear efferents can be found in Warr, Chapter 7, Volume 1 in this series, and Guinan, Chapter 8, this volume.

The medial efferents, arising in the medial portion of the superior olivary complex, provide a substantial innervation to the cochlea. The fibers are large and myelinated and terminate on the OHCs. They use acetylcholine as a neurotransmitter. Most of the effects attributable to efferent stimulation have probably been produced via these medial efferents, since they are easily stimulated electrically (Guinan, Warr, and Norris 1983; Gifford and Guinan 1987).

There is also a population of lateral efferents that originate in the region of the lateral superior olive. These efferents are small and unmyelinated and mostly terminate on the radial afferent fibers that innervate the IHCs. Many of these neurons appear to be cholinergic and also contain CGRP and enkephalins. The role of these efferents is not clear. A second population of lateral efferents may use GABA as a neurotransmitter, since they react to antibodies against the enzyme that makes GAD (glutamate decarboxylase) and do not contain enzymes of the cholinergic system (Vetter, Adams, and Mugnaini 1991). These fibers also generally innervate the radial afferent fibers, although both populations of lateral efferents can provide minor innervation of OHCs, especially in the apical region of the cochlea. The whole picture may not be as simple as portrayed here, since there are considerable differences reported in the distribution of these transmitter markers among different species (for discussion, see Vetter, Adams, and Mugnaini 1991).

3.2 Efferent Synaptic Transmission

The medial efferent fiber terminals contain large numbers of mitochondria and synaptic vesicles. The effects of efferent stimulation are highly dependent on extracellular calcium. Indeed, it is far easier to block efferent transmission by lowering calcium than it is to block afferent transmission (Russell 1971). Directly apposing the efferent terminals, the OHCs contain a complex of subsynaptic cisternae resembling the sarcoplasmic reticulum of muscle. The function of the subsynaptic and subsurface cisternae is not known, although they might be involved in calcium regulation of the efferent effects.

3.3 Acetylcholine

In vivo, efferent stimulation mediated by acetylcholine hyperpolarizes the hair cell and alters its tuning properties (Art, Fettiplace, and Fuchs 1984). One mechanism by which acetylcholine alters hair cell function can be deduced from a number of experiments on isolated hair cells. It is clear that in a variety of hair cells, ranging from chick to turtle to mammal (Art et al. 1982, 1985; Art, Fettiplace, and Fuchs 1984; Housley and Ashmore 1991; Fuchs and Murrow 1992), acetylcholine activates a potassium conductance. Activation of the potassium conductance requires external calcium, suggesting that acetylcholine activates a receptor that allows calcium entry. It is thought that the calcium entering the hair cell could then activate potassium channels. The increase in potassium conductance would tend to bring the resting potential of the hair cell near that of the Nernst potential of potassium, which is around -80 mV. This hyperpolarization of the hair cell is seen in vivo (Art, Fettiplace, and Fuchs 1984). Consistent with this scenario are Fuchs and Murrow's (1992) observations that preceding the potassium conductance there is a rapid inward current consistent with calcium entry into the cell. One consequence of this mechanism is that efferent effects require some time to build up. In isolated hair cells from the chick, the delay was about 75 ms after acetylcholine application (Fuchs and Murrow 1992), whereas OHCs from the mammalian cochlea showed even longer delays (Housley and Ashmore 1991). Similar time constants have been noted in vivo as well (Widerhold and Kiang 1970; Brown, Nuttall, and Masta 1983).

Generally, cholinergic receptors can be characterized as nicotinic or muscarinic. Nicotinic receptors contain an ion channel whereas muscarinic receptors are linked to second messenger systems. The two types of receptors can often be distinguished pharmacologically, since they can be selectively activated and blocked by different pharmacological agents. However, pharmacological characterization of the cholinergic receptor mediating the major effects of efferent stimulation has been difficult, because the receptor is different from many previously characterized receptors.

The major effects of efferent activation are likely to be mediated by an unusual nicotinic receptor. Pharmacological characterization of the action of acetylcholine on isolated OHCs (Housley and Ashmore 1991; Fuchs and Murrow 1992; Erostegui, Norris, and Bobbin 1994), coupled with some pharmacological results in vivo (Bobbin and Konishi 1974; Kujawa et al. 1993, 1994; Sridhar et al. 1995), provide a reasonable description of the pharmacological properties of this receptor. The receptor is blocked by curare, a nicotinic receptor blocker, at concentrations on the order of 10^{-6} to 10^{-7} M, and the blockage is difficult to reverse. But it can also be blocked by strychnine, which is at least as potent as curare. Strychnine is usually thought of as a glycine antagonist, but has been shown to block the effects of acetylcholine elsewhere (Kehoe 1972). The receptor is blocked by α-bungarotoxin (an irreversible antagonist at neuromuscular junction nicotinic receptors), although the block in the cochlea is reversible (Fex and Adams 1978). α-Bungarotoxin-binding sites can be found at the base of hair cells where the efferent terminals are located (Plinkert, Zenner, and Heilbronn 1991). Two other nicotinic blockers, however, are not very effective antagonists. Decamethonium, a neuromuscular nictonic blocker, is more effective than hexamethonium, a neuronal nicotinic blocker. And the receptor is poorly activated by nicotine. Bicuculline, usually thought of as a GABA antagonist, also blocks the effects of acetylcholine on this receptor (Erostegui, Norris, and Bobbin 1994; Kujawa et al. 1994). Nicotinic receptors are coupled to an ion channel and, as mentioned above, the effects of acetylcholine are consistent with activation of an ionotropic receptor, again pointing to a nicotinic receptor, although the mechanism by which the receptor and associated channels exert an effect is markedly different from that of most other nicotinic receptors.

Recently, a novel nicotinic receptor subunit (α9) has been found in hair cells of the organ of Corti (Elgoyhen et al. 1994) which has pharmacological properties similar to that of the acetylcholine receptor described in the cochlea. It also has a relatively high permeability to calcium, making it an extraordinarily good candidate to be a component of the acetylcholine receptor mediating the major efferent effects in the cochlea.

However, there is also evidence to suggest the presence of a muscarinic receptor in hair cells. Muscarinic receptors generally act through second messenger systems and are coupled to G proteins. There is considerable biochemical evidence for the involvement of muscarinic second messenger systems in the action of acetylcholine in the cochlea, and these data are considered in detail by Wangemann and Schacht in Chapter 3. Because of the relatively small amounts of tissue available, it has been difficult to know where in the cochlea the muscarinic receptors are located and what role they might play in generating efferent effects on the cochlea. Other evidence for muscarinic receptors comes from work of Shigemoto and Ohmori (1990), who showed that acetylcholine increases free calcium concentration in isolated chick hair cells. This effect has a much longer time course (tens of seconds to onset) than the effects of ion conductances reported by others (Housley and Ashmore 1991; Fuchs and Murrow 1992) and is more sensitive to block by atropine, a muscarinic antagonist, than to block by curare, a nicotinic antagonist. The increase in free calcium was independent of extracellular calcium, suggesting that the calcium is released from internal stores.

Other evidence exists for second messenger-mediated effects that might be associated with the acetylcholine receptor on OHCs. Kakehata et al. (1993) have found evidence for second messenger-mediated conductance changes in isolated OHCs of the guinea pig associated with acetylcholine application. The conductance change was dependent upon extracellular calcium. Because the effects could be blocked by pertussis toxin, they suggest that acetylcholine effects could be mediated by a G protein that might involve calcium-induced calcium release and activation of calcium-sensitive potassium channels.

Effects of acetylcholine with slow time courses are seen in vivo. Sridhar et al. (1995) demonstrated an effect of olivocochlear bundle (OCB) stimulation that builds up and decays with a time constant of tens of seconds and is superimposed upon the fast (milliseconds) effects of OCB stimulation. The slow effect is apparently mediated by the same acetylcholine receptor as the fast OCB effects.

Although there are reports that acetylcholine can produce slow length changes in OHCs (Brownell et al. 1985), others have shown that these changes may not be specific to acetylcholine (Bobbin et al. 1990). It has been suggested that acetylcholine might affect OHC motility by altering the gain of the voltage-to-movement converter (Sziklai and Dallos 1993a).

3.4 Peptide Transmitter Candidates

Acetylcholine does not appear to be the only neurotransmitter used by efferent fibers. The lateral efferents, which primarily innervate the radial afferent fibers in the IHC region, appear to contain peptide transmitter candidates in addition to acetylcholine. Opioid peptides (Fex and Altschuler 1981; Abou-Madi et al. 1987) and CGRP (Lu et al. 1987) may play a role in inner ear neurotransmission. In addition, a population of efferent fibers may be GABAergic (Fex and Altschuler 1984; Vetter, Adams, and Mugnaini 1991). For all of these substances, there is some evidence that they may be neuroactive in hair cell organs (see below).

Some opioid peptides, the enkephalins and dynorphins, have been immunolocalized in lateral efferent cell bodies (Fex and Altschuler 1981; Altschuler et al. 1984) that project predominantly, although not exclusively, to the radial afferent fibers. Enkephalin appears to be released into perilymph upon stimulation of the cochlea with acoustic stimulation (which might ultimately activate the lateral efferent fibers) (Drescher, Drescher, and Medina 1983; Eybalin et al. 1987). Evidence for functional effects of opioid peptides is, at present, sparse. Eybalin, Pujol, and Bockaert (1987) found that agonists for opioid receptors (specifically the δ and μ receptors), through which enkephalin acts, could inhibit the synthesis of cyclic AMP in homogenates of guinea pig cochlea. Sahley et al. (1991) found that certain antagonists for κ opioid receptors (through which dynorphins act) could lower thresholds for acoustic stimulation, suggesting that neurons containing an endogenous opioid, such as dynorphin, might be tonically active and increase the threshold to acoustic stimulation.

Another peptide present with acetylcholine in lateral efferent neurons is CGRP (Lu et al. 1987). In the lateral line organ of *Xenopus*, afferent fibers that innervate hair cells are excited by CGRP, an effect presumed to be mediated by an increased release of neurotransmitter from the hair cell (Adams, Mroz, and Sewell 1987; Sewell and Starr 1991). In isolated chick hair cells, preincubation with CGRP has been reported to potentiate the amplitude of acetylcholine-induced changes in free calcium concentration, although CGRP produced no response when applied alone (Shigemoto and Ohmori 1990). No action of CGRP has yet been demonstrated in the mammalian cochlea.

3.5 γ-Aminobutyric Acid

A small number of efferent fibers are thought to contain GABA, based upon immunohistochemical localization of glutamate decarboxylase, an enzyme that synthesizes GABA from glutamate (Fex and Altschuler 1984; Eybalin et al. 1988; Vetter, Adams, and Mugnaini 1991). Unlike the peptide neuromodulator candidates, which are co-localized with acetylcholine, those efferent neurons that contain GABA apparently do not contain acetylcholine, at least in the rat (Vetter, Adams, and Mugnaini 1991). These GABAergic terminals end in the IHC region throughout the cochlea and in the OHC region in the apical portion of the cochlea. There has long been evidence for physiological effects of GABA in hair cell organs, such as the lateral line organ (Bobbin et al. 1985; Mroz and Sewell 1989) or the vestibular system (Felix and Ehrenberger 1982). Moreover, isolated OHCs from the cochlea can respond to GABA (Sziklai and Dallos 1993b).

Recently, Kirk and Johnstone (1993) have shown that some of the effects of efferent activation via contralateral sound on cochlear emission could be blocked by bicuculline, a GABA antagonist. This provides some support for a physiological role of GABA in the cochlea. Additional evidence for a role of GABA comes from findings that GABA appears to be released into perilymph during intense acoustic stimulation (Drescher, Drescher, and Medina 1983), possibly as a result of efferent activation by the acoustic stimulation.

4. Neuromodulators and Other Modes of Synaptic Transmission

4.1 Neuromodulators

In addition to the neurotransmitters discussed above, several other substances have been suggested to serve as neuromodulators of synaptic transmission in the inner ear. These include adenosine, histamine, and ATP. Each of these substances can produce effects in the inner ear (see Eybalin 1993 for review), but the specific role that each plays in auditory function is not known.

4.2 Could Potassium Be a Modulator?

Potassium is unlikely to mediate transmission between the hair cell and the afferent fiber, since such effects would more likely be graded than quantal. But if, as Roberts, Jacobs, and Hudspeth (1990) propose, many of the calcium-dependent potassium channels are focally located at the synapse, potassium exiting the hair cell near the synaptic cleft, in principle, could depolarize the afferent fiber, thus contributing to afferent transmission. We know that the transduction current through the IHC is carried mostly by potassium and that all potassium entering the cell via the transduction channel must ultimately exit in the basolateral surface of the cell. A single K_{Ca} channel allows 0.2 pA of potassium current per millivolt change in receptor potential (Hudspeth and Lewis 1988), and there are estimated to be 40 of the channels at each synaptic release site (Roberts, Jacobs, and Hudspeth 1990). Thus, near threshold, a 1-mV receptor potential would increase potassium current by 8 pA or 8×10^{-20} mol of potassium every millisecond. If we assume a diffusional volume (over 1 ms) for the potassium of 6×10^{-16} l,[1] then in 1 ms, the concentration of potassium might change by 1.33×10^{-4} M. Given that extracellular concentrations of

---

[1]Based upon a diffusional radius for K^+ of 3 μm/ms, where the diffusional radius = $\sqrt{4Dt}$, with D, the diffusional coefficient of $K^+ = 2 \times 10^{-5}$ cm^2/s and 2-dimensional spread within the intracellular space, which is 2×10^{-8} m thick.

potassium are at least 20 or 30 times higher, it would seem that these small changes would not likely produce the moment-by-moment changes in afferent activity associated with acoustic stimulation. However, larger (20 mV) changes in hair cell potential might increase extracellular potassium concentration by a few millimolar, which could depolarize surrounding cells, including afferent terminals. Consistent with these arguments, Oesterle and Dallos (1990) have reported slow depolarization of supporting cells with acoustic stimulation that was attributed to the accumulation of potassium in the extracellular space.

5. Ion Channels Involved in Neurotransmitter Release

Normal transmitter release by the hair cell requires calcium entry into the cell via voltage-dependent calcium channels. The voltage-dependent calcium channels in hair cells have appropriate biophysical characteristics to mediate the release of transmitter. In isolated hair cells of the frog saccule (Hudspeth and Lewis 1988), the calcium current was activated at potentials more positive than -60 to -50 mV; that is, the current was activated at the resting potential of the hair cell. Maximal currents were observed at -10 mV, and there was little inactivation during moderate depolarizations of the hair cell. The type of calcium channel involved in the release of transmitter from the hair cell is not firmly established, but it is most similar to the L-type calcium channel (Roberts, Jacobs, and Hudspeth 1990). This assessment is based upon the sensitivity of the channel to dihydropyridine, a lack of inactivation at physiological membrane potentials, a greater conductance to barium than calcium, a large single-channel current, and a propensity for rundown. However, the very low activation potentials of the channels distinguish them from most known L channels.

6. Anatomy of the Afferent Synapse

To appreciate fully the characteristics of afferent synaptic transmission in the cochlea, one must examine the ultrastructural characteristics of the afferent synapse.

6.1 Innervation Patterns

Ninety-five percent of the auditory nerve fibers innervate the IHCs (Spoendlin 1966; Kiang et al. 1982). These myelinated radial afferent fibers carry the temporal information necessary for auditory processing by the central nervous system. The radial afferent fibers are well suited for transmitting rapid information; even the cell bodies are myelinated. Each radial afferent auditory nerve fiber generally makes only one synaptic contact, that is, its entire input consists of only one synaptic site. Each IHC in the cochlea is innervated by from 10 (in the apex of the cochlea) to 30 (in the base) nerve fibers (Liberman 1982). As illustrated in Figure 9.1, this is an interesting configuration, as the hair cell is an enormous presynaptic cell releasing transmitter onto a very small neuronal terminal. One can contrast this with, for example, the neuromuscular junction or many central nervous system cells where there are numerous very small neuronal endings synapsing onto a large cell body. In the latter situation, a relatively large amount of current might be required to depolarize the cell to threshold for spike generation because of the space constants of the postsynaptic cell. The auditory nerve fiber, on the other hand, might be depolarized to threshold by a relatively small amount of current because of its small size. Thus from the simple anatomic relation between the hair cell and the nerve fiber, one might suppose this to be a relatively secure synapse.

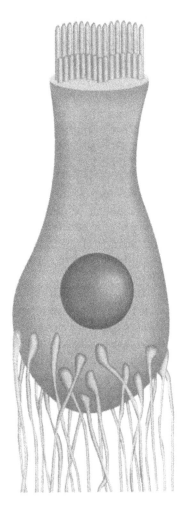

FIGURE 9.1. Each IHC is innervated by 10 to 30 afferent nerve fibers. The hair cell and afferent fibers are drawn approximately to scale based on serial reconstruction of the afferent innervation of an IHC from ultrathin sections through the organ of Corti as published by Liberman (1980).

Only 5% of the afferent fibers innervate the OHCs. These are type II fibers, a separate population of afferent neurons that can be distinguished morphologically from those providing radial afferent projections to the IHCs. Type II fibers are highly branched, with each fiber innervating dozens of OHCs. One might not expect these fibers to be as sharply tuned as radial afferent fibers, since they innervate a broader frequency region of the cochlea. These small fibers are unmyelinated, making it difficult to record their activity. Little is known of their response properties or of their functional role in the auditory system. However, because of their small size and lack of myelination, the conduction velocities of these fibers are likely to be very slow, so that most of the moment-to-moment auditory information must be carried by the radial afferent fibers.

6.2 Synaptic Ultrastructure

The microanatomy of the hair cell synapse is unusual, consisting of a presynaptic electron-dense body that is surrounded by a relatively small number of synaptic vesicles. Similar synaptic structures, also containing electron-dense synaptic structures, are found in retinal rod cells (Sjostrand 1958) (where they are called synaptic ribbons), electroreceptors, taste buds (Jepson 1969), and vestigial light receptors in the pineal of some animals (Wurtman and Axelrod 1968). At the synapse, the afferent fiber and the hair cell are separated by a narrow (10 nm) gap. Postsynaptically, the nerve fiber membrane is described as having a dense spiky coat, which forms an elliptical plaque with dimensions of about 0.4 to 0.7 by 0.6 to 1 μm, with the long axis of the plaque oriented with the long axis of the hair cell. Each afferent fiber has only one synaptic plaque (Liberman 1980). Presynaptically, the hair cell contains an electron-dense synaptic body or bar. In the cochlea, the bar is elongated and oriented parallel to the axis of the postsynaptic plaque. It ranges in length from 0.2 to 0.6 μm and in width and height from 0.05 to 0.1 μm. The synaptic body can appear to be hollow or solid.

The synaptic body is associated with a complex of electron-dense membrane structures that are likely to comprise the voltage-dependent calcium channels needed for transmitter release and a group of calcium-dependent potassium channels. Roberts, Jacobs, and Hudspeth (1990) have shown an organized group of particles clustered presynaptically in an area about the size of the synaptic body complex. Each "active zone" contains a group of approximately 125 particles 12 nm in diameter (in a platinum-carbon replica). The authors provide a reasonable quantitative argument that these particles represent a cluster of approximately 90 voltage-dependent calcium channels and 40 calcium-dependent potassium channels that are involved in transmitter release and in the production of an electrical resonance in these saccular hair cells. Thus one might envision a complex structure where the synaptic body might dock vesicles ready for immediate release in a region directly next to a focal cluster of voltage-dependent calcium channels. Calcium entering through these channels would increase intracellular calcium levels in a highly localized region of the cell and could trigger exocytotic release of the vesicle. Such a system has a number of theoretical advantages for a rapidly operating synapse. It would be a very efficient means of controlling transmitter release and would reduce the metabolic load on the cell (in terms of controlling intracellular calcium). It would allow very rapid reversal of calcium concentrations. It would allow the use of changes in calcium concentrations elsewhere in the hair cell to control other hair cell responses. And it might allow exquisite control of the number of vesicles available for release by retaining highly localized changes in calcium concentration.

Reconstructing the unmyelinated portions of more than 100 radial afferent fibers from serial ultrathin sections, Liberman (1980) characterized two types of synaptic ultrastructures in the cochlea, called simple and complex, based upon the orientation of the synaptic bar relative to the synaptic plaque. Those in which the synaptic bar is located at the center of the synaptic plaque are called simple, whereas those with eccentrically located bodies are called complex. The synaptic ultrastructure and location of the synapse on the hair cell appear to be associated with physiological characteristics of single auditory nerve fibers. Fibers with high spontaneous rates are located on the size of the hair cell away from the modiolus and have simple synaptic bodies. The complex synapses are located on the side of the hair cell facing the modiolus and appear to be associated with units that have either medium or low spontaneous discharge rates and associated elevated thresholds to acoustic stimulation.

One striking aspect of the synaptic structure at the hair cell–afferent fiber synapse is the very small number of apparent stored vesicles near the synapse. For example, Figure 9.2 is based on a three-dimensional reconstruction of serial sections (Liberman 1980) through a synapse in the cochlear IHC. The vesicles present at the synapse probably number fewer than 100. Yet the fiber innervating a synapse such as this can fire spontaneously at rates of up to 100 spikes/s. The implication is twofold—that the turnover of transmitter must be very rapid and that few vesicles may be needed to generate a spike in the afferent fiber. Synaptic membrane recycling does occur and at much higher rates in the IHC than in the OHC (Siegel and Brownell 1986).

The function of the synaptic body is not yet understood, although a number of its properties have been documented. The synaptic body is not likely to be a specialization strictly associated with the high-frequency responsiveness of the synapse, because the synaptic body is present in all hair cells, whether vestibular (which have relatively slow frequency responses) or auditory. One speculation is that it may be associated with the tonic responses of this system, for both vestibular and auditory nerve fibers can discharge at relatively high rates indefinitely. Another is that the

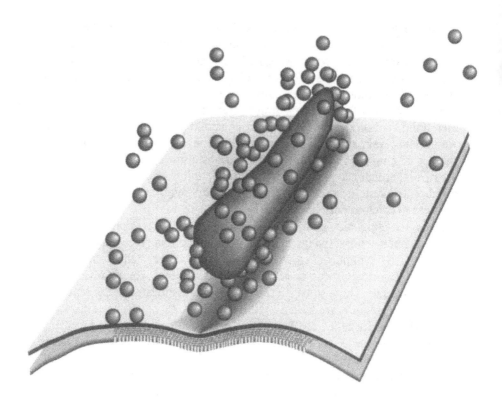

FIGURE 9.2 Three-dimensional depiction of a single afferent synapse as would be seen from the hair cell. The synapse was reconstructed from electron micrographs of eight serial sections in through a single synapse (Liberman 1980, Figure 10). The synaptic body is the elongated mass above a groove in the hair cell membrane and is surrounded by a relatively small number of synaptic vesicles. A synaptic plaque, observed on the postsynaptic membrane, is depicted here as a series of cylinders through the postsynaptic membrane and is the probable location of the neurotransmitter receptors and other associated postsynaptic specializations.

synaptic body serves as a docking site for vesicles in order to align the vesicles near the release site. This idea is supported by the anatomical finding that the synaptic vesicles appear to be tethered to the synaptic body by fine filaments (Takasaka and Smith 1971). Or possibly the synaptic body is involved in the metabolism of transmitter by the hair cell, perhaps containing enzyme complexes necessary to generate hair cell transmitter.

Other properties of the synaptic body are enigmatic as well. For example, it has long been known that the electron density of the synaptic body can be influenced by pharmacological treatment with drugs that interfere with catecholamine storage, although there is no evidence that catecholamines are involved in any way in afferent synaptic transmission (Osborne and Thornhill 1972; Monaghan 1975; Guth et al. 1993).

7. Transfer Function of the Afferent Synapse

7.1 Action Potentials and Synaptic Potentials

A great deal of information can be inferred about afferent synaptic processes by looking at discharge patterns of auditory nerve fibers. It is plausible that, in the mammalian cochlea, the release of a single vesicle (or perhaps the coordinated release of a small group of vesicles) from the hair cell might be capable of triggering an action potential in the afferent fiber. Because afferent terminals are relatively small, they might easily be depolarized to threshold by the release of a single vesicle. Early models of afferent transmission making that assumption were able to describe the discharge properties of afferent fibers reasonably well (Schroeder and Hall 1974; Geisler, Le, and Schwid 1979; Schwid and Geisler 1982). These assumptions by modelers and others (Furukawa 1985) were given more validity when Siegel and Dallos (1986) recorded afferent fiber synaptic potentials from the guinea pig cochlea and found that most action potentials were preceded by excitatory postsynaptic potentials (EPSPs) of similar apparent amplitudes and that EPSPs without action potentials, often seen falling in the refractory period after an action potential, were apparently of unitary amplitude. Thus, the discharge patterns of afferent fibers might be considered to reflect the release of synaptic vesicles by the hair cell, with appropriate consideration of the brief period following an action potential in which the nerve is refractory.

7.2 Spontaneous and Evoked Discharge

In the absence of acoustic stimulation, hair cells are slightly depolarized due to the resting transduction current. The voltage-dependent calcium channels are activated by this slight depolarization (Hudspeth and Lewis 1988), leading to the consequent release of transmitter by the hair cell to produce "spontaneous" discharge in auditory nerve fibers. This discharge in the absence of acoustic stimulation is called "spontaneous," but the fiber is in fact responding to the release of transmitter from a depolarized hair cell. Spontaneous discharge in auditory nerve fibers decreases exponentially when the hair cell is hyperpolarized by decreasing the endocochlear potential (Sewell 1984), implying that transmitter release increases exponentially with depolarization, a behavior known to exist at other chemically mediated synapses (Katz and Miledi 1967). The change in spontaneous activity with membrane potential supports the idea that spontaneous activity is due to voltage-dependent release of transmitter even in the absence of acoustic stimulation.

Spontaneous activity in auditory nerve fibers is described as having an irregular firing pattern (Kiang et al. 1965). Analysis of interspike interval histograms of spontaneous discharge can objectively characterize the patterns of discharge (Walsh et al. 1972). The interspike interval histograms of auditory nerve fibers are clearly asymmetric, having a long tail toward longer intervals. The statistical properties of spontaneous activity are Poisson-like, in that there is an exponential decay in intervals from the mode. However, there are fewer short intervals than expected, probably due to a reduction in the probability of firing during the refractory period following the generation of a spike.

Three populations of auditory nerve fibers can be characterized in the cat based upon spontaneous discharge rate (Liberman 1978). These are low (<2 spikes/s), medium (between 2 and 13 spikes/s), and high (>18 spikes/s). Fibers in these categories exhibit distinct differences in their responses to acoustic stimulation. Fibers with high spontaneous rates have low thresholds to acoustic stimulation. Those with low spontaneous rates have high thresholds, and units with medium spontaneous rates fall in between. Because each IHC is contacted by all three types of fibers (Liberman 1980), it follows that these spontaneous and evoked discharge properties might be determined at the level of the synapse. The high-spontaneous-rate units are thought to comprise the population of mitochondria-rich fibers, and the low- and medium-spontaneous-rate units might correspond to fibers with low mitochondrial content (Liberman 1980).

The discharge rate increases with increasing stimulus intensity. High-spontaneous-rate fibers have about 20 dB of dynamic range before the rate saturates. Medium- and low-spontaneous-rate fibers have higher thresholds and wider dynamic ranges. These three fiber types innervating each hair cell allow for representation, in the afferent discharge patterns, over a wide range of hair cell depolarization voltages.

7.3 Synchrony and Frequency Responsiveness

Afferent nerve fibers can carry timing information about a stimulus at very high frequencies. If, for example, a sinusoidal sound wave is presented, the afferent fiber will fire at a greater probability during one phase of the tone and with less probability during the opposite phase. Indeed, at low sound levels the overall discharge rate may not differ from spontaneous, but the fiber is firing synchronously with the phasic change in the stimulus. The ability of the fiber to synchronize to the stimulus is dependent upon the frequency of the stimulus (Johnson 1980). As the frequency of the stimulus increases, the fiber has more difficulty synchronizing to the stimulus, and at frequencies above 5 kHz there is very little ability to synchronize and the fiber simply increases discharge rate during the presentation of the acoustic stimulus.

The dependence of synchronization of auditory nerve fibers on stimulus frequency behaves as if the cochlea contained several low-pass filters. One of these is the charging of the membrane capacitance of the hair cells in order to produce a voltage change in the hair cell (Sellick and Russell 1980; Palmer and Russell 1986; Weiss and Rose 1988). Two other low-pass filter processes might involve calcium entry into the cell (Kidd and Weiss 1990). The relation between the receptor potential and the activation of the calcium current in the cell is a nonlinear, first-order, low-pass filter whose cutoff frequency decreases with increasing receptor potential magnitude. The relation between calcium current and calcium concentration in the hair cell is a first-order, low-pass filter with constant cutoff frequency. These three processes can account for most, if not all, of the loss of synchronization with increasing stimulus frequency. Because the generation of the action potential is a threshold process, a consequence of these low-pass filtering characteristics is that there is a delay of about 1 ms between receptor potential and action potential generation.

7.4 Adaptation

Auditory nerve fibers adapt to a constant stimulus. The adaptive process gives greater sensitivity to transient stimuli than to steady-state stimuli. The most likely point of occurrence of adaptation is at the afferent synapse (Norris, Guth, and Daigenault 1977; Furukawa and Matsuura 1978). Since the receptor potential of the hair cell is constant (Russell and Sellick 1978), the adaptation must occur after the generation of the receptor potential. Data from the goldfish show that even when the adaptive response to a sound stimulus is present, the amplitude of spontaneous miniature EPSPs is not changed, suggesting that the adaptation is not at the level of the postsynaptic membrane. This implies that adaptation is occurring at the level of the transmitter release process.

The response of auditory nerve fibers to an acoustic stimulus can be described as having three phases (Westermann and Smith 1984). A rapid adaptation occurs with a time constant of a few milliseconds, short-term adaptation occurs with a time constant on the order of 60 ms, and finally the discharge approaches a steady-state rate. The rapid adaptation is difficult to observe experimentally unless the time bins in a poststimulus time histogram are sufficiently small. Rapid adaptation grows with stimulus intensity more rapidly than the steady-state rate and exhibits a greater dynamic range. The time constant of the rapid adaptive process decreases with increasing stimulus intensity. The short-term adaptation is similar to that described by Furukawa and colleagues in goldfish (Furukawa and Matsuura 1978) (and see below). Its time constant is independent of stimulus intensity.

8. Synaptic Transmission

8.1 Overview

It is important to recognize that there are distinct differences between the hair cell synapse and better-studied synapses such as the neuromuscular junction. First, the presynaptic signal is a graded voltage rather than an "all-or-none" action potential, and it is essential that the synapse be able to convey information about the voltage level in the hair cell. Second, the synapse must convey temporal information about auditory signals; it must operate with exceedingly high precision. For example, most mammalian cochleae an convey timing information about a signal for frequencies of up to 5 kHz. Third, the synapse must operate continuously because auditory signals to which the organism must respond are continuously present in the environment and the information must be acquired at a high input rate.

Much of our understanding of synaptic transmission in hair cells has come from the work of Furukawa and colleagues on the goldfish auditory nerve. The goldfish has a group of extraordinarily large auditory nerve fibers. The large size of the fibers has allowed stable microelectrode recordings of synaptic potentials. Some of the fundamental synaptic processes are probably shared between the goldfish and mammalian auditory organs, and an examination of transmission in the goldfish provides insight into cochlear synaptic transmission. However, it is necessary to keep the differences between these organs in mind when attempting to interpret the data obtained from goldfish. Perhaps the most notable difference is that in the mammalian cochlea, each fiber receives only one synaptic input (Liberman 1980), whereas in the goldfish, each large fiber gives off fine dendritic processes that contact as many as 10 hair cells (Sento and Furukawa 1987). In addition, the large size of these goldfish auditory nerve fibers may mean that more EPSPs must sum in order to reach the threshold to generate an action potential than might be necessary in the mammalian auditory nerve fiber.

8.2 Excitatory Postsynaptic Potentials

As is common to all chemical synapses, there is a synaptic delay of about one half ms between the presynaptic signal (as reflected in the extracellular recorded microphonic potential, which reflects current flow through the transduction channels) and the postsynaptic response (the EPSP, which reflects current flow through the postsynaptic receptor channels) (Furukawa and Ishii 1967). This delay is consistent with the amount of time required for calcium to build up in the cytoplasm of the hair cell near the synapse (Kidd and Weiss 1990), a necessary step to evoke release of synaptic vessels. The time required for transmitter to diffuse across the synaptic cleft and bind to the postsynaptic receptor is generally considered to be negligible.

The amplitudes of the EPSPs are graded with stimulus intensity, increasing steeply and ultimately saturating in amplitude. The saturation apparently occurs at the postsynaptic membrane due to nonlinear summation of the EPSPs at large depolarizations, since hyperpolarization of the membrane with current injection through the microelectrode can reduce this nonlinearity (Furukawa 1985). The EPSPs appear to result from an increase in conductance to cations, predominantly sodium. The reversal potential for the EPSP is likely to be somewhere near 0 mV (Furukawa, Ishii and Matsuura 1972). There is a rapid decline in the amplitude of the EPSP with continued stimulation (Furukawa and Matsuura 1978). The decline (adaptation) is initially exponential. The synapse recovers rapidly (15–20 ms) from the adaptation. Both the decline and the recovery are much more rapid than that seen at other synapses (Furukawa 1986).

8.3 Quantal Analysis of Transmitter Release

In intracellular recordings of fibers innervating hair cells in a wide variety of hair cell organs, small spontaneous EPSPs are usually observed (Furukawa and Ishii 1967; Sand 1975; Flock and Russell 1976; Rossi, Valli, and Casella 1977; Crawford and Fettiplace 1980; Kuno 1983; Highstein and Baker 1985; Valli et al. 1985; Kyogoku, Matsuura, and Kuno 1986; Siegel and Dallos 1986; Annoni, Cochran, and Precht 1984; Sewell 1990). These are thought to represent the release of packets of transmitter from the hair cell. Spontaneous EPSPs such as these probably generate the spontaneous action potentials observed in many 8th nerve fibers. In some auditory nerve fibers, such as those in the goldfish, these spontaneous EPSPs are not large enough to trigger a spike because of the large size of the fiber. In auditory organs such as the mammalian cochlea, the size of the fiber may be small enough to allow single spontaneous EPSPs to usually generate a spike (Furukawa 1985; Siegel and Dallos 1986). In other hair cell organs, such as the lateral line organ, in which each nerve fiber contacts large numbers of hair cells, the spontaneously occurring EPSPs are of varying size and duration, often superimposing upon one another to reach threshold for spike initiation (Sewell 1990).

In many chemical synapses, neurotransmitter release can be measured in quantal units, where a quantal unit is thought to reflect the action of the release of a single synaptic vesicle. In general, this appears to be the case at the hair cell synapse. The most extensive examination of the quantal hypothesis for auditory transmission has come from recordings of goldfish auditory nerve fibers. These fibers do not have spontaneous action potential discharge but do have spontaneously occurring miniature excitatory postsynaptic potentials that are of unitary amplitude and time course (Ishii, Matsuura, and Furukawa 1971; Furukawa, Ishii, and Matsuura 1972).

When an acoustic stimulus is applied, the EPSP produced appears to be a summation of many miniature EPSPs. Furukawa and colleagues noticed in early recordings that when low-level sound stimuli were applied, the EPSP amplitudes fluctuated in a stepwise manner, and that sometimes no EPSP was produced (Ishii, Matusuura, and Furukawa 1971). This suggested that transmission was quantal and that the spontaneous miniature EPSP might represent a single quantum of transmitter released, in analogy to what had been observed at the neuromuscular junction. Additional evidence for the idea that each miniature EPSP represents the release of a single packet of transmitter comes from experiments in which voltage-dependent calcium channels, and subsequently the release of transmitter, were blocked with cobalt. In the presence of cobalt, the rate of occurrence of the miniature EPSPs was reduced, eventually to zero, but the amplitudes of the miniature EPSPs did not change (Starr and Sewell 1991).

The quantal hypothesis of transmitter release assumes that the mean number of quanta released (m) is the product of the number (n) of quanta available for release and the probability (p) that any quantum will be released, such that $m = np$. Initial attempts to test the quantal hypothesis at this synapse with the use of Poisson statistics (assuming p is small relative to n) led to a poor agreement between the observed and predicted distributions of EPSP amplitudes, and failures when the sound level was lowered in an attempt to lower the probability of transmitter release. Furukawa eventually demonstrated the surprising finding (see below) that reducing sound level did not substantially lower the probability of the release of a quantum, but rather lowered n, the amount of quanta readily available for release. Later analysis of the fluctuations in the amplitude of EPSPs based on a binomial analysis has proved useful not only in confirming the quantal hypothesis, but also in understanding some of the properties of the transmitter release process in hair cells (Furukawa, Hayashida, and Matsuura 1978).

The binomial analysis quantifies both the probability that any quantum will be released (p) and the number of quanta available for release (n). Furukawa applied binomial statistics to the responses of auditory nerve fibers for a wide range of sound levels and conditions (Furukawa, Hayashida, and Mutsuura 1978; Furukawa and Matsuura 1978; Furukawa, Kuno, and Matsuura 1982; Suzue, Wu, and Furukawa 1987). In summary, they found that the probability that any quantum would be released was generally relatively high and, surprisingly, fairly constant (the probability of release does increase with increasing sound level, but only slightly). The parameter n, the number of quanta available for release, was the parameter most likely to change in conditions such as changing stimulus level. This implies that the amount of transmitter available for release was the main variable that increased with increasing sound intensity.

The binomial analysis of these data has strong implications for understanding the way the synapse operates. Although the final fusion of the synaptic vesicle with the membrane and attendant release of neurotransmitter is due to the entry of calcium through voltage-dependent calcium channels, a process that is assumed to be both probabilistic and voltage-dependent, the binomial analysis suggests that it is the amount of transmitter available for release that is increasing with sound level and with hair cell depolarization. It is not the probability that available transmitter will be released that is changing. Thus it would appear that with increases in sound level, n but not p is voltage-dependent. This is a difficult problem to reconcile with the simple conception that the probability of transmitter release should increase with increasing voltage-dependent calcium entry into the hair cell. Indeed, if calcium entry is blocked with cobalt, transmission at this synapse can be described with Poisson statistics (Starr and Sewell 1991), implying that the probability of transmitter release is lowered when calcium entry is blocked.

It is possible that the microstructure of the synapse can account for its behavior under binomial analysis. Furukawa has proposed multiple release sites, each with a different sensitivity to sound stimulation (Furukawa 1986). Indeed, there might be multiple release sites even within a single synaptic body. This is plausible, given the anatomical relation of the synaptic body to the synaptic vesicles and the calcium channels in the presynaptic membrane. Most of the calcium channels are located in a focal cluster, presumably beneath the synaptic body (Roberts, Jacobs, and Hudspeth 1990). With calcium entry, the highest calcium concentrations would be achieved between the calcium channels and the synaptic body. Calcium concentration might decrease with distance from the calcium channels. With increasing sound stimulation, more calcium would enter, and calcium would be available to evoke release at sites more distant from the presynaptic membrane. Thus n would increase with increasing voltage (calcium entry). One might not see a change in the average overall probability of release of transmitter, since vesicles near the synapse would have a high probability of release and those far from the synapse would have a low probability. However, one would expect fewer failures than predicted by the binomial analysis, since a part of the population would have a higher probability for release than the rest. Indeed, Furukawa did note a failure rate that was lower than expected (Furukawa, Hayashida, and Matsuura 1978).

As mentioned above, auditory nerve fibers exhibit adaptation to a stimulus. The discharge rate drops rapidly following the onset of an acoustic stimulus. The synaptic basis of one component of adaptation has been described by Furukawa from recordings of the goldfish auditory nerve (Furukawa and Ishii 1967; Furukawa and Matsuura 1978). During presentation of a tone burst, the amplitude of the EPSP is decreased in each successive phase of the stimulus. The mean quantal content (m) of the EPSP is successively decreasing. Binomial analysis suggested that this was primarily due to a decrease in the quantal parameter, n, the amount of transmitter available for release, but not in p. This is consistent with the idea that there are multiple release sites with differing sensitivity to sound that are depleted and slowly refilled during continued stimulation. Thus, with a continued stimulation, repletion does not keep up with depletion and the number of vesicles available for release decreases.

9. Summary, Conclusions, and a Few of the Unanswered Questions

Afferent synaptic transmission in the cochlea is a truly remarkable process, capable of transmitting afferent information with exceedingly high temporal resolution. Understanding the molecular and biophysical mechanisms that facilitate this process will present a challenge for neurobiologists in the future. Major questions remain unanswered. What is the identity of the transmitter released by the hair cell? Is there a unique glutamate receptor in the auditory nerve that might contribute to the incredible temporal processing capabilities of these neurons? We are far from understanding efferent neurotransmission in the cochlea. The efferent receptor on hair cells appears to contain an unusual subunit. What are the remaining structural components of the acetylcholine receptor on the hair cells? What are the roles of the various neuropeptides and neuromodulators in cochlear function?

Acknowledgments I thank E. Mroz for helpful discussions during the preparation of this chapter. This work was supported principally by NIH grant DC 00767.

References

Abou-Madi L, Pontarotti P, Tramu G, Cupo A, Eybalin M (1987) Coexistence of putative neuroactive substances in lateral olivocochlear neurons of rat and guinea pig. Hear Res 30:135–146.

Adams JC, Mruz EZ, Sewell WF (1987) A possible neuroransmitter role for CGRP in a hair-cell sensory organ. Brain Res 419:347–351.

Allen CN, Braky R, Swann J, Hori N, Carpenter DO (1988) N-Methyl-D-aspartate (NMDA) receptors are inactivated by trypsin. Brain Res 458:147–150.

Altschuler RA, Hoffman DW, Reeks KA, Fex J (1984) Localization of dynorphin B-like and alpha-neoendorphin-like immunoreactivities in the guinea pig organ of Corti. Hear Res 16:249–258.

Annoni JM, Cochran SL, Precht W (1984) Pharmacology of vestibular hair cells in the frog. J Neurosci 4:2106–2116.

Art JJ, Crawford AC, Fettiplace R, Fuchs PA (1982) Efferent regulation of hair cells in the turtle cochlea. Proc R Soc London B 216:377-384.

Art JJ, Fettiplace R, Fuchs PA (1984) Synaptic hyperpolarization and inhibition of turtle cochlear hair cells. J Physiol 356:525-550.

Art JJ, Crawford AC, Fettiplace R, Fuchs PA (1985) Efferent modulation of hair cell tuning in the cochlea of the turtle. J Physiol 360:397-421.

Bledsoe SC, Chihal DM, Bobbin RP, Morgan DN (1983) Comparative actions of glutamate and related substances on the lateral line organ of *Xenopus laevis*. Comp Biochem Physiol 75C:119-206.

Bobbin RP, Konishi T (1974) Action of cholinergic and anticholinergic drugs at the cross olivocochlear bundle-hair cell junction. Acta Otolaryngol 77:56-65.

Bobbin RP, Bledsoe SCJ, Winbery S, Ceasar G, Jenison GL (1985) Comparative actions of GABA and acetylcholine on the *Xenopus laevis* lateral line. Comp Biochem Physiol 80C:313-318.

Bledsoe SCJ, Bobbin RP, Puel JL (1988) Neurotransmission in the inner ear. In: Jahn AF, Santos-Sacchi J (eds) Physiology of the Ear. New York: Raven Press, pp. 385-406.

Bobbin RP, Ceasar G, Fallon M (1990) Potassium induced release of GABA and other substances from the guinea pig cochlea. Hear Res 46:83-94.

Bobbin RP, Fallon M, Puel JL, Bryant G, Bledsoe SC Jr, Zajic G, Schacht J (1990) Acetylcholine, carbachol, and GABA induce no detectable change in the length of isolated outer hair cells. Hear Res 47:39-52.

Boulter J, Hollman M, Shea-Greenfield A, Hartley M, Deneris E, Maron C, Heinemann S (1990) Molecular cloning and functional expression of glutamate receptor subunit genes. Science 249:1033-1037.

Brown MC, Nuttall AL, Masta RI (1983) Intracellular recordings from cochlear inner hair cells: effects of stimulation of the crossed olivocochlear efferents. Science 222:69-72.

Brownell WE, Bader CR, Bertrand D, de Ribaupierre Y (1985) Evoked mechanical response of isolated cochlear outer hair cells. Science 277:194-196.

Collingridge GL, Lester RA (1989) Excitatory amino acid receptors in the vertebrate central nervous system. Pharmacol Rev 40:143-210.

Crawford AC, Fettiplace R (1980) The frequency selectivity of auditory nerve fibres and hair cells in the cochlea of the turtle. J Physiol (London) 306:79-125.

Devau G, Lehouelleur J, Sans A (1993) Glutamate receptors on type 1 vestibular hair cells of the guinea pig. Eur J Neurosci 5:1210-1217.

Docherty M, Bradford HF, Wu J-Y (1987) Co-release of glutamate and aspartate from cholinergic and GABAergic synaptosomes. Nature 330:64-66.

Doi K, Weinthold RJ, Takahashi Y, Matsunaga T (1993) DNA amplifications of ionotrophic/metabotrophic glutamate receptors and acetylcholine receptors in the rat cochlea. ARO Abstracts 16:92.

Drescher MJ, Drescher DG (1992) Glutamate, of the endogenous primary alpha-amino acids, is specifically released from hair cells by elevated extracellular potassium. J Neurochem 59:93-98.

Drescher MJ, Drescher DG, Medina JE (1983) Effect of sound stimulation at several levels on concentrations of primary amines, including neurotransmitter candidates, in perilymph of the guinea pig inner ear. J Neurochem 41:309-320.

Elgoyhen AB, Johnson DS, Boulter J Vetter DE, Heinemann S (1994) Alpha 9: an acetylcholine receptor with novel pharmacological properties expressed in rat cochlear hair cells. Cell 79:705-715.

Erostegui C, Norris CH, Bobbin RP (1994) In vitro pharmacologic characterization of a cholinergic receptor on guinea pig outer hair cells. Hear Res 74:135-147.

Eybalin M (1993) Neurotransmitters and neuromodulators of the mammalian cochlea. Physiol Rev 73:309-373.

Eybalin M, Pujol R, Bockaert J (1987) Opioid receptors inhibit the adenylate cyclase in guinea pig cochleas. Brain Res 421:336-342.

Eybalin M, Rebillard G, Jarry T, Cupo A (1987) Effect of noise level on Met-enkephalin content of the guinea pig cochlea. Brain Res 418:189-192.

Eybalin M, Parnaud C, Geffard M, Pujol R (1988) Immunoelectron microscopy identifies several types of GABA-containing efferent synapses in the guinea pig organ of Corti. Neuroscience 24:29-38.

Felix D, Ehrenberger K (1982) The action of putative neurotransmitter substances in the cat labyrinth. Acta Otolaryngol 93:311-314.

Fex J, Adams J (1978) Alpha-bungarotoxin blocks reversibly cholinergic inhibition in the cochlea. Brain Res 159:440-444.

Fex J, Altschuler RA (1981) Enkephalin-like immunoreactivity of olivo-cochlear nerve fibers in cochlea of guinea pig and cat. Proc Natl Acad Sci USA 78:1255-1259.

Fex J, Altschuler RA (1984) Glutamic acid decarboxylase immunoreactvity of olivocochlear neurons in the organ of Corti of guinea pig and rat. Hear Res 15:123-131.

Flock A, Russell I (1976) Inhibition by efferent nerve fibers: action on hair cells and afferent synaptic transmission in the lateral line canal organ of the burbot, *Lota lota*. J Physiol 257:45-62.

Fuchs PA, Murrow BW (1992) Cholinergic inhibition of short (outer) hair cells of the chick's cochlea. J Neurosci 12:800-809.

Furukawa T (1985) Mode of operation of afferent synapses between hair cells and auditory fibers. In: Drescher DG (ed) Auditory Biochemistry. Springfield, IL: Charles C Thomas.

Furukawa T (1986) Sound reception and synaptic transmission in goldfish hair cells. Jpn J Physiol 36:1059-1077.

Furukawa T, Ishii Y (1967) Neurophysiological studies on heaing in goldfish. J Neurophysiol 30:1377-1403.

Furukawa T, Matsuura S (1978) Adaptive rundown of excitatory post-synaptic potentials at synapses between hair cells and eighth nerve fibers in the goldfish. J Physiol 276:193-209.

Furukawa T, Ishii Y, Matsuura S (1972) Synaptic delay and time course of postsynaptic potentials at the junction between hair cells and eighth nerve fibers in the goldfish. Jpn J Physiol 22:617-635.

Furukawa T, Hayashida Y, Matsuura S (1978) Quantal analysis of the size of excitatory post-synaptic potentials at synapses between hair cells and afferent nerve fibers in goldfish. J Physiol 276:211-226.

Furukawa T, Kuno M, Katsuura S (1982) Quantal analysis of a decremental response at hair cell-afferent fibre synapses in the goldfish sacculus. J Physiol 322:181-195.

Geisler CD, Le S, Schwid H (1979) Further studies on the Schroeder-Hall hair-cell model. J Acoust Soc Am 65:985-990.

Gifford ML, Guinan JJ (1987) Effects of electrical stimulation of medial olivocochlear neurons on ipsilateral and contralateral cochlear responses. Hear Res 29:179-194.

Gleich O, Johnstone BM, Robertson D (1990) Effects of l-glutamate on auditory afferent activity in view of its proposed excitatory transmitter role in the mammalian cochlea. Hear Res 45:295-312.

Gregor P, Mano I, Maoz I, McKeown M, Teichberg VI (1989) Molecular structure of the chick cerebellar kainate-binding subunit of a putative glutamate receptor. Nature 342:689-692.

Guinan JJ, Warr WB, Norris BE (1983) Differential olivocochlear projections from lateral versus medial zones of the superior olivary complex. J Comp Neurol 221:358-370.

Guth PS, Norris CH, Bobbin RP (1976) The pharmacology of transmission in the peripheral auditory systems. Pharmacol Rev 28:95-125.

Guth P, Norris C, Fermin CD, Pantoja M (1993) The correlated blanching of synaptic bodies and reduction in afferent firing rates caused by transmitter-depleting agents in the frog semicircular canal. Hear Res 66:143-149.

Henley JM, Ambrosini A, Rodriguez-Ithurralde D, Sudan H, Brackley P, Kerry C, Mellor I, Abutidze K, Usherwood PNR, Barnard EA (1992) Purified unitary kainate/α-amino-3-hydroxy-5-methylisooxazolepropionate(AMPA) and kainate/AMPA/N-methyl-D-aspartate receptors with interchangeable subunits. Proc Natl Acad Sci USA 89:4806-4810.

Herb A, Burnashev N, Werner P, Sakmann B, Wisden W, Seeburg PH (1992) The KA2 subunit of excitatory amino acid receptors shows widespread expression in brain and forms ion channels with distantly related subunits. Neuron 8:775-785.

Highstein SM, Baker R (1985) Action of the efferent vestibular system on primary afferents in the toadfish, *Opsanus tau*. J Neurophysiol 54:370-384.

Honore T, Davies SM, Drejer J, Fletcher EJ, Jobobsen P, Lodge D, Nielsen FE (1988) Qunioxalinediones: potent competitive non-NMDA glutamate receptor antagonists. Science 241:701-703.

Housley GD, Ashmore JF (1991) Direct measurement of the action of acetylcholine on isolated outer hair cells of the guinea pig cochlea. Proc R Soc Lond B 244:161-167.

Hudspeth AJ, Lewis RS (1988) Kinetic analysis of voltage- and ion-dependent conductances in saccular hair cells of the bull-frog *Rana catesbieana*. J Physiol 400:237-274.

Huettner JE (1990) Glutamate receptor channels in rat DRG neurons: activation by kainate and quisqualate and blockade of densensitization by Con A. Neuron 5:255-266.

Ishii Y, Matsuura S, Furukawa T (1971) Quantal nature of transmission at the synapse between hair cells and VIII nerve fibers. Jpn J Physiol 21:79-89.

Jepson P (1969) Studies of the structure and function of taste buds. Acta Otolaryngol 55:1-32.

Johnson DH (1980) The relationship between spike rate and synchrony in responses of auditory-nerve fibers to single tones. J Acoust Soc Am 68:1115-1122.

Kakehata S, Nakagawa T, Takasaka T, Akaike N (1993) Cellular mechanism of acetylcholine-induced response in dissociated outer hair cells of guinea pig cochlea. J Physiol 463:227-244.

Katz B, Miledi R (1967) A study of synaptic transmission in the absence of nerve impulses. J Physiol 192:407-436.

Kehoe J (1972) Three acetylcholine receptors in *Aplysia* neurones. J Physiol 225:115-146.

Kiang NY-S, Watanabe T, Thomas EC, Clark LF (1965) Discharge Patterns of Single Fibers in the Cat's Auditory Nerve. Cambridge, MA: MIT Press.

Kiang NY-S, Rho JM, Northrop CC, Liberman MC, Ryugo DK (1982) Hair-cell innervation by spiral ganglion cells in adult cats. Science 217:175–177.

Kidd RC, Weiss TF (1990) Mechanisms that degrade timing information in the cochlea. Hear Res 49:181–208.

Kirk DL, Johnstone BM (1993) Modulation of f_2-f_1: evidence for a GABA-ergic efferent system in apical cochlea of the guinea pig. Hear Res 67:20–34.

Kujawa SG, Glattke TJ, Fallon M, Bobbin RP (1993) Contralateral sound suppresses distortion product otoacoustic emissions through cholinergic mechanisms. Hear Res 68:97–106.

Kujawa SG, Glattke TJ, Fallon M, Bobbin RP (1994) A nicotinic-like receptor mediates suppression of distortion otoacoustic emissions by contralateral sound. Hear Res 74:122–134.

Kuno M (1983) Adaptive changes in firing rates in goldfish auditory fibers as related to changes in mean amplitude of excitatory postsynaptic potentials. J Neurophysiol 50:573–581.

Kuriyana H, Albin RL, Altschuler RA (1993) Expression of NMDA receptor messenger RNA in the rat cochlea. Hear Res 69:215–220.

Kyogoku I, Matsuura S, Kuno M (1986) Generator potentials and spike initiation in auditory fibers of goldfish. J Neurophysiol 55:244–255.

Liberman MC (1978) Auditory-nerve response from cats raised in a low-noise chamber. J Acoust Soc Am 63:442–455.

Liberman MC (1980) Morphological differences among radial afferent fibers in the cat cochlea: an electron-microscopic study of serial sections. Hear Res 3:45–63.

Liberman MC (1982) Single-neuron labeling in the cat auditory nerve. Science 216:1239–1241.

Liberman MC, Oliver ME (1984) Morphometry of intracellularly labeled neurons of the auditory nerve: correlations with functional properties. J Comp Neurol 223:163–176.

Littman T, Bobbin RP, Fallon M, Puel J-L (1989) The quinoxalinediones DNQX and CNQX and two related congeners suppress hair cell-to-auditory nerve transmission. Hear Res 40:45–54.

Lu SM, Schweitzer L, Cant NB, Dawbarn D (1987) Immunoreactivity to calcitonin gene-related peptide in the superior olivary complex and cochlea of cat and rat. Hear Res 31:137–146.

Monaghan P (1975) Ultrastructural and pharmacological studies of the afferent synapse of lateral-line sensory cells of the African clawed toad, *Xenopus laevis*. Cell Tissue Res 163:239–247.

Mroz EA, Sewell WF (1989) Pharmacological alterations of the activity of afferent fibers innervating hair cells. Hear Res 38:141–162.

Nakagawa T, Komune S, Uemura T, Akaike N (1991) Excitatory amino acid response in isolated spiral ganglion cells of guinea pig cochlea. J Neurophysiol 65:715–723.

Niedzielski AS, Wenthold RJ (1995) Expression of AMPA, kainate, and NMDA receptor subunits in cochlear and vestibular ganglia. J Neurosci 15:2338–2353.

Norris CH, Guth PH, Daigneault EA (1977) The site at which peripheral auditory adaptation occurs. Brain Res 123:174–179.

Oesterle EC, Dallos P (1990) Intracellular recordings from supporting cells in the guinea pig cochlea: DC potentials. J Neurophysiol 64:617–636.

Olney JW, Ho OL, Rhee V (1971) Cytotoxic effects of acidic and sulfur containing amino acids on the infant mouse central nervous system. Exp Brain Res 14:61-76.

Osborne MP, Thornhill RA (1972) The effects of monoamine depleting drugs upon the synaptic bars in the inner ear of the bullfrog, *Rana catesbieana*. Z Zellforsch 127:347-355.

Palmer AR, Russell IJ (1986) Phase-locking in the cochlear nerve of the guinea-pig and its relation to the receptor potential of inner hair-cells. Hear Res 24:1-15.

Plinkert PK, Zenner HP, Heilbronn E (1991) A nicotinic acetylcholine receptor-like α-bungarotoxin binding site on outer hair cells. Hear Res 53:123-130.

Pujol R, Lenour M, Robertson D, Eybalin M, Johnstone BM (1985) Kainic acid selectively alters auditory dendrites connected with cochlear inner hair cells. Hear Res 18:145-151.

Raman IM, Trussell LO (1992) The kinetics of the response to glutamate and kainate in neurons of the avian cochlear nucleus. Neuron 9:173-186.

Roberts WM, Jacobs RA, Hudspeth AJ (1990) Colocalization of ion channels involved in frequency selectivity and synaptic transmission at presynaptic active zones of hair cells. J Neurosci 10:3664-3684.

Rossi ML, Valli P, Casella C (1977) Post-synaptic potentials recorded from afferent nerve fibers of the posterior semicircular canal in the frog. Brain Res 135:67-75.

Russell IJ (1971) The pharmacology of efferent synapses in the lateral-line system of *Xenopus laevis*. J Exp Biol 54:643-658.

Russell IJ, Sellick PM (1978) Intracellular studies of hair cells in the mammalian cochlea. J Physiol 284:261-290.

Ryan AF, Brumm D, Kraft M (1991) Occurrence and distribution of non-NMDA glutamate receptor mRNAs in the cochlea. Neuroreport 2:643-646.

Safieddine S, Eybalin M (1992) Co-expression of NMDA and AMPA/kainate receptor mRNAs in cochlear neurones. Neuroreport 3:1145-1148.

Sahley TL, Musiek FE, Hoffman DW (1991) Opiate modulation of N1 and N2 amplitudes in the chinchilla. ARO Abstr 14:132.

Sand O (1975) Effects of different ionic environments on mechanosensitivity of lateral line organs in the mudpuppy. J Comp Physiol 102:27-42.

Schroeder MR, Hall JL (1974) Model for mechanical to neural transduction in the auditory receptor. J Acoust Soc Am 55:1055-1060.

Schwid HA, Geisler CD (1982) Multiple reservoir model of neurotransmitter release by a cochlear inner hair cell. J Acoust Soc Am 72:1435-1440.

Sellick PM, Russell IJ (1980) The responses of inner hair cells to basilar membrane velocity during low frequency auditory stimulation in the guinea pig cochlea. Hear Res 2:439-446.

Sento S, Furukawa T (1987) Intra-axonal labeling of saccular afferents in the goldfish, *Carassius auratus*: correlations between morphological and physiological characteristics. J Comp Neurol 258:352-367.

Sewell WF (1984) The relation between the endocochlear potential and spontaneous activity in auditory nerve fibers of the cat. J Physiol 347:685-696.

Sewell WF (1990) Synaptic potentials in afferent fibers innervating hair cells of the lateral line organ of *Xenopus laevis*. Hear Res 44:71-82.

Sewell WF (1993) Pharmacology of glutamate receptors on afferent fibers innervating hair cells in *Xenopus laevis*: effects of synaptic potentials. ARO Abstr 16:33.

Sewell WF, Mroz EA (1987) Neuroactive substances in inner ear extracts. J Neurosci 17:2465–2475.

Sewell WF, Starr PA (1991) Effects of calcitonin gene-related peptide and efferent nerve stimulation on afferent transmission in the lateral line organ. J Neurophysiol 65:1158–1169.

Shigemoto T, Ohmori H (1990) Muscarinic agonists and ATP increase the intracellular Ca^{++} concentration in chick cochlear hair cells. J Physiol 420:127–148.

Siegel JH, Brownell WE (1986) Synaptic and Golgi membrane recycling in cochlear hair cells. J Neurocytol 15:311–328.

Siegel J, Dallos P (1986) Spike activity recorded from the organ of Corti. Hear Res 22:245–248.

Sjostrand FS (1958) Ultrastructure of retinal rod synapses of the guinea pig as revealed by three dimensional reconstruction from serial sections. J Ultrastruc Res 2:122.

Spoendlin H (1966) The organization of the cochlear receptor. Adv Otorhinolaryngol 13:1–227.

Sridhar TS, Brown MC, Liberman MC, Sewell WF (1995) A novel cholinergic "slow effect" of efferent stimulation on cochlear potentials in the guinea pig. J Neurosci 15:3667–3678.

Starr PA, Sewell WF (1991) Neurotransmitter release from hair cells in its blockade by glutamate-receptor antagonists. Hear Res 52:23–42.

Suzue T, Wu G-B, Furukawa T (1987) High susceptibility to hypoxia of afferent synaptic transmission in the goldfish sacculus. J Neurophysiol 58:1066–1079.

Sziklai I, Dallos P (1993a) Acetylcholine controls the gain of the voltage-to-movement converter in isolated outer hair cells. Acta Otolaryngol 113:326–329.

Sziklai L, Dallos P (1993b) Effect of acetylcholine and GABA on the transfer function of electromotility in isolated OHCs. ARO Abstr 16:83.

Takasaka T, Smith CA (1971) The structure and innervation of the pigeon's basilar papilla. J Ultrastruct Res 35:20–65.

Umekita S, Matsumoto Y, Abe T, Obara S (1980) The afferent neurotransmitter in the ampullary electroreceptors: stimulus-dependent release experiments refute the transmitter role of l-glutamate. Neurosci Lett Suppl 4:S7.

Valli P, Zucca G, Prigioni L, Botta L, Casella C, Guth PS (1985) The effect of glutamate on the frog semicircular canal. Brain Res 330:1–9.

Vetter DE, Adams JC, Mugnaini E (1991) Chemically distinct rat olivocochlear neurons. Synapse 7:21–43.

Wada K, Deschesne C, Shimasaki S, King RG, Kisano K, Buonanno A, Hampson DR, Banner C, Wenthold RJ, Nakatani Y (1989) Sequence and expression of a frog brain complementary DNA encoding a kainate-binding protein. Nature 342:684–689.

Walsh BT, Miller JB, Gacek RR, Kiang NY-S (1972) Spontaneous activity in the eighth cranial nerve of the cat. Int J Neurosci 3:221–236.

Watkins JC, Olverman HJ (1987) Agonists and antagonists for excitatory amino acid receptors. Trends Neurosci 10:265–272.

Weiss TF, Rose C (1898) Stages of degradation of timing information in the cochlea: a comparison of hair-cell and nerve-fiber responses in the alligator lizard. Hear Res 33:167–174.

Werman R (1966) Criteria for identification of a central nervous system transmitter. Comp Biochem Physiol 18:745–766.

Westerman LA, Smith RL (1984) Rapid and short-term adaptation in auditory nerve responses. Hear Res 15:249–260.

Wiederhold ML, Kiang NY-S (1970) Effects of electric stimulation of the crossed olivocochlear bundle on single auditory-nerve fibers in the cat. J Acoust Soc Am 48:950–965.

Wurtman RD, Axelrod JDK (1968) The Pineal. London, New York: Academic Press.

Yamaguchi K, Ohmori H (1990) Voltage-gated and chemically gated ionic channels inthe cultured cochlear ganglion neurone of the chick. J Physiol 420:185–206.

Index

Acetylcholine, 509-511
  cochlea, 148
  efferent synapse, 10, 84, 374
  OHC, 63
Acoustic reflex, MOC, 470ff
Acoustic transformers, 194-195
Acoustic trauma, basilar membrane response, 226
Actin, hair cell, 90ff, 392-393
  OHC, 400
  stereocilia, 56-58, 94
  support cells, 96-97
Actin-binding proteins, cochlea, 94
  support cells, 96-97
Action potential, afferent synapse, 519ff
Active processes, 187
  cochlea, 15ff, 214ff, 228ff
  cochlear models, 281ff
  local, 282ff
  metabolism, 229ff
  organ of Corti, 187-188, 197-198
  otoacoustic emissions, 18
  sensitivity to vibration, 230ff
  two-tone suppression, 233ff
  see Cochlear Amplifier and OHC Motility
Adaptation
  afferent fibers, 365, 521
  efferent effects, 460
  hair bundle stiffness, 241, 342ff
  hair cells, 330ff
  models, 343ff
  motor, 343ff
  temperature, 332-333

tip-links, 241
Adherens junctions, 52-53
Admittance, 199
  OAEs, 242-243, 245
Afferent fibers
  adaptation, 365, 521
  effects of MOC, 461ff, *see also Auditory, Cochlear Afferents, Spiral Ganglion*
  efferent effects on spontaneous activity, 460-461
  efferent effects on synchronization, 461-462
  frequency responsiveness, 520-521
  glutamate receptors, 505-508
  types I and II, 10
Afferent innervation
  cochlea 6, 10, 76-79
  IHC, 78-83, 85-86
  organ of Corti, 79-80
Afferent rate-level functions, MOC effects, 462
Afferent spontaneous activity, MOC effects, 461, 465
Afferent synapse
  anatomy, 514ff
  calcium, 516-517
  hair cell, 77, 85, 320
  OHC, 85
  potentials, 519ff
  synaptic body, 517-518
  ultrastructure, 516-519
African clawed toad, *see Xenopus*
Alligator lizard, *see Gerrhonotus multicarinatus*

536 Index

Aminoglycosides, hair cells, 335–336
Amphibian papilla, electrical resonance, 3
Anesthesia
  efferent spontaneous activity, 471
  sound-driven efferent response, 473
Anion exchangers, OHC, 409–410
Arch of Corti, 200–201
ATP, 132–133
  cochlea, 137
  hair cell, 333
  hair cell currents, 373
  hair cell length change, 391–392
  hair cell motility, 386, 418–419
  as neuromodulator, 513
Attention, efferent effects, 438, 486–487

Basal cells, stria vascularis, 112–113
Basilar membrane, 5, 45ff, 74–76, 186
  active processes, 187–188
  bat, 75
  characteristic frequency, 16–17
  coupling, 190
  DC shifts, 261
  displacement, 13–15
  force from OHC, 417
  frequency response, 17
  gain, 16
  impedance, 285, 290
Basilar membrane impedance
  location of active OHC, 233ff
  motion, 260ff
  resistive vs. reactive, 300
Basilar membrane motion, MOC, 456ff, 466–467
  nonlinearities, 297–298
  phase response, 221ff
Basilar membrane resonance
  models, 265–266, 273ff
  response after trauma, 226
  stiffness and mass, 202, 264–265
  structure, 74–75
  tension, 207–208
  thickness, 75
  transverse coupling, 187
  transverse resonance, 188
  tuning curve, 223–224

Basilar membrane velocity
  cochlear models, 264
  IHC stimulation, 204
Basolateral conductance, MOC, 446ff
Basolateral currents, OHC, 241–242
Basolateral membrane, currents, 352ff
  hair cell, 24ff
Bat
  basilar membrane, 75
  cochlea, 196
  efferent effects, 484
  organ of Corti, 50
  primary afferent tuning, 224
Bidirectional transduction, 193
Binaural efferents, 471–472
Binaural interaction, LOC efferents, 489
Biochemistry, cochlea, 130ff
Bird, electrical resonance, 3
Blood flow, cochlea, 98ff, 133–134
Blood-perilymph barrier, 132, 155–156, 158
BM, *see Basilar Membrane*
Boettcher cells, 71–72
Boltzmann function
  hair cell transduction, 328, 240
  OHC, 369
Bony labyrinth, definition, 44
Boundary layer, cochlear models, 277
Bullfrog, *see Rana catesbieana*

Calcitonin gene-related peptide, efferents, 442, 468–469
Calcium
  afferent synapse, 516–517
  cochlea, 138–139
  effects on slow somatic motility, 419–421
  hair cell adaptation, 330ff
  hair cell length change, 391–392
  hair cell motility, 141
  intracellular regulation, 134ff
  neurotransmitter release, 504
  physiological role, 142–143
  regulation in cochlea, 165–166
Calcium-binding proteins
  hair cell, 95–96, 140
  support cells, 98

Calcium channels, hair cells, 143
Calcium currents
  hair cells, 25-26, 370ff
  phase-locking, 372
  physiological functions, 370ff
Calcium-dependent regulatory proteins, hair cell, 95-96
Calcium homeostasis, cochlea, 138ff
  transduction, 154-155
Calcium stores, cochlea, 140-142
Calcium transport, 139
Calmodulin, hair cell, 96
Carassius auratus, synaptic transmission, 522
Cat, cochlear innervation, 78
  spiral ganglion, 81-83
Catecholaminergic fibers, cochlea, 88
Cerebrospinal fluid, composition, 153, 156
CF, *see Characteristic frequency*
Channels, hair cells, 22-24
  stereocilia, 22-24
Chaos, cochlear models, 308
Cholinergic receptors, organ of Corti, 510-511
Cilia, displacement, 22
  force transmission, 21-22
Ciliary bundle
  mechanical resonance, 3-4
  thermal noise, 2
Claudius cells, 71, 106
Click-evoked OAE (CEOAE), efferent effects, 454-455
CM, *see Cochlear Potentials*
Cochlea, *see also Organ of Corti*
  acetylcholine, 509-511
  actin-binding proteins, 94
  active mechanics, 15ff, 228ff
  afferent innervation, 76-78
  amplification, 4
  anatomy, 44ff
  ATP, 137
  biochemistry, 130ff
  blood flow, 98ff, 133-134, 164-165
  boundaries, 47
  calcium, 138-139
  calcium homeostasis, 138ff
  calcium stores, 140-142
  catecholaminergic fibers, 88
  DC potential, 149
  deoxyglucose uptake, 133
  distortion, 187
  dynamic range, 4, 187
  effects of trauma, 216-217
  efferent innervation, 76-78, 508
  electroanatomy, 446ff
  energy metabolism, 164-165
  fluids, 149ff, 167-168
  frequency analysis, 5
  frequency selectivity, 190
  G proteins, 136-137
  gain control, 187
  glucose metabolism, 132
  guinea pig, 45, 62
  high-pass filtering, 206-207
  history, 44
  homeostasis, 131
  innervation, 76-78
  intermodulation distortion, 237
  ion transport, 143-146, 168
  junctions, 147
  linearity, 187
  longitudinal coupling, 207-208
  micromechanics, 15ff
  modeling description, 262-263
  models, 258ff
  neurobiology, 1ff
  neurotransmitters, 148-149
  nitric oxide, 166
  nonlinearities, 238ff
  partition, 5ff
  passive mechanics, 12ff
  perivascular plexus, 89
  place-frequency map, 186, 214, 265, 274
  potassium composition, 154
  potassium secretion, 160
  protein kinases, 137
  rectification, 237
  reflections, 246
  response characteristics, 1ff
  saturation, 217-218
  second messengers, 137, 165-166
  sensitivity, 187
  silicon model, 259
  SOAE, 187
  spiral, 208
  standing current, 149-150

structure, 5ff
surface wave, 208ff
sympathetic nervous system, 88–90, 134
tubulin, 92
vasculature, 88–90
wave propagation, 12ff
Cochlear admittance
  level-dependence, 247–248
  SFOAE, 247–248
  tympanometry, 243ff
Cochlear afferents
  effects of MOC, 450ff
  rate-level functions, 204–205
  response, 296–297
  tuning, 218ff
Cochlear amplifier, 261ff, 290, 295
  efferents, 438, 484ff
  MOC, 448ff, 456ff, 466–467
  models, 281ff
Cochlear duct, 7
  definition, 45
Cochlea fluids, composition, 150–153
  physiological significance, 153–154
Cochlear gain, distortion products, 305
Cochlear impedance, OHC, 338
Cochlear macromechanics, 205ff
Cochlear mechanics
  efferent effects on, 456ff
  history of study, 189ff
  models, 258ff
Cochlear metabolism, active processes, 229
Cochlear micromechanics, 186, 194ff
Cochlear models
  active vs. passive, 281–282
  anatomy, 262ff
  boundary layer absorption, 277
  cochlear partition, 264, 306
  DC displacements, 308
  distortion products, 296ff, 302ff
  Eikonal equation, 271–272, 276
  feedback, 293–294
  fluid pressures, 262ff
  generalized cross-validation, 284ff
  geometry, 306
  local micromechanics, 287ff
  locally active, 282ff, 287ff, 307

long and short waves, 277
long wave approximation, 266ff, 268ff
longitudinal coupling, 307, 309
nonlinear damping, 298ff
nonlinear local activity, 300ff
nonlinearities, 295ff, 308
OHC, 259ff, 292ff
one-dimensional, 267ff
organ of Corti coupling, 263
panoramic view, 272ff, 278ff, 290
passive, 286, 290
quasi-linear methods, 301–302
rationale, 258ff
relation to psychophysics, 297–298
Siebert equation, 276
SOAE, 307
tectorial membrane, 288ff
three-dimensional, 279ff
time-domain, 299ff
tone interactions, 302ff
traveling wave tube, 295
two-dimensional, 274ff
two-tone suppression, 296–297
undamping, 282, 300
wave equation, 269
wave number, 271
Cochlear nerve fibers, *see also Afferents*
  efferent effects, 459–460
  rate-level functions, 191
Cochlear nonlinearity, 215ff
Cochlear nucleus, efferents, 474–475
  innervation, 81–83
Cochlear partition, models, 262ff
  pressure gradient, 12
Cochlear potentials, 32
  cochlear model, 306
  efferents, 437
  hair cells, 320ff
  MOC, 442–443, 445ff
  OHC active processes, 231ff
  potassium concentration, 154
Cochlear trauma, tuning curves, 224ff
Cochlear vibration, lability, 224ff
Collagen, in basilar membrane, 74–75
Collagen, hair cell, 147
Collagen, in tectorial membrane, 74
Combination tones, cochlear models, 302ff

Communicating junctions, *see Gap Junctions*
Compound action potentials, MOC effects, 443ff
Corti, Alfonse, 189
Cortical lattice, force generation, 401-403
  OHC, 398ff
Corti's organ, *see Organ of Corti*
Cubic difference tones, thresholds, 18-19
Cuticular plate, 56-58
  hair cell motility, 386
  IHC, 59-60
  movement, 425
  myosin, 94
  OHC, 62
Cyclic nucleotide-gated channels, hair cells, 333ff
Cytoskeleton, hair cell, 90ff
  organ of Corti, 90ff
Cytoskeleton proteins, OHC, 93

Damping, 198-199, 272ff
Damping, nonlinear, 298ff
Davis, "battery theory", 131, 132, 149
  transduction model, 131, 132
DC potential, cochlea, 149
DC response, hair cells, 29-33
Deiters cell, cytoskeleton proteins, 93
Deiters cells, 5-6, 69-70
Deoxyglucose, uptake in cochlea, 133
Desmosomes, 53
Development
  hair cell transducer conductance, 329
  OHC currents, 367-368, 372
  potassium currents, 360-361
Discrimination, frequency, 1-2
Dispersion relationship, cochlea, 210ff
Distortion products, 190
  acoustic trauma, 226
  cochlea, 187
  cochlear models, 296ff
  otoacoustic emissions, *see DPOAE*
  psychophysics, 297
Diuretics, hair cells, 336-337
DPOAE (Distortion Products Otoacoustic Emissions)
  cochlear admittance, 248

inhibition by nicotinic receptor, 454
LOC, 468
MOC, 453-454, 458
Dynamic matching network, 198-199
Dynamic range, cochlea, 4
  efferent effects, 484ff

Ear canal, resonance, 11
Efferent effects
  effects of shock rate, 443ff
  level discrimination, 486
  OAE, 458
  OHC motility, 457-458
  signal detection in noise, 486
  vowel discrimination, 486
Efferent fiber, 509
Efferent innervation
  branching pattern, 87
  CNS origin, 10
  cochlea, 6,, 76-78
  IHC, 83-84
  neurotransmitters, 10
  OHC 63, 86-88
  organ of Corti, 79-80
  vestibular and auditory systems, 481ff
Efferent nerve terminals, 87
Efferent neurotransmitter
  acetylcholine, 509-511
  hair cell, 504, 508ff
  IHC, 84
  OHC, 87-88
Efferent synapse
  acetylcholine, 374
  hair cell, 77, 78, 84, 320
  OHC, 87
Efferents, 435ff
  activity during speech, 489
  adaptation, 460
  afferent fiber rate-level functions, 459-460
  central modulation, 478
  cholinergic synapses, 439, 441-442
  cochlear amplifier, 438, 484ff
  cochlear innervation, 440-441
  cochlear nucleus effects, 474-475
  cochlear potentials, 437
  compound action potentials, 437

540   Index

crossed and uncrossed, 435ff
effects of sound level, 439
effects on dynamic range, 484ff
effects on hearing, 480, 491
effect on various OAEs, 454–455
electrical stimulation effects, 440
forward masking, 478
GABA-ergic, 441–442
general anatomy, 439ff
inhibition, 436ff
intense sound protection, 487ff
ipsilateral and contralateral sound, 438
latency, 473
low-spontaneous afferents, 439
masking, 439, 486
medial and lateral bundles, 435ff
medial group, 438ff
myelination, 440
non-mammals, 479–480
OAEs, 438, 487
OHC motility, 438–439
physiology, 435ff
presynaptic facilitation, 480
rate-level functions, 438
response to motion, 483–484
response to sound, 469ff
role in hearing, 484ff
selective attention, 486–487
shocks to brainstem, 443
sound-induced cochlear damage, 439
summary of effects, 489ff
synaptic potentials in hair cells, 479–480
temporary threshold shift, 487ff
unmasking, 459–460
various neurotransmitters, 441–442
vestibular system, 480ff
Eight nerve, OHC effects, 19
Electrical resonance
  amphibians, 3
  hair cells, 3, 362ff, 480
  IHC, 362ff
  potassium currents, 358–359
Electrical tuning, hair cells, 3, 352ff, 362ff, 480
Electrochemical gradients, endolymph, 153
Electromotile responses, OHC, 18–20

Endocochlear potential
  generation, 160ff
  MOC potential, 445ff
  potassium concentration, 154
  stria vascularis, 159
Endolymph, 5, 7, 45, 150ff, 167–168
  electrochemical gradients, 153
  formation, 7
  homeostasis, 156ff
  ionic composition, 157
Endolymph-perilymph barrier, 157–158
Endoplasmic reticulum, OHC, 396–398
EPSP, hair cell, 522–523
Evoked discharge, hair cell, 519–520
Excitatory postsynaptic potentials, see EPSP
External ear, 10–11
External sulcus cells, 72
Extracellular AC potential, 21

Feedback
  basilar membrane, 17ff
  cochlear models, 293–294
  efferents, 438
  micromechanics, 198
  Neely-Kim model, 290
  OHC, 11–12, 18–20
Fibrocytes, ion transport, 106–109
Filaments, organ of Corti, 91
Fimbin, stereocilia, 94
Force transformer, 195ff
Fourier, J., 189
Frequency analysis, 3–4
  history, 5
Frequency discrimination, human, 1–2
Frequency response
  afferent fibers, 520–521
  basilar membrane, 17
  OHC, 18
Frequency-threshold tuning curves, iso-input type, 219ff
Frequency tuning curve, primary afferents, 218ff
Friction, resonator, 199

G proteins, cochlea, 136–137
GABA (γ-aminobutyric acid)
  cochlea, 148
  efferent innervation IHC, 84

efferent synapses, 436, 468
efferents, 441-442, 468-469
  hair cell, 504
  IHC, 512-513
  OHC, 63, 87-88
GAD (glutamate decarboxylase), 508-509
Gain control, cochlea, 187
Gamma-aminobutyric acid, see GABA
Ganglion cells, eighth nerve, 8, 77
Ganglion cells, type I vs type II, 86
Gap junctions, 53-54
  hair cells, 46, 51-52, 147
Gating current, OHC, 406-407, 409
Gating springs
  adaptation, 343ff
  adaptation model, 344-345
  stereocilia, 342ff
  stiffness, 344-345
Gating stiffness, transduction, 240-241
Genes, stretch-activated channels, 338
Gerrhonotus multicarinatus, hair cells, 321
Glia, synaptic function, 149
Glucose metabolism, cochlea, 132, 133
Glutamate, 504-505
  cochlea, 148
Glutamate decarboxylase, see GAB
Glutamate receptors, afferent neurons, 505-508
  types, 505
Glycocalix, OHC, 403-404
Goldfish, see *Carassius auratus*
Group delay, basilar membrane, 221
Guinea pig, cochlea, 62
  cochlea innervation, 78
  OHC, 388

Hair bundle stiffness, adaptation, 241, see also Stereocilia
Hair cell, 54ff, see also Hair Cell Length Change, OHC and IHC
  AC response, 32
  actin, 90ff
  adaptation, 330ff
  adherens junctions, 52-53
  afferent innervation, 6
  afferent synapse, 77, 320, 514ff
  aminoglycosides, 335-336
  amphibian papilla, 3
  ATP, 333
  avian, 3, 319
  basolateral membrane, 24ff
  basolateral membrane currents, 319, 352ff
  birds, 3, 319
  calcium binding proteins, 95-96
  calcium channels, 143
  calcium currents, 25-26, 370ff
  calcium-dependent regulatory proteins, 95-96
  calmodulin, 96
  ciliary channels, 22-24
  collagens, 147
  cuticular plate, 56-58, 425
  cytoskeleton, 90ff
  DC response, 29-33
  desmosomes, 53
  differences between IHC and OHC, 58ff
  displacement sensitivity, 328ff
  distribution in cochlea, 49
  diuretics, 336-337
  efferent innervation, 6
  efferent neurotransmitter, 504, 508ff
  efferent synapse, 77, 78, 320
  electrical resonance, 352ff, 480
  energy requirements, 154
  EPSP, 522-523
  evoked discharge, 519-520
  excitatory stimulus, 22ff
  frequency response, 330, 346ff, 520-521
  function, 7-9
  gating spring stiffness, 342ff
  hair bundle mechanics, 338ff
  high-frequency hearing, 318-319
  innervation, 6, 10, 78ff, 514-516
  input resistance, 34
  intracellular recording, 321ff
  ion balance, 143-146
  ion channels, 323ff
  ionic environment, 324, 335
  junctions, 51ff
  lateral cortex, 395ff
  lizards, 3-4
  low-pass filter, 330

542 Index

mechanical stimulation methods, 325ff
mechanisms of motility, 411–413
membrane channel noise, 334
membrane filtering, 346ff
methods of study, 320ff
microtubule-associated proteins, 95
morphology, 7-9
motility, 141, 386ff
motility and voltage-dependence, 345–346
motility of OHC, 369
motor elements, 392–395
myosin, 94
neurotransmitters, 503ff
non-mammalian, 424–427
number in mammals, 48
organ culture, 322–333
other currents, 372ff
pH balance, 145
physiology, 318ff
potassium channels, 3
potassium currents, 25–26, 354ff
pyrazinecarboxamide block, 336–337
receptors, 352ff, 510–511
resting potentials, 322
saccule, 3
and spontaneous discharge, 519–520
stereocilia, 9–10, 54ff
synaptic transmission, 77, 522ff
tight junctions, 46, 51–52, 147
transducer conductance, 328ff
transducer currents, 326ff, 335
transduction, 22ff, 56
transduction ion selectivity, 333ff
transduction mechanisms, 325ff
transfer function, 326ff
tubulin, 95
type I innervation, 78–83
types, 54
velocity and displacement stimulation, 204
voltage clamp analyses, 323ff
voltage-dependent conductances, 352ff
voltage-gated channels, 25–26
volume regulation, 145–146
Hair cell currents, cations, 373–374
MOC, 446ff

Hair cell length change, 387ff
ATP, 391–392
calcium, 391–392
high frequency mechanism, 414–415
mechanism, 411–414
membrane potential, 389–390
model, 413–414
response to sound, 421–424
voltage dependence, 388–391
Hair cell membrane potential, stereociliary stiffness, 342ff
Head, effects on sound, 11
Hearing, efferent effects, 484ff
Helicotrema, filtering, 206–207
Helmholtz, Hermann, 5, 189
Hensen, first description of stereocilia, 54
Hensens cells, 6, 71
High-frequency hearing, hair cells, 319
Homeostasis, 130ff
cochlea, 131
definition, 130–131
endolymph, 156ff
perilymph, 155–156
stress, 169–170
Horseshoe bat, hair cell innervation, 63
Human
cochlea vasculature, 98
frequency discrimination, 1–2
spiral ganglion cells, 10

IHC, *see also* Hair Cell, OHC
IHC (Inner Hair Cell), 54ff, 187, 200
afferent innervation, 78–83
attachment to tectorial membrane, 72
basilar membrane velocity, 204
basolateral currents, 25ff
calmodulin, 96
conductances, 347ff
coupling to tectorial membrane, 22
cuticular plate, 59–60
differences from OHC, 58ff
distribution, 49
effects of MOC, 453
efferent effects, 444–445
efferent innervation, 83–84, 508
efferent neurotransmitters, 84

efferent synapse, 84
efferent system, 435ff
electrical resonance, 362ff
filtering, 349, 361ff
frequency response, 347ff
GABA, 512–513
hyperpolarization, 348
innervation, 10, 60, 515
input resistance, 34
intracellular recording, 26ff, 192
lateral efferents, 511
LOC effects, 468
membrane time constant, 364–365
MOC effects, 448ff
morphology, 9–10, 59–61
neurotransmitters, 84
opioid receptors, 511–512
organelles, 60
phase-locking, 364–365
physiology, 318ff, 322ff, 334
potassium currents, 354ff, 360–361
potassium currents and receptor potential, 361ff
receptor potential, 29–33
reciprocal synapses, 84
second filter, 298
stereocilia, 54, 59
stereocilia mechanics, 338ff
stiffness, 339ff
support cells, 6, 60
synapse, 78
transduction, 232–233
transfer function, 347ff
tuning curve, 223–224
two-tone suppression, 233ff
Impedance
  cochlea, 195
  cochlear models, 264
  fluid channels, 269–270
  matching, 195
  resonator, 199–200
  transfer, 289–290
Inertial forces, cochlea, 203
Inhibition, efferent, 436ff
Inhibitory post-synaptic potentials, efferents, 480
Inner ear, *see Cochlea and Organ of Corti*
Inner hair cell, *see IHC*

Inner pillar cells, *see Pillar Cells*
Innervation
  cochlea, 76–78
  hair cells, 10
  IHC, 60
  OHC, 85ff
Intense sound
  efferent protection, 487ff
  potassium concentrations, 149
Intercellular communication, organ of Corti, 146ff
Intermediate cells, stria vascularis, 112
Intermediate filaments, support cells, 97–98
Intermodulation distortion, cochlea, 237
Internal spiral sulcus, models, 294
Intrastrial fluid, 150–153
Ion channels, neurotransmitter release, 514
Ion transport
  fibrocytes, 106–109
  Reissner's membrane, 104–105
  stria vascularis, 160ff
Ionic conductances, time-dependent, 353

Junctions, cochlea, 147, 166–167
  hair cells, 51ff

Kanamycin, place-frequency map, 226ff
Kinocilium, 22

Lateral cisternae, OHC, 396–398
Lateral cisternae, role in cell length, 397
Lateral cortex, OHC, 395ff
Lateral efferents, IHC, 511
Lateral line system
  cochlear models, 261
  efferent neurotransmitter, 512
  efferents, 483–484
  hair cells, 318
Lateral olivocochlear efferents, *see LOC*
Lateral plasma membrane, OHC, 403ff

544   Index

Lateral superior olive, 508
  and LOC, 479
Length change, OHC, 387ff, *see also Motility*
Level discrimination, efferent effects, 486
Lizard, mechanical resonance of hair cells, 3-4
LOC (Lateral Olivocochlear Complex), 440ff, 467ff
  acetylcholine, 467-468
  acoustic reflex, 479
  DPOAE, 468
  effects on N1, 444
  excitation of afferents, 468
  GABA, 436
  IHC, 468
  OHC, 468
  possible roles, 489
  summary of effects, 491
Localization, *see Sound Source Localization*
Long wave model
  approximation, 268ff
  history, 270ff
  LG or WKB, 270ff, 276-277, 279
Long waves, cochlear models, 266ff

Macromechanical models, 262
Macromechanics, micromechanics, 228
MAPs, *see Microtubule-Associated Proteins*
Marginal cells, stria vascularis, 110-112
Masking, efferent effects, 439, 478, 486
Mass, organ of Corti, 202
Mechanical nonlinearity, origins, 237ff
Mechanical resonance, hair cells, 3-4
Mechanics, cochlea models, 258ff
Mechanoelectrical transduction, 186-187
Medial olivocochlear efferents, *see MOC*
Medial superior olivary nuclei, 86
Medial superior olivary nucleus, innervation of cochlea, 10

MEM (middle ear muscle reflex), MOC, 477-478, 490-491
Membrane potential, oscillation, 346
Membrane proteins, OHC, 403
Metabolism, 131-133
  and acoustic stimulation, 133
Met-enkephalin, LOC, 479
Micromechanical models, 200ff, 262
Micromechanics
  cochlea, 15ff
  at high frequencies, 197ff
  locally active, 287ff
  at low frequencies, 195
  macromechanics, 228
  morphology, 195-196
Microtubule-associated proteins
  hair cell, 95
  support cells, 97
Microtubules, pillar cells, 69
  support cells, 97
Middle ear, 10-12
  cochlear models, 299
Middle ear muscle reflex, *see MEM*
MOC (Medial Olivary Complex), 440ff
  acoustic reflex, 470ff
  afferent rate level functions, 462
  afferent spontaneous rate, 465
  afferent synchronization, 461-462
  basilar membrane motion, 445, 456ff, 466-467
  basolateral hair cell conductance, 446ff
  best frequency, 473
  binaural efferents, 471-472
  cochlear amplifier, 448ff, 456ff
  DPOAE, 453-454
  effect on afferent spontaneous activity, 460-461
  effect on rate-level functions, 452-453
  effects in human beings, 477-478
  effects on afferent response, 448ff, 461ff
  effects on afferent sensitivity, 451-452
  effects on afferent synchronization, 452-453
  effects on afferent tuning, 450-451

effects on cochlear amplifier, 466–467
effects on cochlear potentials, 442–443
effects on IHC, 448ff
effects on N1, 443–444
effects on SOAE, 455–456
future research directions, 489ff
ipsi- vs contralateral sound, 471ff
MEM, 477–478
noise, 459–460
OAE, 453ff
OAE in human beings, 477–478
OHC conductances, 464
responses to sound, 469ff
slow transient effects, 467
spontaneous activity, 470
summary of effects, 490–491
time course of effects, 463ff
time course over minutes, 464–465
time course over seconds, 463–464
tuning curves, 471–472
MOC potential, 465
afferent spontaneous activity, 461
endocochlear potential, 445ff
mechanisms, 446ff
Models of cochlea, 258ff, *see also Cochlear Models*
Modiolus, 49
blood vessels, 99
capillary bed, 99–100
Mole Rat, *see Spalax ehrenbergi*
Motility, *see also Hair Cell Length Change, Slow Somatic Motility*
cortical lattice, 401–403
cuticular plate, 425
hair cell, 386ff
nonmammalian hair cells, 424–427
OHC, 18–20, 324–325, 350
voltage-dependence, 345–346
Motor element, hair cell, 392–395
Motor processes, OHC, 229ff
Motors, membrane proteins, 403ff
Mouse, hair cell adaptation in neonates, 331
Mudpuppy, *see Necturus nebulosus*
Muscarinic receptors, organ of Corti, 510–511

Myosin
adaptation motor, 345
cuticular plate, 94
hair cell, 94

N-methyl-D-aspartate, *see NMDA*
Necturus nebulosus, hair cells, 321
Neely-Kim model, 288ff
Neuromodulators, 513
Neurotransmitter, *see also Acetylcholine, GABA, Glutamate*
calcium release, 504
cochlea, 148–149
efferent, 10, 504
hair cell, 503ff
removal, 148
Neurotransmitter release
ion channels, 514
quantal analysis, 523–526
Nicotinic receptor
DPOAE inhibition, 454
organ of Corti, 510–511
Nitric oxide, cochlea, 166
NMDA receptors, 505–507
NO, *see Nitric oxide*
Noise
efferent effects, 486
hair cell membrane channels, 334
and MOC, 459–460
Non-NMDA receptors, 505–507
Nonlinear local activity, cochlear models, 300
Nonlinear mechanics, origins, 237ff
Nonlinearity
cochlea, 215ff, 238ff, 295ff
compression, 298, 300
due to potassium conductances, 365

OAE (Otoacoustic Emissions), 18
active influence on maromechanics, 229
cochlear admittance, 242–243, 245
cochlear models, 260–261
efferent effects, 458
efferents, 478, 487

Index

in human beings, MOC, 477–478
MOC effects, 453ff
stimulus frequency, 247–248
tinnitus, 247
transient evoked, 246
OHC (Outer Hair Cell), 54ff, 187, 200, *see also* Hair Cell, IHC
  acetylcholine, 63
  active forces, 307–308
  active processes, 214ff, 228ff
  afferent innervation, 85–86
  afferent synapses, 85
  anchoring by Deiters cells, 6
  anion exchangers, 409–410
  attachment to tectorial membrane, 72
  basolateral currents, 25ff, 241–242
  calmodulin, 96
  cochlear amplification, 11–12
  cochlear impedance, 338
  cochlear models, 259ff
  cochlear sensitivity, 192
  communication between, 62
  connection to tectorial membrane, 22
  cortical lattice, 398ff
  cuticular plate, 62, 425
  cytoskeleton proteins, 93
  Deiters cells, 69–70
  development of currents, 367–368
  differences from IHC, 58ff
  distribution, 49
  effects on eighth nerve, 19
  efferent effects, 445, 464
  efferent innervation, 86–88
  efferent neurotransmitters, 87–88
  efferent specializations, 63
  efferent synapse, 87, 453ff
  electromotile responses, 18–20, 350
  endoplasmic reticulum, 396–398
  feedback, 11–12, 18–20
  filtering, 350ff, 369
  force on basilar membrane 417
  frequency response, 18, 350ff, 388
  GABA, 63, 436
  gating current, 406–407, 409
  glycocalix, 403–404
  guinea pig, 388
  hyperpolarization, 446ff
  innervation, 10, 85ff
  input resistance, 34
  intracellular recording, 26ff
  lateral cisternae, 396–398
  lateral cortex, 395ff
  lateral plasma membrane, 403ff
  length and membrane potential, 242
  length change, 388–391
  LOC effects, 468
  membrane proteins, 403
  morphology, 9–10, 61ff
  motility, 18–20, 65, 324–325, 369, 386ff
  motor, 20, 401, 403ff
  motor element, 392–395
  organelles, 62–63
  physiology, 318ff, 322ff, 329
  plasma membrane lipids, 410–411
  potassium currents, 365ff
  receptor potential, 29–30, 350ff
  reciprocal synapses, 88
  resting potential, 369
  second messengers, 511
  sensory vs. motor processes, 229ff
  stereocilia, 9, 54, 62
  stereocilia mechanics, 338ff
  stiffness, 339ff
  stiffness and voltage, 346
  stretch activation, 414–415
  subsurface cisterns, 63–65
  tectorial membrane, 62
  temporary threshold shift, 488
  in time-domain models, 301–302
  transducer conductances, 350ff
  transfer function, 350
  two-tone suppression, 236ff
  type II ganglion cells, 85–86
  voltage clamp, 366ff
  voltage-gated ion channels, 406
OHC active processes
  cochlear potential, 231ff
  location on BM, 233ff
OHC basolateral conductance, efferent effects, 457–458
OHC damage, 18–20
  place-frequency map, 226ff

OHC motility, 193, 228ff
  cochlear models, 309
  efferent effects, 438-439, 457-458
  mechanisms, 411-414
  models, 292ff
  slow, 417ff
  SOAE, 455-456
Ohm, 189
Olivocochlear efferents, 508
Olivocochlear system, see Efferents
One-dimensional model, cochlea, 267ff
Opioid peptides, organ of Corti, 512
Opioid receptors, IHC, 511-512
Organ of Corti, 46ff, 186, see also
      Cochlea
  afferent innervation, 79
  bat, 50
  Boettcher cells, 71-72
  Claudius cells, 71
  cytoskeleton, 90ff
  Deiters cells, 69-70
  desmosomes, 53
  efferent innervation, 79
  endolymph, 51
  external sulcus cells, 72
  filaments, 91
  gap junctions, 53-54
  Hensen cells, 71
  intercellular communication, 146ff
  intracellular ion balance, 143-146
  junctions, 51ff
  mass, 265
  models, 262ff
  OHC, 415-417
  opioid peptides, 512
  perilymph, 51
  pillar cells, 65ff
  resistance, 265
  reticular lamina, 52
  rigidity, 65ff
  size gradient, 50
  structure, 5-7, 50-51, 415-417
  supporting cells, 96-98, 65ff, 70-72
  tectorial membrane, 72-74
Organ of Corti-specific proteins, 164
Osmotic stress, hair cells, 145-146
Osseous spiral lamina, capillary bed,
      99-100, see also Spiral Lamina

Ossicles, amplification, 11-12
Otoacoustic emissions, see OAE
Outer hair cell, see OHC
Outer pillar cells, see Pillar Cells

Pars pectinata, models, 295
Passive mechanics, cochlea, 12ff
Patch clamp analyses, hair cells, 323ff
Peptide neurotransmitter, 511-512
Perilymph, 5, 45, 47, 150ff, 167-168
  generation, 155
  glucose, 132
  homeostasis, 155-156
  around OHC, 9
Perilymph-endolymph barrier, 157-158
Perivascular plexus, cochlea, 89
pH balance, hair cell, 145
Phase-locking
  and calcium currents, 372
  and IHC, 364-365
Phase response, basilar membrane,
      221ff
Phase velocity, 210ff
Piano, 1-2
Pillar cells, 5, 65ff, 200
  microtubules, 69
  number, 68
Place-frequency map, abnormal, 226ff
  cochlea, 214
Plasma, composition, 153
Plasma membrane lipids, OHC,
      410-411
Positive feedback, resonator, 199-200
Potassium
  effects on motility, 418
  intense sounds, 149
  as neuromodulator, 513-514
Potassium balance, organ of Corti,
      143-144
Potassium channels, hair cells, 3
Potassium concentration
  cochlear potentials, 154
  endocochlear potential, 154
  stria vascularis, 159
Potassium currents
  compressive nonlinearity, 365
  development, 360-361

548 Index

fast and slow, 357ff
hair cell, 25–26, 354ff
IHC, 361ff
OHC, 365ff
time constants in IHC, 364
tuning, 362ff
Potassium secretion, stria vascularis, 160
Presynaptic facilitation, efferents, 480
Protein kinases, 135ff
Protein phosphorylation, 135–136
Pseudemys scripta, basilar papilla hair cells, 319, 322
potassium currents, 358
Psychophysical tuning curves, nonlinearities, 297
Psychophysics, nonlinearities, 297
Pyrazinecarboxamides, hair cells, 336–337

Quality factor
basilar membrane tuning, 223–224
resonator, 199–200
tuning, 215
Quasi-linear method, cochlear models, 301–302

Radial afferent fibers, 81
Radial resonance, tectorial membrane, 187
Rana catesbieana (bullfrog), 319
potassium currents, 357
saccular hair cells, 319, 322ff, 331
Rayleigh, Lord, 189
Receptor potential, hair cells, 29–30
Receptors, cholinergic, 510–511
Reciprocal synapses, IHC, 84
OHC, 88
Recruitment, 218
Rectification, cochlear nonlinearity, 237
Reflections, cochlea, 246, 274
Reissner's membrane, 5, 76
ion transport, 104–105
standing current, 158
structure, 158
Resonance
basilar membrane, 187, 265–266

complex micromechanics, 200
model, 198ff
positive feedback, 199–200
tectorial membrane, 187
time constant, 199
Resting potential, marginal cells, 7–8
Reticular lamina, 6, 21–22, 52, 201, 415
Retzius, Gustav, 44
Rosenthal's canal, 83

Saccular hair cells, bullfrog, 319
electrical resonance, 3
Saimiri sciureus (squirrel monkey), vestibular efferents, 481
Saturation, cochlea, 217–218
Scala media, 5, 45–46
Scala tympani, 5, 45–46
Second filter, 191–192, 298
Second messengers, 136–138, 511
cochlea, 137, 165–166
intracellular regulation, 134ff
SFOAE, cochlear admittance, 247–248
Shear transformer, 195ff
Short-wave model, impedance function, 283–284
Siebert equation, cochlear models, 276
Signal detection in noise, efferent effects, 486
Slow somatic motility
effects of ATP, 418–419
effects of calcium, 419–421
effects of potassium, 418
function, 421
OHC, 417ff
SOAE (Spontaneous Otoacoustic Emissions), 187, 246–247
cochlear models, 307
MOC effects, 455–456
OHC motility, 455–456
SOC (Superior Olivary Complex), efferent system, 8, 436ff, 440ff, 474
Sodium balance, organ of Corti, 143–144
Sodium currents, hair cells, 372ff
Sound, effects on hair cell length changes, 421–424

Sound source localization, LOC efferents, 489
Spalax ehrenbergi, hair cell innervation, 63
Spectral analysis of sound, 3-4
Spectrin, OHC, 400, 401
Speech, efferent activity during, 489
Spiral ganglion, cat, 81-83
 cell composition, 81
Spiral ganglion cells, type I, 10
Spiral lamina, 6
Spiral ligament, 45-47, 105ff
Spiral limbus cells, 50
Spiral prominence, 105ff
Spontaneous activity, 470
 calcium currents, 371
 effects of MOC, 251-452, 460-461, 465
Spontaneous discharge, hair cell, 519-520
Spontaneous otoacoustic emissions, *see* SOAE
Squirrel monkey, *see Saimiri sciureus*
Standing current, cochlea, 149-150
 stria vascularis, 149-150
Stapes, 11-12, 272
Stellate ganglion, innervation of cochlea, 88
Stereocilia, 54ff, 319
 actin, 56-58, 94
 ATP, 424-425
 attachment to tectorial membrane, 203-204
 calcium sensitivity, 58
 channels, 22-24
 crosslinks, 56
 displacement, 326ff
 fimbin, 94
 first description, 54
 hair cell, 9
 IHC, 59
 length, 54
 length changes, 9
 mechanics, 338ff
 OHC, 62
 pattern, 54, 55
 stiffness, 203, 339ff, 343
 structure, 54-56
 threshold of motion, 2-3
 tip link motor, 424-425
 transducer channels, 329, 342ff
 transduction, 56
 voltage-dependent motility, 425-427
Stiffness
 basilar membrane, 202, 264-265
 membrane potential, 342ff
 neonatal hair cells, 341
 rotational, 339ff
 stereocilia, 203, 339ff
 tectorial membrane, 202-203
 voltage-dependence, 345-346
Stimulus-frequency OAE (SFOAE), efferent effects, 454-455
Stress, homeostasis, 169-170
Stretch-activated channels
 diuretics, 337-338
 genes, 337-338
 hair cell transduction, 337-338
 OHC, 414-415
Stria vascularis, 5, 109ff
 blood supply, 101-104
 cell types, 110ff
 Davis transduction model, 132
 endocochlear potentials, 159, 168
 ion transport models, 160ff
 standing current, 149-150, 158
 structure, 159-160
Subsurface cisterns, 63-65
 comparative, 63-65
 pillars, 63, 65
Summating potential, 321
Superior cervical ganglion, innervation of cochlea, 88
Superior olivary complex, *see* SOC
Support cells
 actin, 96-97
 actin-binding proteins, 96-97
 calcium-binding proteins, 98
 without filaments, 70-72
 IHC, 60
 intermediate filaments, 97-98
 microtubule-associated proteins, 97
 microtubules, 97
 organ of Corti, 65ff, 96-98
Surface tension, 210ff
Surface wave, cochlea, 208ff
Sympathetic system, cochlea, 88-90, 134

Synapse, IHC, 78
  ultrastructure, 514ff
Synaptic body, function, 517–518
  ultrastructure, 516
Synaptic potentials, 519ff
Synaptic transmission
  Carassius auratus, 522
  hair cell, 503ff, 522ff
Synaptic vesicles, ultrastructure, 516

Tectorial membrane, 21–22, 72–74
  attachment to stereocilia, 203–204
  collagen, 74
  connection to cilia, 22
  fibers, 72–74
  IHC, 72
  mass and stiffness, 202–203
  models, 288ff
  molecular structure, 74
  movement patterns, 289
  OHC, 62, 72
  radial resonance, 187
  size, 50
  structure, 6
  transduction, 72
Temperature, hair cell adaptation, 332–333
Temporary threshold shift, efferent effects, 487ff
Thermal noise, ciliary bundle, 2
Three-dimensional, cochlear models, 279ff
Threshold, motion of stereocilia, 2–3
Tight junctions, 46, 51–52, 147
  hair cells, 46, 51–52, 147
Time constant, resonator, 199
Time-domain models, 299ff
Time-varying signals, analysis, 1
Tinnitus, OAE, 247
Tip link motor, 424–425
Tip-links, adaptation, 241
  stereocilia, 342
TM, *see Tectorial Membrane*
Tone burst-evoked OAE (TBOAE), efferent effects, 454–455
Transducer channels
  adaptation, 344–345
  dimensions, 334
  ionic selectivity, 333ff
  number, 329, 334
  stiffness, 342ff
  stretch-activation, 337–338
Transducer conductance
  hair cells, 328ff, 350ff
  neonatal hair cells, 329
Transducer current
  adaptation, 330ff
  cilia, 23–24
  hair cells, 319ff, 326ff
  temperature, 332–333
Transduction
  Boltzmann function, 240
  calcium homeostasis, 154–155
  channels, 187
  Davis model, 131
  gating stiffness, 240–241
  hair cell, 22ff
  mechanisms, hair cells, 325ff
  nonlinear OHC, 262ff
  role of tectorial membrane, 72
  stereocilia, 56
Transfer impedance, 289–290, 293, 301
Transformer, middle ear, 11
Transient evoked OAE, 246
Transmitter, *see Neurotransmitter*
Trauma, effects on place-frequency map, 226ff
Traveling wave, 5, 11ff, 21, 261ff
Traveling wave tube, models, 295
TTS, *see Temporary Threshold Shift*
Tubulin, cochlea, 92
  hair cell, 95
Tuning
  in afferents, 218ff
  basilar membrane, 223–224
  efferent effects, 480
  IHC, 223–224, 362ff
  MOC, 473–474
  potassium currents, 362ff
  resonator, 199
Tuning in afferents, effects of MOC, 450–451
Tunnel of Corti, 66–67
Turtle
  hair cell physiology, 24
  hair cell resonance, 352ff
  red-eared, see *Pseudemys scripta*

Two-dimensional, cochlear models, 274ff
Two-tone suppression, 18–19, 190
  active mechanisms, 233ff
  cochlear models, 296–297
  psychophysics, 297
  sensitivity, 237
Tympanic membrane, 11–12
Tympanometry, cochlear admittance, 243ff
Type I afferents, 10
  hair cell innervation, 78–83
Type II afferents, 10
  OHC, 85–86

Undamping, cochlear models, 282, 300
Unmasking, efferents, 459–460

Vasculature, cochlea, 98ff
  modiolus, 99–100
Vestibular hair cells
  potassium currents, 358
  types I and II, 481–482
Vestibular system
  efferent effects, 480ff
  hair cells, 318

Vibration in cochlea, lability, 224ff
Voltage clamp, OHC, 366ff
Voltage clamp analyses, hair cells, 323ff
Voltage-dependence, force, 346
Voltage-dependent calcium channels, 504
Voltage-dependent conductances, hair cells, 352ff
Voltage-gated channels, hair cell, 25–26, 406
Volume regulation, hair cell, 145–146
von Békésy, G., 5, 189–190
  traveling wave, 15ff
Vowel discrimination, efferent effects, 486

Wave equation, cochlear models, 269

Xenopus, lateral line, 512

Zona occludens, see Tight Junction
Zonula adherens, see Adherens junctions

Printed in the USA
CPSIA information can be obtained
at www.ICGtesting.com
LVHW080856230823
755803LV00026B/49